Measuring and assessing the biological health of soils

It is widely recognised that agriculture is a significant contributor to global warming and climate change. Agriculture needs to reduce its environmental impact and adapt to current climate change whilst still feeding a growing population, i.e. become more 'climate-smart'. Burleigh Dodds Science Publishing is playing its part in achieving this by bringing together key research on making the production of the world's most important crops and livestock products more sustainable.

Based on extensive research, our publications specifically target the challenge of climate-smart agriculture. In this way we are using 'smart publishing' to help achieve climate-smart agriculture.

Burleigh Dodds Science Publishing is an independent and innovative publisher delivering high quality customer-focused agricultural science content in both print and online formats for the academic and research communities. Our aim is to build a foundation of knowledge on which researchers can build to meet the challenge of climate-smart agriculture.

For more information about Burleigh Dodds Science Publishing simply call us on +44 (0) 1223 839365, email info@bdspublishing.com or alternatively please visit our website at www.bdspublishing.com.

Related titles:

Advances in measuring soil health
Print (ISBN 978-1-78676-426-3); Online (ISBN 978-1-78676-428-7, 978-1-78676-429-4)

Improving soil health
Print (ISBN 978-1-78676-670-0); Online (ISBN 978-1-78676-672-4, 978-1-78676-673-1)

Managing soil health for sustainable agriculture Volume 1
Print (ISBN 978-1-78676-188-0); Online (ISBN 978-1-78676-190-3, 978-1-78676-191-0)

Managing soil health for sustainable agriculture Volume 2
Print (ISBN 978-1-78676-192-7); Online (ISBN 978-1-78676-194-1, 978-1-78676-195-8)

Understanding and preventing soil erosion
Print (ISBN 978-1-80146-379-9); Online (ISBN 978-1-80146-380-5, 978-1-80146-381-2)

Understanding and utilising soil microbiomes for a more sustainable agriculture
Print (ISBN 978-1-80146-474-1); Online (ISBN 978-1-80146-475-8, 978-1-80146-476-5)

climate•SMART•publishing
IN AGRICULTURAL SCIENCE

BURLEIGH DODDS SERIES IN AGRICULTURAL SCIENCE
NUMBER 173

Measuring and assessing the biological health of soils

Edited by Professor Jeanette Norton, Utah State University, USA, Professor Josh Schimel, University of California – Santa Barbara, USA and Professor Zoë Lindo, University of Western Ontario, Canada

burleigh dodds
SCIENCE PUBLISHING

Published by Burleigh Dodds Science Publishing Limited
82 High Street, Sawston, Cambridge CB22 3HJ, UK
www.bdspublishing.com

Burleigh Dodds Science Publishing, 1518 Walnut Street, Suite 900, Philadelphia, PA 19102-3406, USA

First published 2025 by Burleigh Dodds Science Publishing Limited

Library of Congress Control Number: 2025946834

British Library Cataloguing in Publication Data
A catalogue record for this book is available from the British Library

ISBN 978-1-83545-075-8 (print)
ISBN 978-1-83545-076-5 (ePub)
ISBN 978-1-83545-077-2 (PDF)
ISSN 2059-6936 (print)
ISSN 2059-6944 (online)

DOI: 10.19103/AS.2025.0159

Typeset by Deanta Global Publishing Services, Dublin, Ireland

Contents

Part 1 Advances in DNA-based methods, biological and other indicators to investigate soil microbial activity

Series list

Title	Series number
Achieving sustainable cultivation of maize - Vol 1 From improved varieties to local applications Edited by: Dr Dave Watson, CGIAR Maize Research Program Manager, CIMMYT, Mexico	001
Achieving sustainable cultivation of maize - Vol 2 Cultivation techniques, pest and disease control Edited by: Dr Dave Watson, CGIAR Maize Research Program Manager, CIMMYT, Mexico	002
Achieving sustainable cultivation of rice - Vol 1 Breeding for higher yield and quality Edited by: Prof. Takuji Sasaki, Tokyo University of Agriculture, Japan	003
Achieving sustainable cultivation of rice - Vol 2 Cultivation, pest and disease management Edited by: Prof. Takuji Sasaki, Tokyo University of Agriculture, Japan	004
Achieving sustainable cultivation of wheat - Vol 1 Breeding, quality traits, pests and diseases Edited by: Prof. Peter Langridge, The University of Adelaide, Australia	005
Achieving sustainable cultivation of wheat - Vol 2 Cultivation techniques Edited by: Prof. Peter Langridge, The University of Adelaide, Australia	006
Achieving sustainable cultivation of tomatoes Edited by: Dr Autar Mattoo, USDA-ARS, USA and Prof. Avtar Handa, Purdue University, USA	007
Achieving sustainable production of milk - Vol 1 Milk composition, genetics and breeding Edited by: Dr Nico van Belzen, International Dairy Federation (IDF), Belgium	008
Achieving sustainable production of milk - Vol 2 Safety, quality and sustainability Edited by: Dr Nico van Belzen, International Dairy Federation (IDF), Belgium	009
Achieving sustainable production of milk - Vol 3 Dairy herd management and welfare Edited by: Prof. John Webster, University of Bristol, UK	010
Ensuring safety and quality in the production of beef - Vol 1 Safety Edited by: Prof. Gary Acuff, Texas A&M University, USA and Prof. James Dickson, Iowa State University, USA	011
Ensuring safety and quality in the production of beef - Vol 2 Quality Edited by: Prof. Michael Dikeman, Kansas State University, USA	012
Achieving sustainable production of poultry meat - Vol 1 Safety, quality and sustainability Edited by: Prof. Steven C. Ricke, University of Arkansas, USA	013
Achieving sustainable production of poultry meat - Vol 2 Breeding and nutrition Edited by: Prof. Todd Applegate, University of Georgia, USA	014

Introduction

One of the most important topics in soil ecology is understanding the relationship between soil microbial and faunal biodiversity and soil ecosystem function. Key to this research are techniques for measuring soil organisms and their biological activity. This collection reviews a range of techniques for assessing microbial and faunal abundance and diversity, their activity in soils, and their influence in diverse ecological processes. The conditions and agronomic practices which either inhibit or promote their role in optimizing soil health and ecosystem services are also examined. The book builds on the success of our 2021 title, *Advances in measuring soil health* (ed. Prof. Wilfred Otten, Cranfield University, UK).

An initial overview chapter sets the scene by addressing soil biology and its relationship to soil health and the larger ecosystems affected by soil. Soil's capability to maintain or recover its capacity to continue to provide vital ecosystem services after disturbance is an indicator of the resilience of healthy soils.,The chapter focuses on four essential soil functions: nutrient cycling, water regulation and filtration, carbon cycling and habitat provision for biodiversity. Within these functions are many processes which are mediated by or strongly influenced by soil biodiversity and biological activity. This overview offers a conceptual model that integrates soil biodiversity with its activity into soil health assessments, in the context of soil chemical, physical and environmental parameters. The chapter argues for placing biological indicators on equal footing with chemical, physical, and environmental measures in soil health assessments, ensuring that biological assessment is given appropriate weight in the holistic assessment of soil health.

Part 1 Advances in DNA-based methods, biological and other indicators to investigate soil microbial activity

The chapters of Part 1 explore soil microbial activity and its relationship to soil health. Chapter 2 discusses microbial community characterization in soil. It evaluates different strategies such as amplicon sequencing and shotgun metagenomics. Generating amplicon and metagenomics data requires researchers to make key decisions in the workflow that influence community composition descriptions. The chapter considers these approaches, as well as how sequence data may be used to infer soil health.

Chapter 3 explores quantitative polymerase chain reaction (qPCR) in monitoring soil microbial abundance and activity. Researchers' use of qPCR has allowed them to understand distinct microbial roles in the soil community. This chapter investigates qPCR, including principles, advantages and disadvantages, applications and future directions and its relationship to soil health. Applications of qPCR have, for example, helped to better understand which microbial communities contribute to soil health, the dynamics of pathogens in soil and identify key species in nutrient cycles.

Chapter 4 focuses on soil respiration measurement, outlining its crucial role in providing insight into soil microbial activity. Methodological development and technological advances have allowed for more accurate and sensitive soil respiration assessments. This chapter reviews the different soil respiration techniques, including the historical context and biological bases of different applications for laboratory and field settings. The most common challenges and uncertainties related to respiration-based measurements are addressed. Case studies that use respiration tests to assess how cover cropping, tillage, and fertility management practices affect soil microbial functioning are presented. The chapter concludes by discussing the future of soil respiration research.

Chapter 5 explores soil health through testing the limits and strengths of enzyme assays used as proxies for nutrient cycling and availability in soil. Although soils contain a vast diversity of enzymes, only a limited set of enzyme activities are routinely measured, typically those linked to carbon, nitrogen, phosphorus, and sulphur cycling. These assays nonetheless provide valuable insights into microbial metabolism and nutrient dynamics. Enzyme assays relate to the potential activity under a defined set of conditions. Relationships between enzyme activities and soil health functions require soil environmental context be closely considered. This chapter examines the challenges of interpreting soil enzyme activity, highlighting the use of molecular methods alongside enzyme assays. It concludes with an evaluation of current limitations and outlines future research directions.

As Chapter 6 highlights, soil microorganisms play a critical role in soil health and in delivery of key ecosystem services. However, we still know relatively little about what role individual taxa play in soil processes. Quantifying microbial processes with taxonomic resolution is thus one of the key research challenges in agriculture. By quantifying the incorporation of heavy isotopes into microbial DNA, coupled with taxonomic identifications from DNA sequencing, quantitative stable isotope probing (qSIP) makes it possible to measure taxon-specific activity. This chapter explores how qSIP works and the exciting range of its potential applications in such areas as: increasing soil carbon storage and decreasing carbon emissions, increasing nitrogen retention and decreasing nitrogen losses, increasing plant resistance to environmental stress, supressing pathogens and reducing greenhouse gas emissions. Defining our understanding of and optimizing the role of soil microorganisms in nutrient and carbon cycling has the potential contribute to achieving significant improvements in agricultural sustainability.

Chapter 7 discusses the use of phospholipid fatty acid (PLFA) analysis for assessing microbial community structure and soil health. This technique provides insights into microbial biomass, community composition and physiological status by extracting and analyzing microbial lipids from soil samples. This chapter explores the history of the use of PLFA analysis and recent developments, including advances in gas chromatography-mass spectrometry and statistical tools, that have increased the reliability of PLFA analysis. While molecular techniques based on nucleic acids have gained popularity, this chapter is a reminder that PFLA analysis remains a valuable practice in soil health monitoring and agricultural science in general.

Part 2 Fauna as indicators of soil health

The next chapters explore the effects different fauna have on soil health. Chapter 8 discusses earthworms and enchytraeids and their roles in agroecosystems. Both groups are widely distributed, often occupying different niches, yet together they provide overlapping and complementary soil functions. These faunal groups have recently been reconsidered as biological indicators of soil quality and health. This chapter examines how environmental conditions, soil management practices, and interactions between these faunal groups impact their populations, and what this can then tell us about soil health. The chapter also reviews the various methods used to sample and assess earthworm and enchytraeid activity as well as the future of researching these groups and the knowledge gaps that remain.

Chapter 9 briefly summarizes current knowledge on microarthropod abundance and diversity in soils, their roles in soil processes and ecosystem services, and the agricultural management practices affecting soil microarthropod communities. It also provides an overview of sampling, extraction and identification methods including the importance of accounting for spatial and temporal variability. Potential soil health indicators derived from microarthropod abundance, diversity, and traits are reviewed, with case studies illustrating important aspects of the use of microarthropods as soil health indicators. The chapter concludes with several important challenges and perspectives for future research for the use of these indicators in practice.

Chapter 10 examines soil nematode communities and their role in soil and plant health. Alongside plant-parasitic groups, free-living nematodes contribute to decomposition, nutrient cycling, and microbial regulation, accounting for up to 12% of nitrogen mineralization and serving as valuable indicators of ecosystem function. This chapter explores using nematodes to detect the effects of management practices on soil health measured through colonizer-persister (*cp*) life history traits. This chapter explores the future of nematode research and examines the different knowledge gaps and what would be required to fill in these gaps.

Part 3 Applications

Part 3 examines the applications of the different methods of understanding soil health. As Chapter 11 points out, sustainable soil management requires accurate, standardized and accessible ways to measure and monitor soil health that can be used by managers and farmers. Effective soil health assessment requires integrating biological indicators with other soil properties, functions and ecosystem services in a framework that considers the interconnected nature of soil ecosystems. This requires a sound foundation of scientific knowledge in an appropriate biophysical context (e.g., region or soil type) and set of management factors. In recent years, significant progress has been made towards integrating soil health indicators into commercial testing laboratories and advancing interpretation frameworks. However, further convergence of soil health indicator methodologies and interpretation tools is needed to increase the utility and adoption of soil health testing. This chapter explores ways this may be achieved while stressing the importance of spatial and temporal context.

Chapter 12 examines using biological indicators to assess cropping practices. The concept of agricultural soil health relies on an understanding of soil physical, chemical and biological processes. This chapter explores the wide range of research on the influence farming systems have on biological processes and indicators relevant to soil health. Soil biological health is strongly related to more diverse cropping systems, coupled with high inputs of organic amendments and less frequent and shallow tillage.

Chapter 13 evaluates soil biology in forest soil. Soil health is vital for forest ecosystem health. This chapter traces the shift from traditional soil analysis, focused exclusively on chemical and physical properties, toward an integrated approach that incorporates biological indicators for a more comprehensive assessment of soil health. This chapter assesses the current challenges facing forest soil health and how to evaluate forest soil health using biological indicators. In addition, the aim of this chapter is to help devise simple yet effective biological-based forest soil health indicator systems.

Understanding the relationship of soil health to a soil's biodiversity has recently progressed from cataloging to delving into the multifunctionality of the soil community. Challenges to the interpretation of this realm will depend how we use advanced molecular biological techniques ('omics tools) with advanced data analytics to predict functionality and ecological roles of soil organisms in the context of the soil chemical, physical and environmental matrix.

Chapter 1

Soil biology, soil health and ecosystem services: an overview

C. Vazquez, T. Mulder, L. Chavez Rodriguez, F. David, D. P. Di Lonardo, A. Garsia and R. E. Creamer, Wageningen University and Research, The Netherlands; E.K. Bünemann, Research Institute of Organic Agriculture (FiBL), Switzerland; H. Soinne, Natural Resources Institute Finland (Luke), Finland; P. Cheval, Université de Lorraine, INRAE, LSE, F-54000 Nancy, France; A. Basile, Institute for Mediterranean Agricultural and Forestry Systems, National Research Council of Italy, Italy; A. Bacq-labreuil, Genesis Soil Health, France; J. Nordén, Norwegian Institute for Nature Research (NINA), Norway; L. Cunha, University of Coimbra, Portugal; C. Imbert, Leibniz-Centre for Agricultural Landscape Research (ZALF), Germany; and C. Boix-Fayos, Soil and Water Conservation Research Group, CEBAS-CSIC, Spain

1 Introduction

Soil health refers to the ability of soil, at a specified point in time, to function as a dynamic living system within natural or managed ecosystems. A healthy soil sustains plant, fungal, and animal productivity and health, maintains or enhances water and air quality, and delivers essential ecosystem services over the long term, without increasing trade-offs between ecosystem services (Van den Elsen, et al., 2022). The term is often used interchangeably with soil quality (Bünemann et al., 2018). Soils support a range of ecosystem services and they

http://dx.doi.org/10.19103/AS.2025.0159.01

provide benefits to human societies (Calvaruso et al., 2021) through the delivery of soil functions. Soil functions commonly considered in soil health assessments from an agricultural context are primary productivity and biomass production, climate regulation, water regulation and purification, nutrient cycling and habitat provision for biodiversity (Haygarth & Ritz, 2009; Schulte et al., 2014, Pereira, 2018). Additional functions, such as pest and disease management (Creamer et al., 2022) and pollutant degradation (Vogel et al., 2018) have also been proposed in the literature. Monitoring of soil health aims to assess the resilience of soil to provide continued support in the provision of ecosystem services.

Despite increasing awareness and research, monitoring soil health remains a major challenge due to several factors:

1. Scale mismatch: the spatial scale of the assessment is often not in line with the selected indicators. Indicator selection should consider the purpose of the monitoring programme, who it is for and with which goals in mind monitoring is done (Schreefel et al., 2024).
2. Resource constraints: soil health monitoring is often driven by resources (cost and logistics) and depends primarily on what monitoring systems are common, or already in place or can be expanded upon (Matson et al., 2024).
3. Benchmarking: due to the heterogeneity of soil and the interactions with climate, landscape and land use there is no clear definition of what constitutes healthy soil, as threshold values to benchmark soil health will vary in different pedo-climatic zones and land use systems.
4. Conceptual ambiguity: soil health is a concept that amalgamates several interacting soil functions. This leaves the concept of soil health open to inappropriate use, and indicators are often selected without transparency or pertinence to the functions being assessed.

Since soil health cannot be measured directly, it is typically assessed through two complementary approaches:

1. Threshold-based indicator approach: here, soil health is evaluated by defining thresholds for a range of soil parameters. This approach is often applied in larger-scale assessments (regional, national and continental), where a harmonized minimum dataset is applied to categorize soil as healthy or degraded (Matson et al., 2024).
2. Process-based clustering approach: this approach defines the processes and underlying properties that are relevant to a range of functions. This considers the interaction of soil processes with local conditions and management practices, assessing how they collectively contribute to the delivery of one or more functions.

Published by Burleigh Dodds Science Publishing Limited, 2025.

A well-defined and scientifically rigorous selection of appropriate indicators is critical to ensuring that the selected indicators are both relevant to the processes and overarching functions they are chosen to represent and applicable to the specific land use, management practices and climatic conditions of the ecosystems being evaluated (EEA, 2023).

2 Selecting indicators to assess soil health

Soil functions themselves are challenging to measure directly, as they consist of bundles of soil processes (Bünemann et al., 2018) which continuously interact. Therefore, selected indicators should reflect the processes that are occurring to support the provision of a given function. The first step is the identification of relevant indicators which can directly measure the process. Soil processes can be described as a series of actions or steps that take place that result in a particular outcome (adapted from PROCESS | English meaning - Cambridge Dictionary). Process-based indicator measurements are usually defined as rates which describe the speed at which the process is taking place. As such, process measurements are often conducted either in the laboratory or the field and can be logistically challenging to include in a monitoring programme. In the absence of direct indicator methods for a given process, it is feasible to measure the presence or abundance of (most common) biological drivers and defining (physical or chemical) conditions in which the process will occur. Creamer et al. (2022) presented the main biologically led processes that supported four soil functions (nutrient cycling, water regulation and filtration, carbon regulation and pest and disease management) in a temperate agricultural context through a series of knowledge graphs referred to as cognitive models. The cognitive models identified which processes were relevant for each soil function and how these processes could be bundled together into 'sub-functions' which describe how the co-occurring processes support a specific underlying mechanism of the function. Biological actors were then identified and considered as the main drivers of each process. This facilitated for the first time a transparent methodological approach to describing the role of soil biota in the provision of soil functions, where the soil biological parameters were ranked for their pertinence to supporting the processes underpinning the four soil functions (Zwetsloot et al., 2022). However, the papers of Creamer et al. (2022) and Zwetsloot et al. (2022) did not identify how the biologically mediated soil processes co-occurred with or were mediated by physical and chemical conditions in the soil.

This chapter will present an update on this work by integrating the chemical and physical components of soil health with that of the biologically mediated processes and further define how environmental conditions may contribute. To achieve this, four soil function cognitive models have been developed: (1) carbon and climate regulation, (2) nutrient cycling, (3) water regulation and filtration and (4) habitat provision for biodiversity. These cognitive models define the chemical, physical and biologically mediated processes that occur in soils to support sustainable agricultural production in temperate climate conditions across Europe. These cognitive models provide a first insight into the processes and parameters which support the delivery of each function. Parameters are defined as the chemical, physical or biological properties or climatic or landscape conditions that define a process system or set the conditions of its operation. In many cases, the processes can be measured directly, making them reliable indicators of the functions. However, process-based measurements are often challenging outside of controlled scientific conditions. To address this, the models also incorporate parameters that support these processes. The parameters – soil properties which interact to drive soil processes, either directly, for example, many biological actors are responsible for bioturbation occurring in the soil, or indirectly which can be parameters which define the presence or absence of the biological actors, for example, pH or moisture. These cognitive models facilitate and provide a transparent method for indicator selection of the four functions and overall soil health assessment through a multi-functional approach. Indicators for soil health assessment can then be defined for the processes that are relevant, in a given pedo-climatic context and that are useful in framing the purpose of the monitoring focus. This chapter will provide the detailed scientific underpinning of the processes involved in the delivery of each of the soil functions but will not deliver a set of indicators, as this is dependent on context and purpose.

3 The process of indicator selection

Indicator selection was achieved through a structured interactive approach, where scientists from across Europe engaged in online and in-person workshops. Over 25 soil experts, specializing in different aspects of soil health, collaborated to identify and rank processes relevant to the four specific soil functions. The indicator selection was completed through an iterative process, starting with brainstorming workshops, which provided the first foundation for each of the function cognitive models, integrating biological, chemical and physical soil expertise with landscape and modelling experts. The workshops were followed by comprehensive scientific literature reviews to validate and confirm the role of the identified processes and parameters and uncover any missing elements. After the initial development of the cognitive models, groups were asked to rank the processes and parameters according to their importance for the delivery of a certain function. The ranking system ranged from 1 (very important) to 3 (less important or relevant). This ranking was instrumental in refining the initially highly complex and detailed cognitive models. Processes

Published by Burleigh Dodds Science Publishing Limited, 2025.

having minimal relevance were eventually removed, allowing the models to be simplified without losing critical information. Finally, experts were asked to determine if a parameter had a direct or an indirect effect on the rate of a given process. For example, earthworms have a direct effect on bioturbation (Cunha et al., 2016; Vion-Guibert et al., 2024), whereas their effect on erodibility is indirect, mediated through their impact on aggregation processes (Wen, 2022). However, a direct correspondence does not always equate to a larger effect; indirect effects can also play significant roles depending on the pedo-climatic context and the interplay of multiple factors, such as soil management (Jouquet et al., 2012a; Wen et al., 2022). These indirect or contextual effects have a strong influence on the selection of an appropriate indicator. This systematic and iterative approach ensured that the selected indicators are not only process-specific but also aligned with the broader soil health framework. By tailoring the selection to the unique characteristics of each process, the resulting indicators provide robust tools for monitoring and assessment across diverse systems.

Figure 1 describes the basis of the cognitive model concept. For each soil function cognitive model, the function is first defined. Next, key processes which contribute to the provision of that function are defined. Processes may occur in one or several soil function models. The processes are bundled for the sake of visualization in the model into 'sub-functions' which account for a set of processes that are focused on a specific aspect of a function. Sub-functions are generally not measurable, as they are a visualization of a series of processes. At the process level, it is possible to measure the individual processes, but this often requires specialized or control conditions in which a process is measured over time. Therefore, soil (biological, chemical and/or physical) and environmental parameters are utilized to define the main drivers and conditions under which the process occurs or the rate at which it occurs.

This approach does not imply that all properties involved in all processes need to be monitored continuously. Rather, we propose to follow an approach similar to the logical sieve approach proposed by Ritz et al. (2009) and later adapted to indicator selection for soil biological indicators by Zwetsloot et al. (2022). This method provides a systematic framework for indicator selection by progressively narrowing down a broad list of processes and parameters via specific selection criteria, to define the final set of indicators. The sieve operates through several steps: (1) Pertinence to the soil function: indicators are first evaluated for their relevance to the soil function under consideration. This ensures that each selected indicator provides meaningful information into the processes and sub-functions that underpin the function; (2) Applicability to context: indicators and methods are assessed for the suitability within a specific context of a specific monitoring program (taking into account issues such as land use type and spatial scale of assessment and environmental conditions).

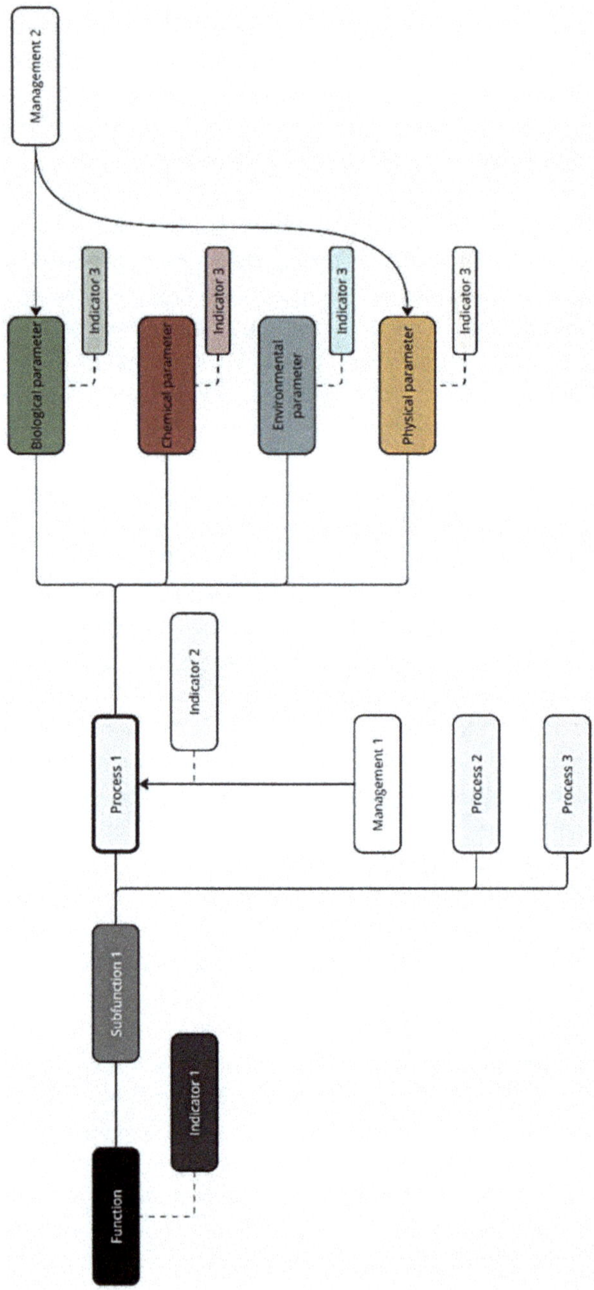

Figure 1 Diagram representing the hierarchical relationship between soil functions, sub-functions, processes, parameters and their associated indicators. At each level of the diagram, indicators could be identified. Indicators used for monitoring purposes should, where possible, be selected at higher levels of the hierarchy. However, indicators at the function or process level are not always available, and indicators at lower levels of the diagram will be needed. In such cases, indicators that represent all parameter types should be selected. Management can affect process rates directly or affect soil functioning through their effect on soil and environmental parameters.

This step ensures that selected indicators better align with the unique characteristics (and constraints) of the monitoring context; and (3) Technical criteria: indicators are further selected based on technical considerations such as the ease of interpretability, throughput, accuracy, cost, and practicality of implementing the associated measurement techniques (Zwetsloot et al., 2022). However, in the context of soil health assessment, the final set of indicators should comprehensively represent all relevant soil functions, ensuring that relevance, applicability and logistical issues have been considered.

In the following sections, each of the soil functions are described in the form of a cognitive model, providing a comprehensive overview of the processes relevant to each of the four functions. As several processes occur across multiple cognitive models, the parameters which define the functions are described in the subsequent section.

4 Carbon and climate regulation

The soil contribution to carbon and climate regulation, i.e. the balance between the uptake of CO_2 and CH_4 by the vegetation and microbes, which is maintained as biomass and carbon compounds in the soil, and the release of greenhouse gas emissions from the soil, encompasses soil processes that contribute to the carbon retention in the soil and regulate the release of major greenhouse gases (BIOSIS, 2023). Within this function, we identified 17 processes underpinning the carbon and climate regulation soil function, which were bundled into four sub-functions: decomposition, organic matter transfer, organic matter stabilization and biochemical transformations (Fig. 2). These sub-functions collectively capture the complex interplay of processes that influence soil carbon dynamics and greenhouse gas emissions. The direct and indirect parameters that support the provision of these processes are described in Table 1.

4.1 Sub-function: decomposition

The breakdown of organic matter (i.e. decomposition) in soil is realized through six processes that result in the production of carbon dioxide (CO_2) and/or methane (CH_4) (Creamer et al., 2022). These processes include: (1) litter photodegradation, (2) fragmentation, (3) faunal respiration, (4) microbial respiration, (5) extracellular depolymerization, and (6) methanogenesis.

Decomposition through these six processes is a careful balance of utilization of existing carbon sources in the soil in combination with new organic matter inputs into the soil. This dual process can result in the release of carbon stored in the soil while simultaneously serving as the basis for the sequestration of new carbon into the soil. Plant material inputs make their way into the soil, where they are fragmented by earthworms, isopods and other macrofauna

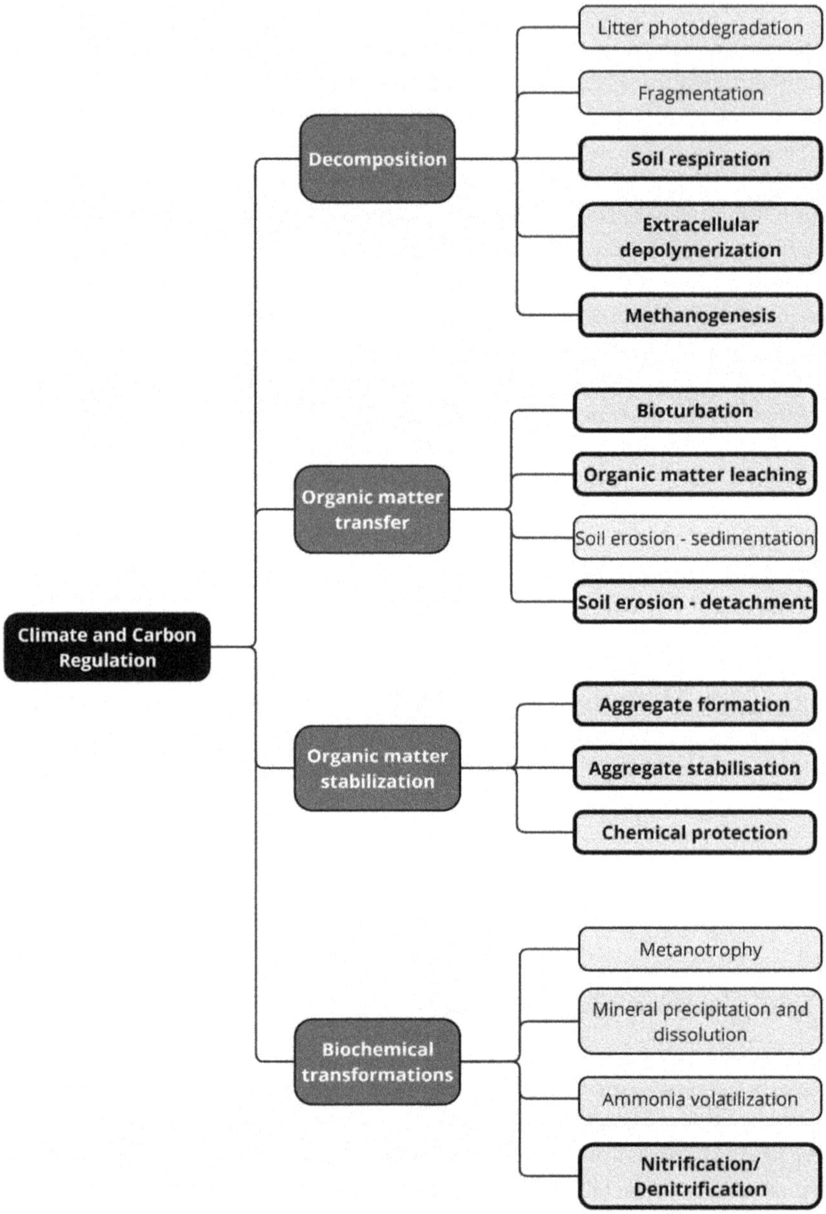

Figure 2 Processes (light grey) and sub-functions (grey) underpinning climate and carbon regulation. Boxes with a thicker outline and bold lettering represent processes deemed as most relevant by the experts.

Published by Burleigh Dodds Science Publishing Limited, 2025.

Table 1 Soil parameters that directly or indirectly mediate the processes that occur in the carbon and climate regulation function. Some parameters or parts of them are expressed with abbreviations or acronyms. Abbreviations and acronyms for chemical parameters: SOM for soil organic matter content, SOM quality for soil organic matter quality, CEC for cation-exchange capacity, oxygen for oxygen concentration, NH_3 for ammonia content, $CaCO_3$ for calcium carbonate, Ca:Mg ratio for calcium to magnesium ratio. Soil functions in which these processes are also present are also present as abbreviations: NC for nutrient cycling and HP for habitat provision for biodiversity

	Soil parameters that underpin each process				Present in other functions*
Process	Biological	Physical	Chemical	Environmental	
Litter photodegradation		Soil moisture	Plant residue quality	Temperature Plant residue layer Solar irradiation	NC
Fragmentation	Acari (Oribatids), Isopods, Collembola, Earthworms, Enchytraeids, Termites, Ants, Millipedes	Soil temperature, Soil moisture,	Soil pH Oxygen Plant residue quality	Temperature Precipitation Plant residue layer	NC; HP
Soil respiration	Archaea, Algae, Bacteria, Fungi, Roots, All soil fauna	Soil temperature, Soil moisture, Soil porosity	Soil pH SOM Nutrient availability Oxygen	Temperature Precipitation	
Extracellular depolymerization	Bacteria, Fungi, Roots	Soil temperature, Soil moisture, Soil porosity	Soil pH, Nutrient availability, SOM, SOM quality		NC; HP

(Continued)

Table 1 (*Continued*)

Process	Soil parameters that underpin each process				Present in other functions*
	Biological	Physical	Chemical	Environmental	
Methanogenesis	Archaea (methanogenic)	Soil temperature, Soil porosity	Soil pH, Plant residue quality, Oxygen		
Bioturbation	Earthworms, Enchytraeids, Termites, Ants	Soil texture, Bulk density, Soil temperature, Soil moisture, Soil depth	Soil pH, Nutrient availability, SOM	Temperature, Precipitation, Plant residue layer	HP
Organic matter leaching	Bacteria, Fungi, Roots, Earthworms, Enchytraeids, Termites, Ants	Soil texture, Soil moisture, Preferential flow path, Percolation, Infiltration, Aggregate stability	Soil pH, SOM, Oxygen	Temperature, Precipitation, Wind speed	
Soil erosion – detachment and runoff	Bacteria, Fungi, Roots, Earthworms, Enchytraeids, Termites, Ants	Soil texture, Soil moisture, Aggregate stability, Impeding layer, Surface stoniness, Surface morphology	$CaCO_3$, SOM	Precipitation, Vegetation density, Vegetation cover, Slope, Position in landscape	NC
Soil erosion – sedimentation		Morphology		Flux velocity, Precipitation, Vegetation density, Vegetation cover, Slope, Position in landscape	NC

Aggregate formation	Earthworms, Enchytraeids, Termites, Ants, Bacteria, Fungi, Roots	Soil temperature, Soil moisture, Soil porosity	Soil pH, CEC, SOM quality, Clay mineralogy	Temperature, Precipitation	HP
Aggregate stabilization	Bacteria, Fungi, Roots, Earthworms, Enchytraeids	Soil texture, Soil temperature, Soil porosity	Soil pH, CEC, Clay mineralogy, Ca:Mg	Freeze-thaw cycles, Dry-rewetting cycles	
Chemical protection	Bacteria, Fungi, Roots	Soil texture	CEC, SOM, SOM quality, Clay mineralogy		
Methanotrophy	Archaea (methanotrophic), Bacteria (methanotrophic)	Soil texture, Soil moisture	Soil pH, Oxygen		
Mineral precipitation and dissolution	Roots	Soil moisture	Soil pH, Oxygen, Clay mineralogy	Freeze-thaw cycles, Bedrock type	
Ammonia volatilization		Soil moisture, Soil temperature, Soil porosity	Soil pH	Temperature, Wind speed	NC
Denitrification	Archaea (denitrifying), Bacteria (denitrifying), Fungi (denitrifying)	Soil moisture, Soil temperature, Soil porosity	Soil pH, Nitrate availability concentration, SOM, Oxygen	Wind speed, Precipitation	

*Where NC is nutrient cycling and HP is habitat provision for biodiversity.

feeding on plant material. Through this process, plant material enters the soil food web and begins transformations into smaller molecules. Root exudates and microbial extracellular enzymes also contribute to the breakdown of larger organic molecules into smaller ones, which then enter the food web. During these processes, both soil fauna and microbes release CO_2 through respiration. When soil conditions are anoxic, decomposition can still take place but it is dominated by methanogenic microorganisms, which release CH_4. Organic matter can remain unprotected, as particulate or dissolved organic matter that is more susceptible to processes of organic matter transfer or can stabilize through physical or chemical protection by associating with mineral surfaces or becoming part of soil aggregates (Cotrufo & Lavallee, 2022).

4.2 Sub-function: organic matter transfer

This sub-function focuses on processes that make resources available, unavailable or that displace them from the site of interest. These processes include bioturbation, organic matter leaching, and soil erosion, including both the accumulation of soil (sedimentation) as well as its loss (detachment).

The overall effect of organic matter leaching on carbon and climate regulation function is difficult to ascertain. Leaching to lower layers of the soil profile could lead to higher levels of protection of organic matter since microbial activity and respiration decrease with depth (Min, 2021) and it would make aggregates less likely to be disturbed by practices such as tillage. However, if dissolved organic matter (DOC) can be leached into groundwater via throughflow to a nearby ditch or stream, this will result in a loss of carbon from the system. A similar problem arises when trying to pinpoint the effect of soil erosion on the carbon and climate regulation function. The impact of soil erosion on the amount and pools of mobilized organic carbon is complex and dependent on the spatial scale. The effect of rain splash and wind on unprotected soil affects detachment and breakdown of aggregates, causing emissions of CO_2. The soil particles can be transported and deposited nearby or further away through the air or stream network. At patch and hillslope scale, selective erosion processes (e.g. sheet runoff) are dominant, mobilizing sediments enriched in organic carbon (Martínez-Mena et al., 2012). At coarser spatial scales (hillslopes, ravines and rivers), non-selective erosion processes (e.g. rill and gully erosion, riverbed and bank erosion) take place, mobilizing sediments from deeper soil layers with lower content of organic carbon (Nadeu et al., 2011; Boix-Fayos et al., 2015). Thus, soil erosion can lead to a loss of both nutrients (see sub-function: nutrient supply) and soil organic carbon at the site and enrichment where the sediment is deposited. At depositional sites the burial of sediments reduces emissions by mineralization, and new soil formation in the upper layers leads to organic carbon enrichment (Martínez-Mena et al., 2019).

Published by Burleigh Dodds Science Publishing Limited, 2025.

In addition, the source of sediments (superficial or deeper soil layers) and the way that organic carbon is transported during the process of erosion (dissolved or protected within stable microaggregates) determine their stabilization in depositional areas (Boix-Fayos et al., 2015, 2017). The stabilization of organic carbon in depositional areas can compensate for organic carbon (OC) losses from eroded sources within catchments (Martínez-Mena et al., 2019).

The erosion potential is strongly determined by the terrain topography and soil physical properties such as aggregate stability, surface roughness and stoniness, soil structure and the presence of a soil crust and seal. Typically, with poor aggregate stability, the presence of stones or a crust can contribute to a higher erosion potential. Transport and sedimentation are also dependent on the topography, especially the terrain roughness and topographic position in the landscape, as they drive the flux velocity and the convergency of fluxes. Furthermore, vegetation cover and vegetation density affect the erosion potential. Vegetation with well-developed root systems typically reduce the erosion potential and positively contribute to the capacity of the soil to retain nutrients in the system. The presence of vegetation is also important for providing soil organic matter to the soil and thereby sustaining soil organisms (bacteria, fungi, earthworms, enchytraeids, termites and ants). Soil organisms can both reduce soil loss, by improving porosity, and increase it, by diminishing soil stability due to mixing activities (Orgiazzi & Panagos, 2018).

4.3 Sub-function: organic matter stabilization

Once organic matter enters the soil, it must be physically or chemically protected to ensure the build-up of organic matter and consequently carbon stocks, which links to the climate regulation component of the function. Processes such as aggregate formation, stabilization and chemical protection are relevant to organic matter protection and stabilization. Aggregate formation and stabilization depend strongly on biological actors that bind soil particles together, such as roots, bacteria or earthworms (Creamer et al., 2022). Environmental variables can also directly affect aggregate formation and stabilization, for example, freeze-thaw cycles have been known to disrupt macroaggregates (Six et al., 2004). Dry-rewetting cycles can also lead to peaks and troughs in decomposition rates. Evidence suggests that soils that undergo dry-rewetting cycles in the long term show higher cumulative soil respiration rates when compared to soils kept under a constant moisture regime (Zhang et al., 2020).

Chemical protection of organic matter leads to the formation of mineral-associated organic matter, which depends strongly on the mineralogy of the soil at hand and the properties of the organic matter (Degens, 1995). The scientific community has recently identified the importance of necromass accumulation to carbon sequestration overall (Liang et al., 2019; Wang et al., 2021; Cotrufo & Lavallee, 2022). Rather than including necromass accumulation as a process relevant to organic matter stabilization, we have included bacterial biomass, which relates to bacterial necromass accumulation in the processes of aggregate formation, stabilization as well as chemical protection.

4.4 Sub-function: biochemical transformations

Biochemical transformations include changes in inorganic molecules that lead to the production of N_2O (denitrification), CO_2 (mineral precipitation and dissolution), NH_3 (ammonia volatilization) as well as the consumption of CH_4 (methanotrophy). In the soil, nitrogen goes through several transformations. Nitrous oxide production is primarily the result of denitrification, which mainly takes place in the presence of NO^-_3 in the soil, which occurs through the process of nitrification. While we include only nitrification in the current function, an assessment of both may be needed to account for the main pathway for N_2O and NOx emissions. The HIP model (hole-in-pipe model) predicts that the production of NOx gases is dependent on the availability of nitrogen in the soil, while the ratio between N_2O and NO production is a function of soil moisture (Davidson, 2000). Other factors, such as soil pH, and soil temperature can also explain N_2O emissions in agricultural systems, particularly when observing emissions during a growing season. Depending on factors like soil diffusivity, temperature and pH, ammonia can also volatilize.

The release of carbon into the atmosphere can be controlled through wider environmental factors. For example, while methanogenesis (see decomposition) is the process by which methanogenic microorganisms acquire energy from the breakdown of organic matter, methanotrophy is the process by which methanotrophic microorganisms oxidize methane. This process happens as soil dries out, or as methane is diffused to higher layers of drier soil. Parent material (geology) of soil can strongly define some processes, such as mineral dissolution, where the release of chemical species from a solid to the surrounding aqueous solution occurs (Noiriel & Daval, 2017). An example of this is the dissolution of carbonates from calcareous parent materials, such as calcite, due to rainwater containing $CaCO_3$ dissolving carbonates from the parent material over time.

5 Nutrient cycling

Nutrient cycling is defined as the capacity of a soil to take up and recycle nutrients from different inputs (e.g. plant residues and manure) and to support the uptake of nutrients from soil minerals and organic matter, water and air by plants and the soil community (BIOSIS, 2023). To understand this function comprehensively, we identified 18 processes allocated to three main sub-functions that determine the delivery of nutrient cycling in soils: nutrient supply, organic matter transformation and nutrient acquisition (Fig. 3). The direct and indirect parameters that support the provision of these processes are described in Table 2.

5.1 Sub-function: nutrient supply

Nutrient supply is determined by processes such as adsorption/desorption, ammonia volatilization, atmospheric deposition, microbial immobilization, nitrification, denitrification, nutrient leaching, precipitation/dissolution, and soil erosion and runoff (specifically soil detachment processes). Plants and soil microorganisms take up dissolved nutrients from the soil solution, primarily in the form of nitrate (NO^-_3) or ammonium (NH_4). For most nutrients, dynamic equilibria exist between the soil solution and various solid forms. In the absence of soil organisms and growing plants, these equilibria are governed by physico-chemical processes such as adsorption/desorption on surfaces as well as precipitation/dissolution. For example, the abiotic retention of phosphate on particles in the soil can be viewed as a continuum of precipitation and adsorption processes (Bünemann, 2015), while phosphate sorption is governed mainly by the concentration, types and surfaces of iron and aluminium oxides (Borggaard et al., 2005). In living soil, these chemical equilibria are additionally affected by biological uptake from the soil solution, transformation into other chemical forms and release upon death and decay. Nitrification and denitrification, as described previously, are transformation processes that are specific to N.

Unutilized nutrients in the soil are at risk of leaching into deeper soil layers, particularly for soils with a sandier texture, through by-pass flow or where groundwater levels are high. When these nutrients are transferred below the rootable soil depth, they can be adsorbed onto mineral surfaces or lost entirely from the soil system, potentially contributing to groundwater contamination in regions with intensive agricultural practices (Lehmann, 2002). Nutrient losses through leaching can be influenced by preferential water flow paths, which may form due to successions of wetting and drying cycles in clayey soils. These pathways can also emerge from structural modifications in the soil, such as biopores created by faunal activity (e.g. earthworms, ants, termites) and plant roots (Zhang et al., 2018). While these structures typically enhance soil aeration and water infiltration, they may, under certain conditions, facilitate deeper nutrient transport beyond the root zone. Vegetation type and density, with its effect on rooting depth and density, can limit nutrient leaching through assimilation (Creamer et al., 2022).

Nutrient mobility in the soil is nutrient-specific and often dependent on their chemical form (speciation). For instance, nitrate (NO_3) is highly mobile and prone to leaching, while phosphate is generally immobile in most soils because

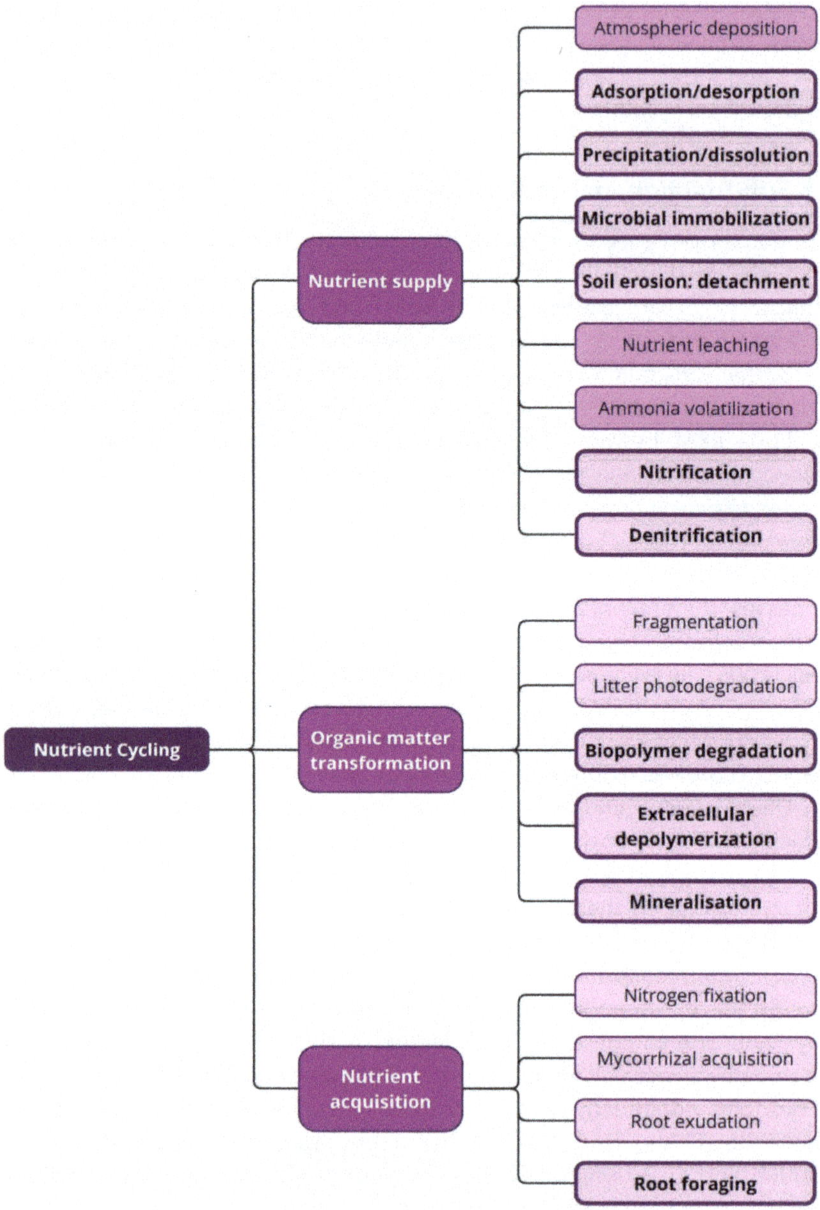

Figure 3 Processes (light purple) and sub-functions (purple) underpinning nutrient cycling. Boxes with a thicker outline and bold lettering represent processes deemed as most relevant by the experts.

Published by Burleigh Dodds Science Publishing Limited, 2025.

Table 2 Soil parameters that directly or indirectly mediate the processes that occur in the nutrient cycling function. Some parameters or parts of them are expressed with abbreviations or acronyms. Abbreviations and acronyms for biological parameters: N-fixing for nitrogen-fixing. Abbreviations and acronyms for chemical parameters: SOM for soil organic matter content, SOM quality for soil organic matter quality, CEC for cation-exchange capacity, $CaCO_3$ for calcium carbonate, MAOM for mineral associated organic matter, oxygen for oxygen concentration, FF for fungivores and BF for bacterivores

	Soil parameters that underpin each process				
Process	Biological*	Physical	Chemical	Environmental	Present in other functions**
Atmospheric deposition				Precipitation Deposit concentration in the air	
Absorption/ desorption	Bacteria Fungi Roots Earthworms	Soil texture Soil moisture Soil temperature Specific surface area Soil porosity	Soil pH CEC Clay mineralogy SOM MAOM	Temperature Precipitation Freeze-thaw cycles	WR
Precipitation/ dissolution	Bacteria Fungi Roots	Soil moisture Soil temperature Soil porosity	Soil pH Chemical saturation Oxygen SOM Clay mineralogy	Temperature Vegetation cover	WR; HP
Microbial immobilization	Bacteria Fungi Nematode (B) Nematode (F) Collembola (B) Collembola (F) Protozoa (B) Protozoa (F) Acari (F)	Soil temperature Soil moisture Soil porosity	Soil pH SOM SOM quality		
Soil erosion – detachment and runoff	Bacteria Fungi Roots Earthworms Enchytraeids Termites Ants	Soil texture Soil moisture Aggregate stability Impeding layer Surface stoniness Morphology	$CaCO_3$ SOM	Precipitation Vegetation density Vegetation cover Slope Position in landscape	CR

(*Continued*)

Table 2 (*Continued*)

Process	Soil parameters that underpin each process				Present in other functions**
	Biological*	Physical	Chemical	Environmental	
Soil erosion – sedimentation				Morphology Flux velocity Precipitation Vegetation density Vegetation cover Slope Position in landscape	CR
Nutrient leachin	Bacteria Fungi Roots Earthworms Enchytraeids Termites Ants	Soil texture Soil moisture Preferential flow path Percolation Infiltration Aggregate stability	Soil pH Sorption/ Desorption Nutrient availability Clay mineralogy	Dry-rewetting cycles Freeze-thaw cycles Vegetation density	
Ammonia volatilization		Soil moisture Soil temperature Soil porosity	Soil pH	Temperature Wind speed	CR
Nitrification	Archaea (nitrifying) Bacteria (nitrifying) Fungi (nitrifying)	Soil moisture Soil temperature Soil porosity	Soil pH Ammonia SOM Oxygen	Temperature Vegetation cover	
Fragmentation	Acari (Oribatids) Isopods Collembola Earthworms Enchytraeids Termites Ants Millipedes	Soil temperature Soil moisture	Soil pH Oxygen Plant residue quality	Temperature Precipitation Plant residue layer depth	CR; HP
Litter photodegradation		Soil moisture	Plant residue quality	Temperature Plant residue layer Solar irradiation	CR

Biopolymer degradation	Acari Collembola Earthworms Nematodes Protozoa Termites Ants Millipedes	Soil texture Soil temperature Soil moisture Soil porosity	Soil pH Nutrient availability SOM Oxygen	Temperature Precipitation	HP
Extracellular depolymerization	Bacteria Fungi Roots	Soil temperature Soil moisture Soil porosity	Soil pH Nutrient availability SOM SOM quality		CR; HP
Mineralization	Bacteria Fungi	Soil texture Soil temperature Soil moisture	Soil pH Nutrient availability SOM SOM quality Oxygen CaCO3	Temperature Precipitation	HP
Nitrogen fixation	Bacteria (N-fixing) Archaea (N-fixing) Legumes	Soil temperature Soil moisture	Soil pH Nutrient availability		
Mycorrhizal acquisition	Plants (mycorrhizal) Fungi (mycorrhizal)	Soil moisture	Soil pH Nutrient availability		
Root exudation	Roots	Soil moisture	Soil pH Nutrient availability Oxygen		HP
Root foraging	Roots	Soil moisture Surface stoniness Effective root depth	Soil pH Nutrient availability Micronutrient availability		

*Where F are fungivores, B are bacterivores.

**Where CR is carbon regulation, WR is water regulation and provision and HP is habitat provision for biodiversity.

of its tendency to precipitate or adsorb onto mineral surfaces (Lehmann, 2002). Another pathway for nitrogen loss is ammonia volatilization, a process that was already discussed under the carbon and climate regulation function model.

Atmospheric deposition is of varying importance, depending on the nutrient. For N, the critical loads theory attributes high nitrogen deposition to the livestock density and type of animals within a local region (Bobbink et al., 2015; Chang et al., 2021). High nitrogen emissions associated with pig and poultry facilities in the local vicinity can significantly increase the rate of nitrogen atmospheric deposition. For sulphur (S), atmospheric deposition has decreased substantially over time, and in the last 2 decades, S deficiency in certain crops has been increasingly observed (Zhao et al., 2003a).

Soil erosion can lead to the transfer of nutrients across the landscape, similar to the erosion processes described in the organic matter transfer sub-function. It is worth highlighting that depositing nutrient-rich sediments may play a dual role. On one hand, it can negatively impact sensitive ecosystems through nutrient enrichment, which may lead to soil or surface water eutrophication. On the other hand, it can benefit agricultural systems that depend on external nutrient inputs to maintain productivity (Nathan et al., 2022).

5.2 Sub-function: organic matter transformation

Through transformations of organic matter, nutrients become available to plants as well as to soil organisms. Processes such as litter photodegradation (Almagro et al., 2015), fragmentation, biopolymer degradation, extracellular depolymerization and mineralization are particularly important in the cycling of nutrients through the soil profile. These processes closely relate to the decomposition sub-function described previously, since as soil organic matter is broken down, nutrients become available to soil organisms as well as to plants after mineralization by biological actors. The parameters that regulate decomposition also play an important part in the release of nutrients.

5.3 Sub-function: nutrient acquisition

Nutrients which are mobile in the soil can reach plant roots via mass flow, while for poorly mobile nutrients, diffusion can replenish depleted nutrients at the root surface (Hinsinger et al., 2011). In addition, the root system often increases to support nutrient acquisition when concentrations are low and/or plant uptake is high. Such root foraging strategies differ between plant species, resulting in below-ground niche complementarity of (e.g.) shallow- vs. deep-rooting species (Zhang, 2014).

Besides growth, many plant species use specific rhizosphere processes such as the exudation of protons or organic ligands to make nutrients more available (Hinsinger et al., 2011). Symbiosis with mycorrhizal fungi can be a possibility to expand the depleted soil volume, partly by growing into smaller soil pores than roots or root hairs. Ectomycorrhizal fungi may also produce exudates that contribute to the mobilization of recalcitrant nutrients such as phosphorus. Biological fixation of nitrogen from the atmosphere mostly occurs in legumes through symbiosis with rhizobia (Buckley, 2007). In addition, free-living diazotrophs have been shown to be present in many soils, even though their contribution to the nitrogen cycle is likely limited compared to that of symbiotic nitrogen fixation.

6 Water regulation and filtration

The water regulation and filtration function includes two main sub-functions: water storage and filtering capacity. We identified 16 main processes that underpin the sub-functions of water storage and filtering capacity (Fig. 4). The direct and indirect parameters that support the provision of these processes are described in Table 3.

6.1 Sub-function: water storage

Water storage is the capacity of the soil to store water as it integrates multiple interrelated processes, influencing one another to determine the overall water storage capacity of the soil. This interconnected nature of processes in the soil-plant-atmosphere continuum justifies their integration under the broader water storage sub-function (Hillel, 2003).

Water enters the soil through infiltration, influenced by rainfall, soil properties, vegetation, land use and temporary barriers like hydrophobic layers or compacted crusts (Coppola et al., 2011). Water that cannot infiltrate may result in runoff or ponding on low-lying ground. Runoff occurs under two conditions: (i) saturation excess runoff (Dunnian process) – when the soil becomes fully saturated, excess water flows over the surface, and (ii) infiltration excess runoff (Hortonian process) – when rainfall exceeds the soil's infiltration capacity, runoff forms even if the soil is not saturated (Blöschl & Sivapalan, 1995). Infiltration and runoff are interconnected – higher runoff reduces infiltration and vice versa (Wang et al., 2016). The rates of infiltration and runoff depend on precipitation intensity, soil properties and landscape features, especially the upper soil horizon and slope. Soil texture plays a crucial role: fine-textured soils have low hydraulic conductivity, slowing water movement and increasing runoff when rainfall exceeds infiltration capacity (Beven and Germann, 2013). In contrast, coarse-textured soils support higher infiltration, reducing runoff under similar conditions. Rock fragments on the surface also influence infiltration, runoff and erosion processes (Cerdà, 2001).

Water returns to the atmosphere through evapotranspiration, which includes evaporation from the soil surface and transpiration through plant

Published by Burleigh Dodds Science Publishing Limited, 2025.

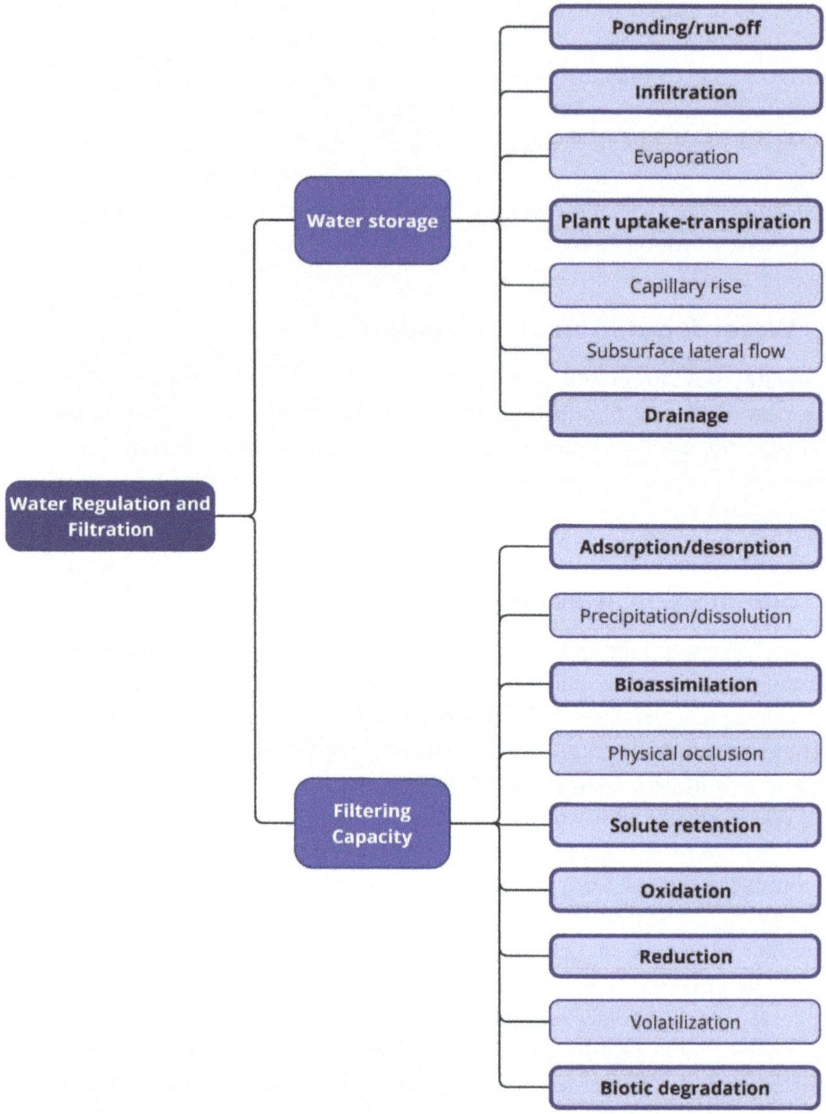

Figure 4 Processes (light blue) and sub-functions (blue) underpinning water regulation and filtration. Boxes with a thicker outline and bold lettering represent processes deemed as most relevant by the experts.

uptake. In crops, over 99% of absorbed water is lost via transpiration, with only a small amount retained in plant tissues (Hillel, 2003). Evapotranspiration is influenced by atmospheric factors (temperature, solar radiation, humidity and wind) and plant traits (root depth, architecture and uptake efficiency) (Feddes

Published by Burleigh Dodds Science Publishing Limited, 2025.

Table 3 Soil parameters that directly or indirectly mediate the processes that occur in the water regulation and filtration function. Some parameters or parts of them are expressed with abbreviations or acronyms. Abbreviations and acronyms for chemical parameters: SOM for soil organic matter content, SOM quality for soil organic matter quality, CEC for cation-exchange capacity, MAOM for mineral-associated organic matter, oxygen for oxygen concentration

	Soil parameters that underpin each process				
Process	Biological	Physical	Chemical	Environmental	Present in other functions*
Infiltration	Fungi Roots Earthworms Enchytraeids Termites Ants	Soil texture Aggregate stability Saturation Soil porosity Impeding layer Surface stoniness Presence of cracks		Temperature Precipitation Vegetation cover Slope Position in landscape	
Evaporation		Soil texture Soil porosity Aggregate Stability Unsaturated hydraulic conductivity Water retention curve Impeding layer Surface stoniness	SOM	Temperature Precipitation Wind speed Vegetation cover Vegetation density Albedo Plant residue layer Relative humidity Latent heat of evaporation	
Plant uptake and transpiration	Roots Plants Mycorrhizal fungi		Soil salinity	Temperature Precipitation Wind speed Vegetation cover Vegetation density Albedo Plant residue layer Relative humidity Latent heat of evaporation Position in landscape Length of growing season	

(*Continued*)

Table 3 (*Continued*)

Process	Soil parameters that underpin each process				Present in other functions*
	Biological	Physical	Chemical	Environmental	
Capillary rise	Bacteria Fungi Roots Earthworms Nematodes Protozoa	Soil texture Soil porosity Saturation Water retention curve Water table depth Clay mineralogy	Water salinity	Temperature Precipitation Position in landscape	
Subsurface lateral flow	Earthworms Roots	Soil porosity Saturation Impeding layer Hydraulic discontinuity		Temperature Precipitation Position in landscape	
Drainage	Roots Earthworms	Soil texture Saturation Soil porosity Soil depth Water table depth Impeding layer Presence of cracks		Temperature Precipitation	
Adsorption/ desorption	Bacteria Fungi Roots Earthworms	Soil texture Soil moisture Soil temperature Specific surface area Soil porosity	Soil pH CEC Clay mineralogy SOM MAOM	Temperature Precipitation Freeze-thaw cycles	
Precipitation/ dissolution	Roots Bacteria Fungi	Soil moisture Soil temperature Soil porosity	Soil pH Chemical saturation SOM Oxygen Clay mineralogy	Temperature Vegetation cover	NC; HP

Bio-assimilation	Bacteria Fungi Roots Earthworms Nematodes Protozoa	Soil temperature Soil moisture	Soil pH SOM	Vegetation
Physical occlusion	Bacteria Fungi Roots Earthworms Protozoa	Soil texture Soil temperature Soil moisture Soil porosity	Soil pH CEC SOM quality Clay mineralogy	Dry-rewetting cycles Freeze-thaw cycles
Solute retention	Bacteria Fungi Roots Earthworms	Soil texture Soil porosity	SOM	
Oxidation	Impeding layer		Soil pH SOM Oxygen	
Reduction	Bacteria (denitrifying) Archaea (denitrifying)	Soil moisture Soil porosity Air permeability	Soil pH SOM Oxygen	
Volatilization	Soil temperature Soil Moisture Soil porosity		Soil pH	Wind speed
Biotic degradation	Bacteria Fungi Collembola Earthworms Protozoa Nematodes Acari	Soil temperature Soil moisture	Soil pH SOM Oxygen Nutrient availability	

*Where NC is nutrient cycling and HP is habitat provision for biodiversity.

et al., 1976). Deeper soils support root growth, providing a larger water reservoir that sustains transpiration during dry periods.

Soil parameters, however, have a complex and sometimes contradictory effect on water availability. For example, fine-textured soils have a higher water-holding capacity, meaning they can store more water. However, their low permeability and high density can impede deep root growth, limiting the plant's ability to access moisture at greater depths. In contrast, coarse-textured soils drain quickly and hold less water but allow for deeper root penetration (Dodd & Lauenroth, 1997).

Water can also move upward from wetter regions (e.g. groundwater) towards drier soil zones due to capillary action. This process is particularly relevant in areas with a shallow groundwater table and is strongly impacted by the capillary porosity and clay content of the soil (Li et al., 2014).

Drainage (or deep percolation) is the downward movement of water out of the soil profile, contributing to groundwater recharge. Soil texture and depth play key roles: sandy soils, with large pores and high conductivity, promote rapid drainage, while clay-rich or compacted soils slow water movement (Keese et al., 2005). Shallow soils over bedrock can also enhance recharge by directing water downward more quickly.

Finally, subsurface lateral flow moves water sideways through the soil towards streams or rivers. This process depends on landscape morphology and soil hydraulic properties. It is most significant when a less permeable layer (e.g. clay-rich or compacted soil) impedes vertical movement, redirecting water laterally along soil horizons (Li et al., 2014).

6.2 Sub-function: filtering capacity

The soil's filtering capacity refers to the capacity of the soil to extract, degrade, transform or contain harmful compounds (BIOSIS, 2023), including the mechanism of their physical transport into the soil.

Each pollutant exhibits a different potential for these processes, and the mechanisms leading to their degradation can be different depending on the pollutant at hand. It is outside of the scope of this chapter to provide the parameters needed to model the degradation of each individual pollutant. Instead, our focus is on describing soils with a high potential for containing, degrading, occluding and transporting pollutants.

Containment refers to soil's capacity to retain elements or compounds within the soil matrix or through biological assimilation, thereby limiting their movement and export. Key containment processes include sorption onto solid surfaces, precipitation, physical occlusion and biological uptake. A soil's ability to retain molecules through chemical reactions depends on the characteristics of sorbent materials and solid surfaces, which can vary with

Published by Burleigh Dodds Science Publishing Limited, 2025.

factors such as redox potential and pH (Naidu, 1998; Sarkar, 2021; Sposito, 1999).

Physical occlusion sequesters elements by making them inaccessible to soil biota and prevents leaching due to reduced water and gas movement, often a result of discontinuous pore structure and aggregation (Chi et al., 2022; Sexstone, 1985). Bio-assimilation relates to the ability of the soil biota to extract elements and molecules from the soil. Fungi and bacteria are particularly relevant to bio-assimilation, as they are the primary consumers of organic pollutants (Saadi, 2012) and micro-plastics (Fojt et al., 2020), and it is through the consumption of these microbes that the pollutants become assimilated at higher levels of the soil food-web (Armitage, 2007). Soil invertebrates can also play an important role in bio-accumulation, both directly through absorption of chemicals through the skin, particularly in the case of soft-bodied soil organisms such as earthworms and molluscs, and indirectly through feeding on contaminated materials (Heikens, 2001; Chao et al., 2023).

Transformation processes include oxidation, reduction and volatilization. Whether a pollutant becomes more or less impactful to the environment through oxidation and reduction processes depends on the nature of the pollutant itself. Therefore, we consider soil with the highest potential for pollutant transformation to be one where both processes can occur over time, allowing various pollutants to undergo different types of transformations. However, we acknowledge that when a specific pollutant is present in the soil, remediation efforts or management practices may be designed to favour either oxidation or reduction processes, depending on which pathway is most effective for pollutant transformation or removal. Ultimately, the most desirable outcome is the complete degradation of pollutants into non-harmful chemical forms. Transformation processes can sometimes contribute to this degradation. While soil organisms, particularly microorganisms, play a key role in the breakdown of organic pollutants, factors such as freeze-thaw cycles have been shown to influence degradation rates indirectly by affecting the microbial community (Ji et al., 2022).

The filtering capacity of soil is also influenced by the processes that govern the fate of pollutants – or, more generally, solutes – as they move through the soil towards groundwater. Among the many mechanisms involved, we focus here on the main ones (Jury & Horton, 2004; Kutílek & Nielsen, 1994): convection (or advection) is the movement of a solute dissolved in soil water, which occurs in direct proportion to the average water flow and solute concentration; diffusion occurs when solute particles move from high-concentration areas to low-concentration areas, following the concentration gradient; mechanical dispersion arises due to variations in water flow within individual pores and among pores of different sizes. Diffusion and mechanical dispersion contribute to solute spreading by smoothing sharp concentration fronts along the main flow direction. Due to their similarities, they are often represented by a combined hydrodynamic dispersion coefficient, determined through miscible displacement experiments in lab and field settings. Given the complexity of these processes, linking measurable soil properties to solute behaviour is challenging. Dispersivity (μ) is a key parameter influencing solute transport. Vanderborght and Vereecken (2007) found that dispersivity increases with transport distance and scale but is unaffected by soil texture, with higher values under saturated conditions. Koestel et al. (2012) identified process scale, water flow level and saturation as critical factors, while Coppola et al. (2011) observed that local heterogeneities, such as stones, also play a role. Godoy et al. (2019) found that while most soil properties showed weak correlations with transport parameters, the adsorption isotherm slope strongly correlated with cation exchange capacity and was significantly influenced by meso- and micro-porosity.

Finally, pollutants can preferentially move through macropores – large structural pores like root channels, earthworm burrows and fissures. Jarvis (2020) found that these biologically and chemically active microsites influence water flow and solute transport, though adsorption and retardation are minimal due to their limited surface area. The macropore network is shaped by soil biota, soil properties, site conditions and land management. High-intensity rainfall can enhance non-equilibrium flow, accelerating solute transport and impacting water quality, especially for low-mobility solutes like pesticides, which pose risks even at low leaching rates.

7 Habitat provision for biodiversity

Habitat provision for biodiversity refers to the capacity of soil to create and sustain suitable habitats for a wide range of organisms, including microorganisms, plants and animals. It encompasses the physical, chemical and biological characteristics of the soil that enable the establishment and maintenance of diverse communities. We present a set of processes and parameters that would, in theory, lead to higher levels of biodiversity (species richness and abundance), both in terms of taxonomic diversity and functional diversity. The habitat provision for biodiversity function assumes that habitats need to be available for colonization and that increased habitat complexity will increase the opportunities for soil organisms to establish and thrive. Habitat provision is influenced by a variety of interconnected processes, grouped into three main sub-functions: spatial availability and complexity, habitat suitability and stability and biotic interactions (Fig. 5). The direct and indirect parameters that support the provision of these processes are described in Table 4.

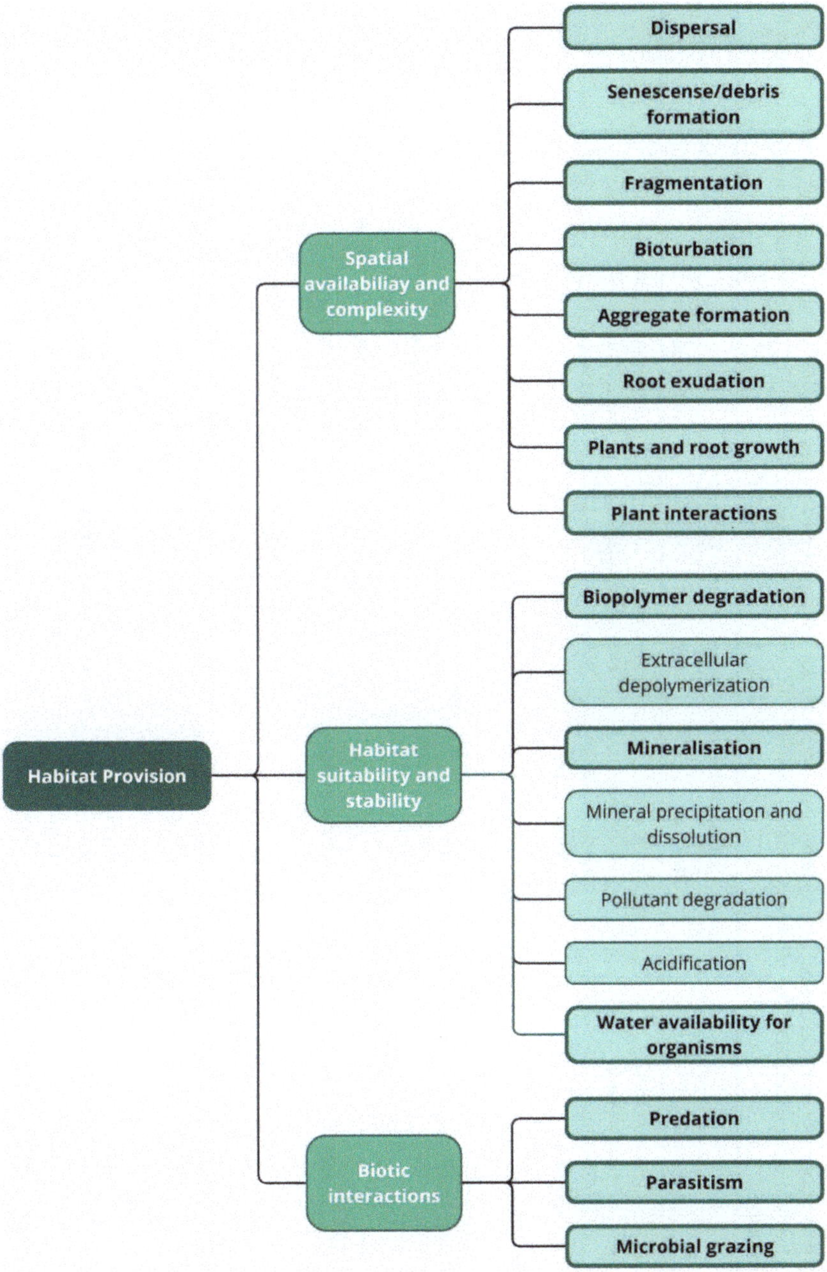

Figure 5 Processes (light green) and sub-functions (green) underpinning habitat provision for biodiversity function. Boxes with a thicker outline and bold lettering represent processes deemed as most relevant by the experts.

Table 4 Soil parameters that directly or indirectly mediate the processes that occur in the habitat provision function. Some parameters or parts of them are expressed with abbreviations or acronyms. Abbreviations and acronyms for biological parameters: N-fixing for nitrogen-fixing. Abbreviations and acronyms for chemical parameters: SOM for soil organic matter content, CEC for cation-exchange capacity, $CaCO_3$ for calcium carbonate, oxygen for oxygen concentration

	Soil parameters that underpin each process				Present in other functions**
Process	Biological*	Physical	Chemical	Environmental	
Dispersal	Earthworms Enchytraeids Termites Ants Nematodes Acari Collembola	Soil texture Bulk density Soil temperature Soil moisture Soil depth	Soil pH Nutrient availability SOM	Temperature Precipitation Wind speed	
Senescence/ debris formation	Plants			Precipitation Wind speed Vegetation cover Vegetation composition	
Fragmentation	Acari (Oribatids) Isopods Collembola Earthworms Enchytraeids Termites Ants Millipedes	Soil temperature Soil moisture	Soil pH Oxygen Plant residue quality	Temperature Precipitation Plant residue layer depth	CR; NC
Bioturbation	Earthworms Enchytraeids Termites Ants	Soil texture Bulk density Soil temperature Soil moisture Soil depth	Soil pH Nutrient availability SOM	Temperature Precipitation Plant residue layer depth	CR

Aggregate formation	Earthworms Enchytraeids Termites Ants Bacteria Fungi Roots	Soil temperature Soil moisture Soil porosity	Soil pH CEC SOM quality Clay mineralogy	Temperature Precipitation	CR
Root exudation	Roots	Soil moisture	Soil pH Nutrient availability Oxygen		NC
Plant and root growth	Roots	Soil moisture Effective root depth Surface stoniness	Soil pH Nutrient availability Micronutrient availability		
Plant interactions	Roots Bacteria (N-fixing) Archaea (N-fixing)	Soil texture Water holding capacity Soil porosity Soil depth	Soil pH Nutrient availability SOM	Temperature Precipitation Topography	
Biopolymer degradation	Acari Collembola Earthworms Nematodes Protozoa Termites Ants Millipedes	Soil texture Soil temperature Soil moisture Soil porosity	Soil pH Nutrient availability SOM Oxygen	Temperature Precipitation	NC
Extracellular depolymerization	Bacteria Fungi Roots	Soil temperature Soil moisture Soil porosity	Soil pH Nutrient availability SOM SOM quality		CR; NC

(Continued)

Table 4 (*Continued*)

	Soil parameters that underpin each process				
Process	Biological*	Physical	Chemical	Environmental	Present in other functions**
Mineralization	Bacteria Fungi	Soil texture Soil temperature Soil moisture	Soil pH Nutrient availability SOM SOM quality Oxygen CaCO3	Temperature Precipitation	NC
Precipitation/ dissolution	Bacteria Fungi Roots	Soil temperature Soil moisture Soil porosity	Soil pH Chemical saturation SOM Oxygen Clay mineralogy	Temperature Vegetation cover	NC; WR
Pollutant degradation	Bacteria Fungi Roots Earthworms	Soil temperature Soil moisture Soil porosity	Soil pH CEC SOM Presence of pollutants Presence of surfactants Oxygen Soil enzyme activity		
Acidification	Roots		SOM CEC	Precipitation Vegetation composition Deposit concentration in air Bedrock type	
Water availability		Soil porosity Soil moisture			

Predation	Acari (O)
	Acari (Pr)
	Collembola (O)
	Collembola (Pr)
	Nematodes (O)
	Nematodes (Pr)
	Spiders
	Insects (Pr)
	Ants (Pr)
Parasitism	Bacteria (Pa)
	Fungi (Pa)
	Insects (Pd)
	Nematodes (Pa)
	Protozoa (Pa)
	Viruses
Microbial grazing	Acari (F)
	Collembola (B)
	Collembola (F)
	Collembola (O)
	Enchytraeids
	Nematodes (B)
	Nematodes (F)
	Nematodes (O)
	Protozoa (B)
	Protozoa (F)
	Protozoa (O)

*Where O are omnivores, Pr are predatory, Pa are parasitic, Pd are parasitoid, F are fungivores, B are bacterivores.
**Where CR is carbon regulation, NC is nutrient cycling and WR is water regulation and provision.

7.1 Sub-function: habitat availability and complexity

This sub-function incorporates the processes that relate to the physical availability of habitats to soil organisms, as well as the complexity of the soil environment. It includes factors that enable organisms to disperse to new habitats and processes that enhance the complexity of the soil habitat. Examples include the variety of soil inputs, for instance, root exudation from a diverse set of plants, or soil organic matter with different properties.

While the habitat provision potential may be high, it is crucial for organisms to physically reach the micro- or macro-sites for these to function as effective habitats. This is also true across scales, from the micro scale (e.g. pore spaces) to the landscape scale. Barriers (whether physical, chemical or anthropogenically derived) at different spatial scales can effectively decrease the habitat provision potential of soil by impeding the dispersal of organisms. Dispersal rates for soil organisms can vary greatly depending on the mode of dispersal, as well as the size of the propagules or dispersal units (*sensu* Kleyer et al., 2008). Smaller dispersal units can be transported long distances by wind or water (Hirst, 1965; Ptatscheck, 2018), but successful (re-)colonization of a habitat patch can still take a long time (e.g. years) (Klein, 2020). Some organisms will also adopt hitchhiking strategies, either on the intestinal tract of other soil organisms or by anchoring themselves to other organisms (Wardle, 2006). However, habitat fragmentation or spatiotemporal discontinuity of the habitat can negatively impact soil biodiversity by limiting the availability of and movement to the habitat, with consequent impacts on the composition and dynamics of local communities as well as biotic interactions (Nordén et al., 2014).

Soil organisms, such as earthworms, ants, enchytraeids and collembola are of great importance to habitat formation through roles such as bioturbation, and the formation of pores and aggregates (Wardle, 2006; Feeney et al., 2006). Microbial activity and plant roots also contribute significantly to these processes by stabilizing aggregates and enriching the soil matrix with exudates and organic matter (Paul, 2016). Through these processes, the physical structure of the soil matrix becomes more complex leading to higher habitat heterogeneity (Coleman et al., 2024). In fact, plants exert a profound influence on the spatial complexity of the soil environment through their root exudates, which alter the composition of microbial communities in their immediate surroundings (Bais et al., 2006; Zhao, et al., 2023b). Different plant species have distinct microbial communities associated with their root system and, therefore, an increase in plant diversity is expected to lead to increasing habitat heterogeneity for associated species (Lamb et al., 2011). Aboveground, different plant species are known to produce litter with varying chemical and physical properties, which can affect detritivore and decomposer communities in the soil. This results in changing patterns of decomposition, fragmentation and organic matter formation, further contributing to habitat heterogeneity.

7.2 Sub-function: habitat suitability and stability

Habitat suitability and stability encompasses the processes that ensure that the soil is a favourable environment for organisms, and remains so over time. These processes are linked to nutrient and water supply, maintenance of soil structure and the mitigation of habitat degradation through mechanisms like acidification control and pollutant degradation. Nutrient cycling is a cornerstone of soil health, driven by the interactions between soil microorganisms, fauna and organic matter inputs. The decomposition of soil organic matter (SOM) is a multi-step process involving fragmentation and depolymerization. Fragmentation involves breaking down plant residues (e.g. leaves and other dead plant material) into smaller, edible pieces for soil fauna and microorganisms (Brown et al., 2018; Frouz, 2018). This process is primarily led by soil fauna such as earthworms, millipedes, isopods, springtails, mites and nematodes. Depolymerization is the further breakdown of complex organic matter into simpler, plant-available elements and is primarily carried out by bacteria and fungi (Kögel-Knabner 2002; Xu et al., 2018; Salvachúa et al., 2016). This leads to the release of plant-available nutrients such as nitrogen and phosphorus (Frouz, 2018; Kögel-Knabner, 2002). The continuous input of SOM, such as crop residues, animal manure and leaf litter, replenishes nutrient pools and maintains microbial activity; in addition, plants release organic compounds as exudates into the rhizosphere, further contributing to microbial nutrient cycling. This creates a feedback loop as plant diversity fosters microbial diversity, which in turn improves nutrient availability for plants (Ehrmann & Ritz, 2014). Additionally, nutrient cycling is linked to biotic and abiotic mineral weathering. Biotic agents, such as bacteria, fungi, lichens and roots, secrete organic acids and enzymes that dissolve primary minerals, releasing essential nutrients into the soil solution. Abiotic factors, such as weathering and erosion, also play a role. These geochemical processes alter the soil's chemical composition over time, influencing parameters like pH, redox potential, and cation exchange capacity (April & Newton, 1992). The release of nutrients depends on intrinsic soil characteristics, such as texture, the presence of coarse fragments, soil depth, mineral composition, and the underlying bedrock (Zhu et al., 2014).

The soil's ability to allow water to infiltrate and to be retained is critical for maintaining habitat stability and ecosystem functions. Soil structure, particularly aggregation and pore formation, plays a central role in water infiltration and retention. Aggregation as the clustering of soil particles into aggregates, stabilizes soil structure and improves the capacity to retain water. Pore formation, involving the creation of spaces within the soil, facilitates the movement of air, water and organisms. Together, these processes enhance the physical structure of soil. Soil texture, SOM content and bioturbation (mixing of soil elements by organisms) significantly influence aggregation and pore formation. Earthworms, for example, create macropores that improve water infiltration and reduce surface runoff during heavy rainfall (Bottinelli et al., 2010, 2015; Bacq-Labreuil et al., 2018; 2019). Similarly, ants, termites, plant roots and fungi contribute to soil porosity and stability by forming and stabilizing soil aggregates (Cammeraat and Risch 2008; Jouquet et al., 2011; Helliwell et al., 2017; Feeney et al., 2006; Martin et al., 2012). Root architecture, mucilage secretion and vegetation density also play critical roles, directly modulating soil structure and hydrological dynamics (Bacq-Labreuil et al., 2019). From a physical perspective, soil porosity, pore size distribution, and pore connectivity are key parameters that determine water retention and gas exchange in soil. These characteristics influence the soil's ability to store water and buffer against drought or flooding (Zhang et al., 2021), thereby ensuring habitat suitability. Keystone species and biological engineers, such as earthworms, arbuscular mycorrhizal fungi (AMF), and plant roots, play critical roles in maintaining soil health and functionality. Via their movements and activity through the soil layers, they play a key role in litter decomposition (Brown et al., 2000), and pore formation (Bottinelli et al., 2015), improving nutrient availability (Fischer et al., 2014) and water infiltration (Capowiez et al., 2015; Jouquet et al., 2012b). Fungal hyphae grow through fissures and pores and form aggregates at a micro scale important for aggregate stability (Tisdall & Oades, 1982; Dorioz et al., 1993). Arbuscular mycorrhizal fungi are symbiotic fungi that facilitate nutrient cycling and enhance plant root access to essential nutrients, such as phosphorus (Miller & Fitzsimons, 2011). Roots also modify soil structure by forming aggregates, increasing porosity, and secreting mucilage and exudates. These processes stabilize soil structure, enhance water retention and improve soil nutrient dynamics (Whalley et al., 2005; Helliwell et al., 2017; Chenu & Cosentino, 2011; Tisdall & Oades, 1982; Read et al., 2003).

7.3 Sub-function: biotic interactions

Biotic interactions encompass the complex relationships among soil organisms and between soil organisms and plants. This sub-function emphasizes the importance of interactions like predation, microbial grazing and parasitism in supporting soil biodiversity and functionality. The primary indicators for the sub-functions here are centred on the diversity and abundance of predators, microbial grazers and parasites, as these metrics provide critical insights into the balance, functionality and resilience of soil ecosystems by reflecting the variety and activity levels of organisms involved in these processes.

Predation dynamics, involving organisms like nematodes, mites, ants, spiders and larger soil fauna (macrofauna), play a pivotal role in regulating soil ecosystems. Predators control populations of prey species, preventing imbalances that could disrupt soil processes (Meyer et al., 2019; Bernardin et al., 2024). By preventing the overpopulation of any single species, predation contributes to a balanced soil community structure, which enhances functional redundancy and strengthens the resilience of the ecosystem (Feit et al., 2019). In addition to regulating prey populations, predators contribute to soil nutrient cycling. Predators indirectly support nutrient turnover and availability for plants through excretion, egestion and translocation of nutrients (Schmitz et al., 2010). For instance, predatory ants and beetles can redistribute organic matter and nutrients across soil profiles, enhancing soil fertility. Predators also play a crucial role in pest control by managing soil-dwelling pest populations, such as root-feeding nematodes and insect larvae. This natural regulation indirectly supports plant health and productivity by reducing the need for chemical pest control and by promoting sustainable land management (Moore et al., 1988; Beretta et al., 2022). The role of predation in pest control highlights its contribution to ecosystem services, such as crop yield enhancement and biodiversity conservation. Microbial grazers, such as protozoa and nematodes, regulate soil microbial populations, contributing to the balance between different microbial groups, and enhancing nutrient cycling. These grazers release nutrients stored in microbial biomass, particularly nitrogen and phosphorus, making them available for plant uptake (Griffiths, 1994; Bonkowski, 2004; Jiang et al., 2017; Camuy-Vélez et al., 2024). The process of microbial grazing is essential for sustaining microbial diversity and maintaining soil fertility. Microbial grazing also influences the size and turnover rate of organic carbon pools, affecting the carbon dynamics in the soil, and its role as a carbon sink (Jiang et al., 2018). While grazing promotes nutrient mineralization, it can also alter the microbial community structure, influencing decomposition processes and overall soil functionality.

Parasitism in soil ecosystems involves interactions where one organism derives nutrients from a host, often regulating host populations. Parasites, such as parasitic nematodes and fungi, can prevent outbreaks of dominant species, maintaining community structure and fostering biodiversity (Bueno-Pallero et al., 2018). Moreover, parasitic nematodes and fungi also directly suppress soil-borne pathogens, indirectly benefiting plant health and reducing the need for chemical interventions (Klingen & Haukeland, 2006). These interactions make parasitism a critical component of soil health, with significant implications for sustainable agriculture.

Other processes like facilitation and competition are equally important for supporting biodiversity. Facilitation occurs when one species creates conditions favourable for another, such as nitrogen-fixing bacteria improving soil fertility for neighbouring plants (van der Heijden et al., 2008). Conversely, competition shapes community composition by determining access to limited resources like nutrients, water and space. While these processes are vital for understanding soil biodiversity, their effects are highly species-specific, making them difficult to generalize in a soil health assessment. As a result, their inclusion is beyond the scope of this framework.

8 Defining soil health through soil functions

Soil health is best assessed through its capacity to sustain multiple essential functions rather than relying on a rigid set of pre-selected indicators. However, the most commonly applied approach in soil health monitoring is a minimum dataset approach, which often prioritizes feasibility over functional relevance. Indicators are typically selected based on:

1. Availability of existing data: whether an existing scheme measured the indicator previously (legacy data);
2. Expertise of the scientists and practitioners: selection is influenced by the knowledge and familiarity of those conducting the assessments;
3. Ease of application – indicators must be measurable across varying spatial scales and monitoring schemes; and
4. Cost considerations – to ensure the cost of the monitoring scheme is kept to a minimum.

While practical, this approach does not always account for the functional significance of selected indicators. It often overlooks whether the indicators effectively represent soil processes, functions and their interactions across different land uses, landscape complexity or climatic conditions.

Monitoring frameworks should select indicators that directly reflect the soil functions considered rather than convenience or availability. This ensures a more accurate and holistic assessment of soil health. Effective soil health monitoring should include a balanced and integrative set of indicators representing the soil's capacity to sustain multiple soil functions (and deliver ecosystem services).

Given the complexity of the soil systems, monitoring should include at least one or more relevant process(es) per sub-function within each soil function. This structured approach ensures that the monitoring system captures the multifaceted nature of soil health, providing a more comprehensive basis for sustainable land management practices.

While the soil function models present a generalized approach, the selection of processes and parameters must be context-specific (Fig. 6). Relevant indicators should be chosen based on:

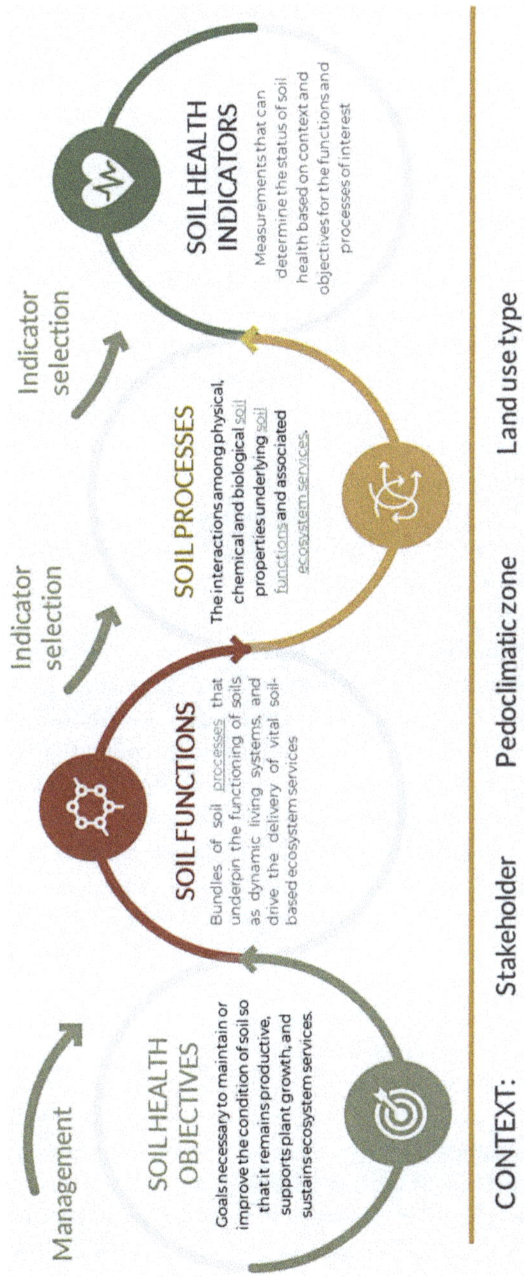

Figure 6 Framework showcasing the relationship between soil health, soil functions, soil processes and soil health objectives.

1 Stakeholder needs: who is the monitoring designed for (e.g. farmers, policymakers)?
2 Pedo-climatic conditions: how do the soil type, climate and environmental constraints affect soil function?
3 Land use: what is the dominant land use (e.g. agriculture, forestry and urban systems), and how does it affect soil health dynamics?

By tailoring indicator selection to these factors, monitoring frameworks can provide more meaningful insights into soil health dynamics and support targeted management strategies.

This research has sought to further clarify the role of soil functions in soil health assessment, by providing a transparent framework from which to select indicators. The soil function cognitive models presented here offer a systematic approach to identifying the most relevant parameters and processes for monitoring the four core functions: carbon and climate regulation, nutrient cycling, water regulation and filtration and habitat provision for biodiversity.

The next steps involve developing an automated system for indicator selection, tailored to context, purpose, and scale of assessment. This would enable more efficient and objective-driven soil health monitoring across different stakeholder levels:

(a) Local scale: at the field level, land managers define the key soil health challenges in their given context and objectives for each function. These objectives guide management decisions to optimize soil health under specific site-specific conditions (local context). The soil processes should align with the objectives enabling the direct assessment of soil functionality. The processes themselves may be measured directly, or a combination of direct and indirect parameters (context-based proxies) can be used to approximate key processes. The measured parameters become indicators, collectively providing a comprehensive picture of soil health at the field scale. This fine-scale assessment allows for a detailed understanding of how soil functions respond to different management practices, supporting adaptive and sustainable land use management strategies.
(b) Regional scale – value chains and agricultural cooperatives: At the value-chain level, soil health monitoring is guided by regional production challenges rather than individual land management decisions. At this level, key soil health challenges are linked to specific crops or livestock systems. Indicators are selected to track the impact of farming practices on soil functions within the region, which enables better alignment of agricultural sustainability strategies across multiple stakeholders, ensuring long-term productivity and resilience.

(c) Broader scales – national and continental monitoring programs: large-scale monitoring efforts focus on assessing the overall soil health trends across diverse landscapes and climatic regions. Standardized indicators must be selected at this scale to ensure comparability across regions. These broad-scale assessments (e.g. European climatic regions) identify the main trends, highlight regions at risk, and inform policy interventions and value-chain support programs. While such assessment provides macro-level insights, they cannot fully capture localized soil health variations, reinforcing the need for nested monitoring approaches that connect local, regional and national assessments.

Structuring soil health monitoring frameworks across multiple scales ensures that assessment remains scientifically robust, stakeholder-relevant and applicable.

9 Conclusion

This chapter has introduced a multi-functional approach to soil health, illustrating the interconnections between key processes and their associated parameters across different soil functions. By integrating this multi-functional perspective, soil health assessment moves beyond isolated measurements towards a more holistic, system-based framework that better reflects soil dynamics and functionality.

The cognitive models presented provide a structured breakdown of function-specific processes, offering deeper insights into the primary drivers and contextual parameters that shape soil health. Recognizing that soil functions are governed by interconnected processes, ensuring that any monitoring efforts capture the complexity of soil systems, rather than relying on individual property measurements in isolation. While some processes are more critical than others, most interact in ways that require a comprehensive, integrated assessment methodology.

By defining appropriate indicators that reflect the underlying mechanisms of soil functions, this work contributes to the development of robust, adaptable and scientifically grounded soil health monitoring frameworks. By identifying the key drivers of soil processes and linking them to relevant, measurable parameters, this work lays the groundwork for a comprehensive and scalable soil health assessment, one that can be tailored to different contexts, land uses and management objectives while ensuring scientific rigour and practical applicability.

10 Acknowledgements

The work presented in this paper is part of and supported by the European Union's Horizon Europe research and innovation program under the Soil Health

BENCHMARKS project, Grant Agreement No. 101091010. We would also like to thank the contributing members of the BENCHMARKS project, who supported the discussions for the development of the cognitive models described in this paper, specifically, Marialaura Bancheri, Graciela Rusch, Geoffroy Sere and David Montague.

11 References

Almagro, M., Maestre, F. T., Martínez-López, J., Valencia, E., & Rey, A. (2015). Climate change may reduce litter decomposition while enhancing the contribution of photodegradation in dry perennial Mediterranean grasslands. *Soil Biology and Biochemistry*, *90*, 214–223. https://doi.org/10.1016/j.soilbio.2015.08.006

April, R., & Newton, R. (1992). Mineralogy and mineral weathering. In D. W. Johnson & S. E. Lindberg (Eds.), *Atmospheric deposition and forest nutrient cycling: A synthesis of the integrated forest study* (pp. 378–425). Springer. https://doi.org/10.1007/978-1-4612-2806-6_10

Armitage, J. M., & Gobas, F. A. P. C. (2007). A terrestrial food-chain bioaccumulation model for POPs. *Environmental Science & Technology*, *41*(11), 4019–4025. https://doi.org/10.1021/es0700597

Bacq-Labreuil, A., Crawford, J., Mooney, S. J., Neal, A. L., Akkari, E., McAuliffe, C., Zhang, X., Redmile-Gordon, M., & Ritz, K. (2018). Effects of cropping systems upon the three-dimensional architecture of soil systems are modulated by texture. *Geoderma*, *332*, 73–83. https://doi.org/10.1016/j.geoderma.2018.07.002

Bacq-Labreuil, A., Crawford, J., Mooney, S. J., Neal, A. L., & Ritz, K. (2019). Cover crop species have contrasting influence upon soil structural genesis and microbial community phenotype. *Scientific Reports*, *9*(1), 7473. https://doi.org/10.1038/s41598-019-43937-6

Bais, H. P., Weir, T. L., Perry, L. G., Gilroy, S., & Vivanco, J. M. (2006). The role of root exudates in rhizosphere interactions with plants and other organisms. *Annual Review of Plant Biology*, *57*, 233–266. https://doi.org/10.1146/annurev.arplant.57.032905.105159

Bernardin, J. R., Gray, S. M., & Bittleston, L. S. (2024). Arthropod prey type drives decomposition rates and microbial community processes. *Applied and Environmental Microbiology*, *90*(7), e00394-24. https://doi.org/10.1128/aem.00394-241734

Beretta, G. M., Deere, J. A., Messelink, G. J., Muñoz-Cárdenas, K., & Janssen, A. (2022). Review: Predatory soil mites as biocontrol agents of above- and below-ground plant pests. *Experimental and Applied Acarology*, *87*(2), 143–162. https://doi.org/10.1007/s10493-022-00723-w

Beven, K. and Germann, P. (2013). Macropores and water flow in soils revisited. *Water Resources Research* 49: 3071–3092.

BIOSIS. (2023). BIOSIS. Retrieved July 10, 2023, from https://biosisplatform.eu/services/soil-biota-and-functions

Blöschl, G., & Sivapalan, M. (1995). Scale issues in hydrological modelling: A review. *Hydrological Processes*, *9*(3–4), 251–290.

Bobbink, R., Tomassen, H., Weijters, M., van den Berg, L., Strengbom, J., Braun, S., & Hettelingh, J. P. (2015). Effects and empirical critical loads of nitrogen for Europe.

Critical Loads and Dynamic Risk Assessments: Nitrogen, Acidity and Metals in Terrestrial and Aquatic Ecosystems, 85–127.

Boix-Fayos, C., Martínez-Mena, M., Cutillas, P. P., de Vente, J., Barberá, G. G., Mosch, W., & Navas, A. (2017). Carbon redistribution by erosion processes in an intensively disturbed catchment. *Catena*, *149*, 799–809. https://doi.org/10.1016/j.catena.2016.08.003

Boix-Fayos, C., Nadeu, E., Quiñonero, J. M., Martínez-Mena, M., Almagro, M., & De Vente, J. (2015). Sediment flow paths and associated organic carbon dynamics across a Mediterranean catchment. *Hydrology and Earth System Sciences*, *19*(3), 1209–1223. https://doi.org/10.5194/hess-19-1209-2015

Bonkowski, M. (2004). Protozoa and plant growth: The microbial loop in soil revisited. *New Phytologist*, *162*(3), 617–631. https://doi.org/10.1111/j.1469-8137.2004.01066.x

Borggaard, O. K., Raben-Lange, B., Gimsing, A. L., & Strobel, B. W. (2005). Influence of humic substances on phosphate adsorption by aluminium and iron oxides. *Geoderma*, *127*(3), 270–279. https://doi.org/10.1016/j.geoderma.2004.12.011

Bottinelli, N., Henry-des-Tureaux, T., Hallaire, V., Mathieu, J., Benard, Y., Duc Tran, T., & Jouquet, P. (2010). Earthworms accelerate soil porosity turnover under watering conditions. *Geoderma*, *156*(1), 43–47. https://doi.org/10.1016/j.geoderma.2010.01.006

Bottinelli, N., Jouquet, P., Capowiez, Y., Podwojewski, P., Grimaldi, M., & Peng, X. (2015). Why is the influence of soil macrofauna on soil structure only considered by soil ecologists? *Soil and Tillage Research*, *146*, 118–124. https://doi.org/10.1016/j.still.2014.01.007

Brown, G., Da Silva, E., Thomazini, M., Niva, C., Decaëns, T., Cunha, L., Nadolny, H., Demetrio, W., Santos, A., Ferreira, T., Maia, L., Conrado, A., Segalla, R., Casadei-Ferreira, A., Pasini, A., Bartz, M., Sautter, K., James, S., Baretta, D., & Lavelle, P. (2018). *The role of soil fauna in soil health and delivery of ecosystem services* (pp. 197–241). Burleigh Dodds Science Publishing. https://doi.org/10.19103/AS.2017.0033.11

Brown, G. G., Barois, I., & Lavelle, P. (2000). Regulation of soil organic matter dynamics and microbial activityin the drilosphere and the role of interactionswith other edaphic functional domains, Paper presented at the 16th World Congress of Soil Science, 20–26 August 1998, Montpellier, France. *European Journal of Soil Biology*, *36*(3), 177–198. https://doi.org/10.1016/S1164-5563(00)01062-1

Buckley, D. H., Huangyutitham, V., Hsu, S.-F., & Nelson Tyrrell, A. (2007). Stable isotope probing with 15N2 reveals novel noncultivated diazotrophs in soil. *Applied and Environmental Microbiology*, *73*(10), 3196–3204. https://doi.org/10.1128/AEM.02610-06

Bueno-Pallero, F. Á., Blanco-Pérez, R., Dionísio, L., & Campos-Herrera, R. (2018). Simultaneous exposure of nematophagous fungi, entomopathogenic nematodes and entomopathogenic fungi can modulate belowground insect pest control. *Journal of Invertebrate Pathology*, *154*, 85–94. https://doi.org/10.1016/j.jip.2018.04.004

Bünemann, E. K. (2015). Assessment of gross and net mineralization rates of soil organic phosphorus – A review. *Soil Biology and Biochemistry*, *89*, 82–98. https://doi.org/10.1016/j.soilbio.2015.06.026

Bünemann, E. K., Bongiorno, G., Bai, Z., Creamer, R. E., De Deyn, G., de Goede, R., Fleskens, L., Geissen, V., Kuyper, T. W., Mäder, P., Pulleman, M., Sukkel, W., van Groenigen, J. W.,

& Brussaard, L. (2018). Soil quality – A critical review. *Soil Biology and Biochemistry, 120*, 105–125. https://doi.org/10.1016/j.soilbio.2018.01.030

Calvaruso, C., Blanchart, A., Bertin, S., Grand, C., Pierart, A., & Eglin, T. (2021). Quels paramètres du sol mesurer pour évaluer les fonctions et les services écosystémiques associés. *Étude et Gestion des Sols, 28*, 3–29.

Cammeraat, E. L. H., & Risch, A. C. (2008). The impact of ants on mineral soil properties and processes at different spatial scales. *Journal of Applied Entomology, 132*(4), 285–294. https://doi.org/10.1111/j.1439-0418.2008.01281.x

Camuy-Vélez, L. A., Banerjee, S., & Sedivec, K. (2024). Grazing intensity alters network complexity and predator–prey relationships in the soil microbiome. *Applied and Environmental Microbiology, 90*(10), e00425-00424. https://doi.org/10.1128/aem.00425-24

Capowiez, Y., Bottinelli, N., Sammartino, S., Michel, E., & Jouquet, P. (2015). Morphological and functional characterisation of the burrow systems of six earthworm species (Lumbricidae). *Biology and Fertility of Soils, 51*(7), 869–877. https://doi.org/10.1007/s00374-015-1036-x

Cerdà, A. (2001). Effects of rock fragment cover on soil infiltration, interrill runoff and erosion. *European Journal of Soil Science, 52*(1), 59–68.

Chang, J., Havlík, P., Leclère, D., de Vries, W., Valin, H., Deppermann, A., Hasegawa, T., & Obersteiner, M. (2021). Reconciling regional nitrogen boundaries with global food security. *Nature Food, 2*(9), 700–711. https://doi.org/10.1038/s43016-021-00366-x

Chao, J., Feng, S., Hao, Y., Lin, J., & Zhang, B. (2023). Pollution characteristics and risk assessment of organophosphate esters in aquaculture farms and natural water bodies adjacent to the Huanghe River delta. *Journal of Oceanology and Limnology, 41*(1), 251–266. https://doi.org/10.1007/s00343-021-1233-4

Chenu, C., & Cosentino, D. (2011). Microbial regulation of soil structure dynamics. In K. Ritz & I. Young (Eds.), *The architecture and biology and soils: life in inner space* (pp. 37–70). CAB International. https://doi.org/10.1079/9781845935320.0037

Chi, J., Fan, Y., Wang, L., Putnis, C. V., & Zhang, W. (2022). Retention of soil organic matter by occlusion within soil minerals. *Reviews in Environmental Science and Bio/Technology, 21*(3), 727–746. https://doi.org/10.1007/s11157-022-09628-x

Coleman, D. C., Geisen, S., & Wall, D. H. (2024). Chapter 5 - Soil fauna: Occurrence, biodiversity, and roles in ecosystem function. In E. A. Paul & S. D. Frey (Eds.), *Soil microbiology, ecology and biochemistry* (5th ed., pp. 131–159). Elsevier. https://doi.org/10.1016/B978-0-12-822941-5.00005-3

Coppola, A., Basile, A., Wang, X., Comegna, V., Tedeschi, A., Mele, G., & Comegna, A. (2011). Hydrological behaviour of microbiotic crusts on sand dunes: Example from NW China comparing infiltration in crusted and crust-removed soil. *Soil and Tillage Research, 117*, 34–43.

Cotrufo, M. F., & Lavallee, J. M. (2022). Chapter one: Soil organic matter formation, persistence, and functioning: A synthesis of current understanding to inform its conservation and regeneration. In D. L. Sparks (Ed.), *Advances in agronomy* (Vol. 172, pp. 1–66). Academic Press. https://doi.org/10.1016/bs.agron.2021.11.002

Creamer, R. E., Barel, J. M., Bongiorno, G., & Zwetsloot, M. J. (2022). The life of soils: Integrating the who and how of multifunctionality. *Soil Biology and Biochemistry, 166*, 108561. https://doi.org/10.1016/j.soilbio.2022.108561

Cunha, L., Brown, G. G., Stanton, D. W. G., Da Silva, E., Hansel, F. A., Jorge, G., McKey, D., Vidal-Torrado, P., Macedo, R. S., Velasquez, E., James, S. W., Lavelle, P., Kille, P.,

& The Terra Preta de Indio, N. (2016). Soil animals and pedogenesis: The role of earthworms in anthropogenic soils. *Soil Science*, *181*(3/4), 110–125. https://doi.org/10.1097/SS.0000000000000144

Davidson, E. A., Keller, M., Erickson, H. E., Verchot, L. V., & Veldkamp, E. (2000). Testing a conceptual model of soil emissions of nitrous and nitric oxides: Using two functions based on soil nitrogen availability and soil water content, the hole-in-the-pipe model characterizes a large fraction of the observed variation of nitric oxide and nitrous oxide emissions from soils. *BioScience*, *50*(8), 667–680. https://doi.org/10.1641/0006-3568(2000)050[0667:TACMOS]2.0.CO;2

Degens, B. P., & Sparling, G. P. (1995). Repeated wet-dry cycles do not accelerate the mineralization of organic C involved in the macro-aggregation of a sandy loam soil. *Plant and Soil*, *175*(2), 197–203. https://doi.org/10.1007/BF00011355

Dodd, M. B., & Lauenroth, W. K. (1997). The influence of soil texture on the soil water dynamics and vegetation structure of a shortgrass steppe ecosystem. *Plant Ecology*, *133*, 13–28.

Dorioz, J. M., Robert, M., & Chenu, C. (1993). The role of roots, fungi and bacteria on clay particle organization. An experimental approach. *Geoderma*, *56*(1), 179–194. https://doi.org/10.1016/0016-7061(93)90109-X

EEA. (2023). Soil monitoring in Europe – Indicators and thresholds for soil health assessments (EEA Report No 08/2022). https://www.eea.europa.eu/en/analysis/publications/soil-monitoring-in-europe

Ehrmann, J., & Ritz, K. (2014). Plant: soil interactions in temperate multi-cropping production systems. *Plant and Soil*, *376*(1), 1–29. https://doi.org/10.1007/s11104-013-1921-8

Feddes, R. A., Kowalik, P., Kolinska-Malinka, K., & Zaradny, H. (1976). Simulation of field water uptake by plants using a soil water dependent root extraction function. *Journal of Hydrology*, *31*(1–2), 13–26.

Feeney, D. S., Crawford, J. W., Daniell, T., Hallett, P. D., Nunan, N., Ritz, K., Rivers, M., & Young, I. M. (2006). Three-dimensional microorganization of the soil–root–microbe system. *Microbial Ecology*, *52*(1), 151–158. https://doi.org/10.1007/s00248-006-9062-8

Feit, B., Blüthgen, N., Traugott, M., & Jonsson, M. (2019). Resilience of ecosystem processes: A new approach shows that functional redundancy of biological control services is reduced by landscape simplification. *Ecology Letters*, *22*(10), 1568–1577. https://doi.org/10.1111/ele.13347

Fischer, C., Roscher, C., Jensen, B., Eisenhauer, N., Baade, J., Attinger, S., Scheu, S., Weisser, W. W., Schumacher, J., & Hildebrandt, A. (2014). How do earthworms, soil texture and plant composition affect infiltration along an experimental plant diversity gradient in grassland? *PLoS One*, *9*(2), e98987. https://doi.org/10.1371/journal.pone.0098987

Fojt, J., David, J., Přikryl, R., Řezáčová, V., & Kučerík, J. (2020). A critical review of the overlooked challenge of determining micro-bioplastics in soil. *Science of The Total Environment*, *745*, 140975. https://doi.org/10.1016/j.scitotenv.2020.140975

Frouz, J. (2018). Effects of soil macro- and mesofauna on litter decomposition and soil organic matter stabilization. *Geoderma*, *332*, 161–172. https://doi.org/10.1016/j.geoderma.2017.08.039

Godoy, V. A., Zuquette, L. V., & Gómez-Hernández, J. J. (2019). Spatial variability of hydraulic conductivity and solute transport parameters and their spatial correlations to soil properties. *Geoderma*, *339*, 59–69.

Griffiths, B. S. (1994). Microbial-feeding nematodes and protozoa in soil: Their effectson microbial activity and nitrogen mineralization in decomposition hotspots and the rhizosphere. *Plant and Soil*, *164*(1), 25–33. https://doi.org/10.1007/BF00010107

Haygarth, P. M., & Ritz, K. (2009). The future of soils and land use in the UK: Soil systems for the provision of land-based ecosystem services. *Land Use Policy*, *26*, S187–S197. https://doi.org/10.1016/j.landusepol.2009.09.016

Heikens, A., Peijnenburg, W., & Hendriks, A. (2001). Bioaccumulation of heavy metals in terrestrial invertebrates. *Environmental Pollution*, *113*(3), 385–393. https://doi.org/10.1016/S0269-7491(00)00179-2

Helliwell, J. R., Sturrock, C. J., Mairhofer, S., Craigon, J., Ashton, R. W., Miller, A. J., Whalley, W. R., & Mooney, S. J. (2017). The emergent rhizosphere: Imaging the development of the porous architecture at the root-soil interface. *Scientific Reports*, *7*(1), 14875. https://doi.org/10.1038/s41598-017-14904-w

Hillel, D. (2003). *Introduction to environmental soil physics*. Elsevier. https://doi.org/10.1016/B978-0-12-348655-4.X5000-X

Hinsinger, P., Brauman, A., Devau, N., Gérard, F., Jourdan, C., Laclau, J.-P., Le Cadre, E., Jaillard, B., & Plassard, C. (2011). Acquisition of phosphorus and other poorly mobile nutrients by roots. Where do plant nutrition models fail? *Plant and Soil*, *348*(1), 29–61. https://doi.org/10.1007/s11104-011-0903-y

Hirst, J. M. (1965). Dispersal of Soil Microorganisms. In F. B. Kenneth & C. S. William (Eds.), *Ecology of soil-borne plant pathogens* (pp. 69–81). University of California Press. https://doi.org/10.1525/9780520407992-014

Jarvis, N. J. (2020). A review of non-equilibrium water flow and solute transport in soil macropores: Principles, controlling factors and consequences for water quality. *European Journal of Soil Science*, *71*(3), 279–302.

Ji, X., Liu, M., Yang, J., & Feng, F. (2022) Meta-analysis of the impact of freeze–thaw cycles on soil microbial diversity and C and N dynamics. *Soil Biology and Biochemistry*, *168*, 108608.

Jiang, Y., Liu, M., Zhang, J., Chen, Y., Chen, X., Chen, L., Li, H., Zhang, X. X., & Sun, B. (2017). Nematode grazing promotes bacterial community dynamics in soil at the aggregate level. *ISME Journal*, *11*(12), 2705–2717. https://doi.org/10.1038/ismej.2017.120

Jiang, Y., Qian, H., Wang, X., Chen, L., Liu, M., Li, H., & Sun, B. (2018). Nematodes and microbial community affect the sizes and turnover rates of organic carbon pools in soil aggregates. *Soil Biology and Biochemistry*, *119*, 22–31. https://doi.org/10.1016/j.soilbio.2018.01.001

Jouquet, P., Huchet, G., Bottinelli, N., Thu, T. D., & Duc, T. T. (2012a). Does the influence of earthworms on water infiltration, nitrogen leaching and soil respiration depend on the initial soil bulk density? A mesocosm experiment with the endogeic species Metaphire posthuma. *Biology and Fertility of Soils*, *48*(5), 561–567. https://doi.org/10.1007/s00374-011-0652-3

Jouquet, P., Janeau, J.-L., Pisano, A., Sy, H. T., Orange, D., Minh, L. T. N., & Valentin, C. (2012b). Influence of earthworms and termites on runoff and erosion in a tropical steep slope fallow in Vietnam: A rainfall simulation experiment. *Applied Soil Ecology*, *61*, 161–168. https://doi.org/10.1016/j.apsoil.2012.04.004

Jouquet, P., Traoré, S., Choosai, C., Hartmann, C., & Bignell, D. (2011). Influence of termites on ecosystem functioning. Ecosystem services provided by termites. *European Journal of Soil Biology*, *47*(4), 215–222. https://doi.org/10.1016/j.ejsobi.2011.05.005

Jury, W. A., & Horton, R. (2004). *Soil physics*. John Wiley & Sons.

Keese, K. E., Scanlon, B. R., & Reedy, R. C. (2005). Assessing controls on diffuse groundwater recharge using unsaturated flow modeling. *Water Resources Research*, *41*(6). http://dx.doi.org/10.1029/2004WR003841.

Klein, A., Eisenhauer, N., & Schaefer, I. (2020). Invasive lumbricid earthworms in North America–different life histories but common dispersal? *Journal of Biogeography*, *47*(3), 674–685. https://doi.org/10.1111/jbi.13744

Kleyer, M., Bekker, R. M., Knevel, I. C., Bakker, J. P., Thompson, K., Sonnenschein, M., Poschlod, P., Van Groenendael, J. M., Klimeš, L., Klimešová, J., Klotz, S., Rusch, G. M., Hermy, M., Adriaens, D., Boedeltje, G., Bossuyt, B., Dannemann, A., Endels, P., Götzenberger, L.,...Peco, B. (2008). The LEDA Traitbase: A database of life-history traits of the Northwest European flora. *Journal of Ecology*, *96*(6), 1266–1274. https://doi.org/10.1111/j.1365-2745.2008.01430.x

Klingen, I., & Haukeland, S. (2006). The soil as a reservoir for natural enemies of pest insects and mites with emphasis on fungi and nematodes. In J. Eilenberg & H. M. T. Hokkanen (Eds.), *An ecological and societal approach to biological control* (pp. 145–211). Springer. https://doi.org/10.1007/978-1-4020-4401-4_9

Koestel, J. K., Moeys, J., & Jarvis, N. J. (2012). Meta-analysis of the effects of soil properties, site factors and experimental conditions on solute transport. *Hydrology and Earth System Sciences*, *16*(6), 1647–1665.

Kögel-Knabner, I. (2002). The macromolecular organic composition of plant and microbial residues as inputs to soil organic matter. *Soil Biology and Biochemistry*, *34*(2), 139–162. https://doi.org/10.1016/S0038-0717(01)00158-4

Kutílek, M., & Nielsen, D. R. (1994). *Soil hydrology: Texbook for students of soil science, agriculture, forestry, geoecology, hydrology, geomorphology and other related disciplines* (p. 370). Catena Verlag.

Lamb, E. G., Kennedy, N., & Siciliano, S. D. (2011). Effects of plant species richness and evenness on soil microbial community diversity and function. *Plant and Soil*, *338*(1), 483–495. https://doi.org/10.1007/s11104-010-0560-6

Lehmann, J., & Schroth, G. (2002). Nutrient leaching. In G. Schroth & F. L. Sinclair (Eds.), *Trees, crops and soil fertility: Concepts and research methods* (pp. 151–166). CABI Publishing. https://doi.org/10.1079/9780851995939.0151

Li, X., Chang, S. X., & Salifu, K. F. (2014). Soil texture and layering effects on water and salt dynamics in the presence of a water table: A review. *Environmental Reviews*, *22*(1), 41–50.

Liang, C., Amelung, W., Lehmann, J., & Kästner, M. (2019). Quantitative assessment of microbial necromass contribution to soil organic matter. *Global Change Biology*, *25*(11), 3578–3590. https://doi.org/10.1111/gcb.14781

Martin, S. L., Mooney, S. J., Dickinson, M. J., & West, H. M. (2012). Soil structural responses to alterations in soil microbiota induced by the dilution method and mycorrhizal fungal inoculation. *Pedobiologia*, *55*(5), 271–281. https://doi.org/10.1016/j.pedobi.2012.06.001

Martínez-Mena, M., Almagro, M., García-Franco, N., de Vente, J., García, E., & Boix-Fayos, C. (2019). Fluvial sedimentary deposits as carbon sinks: Organic carbon pools and stabilization mechanisms across a Mediterranean catchment. *Biogeosciences*, *16*(5), 1035–1051. https://doi.org/10.5194/bg-16-1035-2019

Martínez-Mena, M., López, J., Almagro, M., Albaladejo, J., Castillo, V., Ortiz, R., & Boix-Fayos, C. (2012). Organic carbon enrichment in sediments: Effects of rainfall characteristics under different land uses in a Mediterranean area. *Catena*, *94*, 36–42. https://doi.org/10.1016/j.catena.2011.02.005

Matson, A., Fantappiè, M., Campbell, G. A., Miranda-Vélez, J. F., Faber, J. H., Gomes, L. C., Hessel, R., Lana, M., Mocali, S., Smith, P., Robinson, D. A., Bispo, A., van Egmond, F., Keesstra, S., Saby, N. P. A., Smreczak, B., Froger, C., Suleymanov, A., & Chenu, C. (2024). Four approaches to setting soil health targets and thresholds in agricultural soils. *Journal of Environmental Management*, *371*, 123141. https://doi.org/10.1016/j.jenvman.2024.123141

Meyer, S. T., Heuss, L., Feldhaar, H., Weisser, W. W., & Gossner, M. M. (2019). Land-use components, abundance of predatory arthropods, and vegetation height affect predation rates in grasslands. *Agriculture, Ecosystems & Environment*, *270–271*, 84–92. https://doi.org/10.1016/j.agee.2018.10.015

Miller, R., & Fitzsimons, M. (2011). Fungal Growth in Soils. In *The architecture and biology of soils: Life in inner space* (pp. 149–163). CABI. https://doi.org/10.1079/9781845935320.0149

Min, K., Slessarev, E., Kan, M., McFarlane, K., Oerter, E., Pett-Ridge, J., Nuccio, E., & Berhe, A.A. (2021). Active microbial biomass decreases, but microbial growth potential remains similar across soil depth profiles under deeply-vs. shallow-rooted plants. *Soil Biology and Biochemistry*, *162*, 108401.

Moore, J. C., Walter, D. E., & Hunt, H. W. (1988). Arthropod regulation of micro- and mesobiota in below-ground detrital food webs. *Annual Review of Entomology*, *33*, 419–435. https://doi.org/10.1146/annurev.en.33.010188.002223

Nadeu, E., de Vente, J., Martínez-Mena, M., & Boix-Fayos, C. (2011). Exploring particle size distribution and organic carbon pools mobilized by different erosion processes at the catchment scale. *Journal of Soils and Sediments*, *11*, 667–678. https://doi.org/10.1007/s11368-011-0348-1

Naidu, R., Sumner, M. E., & Harter, R. (1998). Sorption of heavy metals in strongly weathered soils: An overview. *Environmental Geochemistry and Health*, *20*, 5–9.

Nathan, O. O., Monicah, M.-M., Jayne, M. N., Isaya, S., George, N., & Daniel, M. N. (2022). Nutrient and organic carbon losses by erosion, and their economic and environmental implications in the drylands of Kenya. *Environmental Challenges*, *7*, 100519. https://doi.org/10.1016/j.envc.2022.100519

Noiriel, C., & Daval, D. (2017). Pore-scale geochemical reactivity associated with CO(2) storage: New frontiers at the fluid-solid interface. *Accounts of Chemical Research*, *50*(4), 759–768. https://doi.org/10.1021/acs.accounts.7b00019

Nordén, B., Dahlberg, A., Brandrud, T. E., Fritz, Ö., Ejrnaes, R., & Ovaskainen, O. (2014). Effects of ecological continuity on species richness and composition in forests and woodlands: A review. *Écoscience*, *21*(1), 34–45. https://doi.org/10.2980/21-1-3667

Orgiazzi, A., & Panagos, P. (2018). Soil biodiversity and soil erosion: It is time to get married: Adding an earthworm factor to soil erosion modelling. *Global Ecology and Biogeography*, *27*(10), 1155–1167. https://doi.org/10.1111/geb.12782

Paul, E. A. (2016). The nature and dynamics of soil organic matter: Plant inputs, microbial transformations, and organic matter stabilization. *Soil Biology and Biochemistry*, *98*, 109–126. https://doi.org/10.1016/j.soilbio.2016.04.001

Pereira, P., Bogunovic, I., Muñoz-Rojas, M., & Brevik, E. C. (2018). Soil ecosystem services, sustainability, valuation and management. *Current Opinion in Environmental Science & Health*, *5*, 7–13. https://doi.org/10.1016/j.coesh.2017.12.003

Ptatscheck, C., Gansfort, B., & Traunspurger, W. (2018). The extent of wind-mediated dispersal of small metazoans, focusing nematodes. *Scientific Reports*, *8*(1), 6814. https://doi.org/10.1038/s41598-018-24747-8

Read, D. B., Bengough, A. G., Gregory, P. J., Crawford, J. W., Robinson, D., Scrimgeour, C. M., Young, I. M., Zhang, K., & Zhang, X. (2003). Plant roots release phospholipid surfactants that modify the physical and chemical properties of soil. *New Phytologist*, *157*(2), 315–326. https://doi.org/10.1046/j.1469-8137.2003.00665.x

Ritz, K., Black, H. I. J., Campbell, C. D., Harris, J. A., & Wood, C. (2009). Selecting biological indicators for monitoring soils: A framework for balancing scientific and technical opinion to assist policy development. *Ecological Indicators*, *9*(6), 1212–1221. https://doi.org/10.1016/j.ecolind.2009.02.009

Saadi, Z., Rasmont, A., Cesar, G., Bewa, H., & Benguigui, L. (2012). Fungal degradation of Poly(l-lactide) in soil and in compost. *Journal of Polymers and the Environment*, *20*(2), 273–282. https://doi.org/10.1007/s10924-011-0399-9

Salvachúa, D., Katahira, R., Cleveland, N. S., Khanna, P., Resch, M. G., Black, B. A., Purvine, S. O., Zink, E. M., Prieto, A., Martínez, M. J., Martínez, A. T., Simmons, B. A., Gladden, J. M., & Beckham, G. T. (2016). Lignin depolymerization by fungal secretomes and a microbial sink. *Green Chemistry*, *18*(22), 6046–6062. https://doi.org/10.1039/C6GC01531J

Sarkar, B., Mukhopadhyay, R., Ramanayaka, S., Bolan, N., & Ok, Y. S. (2021). The role of soils in the disposition, sequestration and decontamination of environmental contaminants. *Philosophical Transactions of the Royal Society B*, *376*(1834), 20200177. https://doi.org/10.1098/rstb.2020.0177

Schmitz, O. J., Hawlena, D., & Trussell, G. C. (2010). Predator control of ecosystem nutrient dynamics. *Ecology Letters*, *13*(10), 1199–1209. https://doi.org/10.1111/j.1461-0248.2010.01511.x

Schreefel, L., Creamer, R. E., van Zanten, H. H. E., de Olde, E. M., Koppelmäki, K., Debernardini, M., de Boer, I. J. M., & Schulte, R. P. O. (2024). How to monitor the 'success' of agricultural sustainability: A perspective. *Global Food Security*, *43*, 100810. https://doi.org/10.1016/j.gfs.2024.100810

Schulte, R. P. O., Creamer, R. E., Donnellan, T., Farrelly, N., Fealy, R., O'Donoghue, C., & O'hUallachain, D. (2014). Functional land management: A framework for managing soil-based ecosystem services for the sustainable intensification of agriculture. *Environmental Science & Policy*, *38*, 45–58. https://doi.org/10.1016/j.envsci.2013.10.002

Sexstone, A. J., Revsbech, N. P., Parkin, T. B., & Tiedje, J. M. (1985). Direct measurement of oxygen profiles and denitrification rates in soil aggregates. *Soil Science Society of America Journal*, *49*(3), 645–651. https://doi.org/10.2136/sssaj1985.03615995004900030024x

Six, J., Bossuyt, H., Degryze, S., & Denef, K. (2004). A history of research on the link between (micro)aggregates, soil biota, and soil organic matter dynamics. *Soil and Tillage Research*, *79*(1), 7–31. https://doi.org/10.1016/j.still.2004.03.008

Sposito, G., Skipper, N. T., Sutton, R., Park, S.-h., Soper, A. K., & Greathouse, J. A. (1999). Surface geochemistry of the clay minerals. *Proceedings of the National Academy of Sciences*, *96*(7), 3358–3364. https://doi.org/10.1073/pnas.96.7.3358

Tisdall, J. M., & Oades, J. M. (1982). Organic matter and water-stable aggregates in soils. *Journal of Soil Science*, *33*(2), 141–163. https://doi.org/10.1111/j.1365-2389.1982.tb01755.x

Van den Elsen, E., Anton-Sobejano, R., Cousin, I., Fuchs, L., de Haan, J., Teuling, K., Klimkowicz-Pawlas, A., Niedźwiecki, J., Pindral, S., Montagne, D., Scammacca, O., & Weninger, T. (2024). *A framework to assess soil threats, soil functions and soil-based ecosystem services*. Zenodo. https://doi.org/10.5281/zenodo.14056696

Vanderborght, J., & Vereecken, H. (2007). Review of dispersivities for transport modeling in soils. *Vadose Zone Journal*, *6*(1), 29–52.

Van Der Heijden, M. G. A., Bardgett, R. D., & Van Straalen, N. M. (2008). The unseen majority: Soil microbes as drivers of plant diversity and productivity in terrestrial ecosystems. *Ecology Letters*, *11*(3), 296–310. https://doi.org/10.1111/j.1461-0248.2007.01139.x

Vion-Guibert, L., Capowiez, Y., Alavoine, G., Besaury, L., Delfosse, O., Hedde, M., Marsden, C., & Lashermes, G. (2024). The effects of earthworm species on organic matter transformations and soil microbial communities are only partially related to their bioturbation activity. *Soil Biology and Biochemistry*, *199*, 109606. https://doi.org/10.1016/j.soilbio.2024.109606

Vogel, H. J., Bartke, S., Daedlow, K., Helming, K., Kögel-Knabner, I., Lang, B., Rabot, E., Russell, D., Stößel, B., Weller, U., Wiesmeier, M., & Wollschläger, U. (2018). A systemic approach for modeling soil functions. *Soil*, *4*(1), 83–92. https://doi.org/10.5194/soil-4-83-2018

Wang, B., An, S., Liang, C., Liu, Y., & Kuzyakov, Y. (2021). Microbial necromass as the source of soil organic carbon in global ecosystems. *Soil Biology and Biochemistry*, *162*, 108422. https://doi.org/10.1016/j.soilbio.2021.108422

Wang, Y., Fan, J., Cao, L., & Liang, Y. (2016). Infiltration and runoff generation under various cropping patterns in the red soil region of China. *Land Degradation & Development*, *27*(1), 83–91.

Wardle, D. A. (2006). The influence of biotic interactions on soil biodiversity. *Ecology Letters*, *9*(7), 870–886. https://doi.org/10.1111/j.1461-0248.2006.00931.x

Wen, S., Wang, J., Li, Y., & Shao, M. A. (2022). Effects of anecic earthworms on runoff and erosion on the slope with soil from the Loess Plateau under a rainfall simulation experiment. *Agricultural Water Management*, *259*, 107230. https://doi.org/10.1016/j.agwat.2021.107230

Whalley, W. R., Riseley, B., Leeds-Harrison, P. B., Bird, N. R. A., Leech, P. K., & Adderley, W. P. (2005). Structural differences between bulk and rhizosphere soil. *European Journal of Soil Science*, *56*(3), 353–360. https://doi.org/10.1111/j.1365-2389.2004.00670.x

Xu, J., Liu, S., Song, S., Guo, H., Tang, J., Yong, J.W., Ma, Y. and Chen, X. (2018). Arbuscular mycorrhizal fungi influence decomposition and the associated soil microbial community under different soil phosphorus availability. *Soil Biology and Biochemistry*, *120*, 181–190.

Zhang, C., Postma, J. A., York, L. M., & Lynch, J. P. (2014). Root foraging elicits niche complementarity-dependent yield advantage in the ancient 'three sisters' (maize/

bean/squash) polyculture. *Annals of Botany*, *114*(8), 1719–1733. https://doi.org/10.1093/aob/mcu191

Zhang, S., Yu, Z., Lin, J. and Zhu, B. (2020). Responses of soil carbon decomposition to drying-rewetting cycles: a meta-analysis. *Geoderma*, *361*, 114069.

Zhang, Z., Liu, K., Zhou, H., Lin, H., Li, D., & Peng, X. (2018). Three dimensional characteristics of biopores and non-biopores in the subsoil respond differently to land use and fertilization. *Plant and Soil*, *428*, 453–467.

Zhang, Y-W. et al. 2021. Changes in soil water holding capacity and water availability following vegetation restoration on the Chinese Loess Plateau. *Scientific Reports* 11: 9692

Zhao, F. J., Knights, J. S., Hu, Z. Y., & McGrath, S. P. (2003a). Stable sulfur isotope ratio indicates long-term changes in sulfur deposition in the broadbalk experiment since 1845. *Journal of Environmental Quality*, *32*(1), 33–39. https://doi.org/10.2134/jeq2003.3300

Zhao, X., Guo, M., Zhang, T., Bai, S., Meng, Y., Tian, Y., Yang, J., & Ma, F. (2023b). Spatiotemporal dynamics of root exudates drive microbial adaptation mechanisms under day-night alterations in constructed wetlands. *Chemical Engineering Journal*, *477*, 147311. https://doi.org/10.1016/j.cej.2023.147311

Zhu, Y., Duan, G., Chen, B., Peng, X., Chen, Z., & Sun, G. (2014). Mineral weathering and element cycling in soil-microorganism-plant system. *Science China Earth Sciences*, *57*(5), 888–896. https://doi.org/10.1007/s11430-014-4861-0

Zwetsloot, M. J., Bongiorno, G., Barel, J. M., di Lonardo, D. P., & Creamer, R. E. (2022). A flexible selection tool for the inclusion of soil biology methods in the assessment of soil multifunctionality. *Soil Biology and Biochemistry*, *166*, 108514. https://doi.org/10.1016/j.soilbio.2021.108514

Part 1

Advances in DNA-based methods, biological and other indicators to investigate soil microbial activity

Chapter 2

Advances in DNA-based methods for assessing abundance and diversity of soil microbial groups

Dietrich S. Epp Schmidt, Stephanie A. Yarwood and Jared L. Wilmoth, University of Maryland, USA

1 Introduction

Almost a half century ago, Alexander (1977) identified nine genera of bacteria that were important in soils. That list included *Agrobacterium*, *Alcaligenes*, *Arthrobacter*, *Bacillus*, *Flavobacterium*, *Micromonospora*, *Nocardia*, *Pseudomonas*, and *Streptomyces*. Today Alexander's list only represents 2–3% of known soil microorganisms (Janssen 2006) and omits phyla such as *Acidobacteria*, *Planctomycetes*, and *Verrucomicrobia* that are ubiquitous (Delgado-Baquerizo et al. 2018). This discrepancy illustrates the dramatic shift that has taken place in soil microbial ecology due to wide adoption of molecular techniques that has largely replaced culturing to describe communities. Although these techniques have opened entire new lines of inquiry into life within soil, they have also uncovered challenges in our understanding of how microbial communities function within the soil matrix and how we should interpret the vast numbers of organisms that we encounter there.

Technological advancements have resulted in high throughput sequencing that is more affordable and widely available than could have been imagined even in the year 2000. A revolution in sequencing chemistry on microfluidic

http://dx.doi.org/10.19103/AS.2025.0159.02

and nanopore platforms has been accompanied by an ever-increasing computational capacity and rapid development of bioinformatics. This combination allows a broad group of users to sequence the soil microbiome and then evaluate those data for important patterns. This chapter will describe the use of amplicon and shotgun sequencing to describe the microbial community, give a brief overview of current sequencing technologies, offer suggestions about best practices used to collect and analyze the data, and discuss the potential uses of these data in assessing soil health.

2 Amplicon sequencing

The use of amplicon sequencing permeates current soil science research. In 2020, there were 400+ peer-reviewed articles published that included this approach, comprising about 8% of all soil science publications (Alteio et al. 2021). Most of those studies targeted bacterial and archaeal 16S rRNA genes and/or fungal intergenic transcribed spacer (ITS) regions. Both the 16S rRNA and ITS sequencing region provide variable regions sufficient to characterize bacteria and fungi, respectively, to taxonomic levels of family or genus.

General workflows have been established (Fig. 1), but there are a number of choices that researchers make that can introduce variation in sequencing results (reviewed by Alteio et al. 2021). Differences in DNA extraction kits (Galla et al. 2024), primers, and sequencing platforms (Stevens et al. 2023; Manter et al. 2024) may result in differences in estimates of species richness and in determining the relative abundance of identified groups. Papers that have reported these differences recommend the use of a standard soil or mock community that can be used as a reference and point out the need for standard protocols to be established.

2.1 Soil extraction

Today most labs make use of commercially available extraction kits, which use either chemical lyses, physical lyses, or both to release DNA from cells and then purify it by removing potential PCR inhibitors. The choice of extraction kits depends on the number of samples being processed, the amount of microbial biomass, and the concentration of organic material that will need to be removed. For labs processing a few hundred soil samples annually, the most cost-effective options are kits with spin columns where ~30 samples can be extracted at once with use of a bench top centrifuge. Higher throughput methods such as those used by the Earth Microbiome Project (https://earthmicrobiome.org) require vacuum manifolds or magnetic bead capture systems but can accommodate ~100 samples being extracted simultaneously. Commonly used kits are typically appropriate for agricultural soils, but

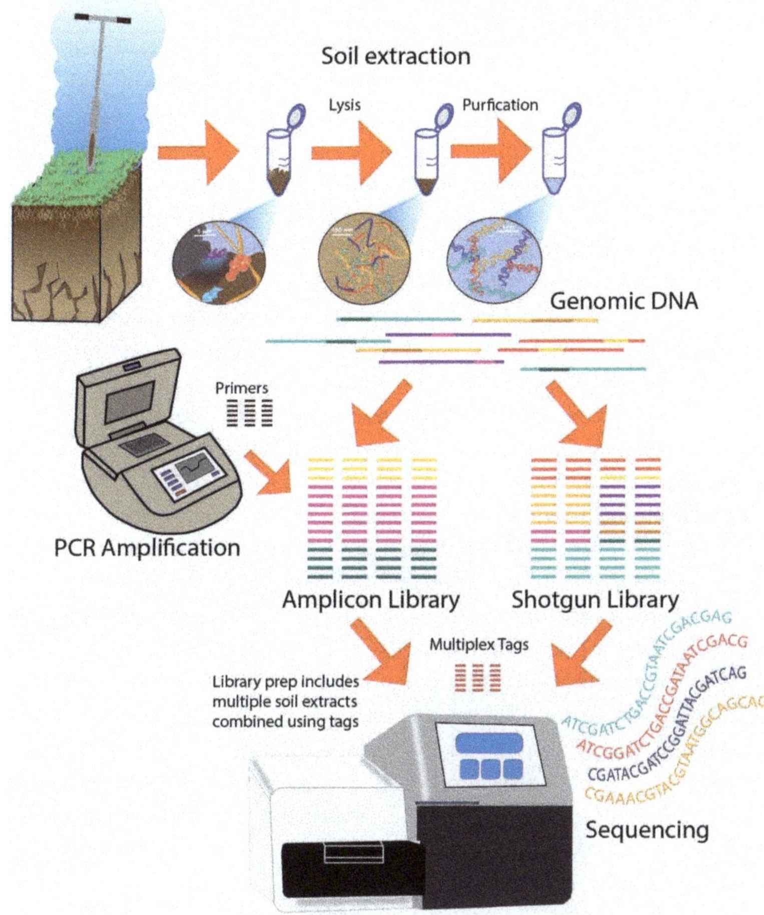

Figure 1 Flow chart showing the workflow used in amplicon and shotgun sequencing. During DNA extraction, cells are lysed. This example illustrates three bacteria and one fungus in the original sample. Soil and cellular material are removed during DNA cleanup, leaving only the genomic DNA. In the case of amplicon sequencing primers are used to target a gene region such as a portion of the rRNA gene (as shown in the contrasting colors on the genomic DNA fragments). Here the PCR reaction results in amplification of the 16S rRNA gene for the three bacteria, excluding the fungal member. In contrast shotgun methods fragment all the genomic DNA so that sequencing input includes both the ribosomal and all other genomic material.

specialized kits are available for soils with low microbial biomass and for soils with high organic matter or clay content that may require additional clean up steps. Quantification of DNA by fluorometry or spectrophotometry is typically used to evaluate the success of individual extractions. Whenever possible, the same extraction method should be used across samples that will later be compared (Galla et al. 2024).

2.2 PCR amplification

DNA extracts are used as templates in PCR reactions, where primer choice and reaction conditions allow researchers to target specific groups of the microbial communities. Most studies target the 16S rRNA gene for bacteria and archaea using primers that flank the V4 region of the gene (F515/R806) (Caporaso et al. 2011). The original primer pair sequences have been improved since their initial design. In the case of the F515 the revision has allowed for better amplification of the Crenarchaeota/Thaumarchaeota (Parada, Needham and Fuhrman 2016) and a revision to R806 improved amplification of SAR11, an ubiquitous marine bacterial clade (Apprill et al. 2015). Studies that include fungal community composition typically target the 18S rRNA gene (Banos et al. 2018) or the ITS1F region between the 18S and the 2.5S rRNA genes (White et al. 1990). Primers targeting fungi tend to vary more across the soil microbial ecology literature than those for 16S rRNA, in part due to concerns about the ability of primer sets to equally amplify all fungal groups (Yu et al. 2022). For example, Liu et al. (2015) observed more fungal richness using ITS compared to 18S primers, but another study found that 18S rRNA primers amplified some soil groups better than ITS targets (Asemaninejad et al. 2016). The ITS sequencing libraries may underestimate arbuscular mycorrhizal (AM) fungi (George et al. 2019). As the number of fungal sequences increase in public repositories, particularly from shotgun libraries, new primer sets are being developed, but the most commonly used primer pair ITSF1/ITS2 targets the ITS1 region (Baldrian et al. 2022; Shaffer et al. 2022) and has been applied to many soils including those under agricultural cultivation.

Aside from primers designed to target large groups of the microbial community, there are also group-specific and functional primers. To address the lack of amplification of AM fungi in ITS sequencing, several AM group-specific primers have been developed and evaluated (Kohout et al. 2014). Group-specific primers have also been developed for Fusarium, an important group of fungal pathogens in agricultural systems (Karlsson et al. 2016). In the case of bacteria, some primers have been developed to target specific taxonomic groups (Abdelmoneim et al. 2023), but those primers have mostly been used in Q-PCR assays.

Numerous primer pairs target functional genes linked primarily to nutrient cycling. Most of these primers amplify genes or gene fragments that code for the active sites of enzymes central to a biogeochemical pathway (Table 1). Today many of these primer pairs are used more in gene quantification than in high throughput sequencing, but functional gene sequencing can provide important

Table 1 Functional gene primer pairs that have been used in high-throughput sequencing with illumina chemistry

Processes	Genes	Primers	Product length	Reference
Denitrification	nirK (bacteria)	F1aCuF/R3CuR	450	(Hallin et al. 1999)
	nirS (bacteria)	cd3aF/R3cdR	423	(Throbäck et al. 2004)
	nosZI (bacteria)	840F/2090R	267	(Henry et al. 2006)
	nosZII (bacteria)	nosZ-II-F/nosZ-II-R	690	(Jones et al. 2013)
	nirK (fungi)	FnirK-F3/FnirK-R2	233	(Chen, Yu and Shi 2016)
Nitrification	amoA (bacteria)	amoA-1F/amoA-2R	491	(Rotthauwe, Witzel and Liesack 1997)
	amoA (archaea)	CrenamoA23f/CrenamoA616r	620	(Tourna et al. 2008)
Nitrogen fixation	nifH (bacteria)	IGK3/DVV	395	(Ando et al. 2005)
Anaerobic methane oxidation	mcrA (archaea)	McrA169F/McrA1360R (nested)	363	(Xu et al. 2018)

new information particularly with respect to biogeochemical pathways such as nitrification, where the discovery of archaeal ammonia oxidizers and comamox bacteria have posed new questions about niche differentiation (Aigle, Prosser and Gubry-Rangin 2019). Similarly, the discovery of a second clade of nosZ bacteria and nirK containing fungi have expanded the population responsible for denitrification (Wan et al. 2023) and high throughput sequencing of nifH has also continued to add knowledge concerning the diversity of diazotrophs (Gaby et al. 2018). Similarly, with respect to methane (CH_4) cycling, sequencing has revealed new archaeal anaerobic CH_4 oxidation pathways (Xu et al. 2018).

2.3 Sequencing platforms

Whether amplicons or total genomic DNA fragments are isolated for analysis, the next important step is to choose the appropriate sequencing technology. Sequencing technologies have advanced rapidly in the last two decades and continue to evolve to meet the demands of ever-larger datasets. The first set of sequencing technologies to move beyond Sanger were pyrosequencing (Ronaghi et al. 1996) and Ion Torrent (Zhang et al. 2015). In the last decade, Illumina, Pacific Biosciences (PacBio), and Oxford Nanopore have led the sequencing market (Rayamajhi, Cheng and Catchen 2022). Of the three, Illumina sequencing platforms like MiSeq and HiSeq have been used most extensively for environmental sequencing studies (Satam et al. 2023).

Illumina technology uses a sequence by synthesis method allowing for the measurement of DNA bases as they are added to complimentary strands synthesized from sample DNA (Quail et al. 2012). Sequences collected from a sample are first prepared by ligating a small complementary oligo sequence that matches one of a multitude of small oligo sequences on an Illumina microfluidic device. When a sample is added to the platform, sample DNA fragments bind to the matching primers and undergo bridge amplification. Bridge amplification allows a single DNA sequence to be replicated locally on the chip (Slatko, Gardner and Ausubel 2018). Following this amplification, the microfluidic device is populated by millions of unique spots of replicated DNA fragments from the sample. These spots of concentrated, single-stranded sample DNA copies are measured with light emitted from each new base addition. The instrument simultaneously tracks each new base addition at each spot to record the sequences.

There are advantages and disadvantages with Illumina sequencing. An advantage is the ability to barcode DNA belonging to a specific samples, such as individual soil extracts that can then be multiplexed and sequenced together (Wong, Jin and Moqtaderi 2013). Following a run, sequences are sorted and traced back to the original sample using the short hexamer barcodes (Fig. 2).

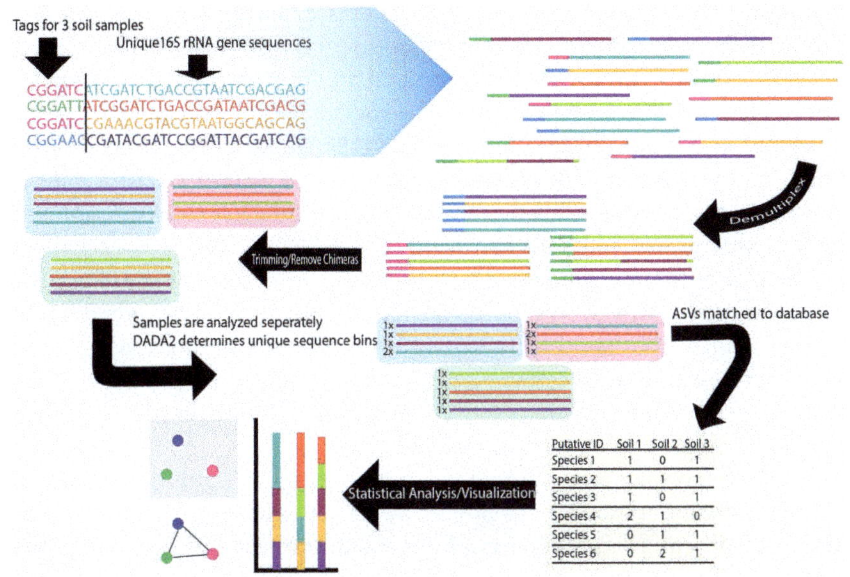

Figure 2 Flow chart showing amplicon sequence analysis. The example includes three soil samples and the distribution of six microbial species.

The Illumina short-read approach generates millions, even billions, of unique DNA fragments (i.e. Gb to Tb of data) from a sample in a short timeframe. Sequences generated by Illumina are often referred to as 'reads'. Although forward-read runs may be used, it is most advantageous to use Illumina paired-end sequencing, where both the forward and reverse reads from a sequence fragment can be joined or merged to generate a longer sequence of up to ~600 bp. A potential disadvantage is this limit on the optimal size of individual DNA fragments. This length does not always allow for adequate coverage to complete a large genomic sequence fragment. Another drawback is that Illumina uses PCR during bridge amplification which can lead to bias in the soil microbial community compositional inference following bioinformatics (Kennedy et al. 2014; Stevens et al. 2023).

Third generation, long-read technologies include PacBio and Oxford Nanopore. They generate sequence up to ~20 kb and do not necessarily require PCR as part of the sequencing platform (Satam et al. 2023). PacBio uses a sequence by synthesis approach akin to Illumina. The difference is that traps or guides single large DNA strands from a sample into a microwell. The added stability within the PacBio microwell allows sequences to reach up to 25 kb with low error rates. Oxford Nanopore platforms (e.g. MinION) allow for long reads up to 30 kb, on par with PacBio. Although error rates can be higher with nanopore technology compared to PacBio when dealing with highly repetitive sequences. The benefits of nanopore are large DNA fragments from a sample are directly sequenced as they pass through a nanopore. Thus no synthesis of a template strand is needed for native DNA molecules, avoiding related bias and errors of template strands. Another advantage is the portability (Sevim et al. 2019). Nanopore technologies like MinION allow users to sequence long reads in the field or lab with minimal prep and turnaround time (Jain et al. 2016). The MinION is small enough to fit into the palm of a researcher's hand and can be linked directly to a computer to display sequence characteristics in real time from virtually anywhere. The nanopore technology works by using nanopores on a chip and enzymes to direct native DNA strands through the pores while each passing base is measured by a change in voltage across the pore (Caldwell and Spies 2017). This allows nanopore sequencing to accurately detect single nucleotide changes compared to other technologies.

Studies can take advantage of coupling datasets from more than one NGS sequencing platform (Sevim et al. 2019). Illumina short-read data can complement long-read data from PacBio to fill in gaps for longer sequences required for metagenomic studies. At present, no single emerging technology is a standalone solution to sequencing research on environmental systems.

Table 2 Open-source software and R packages commonly used in sequence analysis pipelines

Name	Citation	Description
Mothur2	(Schloss Patrick D. et al. 2009)	Open source software package that can be used to process sequence data through statistical analysis
QIIME2	(Bolyen et al. 2019)	Open source software package that can be used to process sequence data through statistical analysis
DADA2 (R package)	(Callahan et al. 2016)	State of the art error inference package used primarily for the Illumina platform but supports MinION data. Accepts raw output from Illumina, is able to join paired ends, has some limited chimera removal and primer/index removal capabilities. Performs taxonomy annotations. Output is a phyloseq object with an OTU table and taxonomy table
Phyloseq (R package)	(McMurdie and Holmes 2013)	An open-source utility package for managing and analyzing sequence data. Uses a standard data structure for storing complete datasets (OTU table, taxonomy table, and sample metadata table). Many common R packages used for microbiome analysis are wrapped into phyloseq, and phyloseq provides support for users to incorporate other software packages as needed
Vegan (R package)	(Oksanen et al. 2024)	Statistical package for community analysis

OTU, Operational Taxonomic Unit.

3 Amplicon sequence analysis

Once sequence data are obtained, early analysis steps involve quality checking the sequences and then matching them to appropriate databases. The use of open-source pipelines has become the norm. For example, open-source pipelines such as QIIME2 and Mothur are very popular (Table 2) and have substantial resources available to guide users through the analysis. Sequence analysis is commonly done in the open-source programming language R using the state of the art error inference algorithm for amplicon sequences, DADA2 (Callahan et al. 2016). The community has also built another package, phyloseq (McMurdie and Holmes 2013), to manage microbiome data for statistical analysis (Table 2). Packages for community ecology such as vegan (Oksanen et al. 2024) are integrated into phyloseq. R has a robust set of packages such as ape (Paradis, Claude and Strimmer 2004) for doing phylogenetic analyses of sequence data. When programming in R, there are two important things to be aware of (1) natively, R uses only one processor, but many functions including those in DADA2 have an option to use additional processors. This can vastly reduce processing time. Settings must be changed by the user. (2) Many of the processes in microbiome analysis are implemented in a way that requires a significant amount of Random Access Memory (RAM). Newer personal computers often have larger RAM storage available; when choosing a machine for sequence analysis using R, it's a good idea to maximize RAM.

Taxonomy in microbial datasets is centered around the concept of an Operational Taxonomic Unit (OTU). OTUs originally were defined by clustering sequences together based on similarity (Schloss and Handelsman 2005), using a number of different clustering algorithms (Bhat, Prabhu and Balakrishnan 2019). It is common to define an arbitrary threshold, such as 97%, for clustering (Dunbar et al. 1999). However, there is evidence to suggest that an arbitrary threshold obscures true variation. For example, Edgar (2018) argues that bacterial taxa should have a 100% match in the 16S rRNA region to be considered the same species. One challenge with clustering methods for defining OTUs is that the results may depend on the complexity and composition of the dataset, where the definition of the OTU depends on sequences that are input. For example, as more sequences are added, OTUs may be split resulting in different apparent community diversity characteristics (He et al. 2015). In general, there is a tradeoff between computational time and cluster accuracy (Bhat, Prabhu and Balakrishnan 2019).

The DADA2 algorithm takes a fundamentally different approach; rather than clumping similar sequences together, DADA2 looks for the number of identical sequences in the dataset to identify sets of sequences that are real signals and develops an error model to predict whether specific single nucleotide polymorphisms should be discarded as an error or included as a true signal (Callahan et al. 2016). This is a powerful approach that outputs true Amplicon Sequence Variants (ASV). It should be noted that ASVs are one potential way to operationally define taxonomic units, and therefore could be considered a constituent category of OTU. However, it is common to find discussions of OTU versus ASV approaches, and when this occurs the author typically is using OTU as a stand-in for several clustering algorithms. When analyzing fungal communities, consider that experts in fungal ecology generally consider the output of ASVs on the ITS locus to be less accurate than traditional OTU clustering methods, in part because ASV approaches tend to filter out rare members of *Glomeromycota* and other non-fungal eukaryotes (Tedersoo et al. 2022). Some fungal genera, for example species in *Glomus*, will maintain multiple distinct genomes within the same organism (Young 2015), so a single organism may have multiple true sequence variants. This makes interpreting the diversity of these taxa somewhat subjective.

3.1 Taxonomic Identification

Once ASVs has been identified, whether through clustering or error inference, there are several databases that can be used to putatively ID sequences (Table 3). The choice of database depends on the amplicons being compared, the size of the database (Balvočiūtė and Huson 2017), and the frequency of updates and curation. All three of the databases listed here have been updated within the last few years and are regularly used for identification of soil amplicon sequences.

3.2 Statistical analysis

Amplicon sequencing data is often used in statistical analysis to assess differences between soil treatments and can be used in network analysis and structural equation modeling. It is important to remember that amplicon sequencing data contain large numbers of zeros, thus the data are not normally distributed. The sequencing mechanisms also results in large differences in sample sequencing depth, thus data should be treated as compositional,

Table 3 Pipelines and taxonomy databases for annotating amplicon sequence data.

Name	Citation	Description
Greengenes	(DeSantis et al. 2006)	Database of 16S ribosomal DNA for taxonomy, recently redesigned
SILVA	(Quast et al. 2012)	Ribosomal gene database including bacteria, archaea, and eukaryotes
UNITE	(Abarenkov et al. 2024)	Database of fungal taxonomy based on the Internal Transcribed Spacer (ITS)

meaning that the data should be interpreted as proportions rather than as accurate abundance data (Gloor et al. 2017). These properties constrain some of the statistics that can be used (Boshuizen and Te Beest 2023), and the inferences that can be made (Barlow et al. 2021). Therefore, researchers must be careful to not overstate differences in communities. These limitations can be mitigated somewhat by combining sequence data with quantitative metrics such as Q-PCR (Barlow et al. 2021).

The quantitative sequencing (QSeq) transformation is done by using a quantitative measure such as QPCR to estimate the total abundance of the community, then multiplying the proportion of each taxon within the community by the total abundance value to make an estimate of abundance for each taxon (Jian et al. 2020). The resulting community data is quantitative, increasing the scope of inference during analysis (Epp Schmidt, Maul and Yarwood 2024) but still retains some characteristics of compositional data. Specifically, QSeq data produced in this way have skewed measurement precision; taxa with low relative abundance have higher associated measurement error than taxa with higher relative abundance in the same sample (Barlow et al. 2021). This implies that researchers should ensure they have sufficient sequencing depth to reliably estimate the abundance of the target taxa in their samples as part of the experimental design.

3.3 Functional inference

Amplicon sequences can also be used to infer the functional capacity of soils using tools such as PICRUSt (Langille et al. 2013; Douglas, Beiko and Langille 2018), Tax4Fun2 (Wemheuer et al. 2020), and FUNGuilds (Nguyen et al. 2016). PICRUSt and Tax4Fun2 can be used to infer bacterial functional attributes by matching putatively identified 16S rRNA gene sequences to known bacterial genomes. The resulting output is a predicted functional gene abundance table like what researchers may directly generate using shotgun metagenomics. In a slightly different approach, FUNGuilds assigns putatively identified fungi into functional guilds, in this case any taxonomic gene marker may be used (e.g.18S or ITS), because the input is identified fungi.

All three tools are limited to the known members of the community and prior knowledge concerning those members' function. Tax4Fun2 allows users to narrow the environment, such as soils, where samples originated which has helped to improve accuracy compared to PICRUSt (Wemheuer et al. 2020). These tools have been applied to understand how management may alter microbial community function. For example, Chen et al. (2020) observed functional differences when *Miscanthus* was grown as a biofuel crop and genes involved in phosphorus metabolism differed in PICRUSt analysis comparing fields with phosphorus biofertilizer additions (Mącik et al. 2023).

The next section will describe shotgun metagenomics in depth, and this might raise the question: Why would researchers use inference tools to assign function rather than shotgun metagenomics directly? Amplicon sequencing is still more broadly accessible to research labs. Amplicon sequencing requires less sequencing depth for individual soil samples, the kits are substantially cheaper per sample, and computation can usually be done on a personal computer with minimal coding knowledge.

4 Shotgun metagenomics

In some ways, shotgun sequencing is a misnomer, as it implies something unique about the sequencing platform, when in fact the differences between amplicon and shotgun sequencing arise during the library preparation process. As discussed earlier, amplicon sequencing targets a specific region of the genome of interest, omitting the rest of the genomic content in the sample. In contrast, shotgun library preparation uses a variety of enzymatic or physical methods (e.g. sonication) to shear DNA into fragments for sequencing (Jones et al. 2015). Preparing a sequence library in this way captures a representative sample of all DNA in the sample, and can be used for both taxonomic classifications, as well as characterizing functional capacity of the community (Fig. 1). Because any DNA can be captured in this method, it also allows researchers to simultaneously collect data on many different domains of the soil community.

One popular library preparation tool is Illumina's Tagmentation kits (Wang et al. 2013), which simultaneously fragments DNA using transposomes and ligates adapters so that any genomic DNA can be sequenced on an Illumina platform. Each of the popular sequencing platform companies offers kits that accomplish both the fragmentation of the DNA, and additional preparations necessary for sequencing such as the ligation of indexes and adaptors. Because each of the platforms work using different sequencing mechanisms, it is important to use a kit that is compatible with the sequencing platform that will be used.

4.1 Sequence analysis

In general, the shotgun sequence analysis pipeline flows from sequence quality filtering, through read mapping and annotation, to statistical analyses. It is in the read mapping and annotation phase that major differences in approach emerge. One standard approach to metagenomics annotation involves the assembly of a Metagenome Assembled Genome or MAG (Setubal 2021). This process mirrors the process of assembling a genome during whole genome sequencing, but instead of assembling a single genome, several genomes are assembled simultaneously (The Genome Standards Consortium et al. 2017). It is relatively rare to achieve a complete assembly. In many cases, MAGs are large fragments of a genome. Assembly into MAGs is done using processes that were first designed to assemble genomes from sometimes error-prone DNA sequences. Assemblers therefore should be thought of as building a representative genomic backbone, but they are prone to generating chimeric MAGs that represent sequences from multiple taxa (Vázquez-Castellanos et al. 2014), which potentially confound phylogenetic analysis.

There is no single best tool for MAG assembly. Some assemblers are better at assembling genes from the sequence data, while others are better at assembling whole genomes. Assembly is a memory-intensive process that in some cases requires greater than 500 GB of RAM (current Mac laptops offer up to 128 GB of RAM). Therefore, the selection of the assembler should reflect the available computing resources and the objective of the study. Van Der Walt et al. (2017) presents a useful decision tree for selecting an assembler. Some of the more popular assemblers include Spades (Bankevich et al. 2012), metaSpades (Nurk et al. 2017), and MegaHit (Li et al. 2015). MetaSpades is good at identifying genes from highly diverse samples, while megaHit excels at differentiating genes and genomes in very similar communities (Greenwald et al. 2017). If phylogenetic analysis is of interest, then a user should select megaHit. If, however, the study is focused on identifying functional capacity of a highly diverse community, metaSpades may be a better choice.

Although MAG generation is sometimes viewed as a critical step in the annotation pipeline, it is not required to answer questions that focus on which genes are in the sample, rather than which genomes (Quince et al. 2017). A benefit of assembling a MAG is that assembly ensures a high confidence that the genes in the sequence library are real genes in the sample or genome. There are drawbacks, however. No assembler is perfect, and all assemblers therefore drop a certain percentage of genes that are represented in the sequence library. For example, MAGs assembled from short reads disproportionately omit genes found in plasmids and genomic islands (Maguire, Jia and Gray, 2020). Assembly is generally seen as a quality control mechanism, but in the case of environmental samples it may exclude the very genes that are of interest to the researchers, especially if those genes are less prevalent in the population.

An alternative approach is to focus on obtaining high quality DNA and high quality annotations of DNA fragments. This can be accomplished by setting an extremely conservative E-value when annotating the dataset (Epp Schmidt et al. 2019). The E-value is the expected probability that a sample sequence matches the reference at random (González-Pech, Stephens and Chan 2019). Setting an E-Value threshold determines the likelihood that any specific annotation within the dataset is erroneous. A researcher can mitigate

Table 4 Gene annotation algorithms

Name	Citation	Description
Kraken2	(Lu et al. 2022)	Bacterial read classifier
Bracken	(Lu et al. 2017)	Bacterial read classifier that is used to refine results from Kraken2
Clark	(Ounit et al. 2015)	K-mer mapping and classification tool
K-Slam	(Ainsworth et al. 2016)	K-mer mapping and classification tool, including pseudo assembly and variant calling
Centrifuge	(Kim et al. 2016)	Memory efficient sequence classifier
Diamond	(Buchfink, Xie and Huson 2015)	Protein alignment tool
Kaiju	(Menzel, Ng and Krogh 2016)	Rapid classification tool for metagenomic sequences
MetAnnotatorX2	(Milani et al. 2021)	Bash script that may be used to generate MAG or directly annotate shotgun sequence data from short read or long read platforms
MetaPhLan 4	(Blanco-Míguez et al. 2023)	Taxanomic profiler for metagenomics datasets
PhiSiGns	(Dwivedi et al. 2012)	Pipeline to identify phages and design primers
FragGeneScan	(Rho, Tang and Ye 2010)	Gene prediction from short and error prone reads
Glimmer-MG	(Kelley et al. 2012)	Gene prediction and clustering
Prodigal/metaprodigal	(Hyatt et al. 2012)	Gene and translation initiation site prediction
Prokka	(Seemann 2014)	Rapid prokaryote genome annotation

the likelihood of misannotation by setting the E-value as several orders of magnitude higher than the size of the library. For example, Epp Schmidt et al. (2019) used an E-value of 1.0e-15 for a total library size of 1.2e9 sequences. The result was an expected probability of an erroneous annotation in the dataset was approximately one in a million. There are a wide range of algorithms available for gene or taxonomy annotation that range in their approach and target organisms (Tables 4 and 5).

A significant barrier to shotgun metagenomics analysis has been the technical capacity to build and manage a complex sequence analysis pipeline. This is no longer necessary, however. There are several open-source pipelines that assemble and annotate without requiring user programming (Table 6). Mg-RAST (Keegan, Glass and Meyer 2016) is a web-based platform for QC, assembly, and annotation of shotgun metagenomics sequences. It references MD5nr non-redundant protein database (Table 5), an indexed and

Table 5 Taxonomy and metabolism databases that can be used with shotgun sequencing data

Name	Citation	Description
KEGG	(Kanehisa et al. 2012)	Network of gene function from genomic sequence data
MetaCyc	(Caspi et al. 2014)	Curated database of experimentally determined metabolic pathways, including reactions, enzymes, and genes
SEED	(Overbeek 2005)	Genome annotation system based on curated database of protein families and their functions
MD5nr	(Wilke et al. 2012)	Uses MD5 checksums to generate a non-redundant sequence database from many different sources
EBI	(Brooksbank et al. 2014)	The European Bioinformatics Institute offers a wide range of genomics tools including for genomic annotation
GO	(The Reference Genome Group of the Gene Ontology Consortium 2009)	A model of biological systems across species and levels of organization; provides annotation for gene functions
JGI	(Markowitz et al. 2007)	Joint Genome Institute dataset and sequence QA/QC and annotation pipeline
NCBI	(Benson et al. 2012)	Publicly available database of nucleotide sequences.
Uniprot	(Magrane and Consortium 2011)	Database on protein sequence and function
PATRIC/VBI	(Snyder et al. 2007)	National Institute of Allergies and Infectious Disease resource center dedicated to selected pathogens; good resource for environmental monitoring
eggNOG	(Jensen et al. 2007)	Automated construction and annotation of orthologous groups of genes
CAZY	(Drula et al. 2022)	The carbohydrate-active enzyme database
CARD	(McArthur et al. 2013)	Comprehensive Antibiotic Resistance Database
RDP	(Wang and Cole 2024)	16S ribosomal DNA taxonomy classifier

searchable database that combines multiple other gene databases. KBase (Arkin 2018) is an open-source platform for genomics analysis maintained by the US Department of Energy, via the Joint Genome Institute. Galaxy (Schiml et al. 2023) is a platform for meta-omics bioinformatics and analysis. The European Bioinformatics Institute (EBI) provides a suite of open-sourced bioinformatics tools for metagenome assembly and annotation (Brooksbank et al. 2014).

Table 6 Open-sourced whole genome pipelines and platforms

Name	Citation	Description
Mg RAST	(Keegan, Glass and Meyer 2016)	Web-based platform for QC, assembly and annotation of shotgun metagenomics sequences. It references MD5nr non-redundant protein database, an indexed and searchable database that combines multiple other gene databases
EBI	(Brooksbank et al. 2014)	The European Bioinformatics Institute offers a wide range of genomics tools including for genomic annotation
KBase	(Arkin 2018)	US Department of Energy platform for genomics analysis; open source
IMG/M	(Chen et al. 2020)	Reference database built by the Joint Genome Institute for studying isolates and metagenomes
Galaxy	(Schiml et al. 2023)	Platform for meta-omics bioinformatics and analysis

Interpreting the functional implications of metagenome sequence data is not straightforward. Soil metagenomes tend to map poorly to reference databases leading to incomplete annotation of the dataset and unknown biases are inherent in the data (Anthony et al. 2024). Even when genes are annotated, interpreting community function from the annotations alone may not be possible. For example, trace metal efflux pumps are important for microbial adaptation to the soil environment because they regulate osmotic pressure, mitigate metal toxicity in cells, and capture nutrients. Yet there is now compelling evidence of an association between trace metal resistance and antibiotic resistance mechanisms (e.g. Seiler and Berendonk 2012; Knapp et al. 2017; Heydari et al. 2022). This complicates data interpretation: if the metagenome is annotated as being enriched with both trace metal and antibiotic resistance genes, how does one distinguish the underlying evolutionary driver maintaining these genes in a particular environment? These are ongoing challenges that researchers should be aware of as they interpret their data and evaluate other studies.

5 DNA sequencing and soil health

Soil health has been defined as 'the continued capacity of soil to function as a vital living ecosystem that sustains plants, animals, and humans' (Lehmann et al. 2020). The importance of the soil microbial community in healthy functioning of soil is widely acknowledged, but whether certain compositional metrics are indicative of soil health remains an open question (Wilhelm et al. 2023). There are many ways in which soils must function to sustain life, and large sequencing efforts, such as a recent survey across the European Union, underscore that soils function with considerable variation in microbial community composition (Labouyrie et al. 2023). Although microbial taxonomic and functional diversity measures are sometimes used as proxies for soil health (Maurya et al. 2020), to get a detailed understanding of specific functional capacities, we need specific data about the functional attributes of members of the microbial community. These data exist for some functions but are sorely lacking for others.

Classifying organisms by taxonomy may not be enough to understand how they are functioning. Differences of only a few genes can transform a commensal organism into a potent pathogen (Denamur et al. 2021). In bacteria, differences of a single nucleotide in the 16S ribosomal gene sequence (<1%) is known to potentially indicate substantial differences in genotype (Ward et al. 1998; Thompson et al. 2005; Eren et al. 2013), meaning closely related taxa may have markedly different ecology.

Similar patterns can be found in fungal pathogens as well, e.g. Fusarium wilt is a potent agricultural disease caused by a strain of *Fusarium oxysporum*. Yet, there is good evidence that suppressive soils for Fusarium wilt can develop because of competition over nutrients between commensal and pathogenic strains of *F. oxysporum* (Alabouvette 1986). Shotgun metagenomics and whole genome sequencing continue to be used to discover genes associated with *Fusarium* pathogenicity. Unique sets of genes appear to be associated with pathogenicity toward specific hosts, and in some cases, pathogenicity appears to be conferred by just a few genes (Edel-Hermann and Lecomte 2019). This means that identification of specific virulence factors would be necessary to predict if a *Fusarium* strain will be pathogenic.

Across the entire microbial community, metagenomic datasets can therefore provide significantly more insight into the function and dynamics of the soil community compared to amplicon sequencing. For this reason, metagenomics is useful for understanding dynamics in soil microbial community function and are being used to uncover fundamental mechanisms by which soil microbiomes interact with agricultural production. Soil microbial metagenome-wide association studies have sought to identify functional characteristics of microbial communities that improve yield (Iquebal et al. 2022). The ultimate hope is that one day agricultural microbiomes could be engineered to maximize crop performance (Toju et al. 2018).

Identifying useful compounds from soil is not limited to improving crops, however. Many antibiotics have been isolated from soil microbes (Davies and Davies 2010). Shotgun metagenomics are now being leveraged for bioprospecting new antibiotic compounds (Mantri et al. 2021).

The same sequencing techniques that are being used to mine for antibiotics are also being used to uncover antibiotic resistance dynamics. For example, livestock are important vectors of antibiotic resistance and soil is thought to be an important reservoir. Using shotgun metagenomics, Li et al. (2022) found that applying manure in a row cropping enriched antibiotic resistance genes in the soil microbiome. In contrast, shotgun metagenomics data from Zheng et al.

(2021) suggest that soils under grazing management, therefore, also influenced by animal manure, had no impact on antibiotic resistance profiles in soils. These differences could be emergent properties of the specific farming systems. For example, the effective rate of manure application may be much lower in the grazing system. Furthermore, the perennial plant growth in the grazing system could also have a stabilizing effect on the microbial community, making it more resistant to introduced microbes. Another possibility is that manure used in the fertilizer trial may inherently have a higher abundance of antibiotic resistance genes due to animal management. Perhaps all these mechanisms play a role.

Metagenomics investigation has found some surprising mechanisms that maintain selection for antibiotic resistance. For example, application of several widely used herbicides (e.g. glyphosate, glufosinate, and dicamba) selected for microbes carrying antibiotic-resistant genes in soils (Liao et al. 2021). This selection appeared to be due to mobile elements that contain antibiotic resistance mechanisms, rather than changes in the abundance of microbial taxa. Taxonomic data alone could not uncover this pattern. In this case, it was necessary to get data on both taxonomy and function.

These findings align well with new research on microbial genomic evolution in soils. Whole genomic sequencing of soil microbes has revealed that as much as 35% of loci genome-wide in soil bacteria show evidence of recombination (Crits-Christoph et al. 2020). This may indicate hyper-local adaptation in biosynthesis and excretion pathways by soil bacteria. Another study using shotgun sequencing found trait-specific adaptation by bacteria to soil conditions (Chase, Weihe and Martiny 2021). This agrees with observations that bacterial genomes tend to be dominated by locally adapted genes (Coelho et al. 2022), especially in soils (Mantri et al. 2021). These insights arise from analyzing many genes simultaneously and analyzing whole microbial genomes. Whole genome metagenomics approaches are also being applied to better understand biotic stress mitigation through Plant Growth Promoting Bacteria and other beneficial microbes (Trivedi et al. 2021; Sharuddin et al. 2022); greenhouse gas production dynamics (Li et al. 2024); and carbon and nutrient cycling (Long et al. 2016).

6 Challenges and future trends

As DNA sequencing has become ubiquitous in soil science research, it has become easier for researchers to ignore some of its limitations. For example, it is challenging to adequately assess whether a systemic bias occurred during sampling (Nesme et al. 2016). *In situ* (Meslier et al. 2022) and *in silico* (Bokulich and Mills 2013; Zhang et al. 2023) benchmarking have focused respectively on sequencing platform and bioinformatics pipeline performance. Studies on DNA extraction have focused on obtaining sufficient DNA yield (Dineen et al.

2010; Young et al. 2014). Although these are worthy goals, they do not address potential bias in extraction efficiency of microbes from soils, and it has been observed that extraction methods affect microbial community data (Iturbe-Espinoza et al. 2021).

There is not yet a gold standard for determining which extraction method leads to the most representative dataset across the variable range of soil characteristics, and as described throughout this chapter, there are a wide variety of choice that researchers in these workflows make that impact the final data outcomes. There are currently efforts to establish *Standards for Technical Reporting in Environmental and Host-Associated Microbiome Studies* (STREAMS) (Kelliher et al. 2024), allowing for better comparability across studies. Ultimately, the ability to describe the microbial community via taxonomy or function should be considered just one facet to holistically understand the health status of a soil. As the rest of this book describes other methods, those can all compliment the data gained through sequencing, and sequencing can also serve as a first step to ask more detailed questions.

Sequencing technology has revolutionized our understanding of environmental microbiomes. From high-throughput taxonomic classifications to meta-genome reconstruction, our ability to access information encoded in DNA is helping us to probe a previously invisible dimension of our ecosystem. But the sequencing platforms and the downstream data analysis approaches are not one size fits all. There are trade-offs between short-read and long-read platforms and between amplicon and shotgun approaches. The choice for the best sequencing platform depends on the specific question that needs to be answered. Similarly, the bioinformatics method to analyzing the sequence data needs to be tailored to the question. Properly applying sequencing technology in an experimental context requires a complex set of expertise spanning ecology, molecular biology, and computer science. In this chapter, we have set out to provide guideposts to help the reader navigate the common decision points in environmental microbiome research and avoid common pitfalls.

7 Where to look for further information

A great resource for those just beginning their journey in the diverse field of environmental microbiology is https://microbenotes.com/. This website offers introductory material on a wide range of topics in microbiology, from fundamental biology to common methodologies used for studying microbes, including non-sequencing approaches for investigation. For additional information on specific environmental microbiomes, tailored methods for sampling, and emerging research in soil and environmental microbial ecology, there are a few additional resources: The Earth Microbiome Project is a fantastic resource, including standardized protocols and code for processing samples across a wide range of environments (https://earthmicrobiome.ucsd.edu/protocols-and-standards/). The Fungal Diversity Survey (FunDiS) is an effort to document fungal biodiversity in the United States, with the goal of supporting conservation efforts in mycology. More information can be found at https://www.fundis.org/about-fundis. The Society for the Protection of Underground Networks (SPUN) is a non-profit focused on the conservation of soil fungi. Their website includes information about the ecology of many species of mycorrhizal fungi.

8 References

Abarenkov K, et al. The UNITE database for molecular identification and taxonomic communication of fungi and other eukaryotes: sequences, taxa and classifications reconsidered. Nucleic Acids Research 2024;52:D791–7.

Abdelmoneim TK, et al. Development of rapid and precise approach for quantification of bacterial taxa correlated with soil health. *Front Microbiol* 2023;13:1095045.

Aigle A, et al. The application of high-throughput sequencing technology to analysis of amoA phylogeny and environmental niche specialisation of terrestrial bacterial ammonia-oxidisers. *Environ Microbiome* 2019;14:3.

Ainsworth D, et al. k-SLAM: accurate and ultra-fast taxonomic classification and gene identification for large metagenomic data sets. *Nucleic Acids Res* 2016:44:1–12.

Alabouvette C. Fusarium-wilt suppressive soils from the Châteaurenard region: review of a 10-year study. *Agronomie* 1986;6:273–284.

Alexander M. *Introduction to soil microbiology*. New York, NY: Wiley, 1977:478–478.

Alteio LV, et al. A critical perspective on interpreting amplicon sequencing data in soil ecological research. *Soil Biol Biochem* 2021;160:108357.

Ando S, et al. Detection of *nifH* Sequences in Sugarcane (*Saccharum officinarum* L.) and Pineapple (*Ananas comosus* [L.] Merr.). *Soil Sci Plant Nutr* 2005;51:303–308.

Anthony WE, et al. From soil to sequence: filling the critical gap in genome-resolved metagenomics is essential to the future of soil microbial ecology. *Environ Microbiome* 2024;19:56.

Apprill A, et al. Minor revision to V4 region SSU rRNA 806R gene primer greatly increases detection of SAR11 bacterioplankton. *Aquat Microb Ecol* 2015;75:129–137.

Arkin AP. KBase: The United States department of energy systems biology knowledgebase. *Nat Biotechnol* 2018;36:566–569.

Asemaninejad A, et al. New primers for discovering fungal diversity using nuclear large ribosomal DNA. Pöggeler S (ed.). *PLoS One* 2016;11:e0159043.

Baldrian P, et al. High-throughput sequencing view on the magnitude of global fungal diversity. *Fungal Divers* 2022;114:539–547.

Balvočiūtė M and Huson DH. SILVA, RDP, greengenes, NCBI and OTT – how do these taxonomies compare? *BMC Genomics* 2017;18:114.

Bankevich A, et al. SPAdes: a new genome assembly algorithm and its applications to single-cell sequencing. *J Comput Biol* 2012;19:455–477.

Banos S, et al. A comprehensive fungi-specific 18S rRNA gene sequence primer toolkit suited for diverse research issues and sequencing platforms. *BMC Microbiol* 2018;18:190.

Barlow JT, et al. Quantitative sequencing clarifies the role of disruptor taxa, oral microbiota, and strict anaerobes in the human small-intestine microbiome. *Microbiome* 2021;9:214.

Benson DA, et al. GenBank. *Nucleic Acids Res* 2012;41:D36–D42.

Bhat AH, et al. A critical analysis of state-of-the-art metagenomics OTU clustering algorithms. *J Biosci* 2019;44:148.

Blanco-Míguez A, et al. Extending and improving metagenomic taxonomic profiling with uncharacterized species using MetaPhlAn 4. *Nat Biotechnol* 2023;41:1633–1644.

Bokulich NA and Mills DA. Improved selection of internal transcribed spacer-specific primers enables quantitative, ultra-high-throughput profiling of fungal communities. *Appl Environ Microbiol* 2013;79:2519–2526.

Bolyen E, et al. Reproducible, interactive, scalable and extensible microbiome data science using QIIME 2. *Nat Biotechnol* 2019;37:852–857.

Boshuizen HC and Te Beest DE. Pitfalls in the statistical analysis of microbiome amplicon sequencing data. *Mol Ecol Resour* 2023;23:539–548.

Brooksbank C, et al. The European Bioinformatics Institute's data resources 2014. *Nucleic Acids Res* 2014;42:D18–D25.

Buchfink B, et al. Fast and sensitive protein alignment using DIAMOND. *Nat Methods* 2015;12:59–60.

Caldwell CC and Spies M. Helicase SPRNTing through the nanopore. *Proc Natl Acad Sci* 2017;114:11809–11811.

Callahan BJ, et al. DADA2: High-resolution sample inference from Illumina amplicon data. *Nat Methods* 2016;13:581–583.

Caporaso JG, et al. Global patterns of 16S rRNA diversity at a depth of millions of sequences per sample. *Proc Natl Acad Sci* 2011;108:4516–4522.

Caspi R, et al. The MetaCyc database of metabolic pathways and enzymes and the BioCyc collection of Pathway/Genome Databases. *Nucleic Acids Res* 2014;42:D459–D471.

Chase AB, et al. Adaptive differentiation and rapid evolution of a soil bacterium along a climate gradient. *Proc Natl Acad Sci* 2021;118:e2101254118.

Chen H, et al. Global meta-analyses show that conservation tillage practices promote soil fungal and bacterial biomass. *Agric Ecosyst Environ* 2020;293:106841.

Chen H, et al. Detection of N2O-producing fungi in environment using nitrite reductase gene (nirK)-targeting primers. *Fungal Biol* 2016;120:1479–1492.

Coelho LP, et al. Towards the biogeography of prokaryotic genes. *Nature* 2022;601:252–256.

Crits-Christoph A, et al. Soil bacterial populations are shaped by recombination and gene-specific selection across a grassland meadow. *ISME J* 2020;14:1834–1846.

Davies J and Davies D. Origins and evolution of antibiotic resistance. *Microbiol Mol Biol Rev* 2010;74:417–433.

Delgado-Baquerizo M, et al. A global atlas of the dominant bacteria found in soil. *Science* 2018;359:320–325.

Denamur E, et al. The population genetics of pathogenic Escherichia coli. *Nat Rev Microbiol* 2021;19:37–54.

DeSantis TZ, et al. Greengenes, a Chimera-Checked 16S rRNA Gene Database and Workbench Compatible with ARB. *Applied and Environmental Microbiology* 2006;72:5069–72.

Dineen SM, et al. An evaluation of commercial DNA extraction kits for the isolation of bacterial spore DNA from soil. *J Appl Microbiol* 2010;109:1886–1896.

Douglas GM, et al. Predicting the functional potential of the microbiome from marker genes using PICRUSt. In: Beiko RG, et al. (eds.). *Microbiome analysis: methods and protocols*. New York, NY: Springer New York, 2018:169–177.

Drula E, et al. The carbohydrate-active enzyme database: functions and literature. *Nucleic Acids Res* 2022;50:D571–D577.

Dunbar J, et al. Levels of bacterial community diversity in four arid soils compared by cultivation and 16S rRNA gene cloning. *Appl Environ Microbiol* 1999;65:1662–1669.

Dwivedi B, et al. PhiSiGns: an online tool to identify signature genes in phages and design PCR primers for examining phage diversity. *BMC Bioinformatics* 2012;13:37.

Edel-Hermann V and Lecomte C. Current status of *Fusarium oxysporum Formae Speciales* and races. *Phytopathology®* 2019;109:512–530.

Edgar RC. Taxonomy annotation errors in 16S rRNA and fungal ITS sequence databases. PeerJ 2018;6:e5030.

Epp Schmidt D, et al. Quantitative amplicon sequencing is necessary to identify differential taxa and correlated taxa where population sizes differ. *Microb Ecol* 2024;86:2790–2801. https://doi.org/s00248-023-02273-z.

Epp Schmidt DJ, et al. Metagenomics reveals bacterial and archaeal adaptation to urban land-use: N catabolism, methanogenesis, and nutrient acquisition. *Front Microbiol* 2019;10:02330.

Eren AM, et al. Oligotyping: differentiating between closely related microbial taxa using 16S rRNA gene data. Freckleton R (ed.). *Methods Ecol Evol* 2013;4:1111–1119.

Gaby JC, et al. Diazotroph community characterization via a high-throughput *nifH* amplicon sequencing and analysis pipeline. Pettinari MJ (ed.). *Appl Environ Microbiol* 2018;84:e01512-17.

Galla G, et al. Comparison of DNA extraction methods on different sample matrices within the same terrestrial ecosystem. *Sci Rep* 2024;14:8715.

George PBL, et al. Primer and database choice affect fungal functional but not biological diversity findings in a national soil survey. *Front Environ Sci* 2019;7:173.

Gloor GB, et al. Microbiome datasets are compositional: and this is not optional. *Front Microbiol* 2017;8:2224.

González-Pech RA, et al. Commonly misunderstood parameters of NCBI BLAST and important considerations for users. Hancock J (ed.). *Bioinformatics* 2019;35:2697–2698.

Greenwald WW, et al. Utilization of defined microbial communities enables effective evaluation of meta-genomic assemblies. *BMC Genomics* 2017;18:296.

Hallin S and Lindgren P-E. PCR detection of genes encoding nitrite reductase in denitrifying bacteria. *Appl Environ Microbiol* 1999;65:1652–1657.

He S, et al. Patterns in wetland microbial community composition and functional gene repertoire associated with methane emissions. *mBio* 2015;6:e00066-15.

Henry S, t al. Quantitative detection of the nosZ gene, encoding nitrous oxide reductase, and comparison of the abundances of 16S rRNA, narG, nirK, and nosZ genes in soils. *Appl Environ Microbiol* 2006;72:5181–5189.

Heydari A, et al. Co-selection of heavy metal and antibiotic resistance in soil bacteria from agricultural soils in New Zealand. *Sustainability* 2022;14:1790.

Hyatt D, et al. Gene and translation initiation site prediction in metagenomic sequences. *Bioinformatics* 2012;28:2223–2230.

Iquebal MA, et al. Potential use of microbial community genomes in various dimensions of agriculture productivity and its management: a review. *Front Microbiol* 2022;13:708335.

Iturbe-Espinoza P, et al. Effects of DNA preservation solution and DNA extraction methods on microbial community profiling of soil. *Folia Microbiol (Praha)* 2021;66:597–606.

Jain M, et al. The Oxford Nanopore MinION: delivery of nanopore sequencing to the genomics community. *Genome Biol* 2016;17:239.

Janssen PH. Identifying the dominant soil bacterial taxa in libraries of 16S rRNA and 16S rRNA genes. *Appl Environ Microbiol* 2006;72:1719–1728.

Jensen LJ, et al. eggNOG: automated construction and annotation of orthologous groups of genes. *Nucleic Acids Res* 2007;36:D250–D254.

Jian C, et al. Quantitative PCR provides a simple and accessible method for quantitative microbiota profiling. Vaz-Moreira I (ed.). *PLoS One* 2020;15:e0227285.

Jones CM, et al. The unaccounted yet abundant nitrous oxide-reducing microbial community: a potential nitrous oxide sink. *ISME J* 2013;7:417–426.

Jones MB, et al. Library preparation methodology can influence genomic and functional predictions in human microbiome research. *Proc Natl Acad Sci* 2015;112:14024–14029.

Kanehisa M, et al. KEGG for integration and interpretation of large-scale molecular data sets. *Nucleic Acids Res* 2012;40:D109–D114.

Karlsson I, et al. Genus-specific primers for study of fusarium communities in field samples. Cullen D (ed.). *Appl Environ Microbiol* 2016;82:491–501.

Keegan KP, et al. MG-RAST, a metagenomics service for analysis of microbial community structure and function. In: Martin F and Uroz S (eds.). *Microbial environmental genomics (MEG)*. New York, NY: Springer New York, 2016:207–233.

Kelley DR, et al. Gene prediction with Glimmer for metagenomic sequences augmented by classification and clustering. *Nucleic Acids Res* 2012;40:e9–e9.

Kelliher J, et al. *Microbiome data management in action workshop*. Atlanta, GA, USA, 2024.

Kennedy K, et al. Evaluating bias of illumina-based bacterial 16S rRNA gene profiles. Wommack KE (ed.). *Appl Environ Microbiol* 2014;80:5717–5722.

Kim D, et al. Centrifuge: rapid and sensitive classification of metagenomic sequences. *Genome Res* 2016;26:1721–1719.

Knapp CW, et al. Relationship between antibiotic resistance genes and metals in residential soil samples from Western Australia. *Environ Sci Pollut Res* 2017;24:2484–2494.

Kohout P, et al. Comparison of commonly used primer sets for evaluating arbuscular mycorrhizal fungal communities: Is there a universal solution? *Soil Biol Biochem* 2014;68:482–493.

Labouyrie M, et al. Patterns in soil microbial diversity across Europe. *Nat Commun* 2023;14:3311.

Langille MGI, et al. Predictive functional profiling of microbial communities using 16S rRNA marker gene sequences. *Nat Biotechnol* 2013;31:814–821.

Lehmann J, et al. The concept and future prospects of soil health. *Nat Rev Earth Environ* 2020;1:544–553.

Li H, et al. The vertical migration of antibiotic-resistant genes and pathogens in soil and vegetables after the application of different fertilizers. *Environ Res* 2022;203:111884.

Li J, et al. Urbanization-driven forest soil greenhouse gas emissions: insights from the role of soil bacteria in carbon and nitrogen cycling using a metagenomic approach. *Sci Total Environ* 2024;923:171364.

Li M, et al. Genomic and transcriptomic evidence for scavenging of diverse organic compounds by widespread deep-sea archaea. *Nat Commun* 2015;6:8933.

Liao H, et al. Herbicide selection promotes antibiotic resistance in soil microbiomes. Falush D (ed.). *Mol Biol Evol* 2021;38:2337–2350.

Liu J, et al. Comparison of ITS and 18S rDNA for estimating fungal diversity using PCR–DGGE. *World J Microbiol Biotechnol* 2015;31:1387–1395.

Long PE, et al. Microbial metagenomics reveals climate-relevant subsurface biogeochemical processes. *Trends Microbiol* 2016;24:600–610.

Lu J, et al. Bracken: estimating species abundance in metagenomics data. *PeerJ Comput Sci* 2017;3:e104.

Lu J, et al. Metagenome analysis using the Kraken software suite. *Nat Protoc* 2022;17:2815–2839.

Mącik M, et al. New insight into the soil bacterial and fungal microbiome after phosphorus biofertilizer application as an important driver of regenerative agriculture including biodiversity loss reversal and soil health restoration. *Appl Soil Ecol* 2023;189:104941.

Magrane M and Consortium U. UniProt Knowledgebase: a hub of integrated protein data. *Database* 2011;2011:bar009–bar009.

Maguire F, et al. Metagenome-a- ssembled genome binning methods with short reads disproportionately fail for plasmids and genomic Islands. *Microb Genomics* 2020.

Manter DK, et al. Unveiling errors in soil microbial community sequencing: a case for reference soils and improved diagnostics for nanopore sequencing. *Commun Biol* 2024;7:913.

Mantri SS, et al. Metagenomic sequencing of multiple soil horizons and sites in close vicinity revealed novel secondary metabolite diversity. Yang Y-L (ed.). *mSystems* 2021;6:01018-21.

Markowitz VM, et al. IMG/M: a data management and analysis system for metagenomes. *Nucleic Acids Res* 2007;36:D534–D538.

Maurya S, et al. Indicators for assessment of soil quality: a mini-review. *Environ Monit Assess* 2020;192:604.

McArthur AG, et al. The comprehensive antibiotic resistance database. *Antimicrob Agents Chemother* 2013;57:3348–3357.

McMurdie PJ and Holmes S. phyloseq: an R package for reproducible interactive analysis and graphics of microbiome census data. Watson M (ed.). *PLoS One* 2013;8:e61217.

Menzel P, et al. Fast and sensitive taxonomic classification for metagenomics with Kaiju. *Nat Commun* 2016;7:11257.

Meslier V, et al. Benchmarking second and third-generation sequencing platforms for microbial metagenomics. *Sci Data* 2022;9:694.

Milani C, et al. METAnnotatorX2: a comprehensive tool for deep and shallow metagenomic data set analyses. Arumugam M (ed.). *mSystems* 2021;6:e00583-21.

Nesme J, et al. Back to the future of soil metagenomics. *Front Microbiol* 2016;7:00073.

Nguyen NH, et al. FUNGuild: an open annotation tool for parsing fungal community datasets by ecological guild. *Fungal Ecol* 2016;20:241–248.

Nurk S, et al. metaSPAdes: a new versatile metagenomic assembler. *Genome Res* 2017;27:824–834.

Oksanen J, et al. vegan: community ecology package. 2024.

Ounit R, et al. CLARK: fast and accurate classification of metagenomic and genomic sequences using discriminative k-mers. *BMC Genomics* 2015;16:236.

Overbeek R. The subsystems approach to genome annotation and its use in the project to annotate 1000 genomes. *Nucleic Acids Res* 2005;33:5691–5702.

Parada AE, et al. Every base matters: assessing small subunit rRNA primers for marine microbiomes with mock communities, time series and global field samples. *Environ Microbiol* 2016;18:1403–1414.

Paradis E, et al. APE: analyses of phylogenetics and evolution in R language. *Bioinformatics* 2004;20:289–290.

Quail M, et al. A tale of three next generation sequencing platforms: comparison of Ion torrent, pacific biosciences and illumina MiSeq sequencers. *BMC Genomics* 2012;13:341.

Quast C, et al. The SILVA ribosomal RNA gene database project: improved data processing and web-based tools. Nucleic Acids Research 2012;41:D590–6.

Quince C, et al. Shotgun metagenomics, from sampling to analysis. *Nat Biotechnol* 2017;35:833–844.

Rayamajhi N, et al. Evaluating illumina-, nanopore-, and PacBio-based genome assembly strategies with the bald notothen, *Trematomus borchgrevinki*. Rokas A (ed.). *G3 GenesGenomesGenetics* 2022;12:jkac192.

Rho M, et al. FragGeneScan: predicting genes in short and error-prone reads. *Nucleic Acids Res* 2010;38:e191–e191.

Ronaghi M, et al. Real-time DNA sequencing using detection of pyrophosphate release. *Anal Biochem* 1996;242:84–89.

Rotthauwe J-H, et al. The ammonia monooxygenase structural gene amoA as a functional marker: molecular fine-scale analysis of natural ammonia-oxidizing populations. *Appl Environ Microbiol* 1997;63:4704–4712.

Satam H, et al. Next-generation sequencing technology: current trends and advancements. *Biology* 2023;12:997.

Schiml VC, et al. Integrative meta-omics in Galaxy and beyond. *Environ Microbiome* 2023;18:56.

Schloss Patrick D., et al. Introducing mothur: open-source, platform-independent, community-supported software for describing and comparing microbial communities. *Appl Environ Microbiol* 2009;75:7537–7541.

Schloss PD and Handelsman J. Introducing DOTUR, a computer program for defining operational taxonomic units and estimating species richness. *Appl Environ Microbiol* 2005;71:1501–1516.

Seemann T. Prokka: rapid prokaryotic genome annotation. *Bioinformatics* 2014;30:2068–2069.

Seiler C and Berendonk TU. Heavy metal driven co-selection of antibiotic resistance in soil and water bodies impacted by agriculture and aquaculture. *Front Microbiol* 2012;3:00399.

Setubal JC. Metagenome-assembled genomes: concepts, analogies, and challenges. *Biophys Rev* 2021;13:905–909.

Sevim V, et al. Shotgun metagenome data of a defined mock community using Oxford Nanopore, PacBio and Illumina technologies. *Sci Data* 2019;6:285.

Shaffer JP, t al. Standardized multi-omics of Earth's microbiomes reveals microbial and metabolite diversity. *Nat Microbiol* 2022;7:2128–2150.

Sharuddin SS, et al. Advancement of metatranscriptomics towards productive agriculture and sustainable environment: a review. *Int J Mol Sci* 2022;23:3737.

Slatko BE, et al. Overview of next-generation sequencing technologies. *Curr Protoc Mol Biol* 2018;122:e59.

Snyder EE, et al. PATRIC: The VBI pathosystems resource integration center. *Nucleic Acids Res* 2007;35:D401–D406.

Stevens BM, et al. Comparison of Oxford nanopore technologies and illumina MiSeq sequencing with mock communities and agricultural soil. *Sci Rep* 2023;13:9323.

Tedersoo L, et al. Best practices in metabarcoding of fungi: from experimental design to results. *Mol Ecol* 2022;31:2769–2795.

The Genome Standards Consortium, et al. Minimum information about a single amplified genome (MISAG) and a metagenome-assembled genome (MIMAG) of bacteria and archaea. *Nat Biotechnol* 2017;35:725–731.

The Reference Genome Group of the Gene Ontology Consortium. The gene ontology's reference genome project: a unified framework for functional annotation across species. Bourne PE (ed.). *PLoS Comput Biol* 2009;5:e1000431.

Thompson JR, et al. Genotypic diversity within a natural coastal bacterioplankton population. *Science* 2005;307:1311–1313.

Throbäck IN, et al. Reassessing PCR primers targeting nirS, nirK and nosZ genes for community surveys of denitrifying bacteria with DGGE. *FEMS Microbiol Ecol* 2004;49:401–417.

Toju H, et al. Core microbiomes for sustainable agroecosystems. *Nat Plants* 2018;4:247–257.

Tourna M, et al. Growth, activity and temperature responses of ammonia-oxidizing archaea and bacteria in soil microcosms. *Environ Microbiol* 2008;10:1357–1364.

Trivedi P, et al. Enabling sustainable agriculture through understanding and enhancement of microbiomes. *New Phytol* 2021;230:2129–2147.

Van Der Walt AJ, et al. Assembling metagenomes, one community at a time. *BMC Genomics* 2017;18:521.

Vázquez-Castellanos JF, et al. Comparison of different assembly and annotation tools on analysis of simulated viral metagenomic communities in the gut. *BMC Genomics* 2014;15:37.

Wan Z, et al. nirS and nosZII bacterial denitrifiers as well as fungal denitrifiers are coupled with N2O emissions in long-term fertilized soils. *Sci Total Environ* 2023;897:165426.

Wang Q and Cole JR. Updated RDP taxonomy and RDP Classifier for more accurate taxonomic classification. Pritchard L (ed.). *Microbiol Resour Announc* 2024;13:e01063-23.

Wang Q, et al. Ecological patterns of nifH genes in four terrestrial climatic zones explored with targeted metagenomics using framebot, a new informatics tool. *mBio* 2013;4:e00592-13-e00592-13.

Ward DM, et al. A natural view of microbial biodiversity within hot spring cyanobacterial mat communities. *Microbiol Mol Biol Rev* 1998;62:1353–1370.

Wemheuer F, et al. Tax4Fun2: prediction of habitat-specific functional profiles and functional redundancy based on 16S rRNA gene sequences. *Environ Microbiome* 2020;15:11.

White TJ, et al. Amplification and direct sequencing of fungal ribosomal RNA genes for phylogenetics. In: Innis MA, et al. (eds.). *PCR protocols*. San Diego: Academic Press, 1990:315–322.

Wilhelm RC, et al. Ecological insights into soil health according to the genomic traits and environment-wide associations of bacteria in agricultural soils. *ISME Commun* 2023;3:1.

Wilke A, et al. The M5nr: a novel non-redundant database containing protein sequences and annotations from multiple sources and associated tools. *BMC Bioinformatics* 2012;13:141.

Wong KH, et al. Multiplex illumina sequencing using DNA barcoding. *Curr Protoc Mol Biol* 2013;101:Unit 7.11. https://doi.org/10.1002/0471142727.mb0711s101.

Xu S, et al. Different clusters of Candidatus 'Methanoperedens nitroreducens'-like archaea as revealed by high-throughput sequencing with new primers. *Sci Rep* 2018;8:7695.

Young JM, et al. Limitations and recommendations for successful DNA extraction from forensic soil samples: a review. *Sci Justice* 2014;54:238–244.

Young JPW. Genome diversity in arbuscular mycorrhizal fungi. *Curr Opin Plant Biol* 2015;26:113–119.

Yu Y, et al. ITS3 / ITS4 outperforms other ITS region and 18S rRNA gene primer sets for amplicon sequencing of soil fungi. *Eur J Soil Sci* 2022;73:e13329.

Zhang B, et al. Evaluation of the ion torrent personal genome machine for gene-targeted studies using amplicons of the nitrogenase gene *nifH*. Lovell CR (ed.). *Appl Environ Microbiol* 2015;81:4536–4545.

Zhang Z, et al. Benchmarking genome assembly methods on metagenomic sequencing data. *Brief Bioinform* 2023;24:bbad087.

Zheng Z, et al. The composition of antibiotic resistance genes is not affected by grazing but is determined by microorganisms in grassland soils. *Sci Total Environ* 2021;761:143205.

Chapter 3

Advances in soil quantitative polymerase chain reaction (qPCR) techniques for identifying and studying soil microbial communities

M. Giles, The James Hutton Institute, UK; A. Cotton, The University of Manchester, UK; and C. Beukes, The James Hutton Institute, UK and University of Pretoria, South Africa

1 Introduction

For over 25 years, real-time quantitative polymerase chain reaction (qPCR), also known as quantitative PCR or real-time PCR (RT-PCR or RT-qPCR), has offered a comparatively cheap and rapid means of quantifying gene abundances in a diverse range of samples (VanGuilder et al., 2008) including environmental samples such as soils. This has allowed unprecedented insight into important microbial metrics such as the abundance of different bacterial taxa (Fierer et al., 2005, Zhang et al., 2017) and the abundance of process-specific functional genes, enabling their associations with functional measures, e.g. volatile organic compounds or greenhouse gas emissions (Shahsavari et al., 2016; Giles et al., 2017). In addition, it has allowed the creation of diagnostic and monitoring tools for both soil prokaryotes and eukaryotes (Jiménez-Fernández et al., 2010). This has revolutionised the ability of microbial ecologists to both examine changes

http://dx.doi.org/10.19103/AS.2025.0159.03

in soil communities, as well as identify and quantify populations of beneficial and pathogenic organisms, thereby generating a better understanding of the role microbial soil communities play in supporting soil function and broader ecosystem services. Using qPCR to investigate soil communities, however, has a wide range of advantages, disadvantages, and potential limitations that need to be accounted for when designing, implementing, and interpreting qPCR-based studies. The continued improvement of qPCR approaches, together with their integration into the growing range of other molecular methods targeted at exploring soil communities, means that qPCR remains one of our most powerful tools for studying soil microbial communities and dissecting their various roles in soil health.

2 Principles of qPCR

The polymerase chain reaction (PCR) is the fundamental basis of many microbial community analyses, allowing the amplification of a specific gene region through *in vitro* enzymatic replication. The product of a successful PCR is known as an 'amplicon'. In traditional PCR, only the presence of an amplicon at the end of the PCR protocol is assessed. In contrast, qPCR can measure the increase in amplicon abundance throughout the PCR reaction in real time. qPCR works by harnessing the PCR reaction in a quantifiable way using the linearity of DNA amplification to allow quantification of the target gene in a sample. This is achievable through the predictable nature of the PCR reaction, which goes through three distinct phases (Figure 1): (1) the exponential phase, where the target gene will in theory double with every cycle, (2) the linear phase, where amplicons no longer double because PCR reagents are becoming limited, and (3) the plateau phase, where PCR reagents are exhausted and no more replication can occur (McKew and Smith, 2015). In qPCR, by introducing a fluorescent dye or probe, the PCR reaction can be tracked by measuring fluorescence across the reaction period. During the exponential phase of the PCR reaction, levels of fluorescence will be proportional to the copies of the amplicon at any given cycle and follow a definable log-linear relationship. qPCR machines can measure this fluorescence and routinely do so after every PCR cycle. The cycle number at which a threshold level of fluorescence is reached is recorded and used in downstream data analysis. This is the foundation on which qPCR operates, though the exact dyes, reaction chemistries, fluorescence measures, and threshold values will vary between the wide range of qPCR machines, and DNA polymerases available.

Despite this wide range of qPCR machines, reagents and protocols, the general principles and steps involved in carrying out a qPCR study (Figure 2) remain much the same. This involves one of two methods: (1) an intercalating dye such as SYBR Green that has low fluorescence when not bound to DNA and

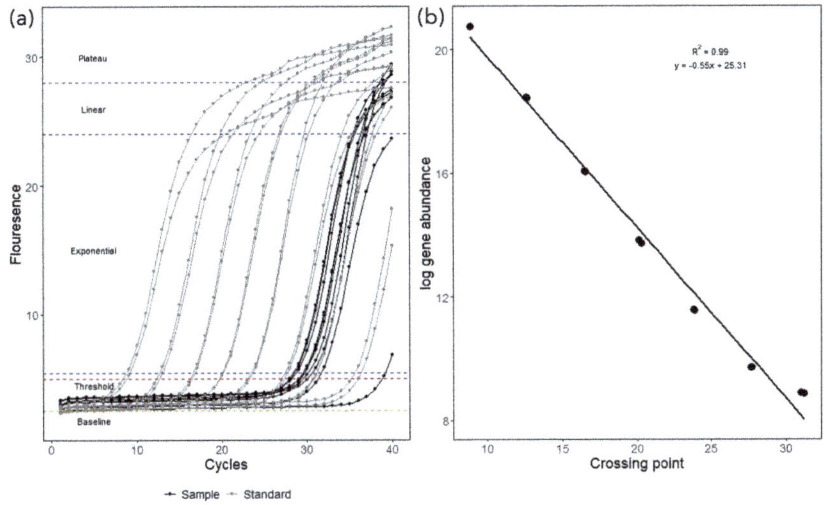

Figure 1 (a) Fluorescence ~ cycle curves for qPCR with bacterial *amoA* target and using DNA extracted from soils. Estimates of the baseline and threshold as well as the exponential, linear and plateau phases are denoted by dotted lines. Light grey lines denote standards which were used to produce (b) a qPCR standard curve.

which cannot bind to single-stranded DNA but which will bind to and fluoresce when intercalated into double-stranded DNA, and (2) a fluorescent probe specific to the target being investigated, where binding of the probe to target DNA allows the probe to fluoresce, for example TaqMan based qPCR assays. In both cases fluorescence increases as DNA replicates during the PCR reaction. The key difference between the two methods is the level of specificity with probes being more specific than intercalating dyes as probes are designed for specific targets, while intercalating dyes are able to bind to any double stranded DNA (Adams, 2020). Due to these differences, the two forms of dye are often used for slightly different purposes. Intercalating dyes are mostly used when a study needs a cost-effective way to quantify multiple genes for example

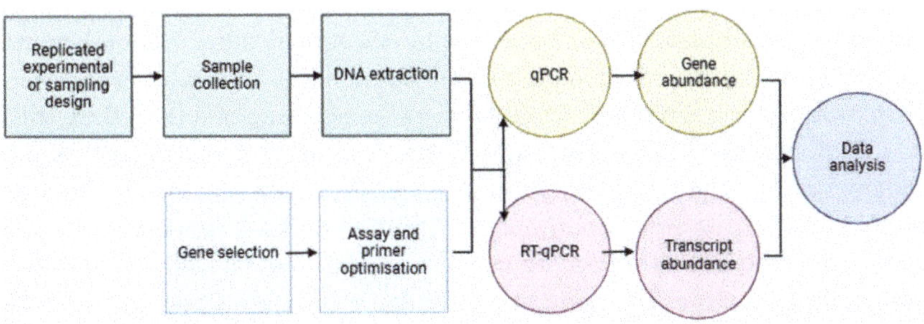

Figure 2 Diagrammatic representation of the steps involved in generating data by qPCR.

soil Nitrogen (N) cycling studies such as that by Langarica-Fuentes et al. (2018) where multiple genes are needed to characterise a process. In contrast, probe-based dyes are mainly used where the goal is accurate quantification of a single soil organism for example the study by Rocha et al. (2023) which monitored soybean soil-borne pathogens. Both methods however will lead to a predictable increase in fluorescence during the exponential phase of the PCR, though care should be taken when interpreting the results of studies using intercalating dyes, as non-target DNA might also be quantified as a consequence of poor primer specificity (Adams, 2020).

2.1 Quantification

Quantification by qPCR can be either relative or absolute, with absolute quantification more common in environmental studies (Brankatschk et al., 2012). Both require determination of an appropriate threshold value that can be referred to as the threshold cycle (C_t), crossing point (C_p), take-off point or quantification cycle (C_q) (Bustin et al., 2009). This is critical for an accurate qPCR assay. The C_q value is set at the level where sample fluorescence becomes higher than background fluorescence. This is normally during the exponential phase of the PCR reaction and at a point where sample curves are parallel; otherwise, comparison of C_q values between samples becomes invalid (Figure 1). C_q values may therefore differ between qPCR assays, machines and dyes and are one of the additional factors that need optimising above what is needed to optimise a conventional PCR.

Relative quantification uses the differences in C_q values (ΔC_q) between a target and reference gene to determine the relative concentration of the target gene in a sample. It is commonly used in gene expression studies but is challenging to use in environmental samples where there may not be an appropriate reference gene available. In contrast, absolute quantification uses a standard curve that can be created from serial dilutions of a linearised plasmid that contains an insert of the target of interest. These can be created from both environmental samples or cultures which possess the gene of interest. The C_q values from this standard curve can be used to quantify the target gene in samples. While absolute quantification is often more relevant to environmental samples than relative quantification, it is important to recognise that in most cases it is not actually definitively an absolute quantification. This is due to both the multi-organism nature of environmental samples and the high probability of co-purifying inhibitors. A standard curve will usually be made from either a culture or cleaned up PCR product and will represent the gene of interest as found in a single organism. In environmental samples a gene of interest can sometimes be found across multiple organisms and can therefore vary in its sequence and GC content leading to biases in amplification between different organisms and variation in PCR efficiencies (Brankatschk et al., 2012). Additionally, unless intentionally added into the PCR reactions, standard curves will contain none of the inhibitors that may affect PCR efficiency in an environmental sample. Consequently, in any given qPCR run, the PCR efficiency of the standard curve will vary from that of the samples and will therefore not allow a perfect quantification of the gene of interest. Despite this, absolute quantification has been used successfully in a wide range of soil-based qPCR studies (e.g. Giles et al., 2017).

3 Factors affecting the reproducibility of qPCR assays

A range of factors can affect the reproducibility of qPCR assays. These can be specific to the assay itself and/or a consequence of carrying out qPCR on environmental samples where factors such as inhibition can play a role. To ensure that qPCR studies which investigate soil microbial communities are reproducible, it is important to understand both sets of factors, the metrics that can be used to identify and quantify them, and how they can be mitigated. As so many factors can affect the efficacy of a qPCR assay, accurate reporting of methods and quality control of data has been recognised as key to creating reproducible and comparable qPCR results. Reporting DNA extraction methods, yields, reaction conditions, primers, standards, reference genes, and PCR efficiencies are extremely important, and guidelines on the minimum publishable information required for qPCR studies have been laid out in papers such as Bustin et al. (2009).

Systematic errors are those introduced by the assay itself and can include variation in PCR efficiencies – a key, though often under-reported consideration in qPCR studies (Bustin et al., 2009). PCR efficiency is defined as the ratio of the number of target gene molecules at the end of a PCR cycle to the number at the beginning and is a metric of how close to the theoretical optimum a PCR reaction is. In optimal reactions, PCR efficiencies will be 100%, i.e. copies of the target gene will double with each cycle. However, in reality, this is often not the case, and it is common to achieve PCR efficiencies above and below 100%. PCR efficiencies below the optimum often indicate a poorly optimised reaction with suboptimal reagent concentrations or reaction conditions. In addition, in many environmental samples, the range of organisms present necessitates the use of degenerate primers which can also impact PCR efficiency (Brankatschk et al., 2012). PCR efficiencies over 100% can be caused by inhibition which changes the nature of the log – linear relationship defining cycles and gene amplification. PCR efficiencies are therefore a critical piece of information for both optimising a PCR assay and assessing how comparable qPCR data may be between qPCR runs, different targets, and different qPCR-based studies. In theory, a well optimised PCR will have an efficiency of between 90 and 110%.

While PCR efficiencies are one of the most well-known factors affecting a qPCR, as it is defined by a wide range of variables, it is important to also consider more specific technical factors that may affect the efficacy and reproducibility of qPCR assays. These include the linear dynamic range, limit of detection and precision of the assay itself (Bustin et al., 2009). Collectively these factors describe the range of concentrations over which target genes can be reproducibly quantified. The linear dynamic range describes the range of gene target concentrations that have a linear relationship with fluorescence, and both sample and standard curve concentrations should fall within this range. The limit of detection describes the lowest number of copies at which 95% of the positive samples are detected and is of importance because PCR reactions are stochastic at low concentrations (Bustin et al., 2009). 'Precision' describes how far removed from the actual value any qPCR-derived quantification of the target is. This is important to understand within and between qPCR runs because it will form part of the variability in any set of results. In a successful assay, the technical precision will be high enough that the predominant source of variation in a set of results will come from biological variability and not variability introduced by the qPCR assay itself.

3.1 DNA extractions

When interpreting qPCR data, it is also important to recognise that factors beyond the PCR reaction itself can ultimately affect the validity of the results of any given qPCR assay. One important source of introduced variability comes from DNA extractions where both small errors in pipetting and differences in extraction efficiency between different soils and extraction methods will affect how much template DNA is present in any given sample. For example, a study on soil protists found variation in 18S rRNA gene copies which was due to the DNA extraction method used instead of being a true reflection of the environmental samples (Santos et al., 2015).

In order to provide an absolute quantification of any given gene target in a sample, DNA extractions would need to be 100% efficient which is unfortunately not achievable. In soils, factors such as the co-extraction of inhibitors and inefficiencies in the lysis step can affect DNA yield. Many commercial kits prioritise inhibitor removal over yields while non-kit-based methods tend to give higher DNA yields but suffer from inhibition. While there have been attempts to account for varying DNA yields in soil health based studies such as standardising output from qPCR as copies per ng DNA (e.g. Duff et al., 2022), normalising template DNA to a fixed value (e.g. Pagaling et al., 2023) or including a standard of known concentration into samples at the DNA extraction stage (e.g. Daniell et al., 2012), there is no perfect way to account for variation introduced by DNA extractions. All of the previous methods are important to consider as options for normalising data but ultimately the best way to ensure that variation in extraction efficiency does not affect the reproducibility of qPCR results is to randomise the order in which samples are extracted to prevent batch effects (i.e. treatments are randomly distributed through DNA extraction batches so that 'batch' does not equal 'treatment') and to ensure that the DNA extraction method remains constant throughout a project, though there will be differences in DNA extraction efficiencies between soils (Park and Crowley, 2005). While extraction efficiency remains a factor that needs to be accounted for, ever improving methods of soil DNA extraction that reduce the co-extraction of inhibitors have gone some way to helping improve reliability of downstream PCRs from soil samples.

4 qPCR targets

It is possible to quantify almost any gene within a soil community as long as the gene in question contains two conserved regions from which to design primers and the resulting amplicon length is generally less than 200 base pairs. This means that the key constraint for using qPCR to characterise soil communities is often not just the creation of an optimised qPCR assay that can amplify the gene of interest without off-target amplification, but also the identification of a gene that can characterise an organism or process of interest within the myriad of microorganisms and functional pathways that constitute a soil community. For genes used to quantify processes that are carried out by phylogenetically diverse organisms, such as denitrification (the sequential reduction of nitrate to either nitrous oxide (N_2O) or dinitrogen gas (N_2)), it is often not possible to find a conserved region across all the organisms capable of that process. qPCR therefore often underestimates the functional capabilities of soil communities. A recent study by Ma et al. (2019) compared the efficacy of commonly used qPCR primers for genes associated with denitrification (nitrate reduction: *narG* and *napA*; nitrite reduction: *nirK* and *nirS*; nitric oxide reduction: *norB*; nitrous oxide reduction: *nosZ*). The authors found that primer coverage could range anywhere from 4% to 99% depending on the gene being quantified and primer set used, and for *nirK* and *nosZ*, for instance, multiple primer sets may be needed to cover all possible clades. In addition, for processes such as soil organic matter mineralisation where a diverse range of genes are involved in the process and where no one gene can reliably characterise a rate-limiting step, it may not be possible to use qPCR to characterise microbial potential for the process. Nonetheless, despite these limitations, qPCR remains one of the only means of quantifying DNA and a useful measure, particularly when comparing relative changes between different systems or treatments. As described later, there are many examples of where qPCR has provided useful insights into soil health, including the ecology of C and N cycling bacteria, pathogens, and rhizobia.

5 Linking soil communities to measured processes

As many of the beneficial functions soil provides are underpinned by microbial processes, e.g. organic matter mineralisation, there is a need to understand the relationship between functional genes and process rates. However, previous studies have found mixed results when examining associations between the two. For example, Shrewsbury et al. (2016) found relationships between enzyme activity and the genes *nirS* and *nirK*. In contrast, Giles et al. (2017) did not identify any links between the production of N_2O and genes involved in denitrification (*nirS, nirK, nosZI*). Consequently, it should not be assumed that gene abundance is representative of what processes are occurring and their respective magnitudes. Gene abundance is not a proxy for process rates. Despite this, qPCR still offers vital insights into soil health as no microbially mediated process can occur without the presence of the gene that regulates it and it is possible to draw conclusions regarding functional potential between treatments and systems (e.g. Samad et al., 2016).

When linking qPCR to measured processes it's important to consider the temporal scale at which changes in process rates and gene abundances occur. This is fundamental to understanding the variability in successfully identifying associations between the two measures. A process rate is not just a function of whether or not the gene needed to regulate it occurs in a soil but also the efficiency with which a given community member can transcribe and translate the gene in question as well as the downstream efficiency of the resulting enzyme. Biotic and abiotic soil conditions can select for organisms with a particular functional gene, e.g. Herold et al. (2018), but this will be a function of soil conditions over a longer time period than those that affect transcription, translation and enzyme efficiency. Process rates therefore represent a snapshot of a moment in time whereas gene abundances represent a short-term accumulation of these snapshots. Additionally, the presence of a gene does not mean that it is actively being expressed and qPCR targeted at DNA (rather than mRNA, discussed next) will capture all organisms that possess the genes, i.e. those that both are and aren't actively expressing at the time of sampling.

One way to tackle some of these challenges in linking functional gene abundance to process rates is to investigate the expression of the gene of interest rather than simply its abundance. This can be done by quantifying messenger RNA (mRNA) rather than DNA. mRNA is a single-stranded RNA molecule created when a gene is transcribed and, through its translation, creates the protein responsible for a given process. Quantitative reverse transcription PCR (RT-qPCR) can be used to quantify RNA (Adams, 2020). It differs from conventional qPCR by including a step where mRNA is transcribed to complementary DNA (cDNA). This can be either a one-step process where reverse transcription and cDNA quantification occur in the same reaction tube or a two-step process where reverse transcription and cDNA quantification are carried out in separate tubes. In either approach, qPCR can quantify the resulting cDNA to give an indication of the expression levels of a functional gene. There are a number of challenges in studying mRNA in soil systems. These are important to understand when designing and implementing an mRNA-based study. Firstly, at the point of sample collection, it is important to capture an accurate representation of the mRNA present in a soil sample. As mRNA levels can change rapidly, samples need to be snap-frozen in liquid nitrogen. mRNA is also particularly susceptible to degradation, with the half-life of mRNA being just minutes (Steiner et al., 2019), therefore all processing steps, from RNA extraction to qPCR set-up, need to be performed on ice in an RNase-free environment. Additionally, any quantified cDNA will be dependent not only on the efficiency of the qPCR reaction but also the transcription step preceding it (Wong and Medrano, 2005). This can make it difficult to get sufficient cDNA to produce an accurate quantification of mRNA levels and to compare results between soils.

Despite these technical challenges, using RT-qPCR has led to a number of insights into the relationship between processes and soil communities e.g. Henderson et al. (2010). As with gene abundance, though, it will not always be directly related to process rates as enzyme efficiency can vary post translation. This is often linked to factors such as soil pH or soil moisture, which can impact how an enzyme operates in the environment (Knowles, 1982). In many cases, the most meaningful insights into soil health using qPCR will come from the combination of methods that quantify community attributes in a range of ways along with direct quantification of process rates.

6 Case studies of the use of qPCR for measuring soil health

6.1 qPCR to characterise N cycling potential

N cycling is intimately linked to soil health, with this multi-process cycle capable of releasing ammonium from organic matter through mineralisation, turning ammonium to nitrate through nitrification, a process that produces the greenhouse gas N_2O, and releasing nitrogen from soils through denitrification, a process that sequentially reduces nitrate to either N_2O or N_2 depending on soil conditions. Consequently, optimising the balance and timing of the different N cycling processes is key to the management of low input agricultural systems. This will allow maximising the release of nitrogen from organic matter when plants require it for growth and minimising nitrogen loss through nitrate leaching and nitrogenous gas production. As the processes that underpin N cycling are predominately microbially mediated, qPCR methods have been key to understanding and characterising N cycling in agronomic systems, providing insights into the mechanisms needed to achieve low input agriculture. As detailed earlier, a range of different genes are used to characterise nitrification and denitrification. These include, but are not limited to, bacterial and archaeal ammonium monooxygenase (*amoA*, nitrification), nitrate reductase (*narG*, nitrate reduction), nitrite reductase, and N_2O reductase (*nirK*, *nirS*, and *nosZ*, denitrification). Studies have used qPCR to quantify these genes under a range of conditions, which has been critical to understanding how microorganisms control the different N cycling processes and what soil conditions select for these functional genes (Tang et al., 2024).

With nitrification, the quantification of ammonium monooxygenase has helped to identify that soil pH controls the balance of archaeal and bacterial *amoA* (Nicol et al., 2008) due to the different pH requirements of the nitrifying organisms found in these two kingdoms (Gubry-Rangin, 2010). This determination has subsequently been found to have implications for the sustainability of agricultural systems, with the balance between these two kingdoms affecting nitrification-driven N_2O emissions (Hink et al., 2017).

In contrast to nitrification, denitrification is a multi-stage process carried out by phylogenetically diverse organisms, and the process might not necessarily always reach completion. This process is affected by a wide range of factors including soil oxygen content, carbon to nitrogen ratios, pH and carbon quality (the range and type of carbon-containing compounds available to microorganisms). The final step of the process, the reduction of N_2O to N_2, is a key control point for the emission of N_2O from soils. Increased reduction of N_2O to N_2 can reduce the emissions of N_2O. qPCR has been used to understand both how denitrifying organisms respond to different controlling factors and how they relate to the emissions of N_2O. Carbon to nitrogen ratios can be key to controlling the production-to-reduction balance of N_2O. Consequently, studies seeking to understand the microbial basis of this have investigated the effect of different forms of carbon on denitrifiers. Studies have been able to show that carbon additions can affect the abundance of denitrification genes with *nirS* and *nosZI* showing an increase in abundance with labile carbon addition (Langerica-Fuentes et al., 2018). The form carbon takes was also found to be important with *nosZ* and *nirS* responding to the addition of an amino acid to soil but not to the addition of a sugar or an organic acid (Giles et al., 2017). This has implications for the selection of denitrifiers in the rhizosphere where carbon is one of the factors leading to the selection of rhizosphere microbial communities. This is reflected in the greater prevalence of *nosZ* from clade I than that of clade II in plant rhizospheres and is thought to have an impact on N_2O emissions because of clade I's ability to both produce and reduce N_2O (Graf et al., 2022). While carbon addition may affect the ratio of N_2O to N_2, the role *nosZ* plays in this is not always clear with some studies finding no associations between carbon addition, the copies of *nosZ* as quantified by qPCR and N_2O emissions (Highton et al., 2020). The quantification of mRNA transcripts can sometimes find relationships between denitrifiers and the N_2O emissions that aren't apparent in DNA-based qPCR data such as the association between waterlogging, *nosZ* and N_2O found by Uchida et al. (2014).

Despite the sometimes contradictory results, qPCR data have collectively enhanced our understanding of the forces that select nitrifiers and denitrifiers in soil systems. It has shown management-driven selection of the different N cycling functional genes such as the association between organic amendments and *nosZ* (Wang et al., 2017; Yang et al., 2022; Tang et al., 2024). Moreover, it has allowed the tracking of inoculant treatments aimed at manipulating N cycling (Hiis et al., 2024).

6.2 qPCR for pathogen quantification

Quantitative PCR is also increasingly being used to measure the levels of deleterious organisms in soil such as pathogens. Soil is a habitat for many plant and human pathogens (Samaddar et al., 2021) and the occurrence and severity of disease they can cause may be influenced by their populations in the soil (Singh et al., 2014; Chatterton et al., 2023). Therefore, accurate quantification is necessary for both estimating the risk of disease and assessing the efficacy of management strategies such as non-susceptible host planting, soil disinfestation and biocontrol (Katan, 2017; Abdelaziz et al., 2023). Furthermore, methods of measuring population sizes are also extremely valuable for conducting research elucidating the mechanisms of pathogenicity.

Culture-based techniques for assessing pathogens are time-consuming and labour-intensive. Moreover, they can lack sensitivity as they can lead to underestimation of population sizes and false negatives due to overgrowth of the pathogen, competition by other microbes and the presence of the pathogen in unculturable yet viable states (Um et al., 2013; Pradhanang et al., 2000; Göhler et al., 2017). Serological methods such as enzyme-linked immunosorbent assays (ELISAs) have been developed for several pathogens, but they can have relatively low sensitivity (Chen et al., 2023).

qPCR therefore offers a promising tool for avoiding these issues and protocols have now been developed for many major pathogens including the *Ralstonia solanacearum* species complex (Chen et al., 2023), the *Fusarium oxysporum* species complex (Jiménez-Fernández et al., 2010), *Phytophthora nicotianae* (Huang et al., 2010) and *Rhizoctonia solani* (Woodhall et al., 2013).

However, qPCR for pathogen assessment is not without its own limitations. Both false positives and negatives for detection are likely and can have profound consequences. PCR inhibitors create the need to dilute samples that can potentially lead to false negative results as pathogen abundances may fall below the sensitivity threshold. Interestingly, digital droplet PCR protocols (see Section 7.1) are being developed for the detection of some soil pathogens and these have been found to be both more sensitive than real-time PCR while also eliminating inhibition issues (Ren et al., 2022). However, even with this method, false negatives could still occur due to problems with pathogen DNA extractions. This can be partially overcome by incorporating internal sample process controls. These can be cells of organisms, rather than extracted nucleic acids, thus enabling assessment of the likelihood of difficulties in DNA extraction resulting from the physical properties of pathogens that make them resistant to lysis (e.g. resting spores with particularly thick wall cells; Chen et al., 2023). Ideally, internal controls should be as similar to the target organism as possible (Rossmanith et al., 2011) and have already been developed for some pathogens (Chen et al., 2023).

The potential for cross-hybridisation of qPCR primers and probes designed for pathogenic taxa against non-pathogenic taxa could also potentially lead to false positives and unreliable population estimation (Lievens et al., 2006). Further sequencing of PCR products offers a potential strategy to check for this specificity issue. False positives may also occur due to the presence of DNA from non-viable microbes. It has been estimated that an average of 40% of fungal and bacterial DNA in soil is either extracellular or from non-intact cells (Carini et al., 2016). The use of propidium monoazide, which prevents the amplification of DNA that is not present in intact cells, is a promising development for reducing this problem and has recently been incorporated into the assessment of soil-borne pathogens (Fu et al., 2020).

In some cases, diseases can be caused by closely related species and differentiating between them is not necessary to determine the risk of disease (e.g. bacterial wilt caused by the *R. solanacearum* species complex; Chen et al., 2023). However, in other cases there can be intra-specific variation in pathogenicity, making it difficult or even impossible to identify pathogens based on commonly used DNA markers such as 16S rRNA gene sequences. In these cases, targeting genes involved with virulence may be a better strategy (Lievens and Thomma, 2005).

Finally, careful consideration must be given to soil sampling strategies when monitoring pathogens. Microbial communities in soils can be extremely spatially heterogenous (Franklin and Mills, 2003). Given that DNA extraction kits often only utilise very small quantities of soil, it is contentious whether DNA extractions can be usefully representative of whole fields (Clarkson et al.,

2015). It is therefore recommended to take many soil samples in the field and homogenise them before analysis or ideally analyse them separately. Exact sampling recommendations will be determined by the pathogen of interest and specific field conditions.

6.3 qPCR for detection and monitoring of beneficial bacteria in soil

6.3.1 Rhizobia and Plant Growth-Promoting Rhizobacteria (PGPRs)

The term 'rhizobia' refers to a polyphyletic group of soil bacteria that possess the ability to form nodules in association with legumes (such as beans, peas and soya). In these structures, predominantly formed on the plant roots, the bacteria can fix atmospheric nitrogen for their hosts while being provided with photosynthates (Checcucci et al., 2016; Maluk et al., 2023). This mutualistic interaction is a form of biological nitrogen fixation (BNF), and legumes capable of nodulation are not dependent upon synthetic nitrogen fertilisers as long they have access to compatible rhizobia under nitrogen-limiting conditions (Maluk et al., 2023). Legumes are known for their high protein content, making them important to animal and human nutrition. Their incorporation into agricultural systems (for instance as cover crops) has also been shown to have a positive impact on soil health. This has been mediated by improving hydraulic and structural traits, as well as increasing both the activity of soil microbes and the organic matter content of the soil while reducing erosion and compaction (Kocira et al., 2020).

The most effective way to determine whether rhizobial inoculation of a legume crop is necessary is to measure the indigenous rhizobial population, as the response of yield to inoculation is already variable and predictions from field trials and historical studies cannot effectively predict the outcome (Furseth et al., 2010). The most probable number (MPN) method has been routinely used to determine rhizobial populations (Furseth et al., 2010). This method involves the inoculation of a set of legumes with a soil dilution series prepared from samples collected from the prospective site and subsequently counting the nodulation incidence (Furseth et al., 2010). There are several drawbacks to the MPN. Firstly, the bacteria need to remain viable, which limits the number of samples that can be stored and investigated at any given time. Furthermore, allowing 4 to 6 weeks for plant growth significantly adds to the time before results are available (Furseth et al., 2010). However, good correlation has been found between quantification of soybean inoculants with qPCR and the MPN method (Furseth et al., 2010).

Factors that could influence both the results of the MPN method as well as yield response in the field are compatibility (with either the rhizobium or plant being either very specific or promiscuous in their symbiont interactions; Da Silva et al., 2024) and efficiency (as nodules can host multiple rhizobial strains, with not all being equally efficient at fixing nitrogen (Checcucci et al., 2016). The application of qPCR focused on taxonomic markers could be used to determine whether nodules contain multiple strains, the fitness of the various strains (determined based on the number of cells for a specific strain being present in the nodules), as well as the competitiveness of the different strains determined as the log of the ratio of qPCR estimates between two strains (Checcucci et al., 2016). These factors are also considered when deciding whether rhizobial strains should be viewed as 'elite' strains (Checucci et al., 2016; Da Silva et al., 2024). Whether higher nitrogen-fixation could be achieved with co-inoculation (as opposed to single-strain inoculants) is also an approach that could be tested by utilising qPCR (Da Silva et al., 2024).

The tracking and quantification of specific rhizobial symbionts provides the opportunity to study the effect of crop rotations and different agricultural management approaches on the applied inoculant/s or even just on the indigenous rhizobial community (Furseth et al., 2010). It could also be used as a diagnostic tool, should nodulation of the crop be suboptimal or to determine whether inoculation would be necessary when a new legume crop is introduced (Yudistira et al., 2021). Moreover, it could also be used to measure whether the strain in an inoculant is becoming established, as has been shown for the introduction of soybean in Scotland where the inoculant *Bradyrhizobium* strain could still be detected after winter (Maluk et al., 2023). The accurate quantification of specific inoculant strains could even enable testing various agronomic practices – such as whether plant density (as determined by row spacing) has an influence on rhizobial populations (Yudistira et al., 2021).

qPCR can also be applied to studies focusing on plant growth-promoting rhizobacteria (PGPRs). These are bacteria that are able to improve soil health in various ways such as phosphate solubilisation, production of phytohormones, sequestering heavy metals, assisting in the breakdown of plant debris as well as the mineralisation of soil organic matter (Khatoon et al., 2020). As for rhizobial inoculants, qPCR can be used to quantify and monitor different PGPRs providing the basis for quality control (Hale et al., 2014). The method can even be used to verify the impact of novel inoculum carriers such as different types of biochar on the survival and effectiveness of the PGPRs and the effect their application has on indigenous rhizosphere communities (Hale et al., 2014).

6.3.2 Bioremediation of soils

Bioremediation has been defined as an eco-friendly and low-cost treatment based on the capability of bacteria, fungi, archaea, and plants to remove or transform compounds or elements into less toxic forms (Bravo et al., 2020). The presence of toxins such as heavy metals (e.g. mercury), in soils can have serious deleterious effects on soil health, resulting in large-scale changes in both soil microbiome activity and composition (Bravo et al., 2020). Due to co-selection, it is also possible for soil metal pollution to drive anti-microbial resistance (Zhu et al., 2022). One bacterium which can be used for mercury bioremediation of soil is *Cupriavidus metallidurans* MSR33. Bravo et al. (2020) investigated the use of this bacterium for the bioremediation of agricultural soil and determined the effect of this process on nitrogen-fixing and nitrifying bacteria by quantifying the *nifH* (associated with the fixation of N) and ammonia oxidising bacteria *amoA* gene (encoding ammonia monooxygenase) copy numbers, respectively. Both groups of bacteria were negatively affected by mercury exposure, although the nitrogen-fixing bacteria to a much larger degree. One week after bioremediation, however, both groups showed a significant increase in function (Bravo et al., 2020). Quantification of the bioremediation strain showed it was maintained throughout the treatment process and even increased in abundance after treatment, suggesting that it might have adapted to the soil.

Zhu et al. (2022) recently designed a metal resistance gene chip, which is a high throughput qPCR-based chip containing the 16S rRNA locus as a reference but focusing on 56 metal resistance genes, and with which the homeostasis of nine metals can be determined. Its utilisation has enabled assessment of the preferred resistance strategies and the distribution of resistance across various habitats.

7 Recent developments

As the case studies above highlight, traditional qPCR has been successfully used in many studies. Despite this, it is continually being developed to improve its accuracy, precision and scope. This includes the development of digital PCR, microfluidics, and integrating qPCR with sequencing.

7.1 Droplet Digital PCR

First developed in the 1990s by Vogelstein and Kinzler (Vogelstein and Kinzler, 1999) droplet digital PCR (ddPCR) works by measuring the presence or absence of a fluorescence signal in droplets containing a single molecule of DNA (Morley, 2014). These droplets are created by partitioning out reactions containing sample template into many thousands of mini droplet reactions. The number of target molecules is then calculated using a Poisson distribution based on the presence or absence of fluorescence in a droplet (Pinheiro et al., 2011). Digital PCR has a number of advantages over qPCR, though tends not to be as cost-effective. As the method works by identifying target molecules in the different droplets, it does not require a standard curve, is not dependent on reaction efficiencies, and is not as impacted by inhibitors (Voegel et al., 2021).

While use of digital PCR in soil systems remains limited, there are examples of where it has been used successfully. For example, Voegel et al. (2021) were able to use digital PCR to track the changes in the abundance of nitrification and denitrification functional genes in vineyard soils over the course of a growing season. In addition, they compared qPCR and ddPCR and found that the upper limit of the dynamic range was lower with ddPCR than qPCR. However, for ddPCR, both the limit of detection and limit of quantification were equal to or better than those achieved by qPCR, depending on which of the functional genes were being quantified. While ddPCR may not be appropriate for genes present in high abundance, it offers the ability to detect low abundance genes with a greater degree of reliability than qPCR and may allow tracking of low abundance organisms that have to date been difficult to quantify.

7.2 qPCR and NGS

The development of Next Generation Sequencing (NGS) technology has allowed soil communities to be characterised both taxonomically (amplicon libraries) and functionally (shotgun metagenomics) in greater detail than ever before. However, despite falling costs, it remains time-consuming and expensive and has the limitation that it produces relative quantifications (Gloor et al., 2017). Moreover, it also may not have great enough sequencing depth to appraise the relative quantities of low abundance genes, meaning it is not always possible to investigate associations between gene abundances and soil processes.

Consequently, qPCR has the advantage of being both more targeted and cheaper per sample than either shotgun metagenomics or amplicon sequencing and will remain useful in cases where there are a low number of genes of interest. However, the combination of qPCR and NGS offers the potential for novel insights into soil communities. For example, Tang et al. (2024) used a combination of qPCR and NGS on functional genes involved in the N cycle to show that agricultural management treatments caused both changes in gene abundances and the functional communities likely to be performing denitrification. Furthermore, NGS data can be used to identify organisms or genes of interest with associated sequence data that can be used to develop qPCR assays. This offers both a cost-effective way of validating initial findings from sequencing studies and developing monitoring methods, and has the advantage that it will be able to identify important organisms or genes that are not traditionally used in soil-based qPCR studies and that may have been missed to date.

Recently the bioinformatics pipeline, SpeciesPrimer, has been developed to assist in the identification of species-specific target regions and the design of appropriate primers that can potentially be used for qPCR (Dreier et al., 2020).

This pipeline is automated and uses downloaded bacterial genomes to assess the quality of proposed primers. The pipeline is designed to be especially user-friendly as it can be run on a standard desktop PC and does not require high-performance computer clusters nor time-consuming manipulation of data. It consists of three main sections: (1) downloading genome assemblies, which are then annotated and subjected to quality control measures; (2) searching for single-copy core genes shared among the genomes, with conserved sequences being extracted and their specificity for a desired target assessed; and (3) lastly specific primers are designed and tested in a quality control assessment (Dreier et al., 2020).

7.3 Microfluidic qPCR (MFQPCR)

The SpeciesPrimer pipeline is thought to be especially useful for the design of primers for use in a relatively new technology – high-throughput microfluidic qPCR (MFQPCR; Crane et al., 2018) where different primer sets should be able to bind to template DNA under the same PCR conditions (Dreier et al., 2020). The basis of MFQPCR is that multiple qPCR reactions occur simultaneously on a single chip at nanolitre volumes, where a system of microfluidic channels and pressurised valves is used for mixing and thermocycling of multiple reactions (Spurgeon et al., 2008; Crane et al., 2018). The multiple reactions can include different samples, different loci of interest or technical replicates. The advantages of MFQPCR include reducing the cost per sample, the time required for results and errors due to the automated distribution of reagents and samples (Crane et al., 2018; Dreier et al., 2020). MFQPCR is however less adaptable than 'traditional' qPCR as it requires that amplification occurs at identical PCR conditions with the same reagents (Crane et al., 2018; Dreier et al., 2020). Although initially developed for quantification of human gene expression via RT-qPCR (Spurgeon et al., 2008), MFQPCR has featured in the study of soil microbial communities with the adaptation of primer design and the addition of reagents to manage PCR inhibitors found in DNA extracted from soil (such as humic acids; Crane et al., 2018). This publication highlights ten parameters vital to MFQPCR success in soil microbial ecology studies. In particular, specific primer parameters were found to be essential in determining the accuracy of studies quantifying functional genes in soil environments. This approach has allowed the reliable quantification of taxonomically informative genes as well as those involved in hydrocarbon degradation and nitrogen cycling, concurrently across 200 soil DNA extracts (Crane et al., 2018).

8 Conclusion

Using qPCR to quantify microorganisms in soils has a number of caveats, but despite this, it has been key to understanding changes in microbial communities and their functional traits. In addition, qPCR is a valuable tool for monitoring important organisms such as pathogens, plant growth–promoting bacteria and rhizobia, as well as genes that determine important processes such as N cycling. In the future, new technologies may offer alternative means of quantifying genes in soil systems, but qPCR is likely to remain a key tool, especially as it is continually being developed and refined to produce new capabilities.

9 References

Abdelaziz, A. M., et al. (2023) Biocontrol of soil borne diseases by plant growth promoting rhizobacteria. *Tropical Plant Pathology* 48(2): 105–127. https://doi.org/10.1007/s40858-022-00544-7

Adams, G. (2020) A beginner's guide to RT-PCR, qPCR and RT-qPCR. *The Biochemist (Lond)* 42(3): 48–53. https://doi.org/10.1042/BIO20200034

Brankatschk, R., et al. (2012) Simple absolute quantification method correcting for quantitative PCR efficiency variations for microbial community samples. *Applied and Environmental Microbiology* 78(12): 4481–4489. https://doi.org/10.1128/AEM.07878-11

Bravo, G., et al. (2020) Bioremediation by *Cupriavidus metallidurans* strain MSR33 of mercury-polluted agricultural soil in a rotary drum bioreactor and its effects on nitrogen cycle microorganisms. *Microorganisms* 8: 1952. https://doi.org/10.3390/microorganisms8121952

Bustin, S.A., et al (2009). The MIQE guidelines: minimum information for publication of quantitative real-time PCR experiments. *Clinical Chemistry* 55(4): 611–622. https://doi.org/10.1373/clinchem.2008.112797

Carini, P., et al. (2016) Relic DNA Is abundant in soil and obscures estimates of soil microbial diversity. *Nature Microbiology* 2(3): 1–6. https://doi.org/10.1038/nmicrobiol.2016.242

Chatterton, S., et al. (2023) Inoculum dose–disease response relationships for the pea root rot pathogen, *Aphanomyces euteiches*, are dependent on soil type and other pathogens. *Frontiers in Plant Science* 14: 1115420. https://doi.org/10.3389/fpls.2023.1115420

Checcucci, A., et al. (2016) Mixed nodule infection in *Sinorhizobium meliloti-Medicago sativa* symbiosis suggest the presence of cheating behaviour. *Frontiers in Plant Science* 7: 835. https://doi.org/10.3389/fpls.2016.00835

Chen, W., et al. (2023) Quantitative detection of the *Ralstonia solanacearum* species complex in soil by qPCR combined with a recombinant internal control strain. *Microbiology Spectrum* 11: e00210-23. https://doi.org/10.1128/spectrum.00210-23

Clarkson, J., et al. (2015) Mini-Paper - Monitoring of Soil-Borne Pathogens (Fungi, Protists and Nematodes) and Soil Tests. Focus Group SOIL-BORNE DISEASES. https://ec.europa.eu/ep/agricultre/en/content/ipm-practices-soil-borne-diseases-suppression-vegetables-and-arable-crops

Crane, S.L., et al. (2018) Microfluidic qPCR enables high throughput quantification of microbial functional genes but requires strict curation of primers. *Frontiers in Environmental Science* 6: 145. https://doi.org/10.3389/fenvs.2018.00145

Da Silva, C.G.N., et al. (2024) Specific primers for the rapid detection and quantitation of Rhizobium elite strains of common beans in the plant and environment. *Applied Soil Ecology* 193: 105156. https://doi.org.10.1016/j.apsoil.2023.105156

Daniell, T.J., et al. (2012) Improved real-time PCR estimation of gene copy number in soil extracts using an artificial reference. *Journal of Microbiological Methods* 91(1): 38–44. https://doi.org/10.1016/j.mimet.2012.07.010

Dreier, M., et al. (2020) SpeciesPrimer: a bioinformatics pipeline dedicated to the design of qPCR primers for the quantification of bacterial species. *PeerJ* 8: e8544. http://doi.org/10.7717/peerj.8544

Duff, A.M., et al. (2022). Assessing the long-term impact of urease and nitrification inhibitor use on microbial community composition, diversity and function in grassland soil. *Soil Biology and Biochemistry* 170: 108709.

Fierer, N., et al. (2005) Assessment of soil microbial community structure by use of taxon-specific quantitative PCR assays. *Applied and Environmental Microbiology* 71(7): 4117–4120. https://doi.org/10.1128/AEM.71.7.4117-4120.2005

Franklin, R.B. and Mills, A.L. (2003) Multi-scale variation in spatial heterogeneity for microbial community structure in an eastern Virginia agricultural field. *FEMS Microbiology Ecology* 44(3): 335–346. https://doi.org/10.1016/S0168-6496(03)00074-6

Fu, Y., et al. (2020) An optimized method to assess viable *Escherichia coli* O157:H7 in agricultural soil using combined propidium monoazide staining and auantitative PCR. *Frontiers in Microbiology* 11: 1809. https://doi.org/10.3389/fmicb.2020.01809

Furseth, B.J., et al. (2010) Enumeration of soybean-associated rhizobia with quantitative real-time polymerase chain reaction. *Crop Science* 50: 2591–2596. https://doi.org/10.2135/cropsci2010.03.0141

Giles, M.E., et al. (2017) Compound driven differences in N_2 and N_2O emission from soil the role of substrate use efficiency and the microbial community. *Soil Biology and Biochemistry* 106: 90–98. https://doi.org/10.1016/j.soilbio.2016.11.028

Gloor, G.B., et al. (2017) Microbiome datasets are compositional: and this ss not optional. *Frontiers in Microbiology* 8: 2224. https://doi.org/10.3389/fmicb.2017.02224

Graf, D.R.H., et al. (2022) Assembly of root-associated N_2O-reducing communities of annual crops is governed by selection for *nosZ* clade I over clade II. *FEMS Microbiology Ecology* 98(9): fiac092. https://doi.org/10.1093/femsec/fiac092

Gubry-Rangin, C., et al. (2010) Archaea rather than bacteria control nitrification in two agricultural acidic soils. *FEMS Microbiology Ecology* 74(3): 566–574. https://doi.org/10.1111/j.1574-6941.2010.00971.x

Göhler, A., et al. (2017) Multitarget quantitative PCR improves detection and predicts cultivability of the pathogen *Burkholderia pseudomallei*. *Applied and Environmental Microbiology* 83(8): e03212-16. https://doi.org/10.1128/AEM.03212-16

Hale, L., et al. (2014) Evaluation of pinewood biochar as a carrier of bacterial strain *Enterobacter cloacae* UW5 for soil inoculation. *Applied Soil Ecology* 84: 192–199. http://dx.doi.org/10.1016/j.apsoil.2014.08.001

Henderson, S.L., et al. (2010) Changes in denitrifier abundance, denitrification gene mRNA levels, nitrous oxide emissions, and denitrification in anoxic soil microcosms amended with glucose and plant residues. *Applied and Environmental Microbiology* 76(7): 2155–2164. https://doi.org/10.1128/AEM.02993-09

Herold, M.B., et al. (2018) Variable response of *nirK* and *nirS* containing denitrifier communities to long-term pH manipulation and cultivation. *FEMS Microbiology Letters* 365(7): fny035. https://doi.org/10.1093/femsle/fny035

Highton, M.P., et al. (2020) Soil N2O emission potential falls along a denitrification phenotype gradient linked to differences in microbiome, rainfall and carbon availability. *Soil Biology and Biochemistry*, 150, p.108004.

Hiis, E.G., et al. (2024) Unlocking bacterial potential to reduce farmland N_2O emissions. *Nature* 630: 421–428. https://doi.org/10.1038/s41586-024-07464-3

Hink, L., et al. (2017) Archaea produce lower yields of N2O than bacteria during aerobic ammonia oxidation in soil. *Environ Microbiol*, 19: 4829–4837. https://doi.org/10.1111/1462-2920.13282

Huang, J., et al. (2010) Detection of phytophthora nicotianae in soil with real-time quantitative PCR. *Journal of Phytopathology* 158(1): 15–21. https://doi.org/10.1111/j.1439-0434.2009.01554.x

Jiménez-Fernández, D., et al. (2010) Identification and auantification of *Fusarium xxysporum in planta* and soil by means of an improved specific and quantitative PCR assay. *Applied Soil Ecology* 46(3): 372–382. https://doi.org/10.1016/j.apsoil.2010.10.001

Katan, J. (2017) Diseases caused by soilborne pathogens: biology, management and challenges. *Journal of Plant Pathology* 99(2): 305–315.

Khatoon, Z., et al. (2020) Unlocking the potential of plant growth-promoting rhizobacteria on soil health and the sustainability of agricultural systems. *Journal of Environmental Management* 273: 111118. https://doi.org/10.1016/j.jenvman,2020.111118

Knowles, R. (1982) Denitrification. *Microbiological Reviews* 46: 43–70. https://doi.org/10.1128/mr.46.1.43-70.1982

Kocira, A., et al. (2020) Legume cover crops as one of the elements of strategic weed management and soil quality improvement: a review. *Agriculture* 10: 394. https://doi.org/10.3390/agriculture10090394

Langarica-Fuentes, A., et al. (2018) Effect of model root exudate on denitrifier community dynamics and activity at different water-filled pore space levels in a fertilised soil. *Soil Biology and Biochemistry* 120: 70–79. https://doi.org/10.1016/j.soilbio.2018.01.034

Lievens, B., et al. (2006) Real-time PCR for detection and quantification of fungal and oomycete tomato pathogens in plant and soil samples. *Plant Science* 171(1): 155–165. https://doi.org/10.1016/j.plantsci.2006.03.009.

Lievens, B. and Thomma, B.P.H.J. (2005) Recent developments in pathogen detection arrays: implications for fungal plant pathogens and use in practice. *Phytopathology* 95(12): 1374–1380. https://doi.org/10.1094/PHYTO-95-1374

Ma, Y., et al. (2019) An evaluation of primers for detecting denitrifiers via their functional genes. *Environmental Microbiology* 21(4): 1196–1210. https://doi.org/10.1111/1462-2920.14555

Maluk, M., et al. (2023) Biological nitrogen fixation by soybean (*Glycine max* [L.] Merr.), a novel, high protein crop in Scotland, requires inoculation with non-native bradyrhizobia. *Frontiers in Agronomy* 5: 1196873. https://doi.org/10.3389/fagro.2023.1196873

Morley, A.A. (2014) Digital PCR: a brief history. *Biomolecular Detection and Quantification* 15(1): 1–2. https://doi.org/10.1016/j.bdq.2014.06.001

McKew, B.A. and Smith, C.J. (2015) Real-Time PCR Approaches for Analysis of Hydrocarbon-Degrading Bacterial Communities. In: McGenity, T., et al. (eds) *Hydrocarbon and Lipid Microbiology Protocols. Springer Protocols Handbooks*. Springer, Berlin, Heidelberg. https://doi.org/10.1007/8623_2015_64

Nicol, G.W., et al. (2008) The influence of soil pH on the diversity, abundance and transcriptional activity of ammonia oxidising archaea and bacteria. *Environmental Microbiology* 10(11): 2966–2978. https://doi.org/10.1111/j.1462-2920.2008.01710.x

Pagaling, E., et al. (2023) Antibiotic resistance patterns in soils across the Scottish landscape. *Communications Earth Environment* 4: 403. https://doi.org/10.1038/s43247-023-01057-0

Park, J.W. and Crowley, D.E. (2005) Normalization of Soil DNA extraction for accurate quantification of target genes by real-time PCR and DGGE. *BioTechniques* 38(4): 579–586. https://doi.org/10.2144/05384ST04

Pinheiro, L.B., et al. (2011) Evaluation of a droplet digital polymerase chain reaction format for DNA copy number auantification. *Analytical Chemistry* 84(2): 1003–1011. https://doi.org/10.1021/ac202578x

Pradhanang, P.M., et al. (2000) Sensitive detection of *Ralstonia solanacearum* in soil: a comparison of different detection techniques. *Plant Pathology* 49(4): 414–422. https://doi.org/10.1046/j.1365-3059.2000.00481.x

Ren, Z., et al. (2022) Development of real-time PCR and droplet digital PCR based marker for the detection of *Tilletia caries* inciting common bunt of wheat. *Frontiers in Plant Science* 13: 1031611. https://doi.org/10.3389/fpls.2022.1031611

Rocha, L.F., et al. (2023) A panel of qPCR assays to detect and quantify soybean soil-borne pathogens. *Letters in Applied Microbiology* 76: 1. https://doi.org/10.1093/lambio/ovac023

Rossmanith, P., et al. (2011) Proof of concept for recombinant cellular controls in quantitative molecular pathogen detection. *Applied and Environmental Microbiology* 77(7): 2531–2533. https://doi.org/10.1128/AEM.02601-10

Samad, M., et al. (2016) Phylogenetic and functional potential links pH and N2O emissions in pasture soils. *Scientific Reports* 6: 35990. https://doi.org/10.1038/srep35990

Samaddar, S., et al. (2021) Role of soil in the regulation of human and plant pathogens: soils' contributions to people'. *Philosophical Transactions of the Royal Society B: Biological Sciences* 376(1834): 20200179. https://doi.org/10.1098/rstb.2020.0179

Santos. S.S., et al. (2015) Comparison of three DNA extraction methods for recovery of soil protist DNA. *Journal of Microbiological Methods* 115: 13–19. https://doi.org/10.1016/j.mimet.2015.05.011

Shahsavari, E., et al. (2016) A quantitative PCR approach for quantification of functional genes involved in the degradation of polycyclic aromatic hydrocarbons in contaminated soils. *MethodsX* 3: 205–2011. https://doi.org/10.1016/j.mex.2016.02.005

Shrewsbury, L.H., et al. (2016) Denitrifier abundance has a greater influence on denitrification rates at larger landscape scales but is a lesser driver than environmental variables. *Soil Biology and Biochemistry* 103: 221–231. https://doi.org/10.1016/j.soilbio.2016.08.016

Singh, D., et al. (2014) Effect of temperature, cultivars, injury of root and inoculums load of *Ralstonia solanacearum* to cause bacterial wilt of tomato. *Archives of Phytopathology and Plant Protection* 47(13): 1574–1583. https://doi.org/10.1080/03235408.2013.851332

Spurgeon, S.L., et al. (2008) High throughput gene expression measurement with real time PCR in a microfluidic dynamic array. *PLoS One* 3(2): e1662. https://doi.org/10.1371/journal.pone.0001662

Steiner, P.A., et al. (2019) Highly variable mRNA half-life time within marine bacterial taxa and functional genes. *Environmental Microbiology* 21(10): 3873–3884. https://doi.org/10.1111/1462-2920.14737

Tang, Q., et al. (2024) Organic fertilization reduces nitrous oxide emission by altering nitrogen cycling microbial guilds favouring complete denitrification at soil aggregate scale. *Science of the Total Environment* 10(946): 174178. http://doi.org/10.1016/j.scitotenv.2024.174178

Uchida, Y., et al. (2014) Expression of denitrification genes in response to a waterlogging event in a Fluvisol and its relationship with large nitrous oxide pulses. *FEMS Microbiology Ecology* 88(2): 407–423. https://doi.org/10.1111/1574-6941.12309

Um, H.Y., et al. (2013) Altered gene expression and intracellular changes of the viable but nonculturable state in *Ralstonia solanacearum* by copper treatment. *The Plant Pathology Journal* 29(4): 374–385. https://doi.org/10.5423/PPJ.OA.07.2013.0067

VanGuilder, H.D., et al. (2008) Twenty-five years of quantitative PCR for gene expression analysis. *BioTechniques* 44(5): 619–626. https://doi.org/10.2144/000112776

Voegel, T.M., et al. (2021) Development of droplet digital PCR assays to quantify genes involved in nitrification and denitrification, comparison with quantitative real-time PCR and validation of assays in vineyard soil. *Canadian Journal of Microbiology* 67(2): 174–187. https://doi.org/10.1139/cjm-2020-0033

Vogelstein, B. and Kinzler, K.W. (1999) Digital PCR. *Proceedings of the National Academy of Sciences of the United States of America* 96(16): 9236–9241. https://doi.org/10.1073/pnas.96.16.9236

Wang, Y.Y., et al. (2017) Responses of N_2O reductase gene (*nosZ*)-denitrifer communities to long-term fertilization follow a depth pattern in calcareous purplish paddy soil (2017). *Journal of Integrative Agriculture* 16(11): 2597–2611. https://doi.org/10.1016/S2095-3119(17)61707-6

Wong, M.L. and Medrano, J.F. (2005) Real-Time PCR for mRNA Quantitation. *BioTechniques* 39(1): 75–85. https://doi.org/10.2144/05391RV01

Woodhall, J.W., et al. (2013) A new quantitative real-time PCR assay for *Rhizoctonia solani* AG3-PT and the detection of AGs of *Rhizoctonia solani* associated with potato in soil and tuber samples in Great Britain. *European Journal of Plant Pathology* 136(2): 273–280. https://doi.org/10.1007/s10658-012-0161-8

Yang, Y., et al. (2022) Response of N2O emission and denitrification genes to different inorganic and organic amendments. *Sci Rep* 12, 3940. https://doi.org/10.1038/s41598-022-07753-9

Yudistira, H., et al. (2021) qPCR assay targeting *Bradyrhizobium japonicum* shows that row spacing and soybean density affects *Bradyrhizobium* population. *Canadian Journal of Microbiology* 67(7): 529–536. https://doi.org/10.1139/cjm-2020-0334

Zhang, Z., et al. (2017) Soil bacterial quantification approaches coupling with relative abundances reflecting the changes of taxa. *Scientific Reports* 7: 4837. https://doi.org/10.1038/s41598-017-05260-w

Zhu, J., et al. (2022) MRG chip: a high-throughput qPCR-based tool for assessment of the heavy metal(loid) resistome. *Environmental Science & Technology* 56: 1056–10667. https://doi.org/10.1021/acs.est.2c00488

Chapter 4

Advances in soil respiration techniques for assessing soil microbial activity

Ryan D. Stewart, Virginia Tech, USA; and Michael S. Strickland, University of Idaho, USA

1 Introduction

Soil represents the most biodiverse terrestrial ecosystem on the planet, with millions to billions of microbes present in a single gram of soil (Nielsen et al., 2015, Trevors, 2010). Only a few of these organisms have been identified, with little known about most microbes that exist in soil. Respiration measurements offer the ability to quantify and understand soil microbial activity in different ecosystems and contexts. For this reason, respiration-based procedures are gaining broad interest as part of comprehensive soil health analyses. New and adapted methodologies are likewise being developed to make use of readily available and inexpensive tools, as opposed to specialized laboratory equipment. Nonetheless, there is continued discussion regarding the underlying mechanisms and interpretation of many of these tests. Moreover, even when performed under controlled laboratory conditions, some of these tests have greater experimental uncertainty than other soil testing methods (e.g. nutrient analyses), with intra-laboratory variability identified as a major challenge (Wade et al., 2018). New advances in application and knowledge are therefore needed to fully realize the potential of these techniques. In this chapter, we review the history and basis of many of the most commonly used respiration-based methods to assess soil microbial characteristics, address

http://dx.doi.org/10.19103/AS.2025.0159.05

some of the most critical challenges and uncertainties, and discuss future needs and trends in soil respiration research.

2 Basis and application of soil respiration-based measurements

Soil respiration has been used as an indicator of microbial activity for more than a century, dating back to observations of carbon dioxide (CO_2) pulses occurring after moderate heating (Dehérain and Demoussy, 1896), fumigation (Darbishire and Russell, 1907), or rapid rewetting of dried soil (Birch and Friend, 1956). These initial studies led to much speculation and subsequent investigations about the mechanisms underlying such behaviors (Jenkinson et al., 2004). For example, an early hypothesis was that the CO_2 flush after fumigation arises from surviving microbes quickly decomposing others that were killed (Störmer, 1908). This idea was initially rebutted, with alternative theories including that fumigation reduces competition between different microbe types or increases substrate availability, until isotope labeling experiments provided new evidence that much of the excess respiration does indeed come from deceased soil microbes (Jenkinson, 1966). Similar uncertainties and contradictions have emerged regarding the source of elevated respiration when dry soils are rewetted (Barnard et al., 2020, Unger et al., 2010, Fierer and Schimel, 2003), with increasingly sophisticated efforts aimed at untangling cellular sources (e.g. osmo-protectants such as trehalose) from extracellular compounds made available during rewetting (Slessarev and Schimel, 2020).

Contemporary methods assess soil respiration under both laboratory and field settings (Fig. 1). Most laboratory measurements are designed to isolate one or more aspects of soil microbiological functioning and are typically performed as either (1) incubations under idealized environmental conditions or (2) assessments done after imposing one or more types of disturbance on the samples. Field measurements quantify soil respiration rate (R_s) under actual environmental conditions and serve as a standard measure of microbial activity (Haney et al., 2008). Most field-based assessments are conducted under ambient environmental conditions, though some experiments have included perturbations such as warming (Schindlbacher et al., 2012). Soil respiration measurements can also be done under greenhouse conditions, which can be designed to impose controlled environmental conditions (Bryla et al., 2001), more similar to laboratory tests, or variable environmental conditions (Bouma et al., 1997), more similar to field settings.

2.1 Laboratory-based assessments based on respiration

One common laboratory measurement is basal respiration, in which CO_2 fluxes are quantified from non-amended soil (Anderson and Domsch, 2010).

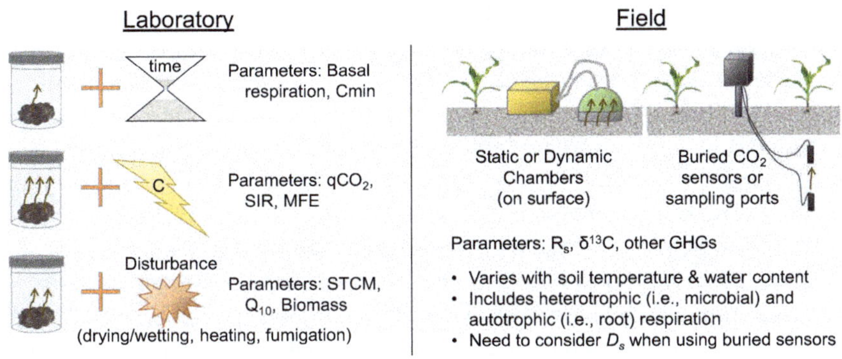

Figure 1 Soil respiration measurements that can be done in laboratory and field settings, with varying conditions in terms of time of measurement, carbon (C) allocations, imposed disturbance types, and other environmental variables. Relevant parameters that can be determined include basal respiration, mineralizable carbon (Cmin), metabolic quotient (qCO_2), substrate-induced respiration (SIR), microbial functional evenness (MFE), short-term carbon mineralization (STCM), temperature sensitivity of soil respiration (Q_{10}), microbial biomass, soil respiration (R_s), stable carbon isotopic signature ($\delta^{13}C$), and fluxes of other greenhouse gases (GHGs). Note that D_s represents the soil gas diffusivity constant.

This parameter is used to characterize microbial functions such as yield (i.e. the conversion of carbon resources into microbial biomass) and the amount of carbon required for maintenance of living soil microorganisms. Basal respiration measurements collected over longer periods (e.g. 60 days) are used to quantify the amount of mineralizable carbon (Cmin) in a soil (Strickland et al., 2017). This parameter, also known as bioavailable carbon, represents carbon that is preferentially used by microbes and is associated with longer-term stabilization in the subsurface (Bradford et al., 2013, Strickland et al., 2015).

Basal respiration measurements also serve for comparison with soils that have been incubated with energy sources such as glucose. The qCO_2 of basal respiration (also known as metabolic coefficient or specific respiration) is a metric that quantifies CO_2 efflux per unit biomass and is assessed by taking the ratio of respiration from non-amended versus glucose-amended soil samples (Anderson, 1990). Since it normalizes respiration based on the amount of active microbes present in a sample, this parameter can be compared between different soils and be used to identify microbial stress (Wepking et al., 2017). Basal respiration measurements from samples incubated with different glucose levels can help to identify the stable respiration rate for a given soil. This stable rate is presumed to represent the amount of energy required for optimal microbial activity without growth (Anderson and Domsch, 1985). Though other non-growth aspects such as cell leakage and death contribute to this steady rate respiration, conceptual models continue to be refined that differentiate physiological maintenance from other processes (van Bodegom, 2007).

Active microbial biomass is quantified using a method known as substrate-induced respiration (SIR), in which respiration is measured for soil incubated with ample energy and carbon available to the microbes (Anderson and Domsch, 1978). Samples are typically pre-incubated with a glucose or autolyzed yeast substrate, and soil respiration is then analyzed after a short incubation (e.g. 4 h at 20°C). The active microbial biomass is quantified by assuming 1 mL CO_2 h^{-1} is equivalent to 40 mg of biomass. This method builds on earlier evidence of links between microbial biomass and respiration rates (Drobnik, 1960, Wright and Hobbie, 1966). A variation of the SIR test, called the catabolic response profile, uses multiple easily degraded substrates, such as glycine, cellulose, and chitin, to assess the functional diversity of soil microbes (Degens and Harris, 1997, Strickland et al., 2017). Results from this method can be applied to calculate microbial functional evenness (MFE) using the Simpson-Yule index (Magurran, 1988, Gyawali et al., 2022). Functional evenness provides a measure of ecosystem diversity and resilience to stress (Escalas et al., 2019, Wittebolle et al., 2009) and can vary due to vegetation or other management factors (Schipper et al., 2001). This parameter is often well-correlated with measures of microbial functionality such as qCO_2 (Schipper et al., 2001).

Respiration measurements after disturbance also reveal important microbial characteristics and functionality. The classic fumigation-incubation method determines soil microbial biomass by measuring soil respiration for some time (e.g. 10 days) after fumigation with chloroform ($CHCl_3$). The total amount of respiration is adjusted based on respiration from a non-fumigated control. A conversion factor, k_c, is then applied to translate the amount of CO_2 evolved into equivalent microbial biomass (Jenkinson and Powlson, 1976). A k_c value of 0.45 is most commonly used for incubations at 25°C, but other values have been proposed based on the type and pH of the soil (Jenkinson et al., 2004). The fumigation-incubation method can be inaccurate when applied to low pH soils and those with large quantities of fresh substrate (Martens, 1995), leading many to instead use a fumigation-extraction procedure, where the amount of dissolved carbon is measured on liquid extractions from the sample (Jenkinson et al., 2004). The fumigation-incubation and fumigation-extraction methods provide similar results for many soils (Vance et al., 1987), meaning that the former may continue to be used when assessing soil microbial characteristics. It also can find application when re-analyzing historic datasets.

The decomposition potential of carbon in soil can be determined by adjusting the temperature and moisture conditions of the incubations. Short-term (e.g. 1-d) incubations performed at different temperatures are used to quantify the temperature sensitivity of soil respiration, i.e. Q_{10} (Fierer et al., 2003). This parameter depicts how temperature changes affect decomposition processes in different soils and has important implications for carbon fluxes and biogeochemical cycling at local and global scales, including how soil organic matter stores will shift in different ecosystems under future warming conditions (Warner et al., 2019, Meyer et al., 2018b). Multi-month incubations can be used to determine decomposition potential under different moisture and temperature regimes (Possinger et al., 2021). Such work has identified how climatic drivers interact with biological and mineralogical factors to affect carbon persistence in subsoils.

Another disturbance-based measurement is the short-term carbon mineralization (STCM) test, which quantifies the flush of CO_2 following rapid rewetting of a dried soil sample. In practice, many variations of the test have been developed (NRCS, 2019). Carbon dioxide efflux can be measured using alkaline traps that are titrated or analyzed for changes in electrical conductivity (Franzluebbers and Veum, 2020) or by collecting headspace samples through time, which are passed through a gas chromatograph (McGowen et al., 2018) or infrared analyzer (Gyawali et al., 2019). Gel paddles that change color with the amount of absorbed CO_2 can also be used, albeit with a more limited measurement range (McGowen et al., 2018). Incubation times range from 1 to 4 days, and rewetting procedures vary from capillary wetting from below to adding 25–100% of water-holding capacity to the soils; each of these variations can affect measured values (Wade et al., 2018). Drying temperatures also influence the amount of CO_2 produced during rewetting, with oven-dried samples typically producing a greater respiration flux than cooler temperatures (Birch and Friend, 1956). Our own work has confirmed that drying conditions affect respiration rates, with samples dried for 2 days at 55°C having a greater flush of CO_2 when rewetted compared to samples that were dried for 21 days at 22°C (Fig. 2). While more work is needed to understand the biological causes of this response, this result emphasizes how interpretation of tests such as the STCM assay hinges on identifying methodologies that induce consistent and interpretable responses.

2.2 Field-based assessments of respiration

Agricultural management practices have both short- and long-term effects on the biological composition and function of field soils. Because microbial populations tend to respond much more rapidly than other soil properties to changes in management, monitoring of soil biological composition and function has become increasingly important (Bossio et al., 2005, Le Roux et al., 2008). As such, respiration-based methods have been widely used to understand how agricultural practices affect soil microbiology. Studies comparing conventional tilled versus no-till systems have revealed that tillage is often associated with short-term increases in respiration (Reicosky

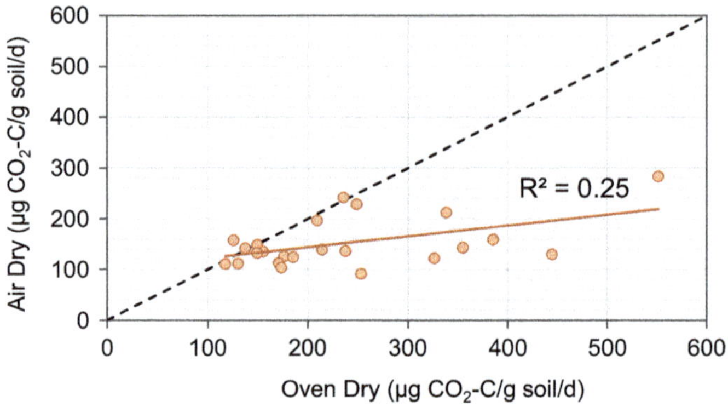

Figure 2 Comparison of short-term carbon mineralization (STCM) test results (in μg CO_2-C/g soil/day) for samples prepared via oven-drying versus air-drying. Samples came from four different areas of a row-cropping field located in a Ross-series silt loam (fine-loamy, mixed, superactive, mesic Cumulic Hapludoll) near Blacksburg, VA, U.S. The black dashed line represents the 1:1 relationship; the orange line represents the best fit from linear regression.

et al., 1997), due to physical rearrangements of particles and destruction of aggregates that can temporarily boost substrate and oxygen availability to microbes (Al-Kaisi and Yin, 2005. These effects become muted through time as tilled soils recompact – thus reducing gas exchange – and labile carbon is consumed. Over longer time periods, reduced or no-till systems may have similar or greater cumulative CO_2 emissions than tilled soils (Hendrix et al., 1988, Franzluebbers et al., 1995).

Soil respiration is most commonly measured using chambers installed at the soil surface; chambers can have open or closed configurations (Knoepp and Vose, 2002). After installation, discrete gas samples collected from the chamber headspace can be analyzed for CO_2 concentration back in the laboratory (e.g. using gas chromatography), or the headspace gas can be passed through a non-dispersive infrared (NDIR) sensor to determine concentrations in near-real time. Changes in CO_2 concentration per time are used to calculate the soil respiration flux, typically accounting for the volume of the chamber and the area of surface that it covers (Gyawali et al., 2019). An alternative approach is to measure CO_2 concentrations at multiple depths in the soil, and thereby calculate the concentration gradient that exists in the profile. Subsurface CO_2 concentrations are measured using buried NDIR sensors (Tang et al., 2003) or air sampling ports (Buyanovsky and Wagner, 1983). Soil respiration is then calculated using Fick's first law of gas diffusion based on the measured concentration gradient and an estimate of the soil gas diffusivity constant, D_s (Thompson et al., 2023). This approach has several advantages, including limited disturbance of the soil surface (beyond the initial sensor install), relatively low costs, the ability to link with simultaneous oxygen readings, and the capability to monitor soil respiration at high frequencies (e.g. minutes to hours) over multi-month to multi-year periods (Turcu et al., 2005). Disadvantages include uncertainty about the magnitude of D_s and possible sensor failures (Thompson et al., 2023).

Field-based assessments of soil respiration include some other caveats and cautions. If using chambers to measure R_s, it is best practice to install the containment ring or barrier at least 24 hours prior to measurement to minimize short-term artifacts resulting from that disturbance (Collier et al., 2014). Chambers should be vented and insulated if possible. Fans can be used to mix the headspace, though the effects on measurements are still not fully understood (Clough et al., 2020). Many studies rely on one or only a few discrete time points to compare how management practices affect soil respiration and generation of other greenhouse gases. Soil CO_2 emissions vary widely with water content and temperature of the soil (MacDonald et al., 2010, Yonemura et al., 2014), meaning that time of day and environmental conditions at the time of measurement can induce variability that supersedes any management-related indications (Jian et al., 2018). Gas diffusivity varies with soil wetness, since rapid gas exchange is facilitated by larger air-filled pores in the medium (McCourty et al., 2018, Moldrup et al., 2005). Conversely, gas exchange is inhibited when such pores become water-filled (e.g. after a rainstorm), allowing for temporary decoupling of CO_2 production in the subsurface versus CO_2 fluxes leaving the surface (Chen et al., 2005). Soil respiration measurements also do not typically differentiate between heterotrophic respiration associated with soil microbes versus autotrophic respiration from living plants, thus necessitating the use of methods such as (1) managing fallow plots for comparison, (2) destructively isolating portions of the soil from roots with trenching or collars, (3) using carbon isotope labeling approaches, or (4) collecting samples and relying on laboratory analyses (e.g. separate incubations for plant versus root-free soil) to distinguish between the two (Kuzyakov and Larionova, 2005, Brown and Markewitz, 2018, Savage et al., 2013).

Field-based R_s measurements can be accompanied by stable isotope measurements to infer the fate of specific carbon compounds or recently fixed carbon. The relative concentration of ^{13}C ($\delta^{13}C$) in CO_2 can be measured from discrete gas samples collected in field or greenhouse settings or by using field-deployable spectrometers (McCloskey et al., 2020). Respiration associated with ^{13}C-labeled glucose has been tied to land-use history and can be used to help understand the response of soil microbial communities to nutrient additions (Bradford et al., 2013, Strickland et al., 2010). Additionally, the use of stable isotopes has allowed researchers to track the fate of elements through soil food webs and their incorporation into soil organic matter (Strickland et al., 2013, Wepking et al., 2019). Similar work has assessed how microbes consume specific compounds present in soil, revealing that microbial communities shift both in composition and in cumulative respiration depending on the type of volatile organic carbons released from leaf litter (McBride et al., 2020).

2.3 Use in soil health assessment

Soil health, which emphasizes the ability of soils to sustain ecosystem and societal functions, is often framed in terms of a combination of physical, chemical, and biological properties (Stewart et al., 2018). Interpretations of soil health are well established for many physical properties (e.g. permeability, bulk density) and chemical indicators (e.g. organic matter, macro- and micro-nutrients). The applicability of soil biological characteristics is far less agreed upon. Field-based measurements of R_s have long been used as part of soil health assessments and are typically performed after infiltration tests by capping the infiltration ring and measuring CO_2 concentration using a Draeger tube or color-changing gel paddles (USDA, 1999). Robust interpretations of these measurements are still lacking, however. Other studies and laboratories have identified the STCM test as a relatively quick and easy indicator to measure for soil health assessment (NRCS, 2019). Nonetheless, there is still a need for more work to understand the implications of different configurations and interpretations of this test.

Recent studies have included other biological characteristics into soil health assessment, including respiration-based parameters such as SIR, Cmin, and catabolic response profile. These measurements have clear responses when comparing contrasting land covers, such as grasslands versus croplands (Congreves and Wu, 2024) or forests versus croplands (Nogueira et al., 2006). When comparing different management practices within cropping systems, however, the relatively high variability of respiration-based measurements limits their effectiveness (Gyawali et al., 2022, Bierer et al., 2020). There is potential to use CO_2 evolution as an early indicator of soil health improvements due to perennialization (Wu et al., 2024), but differences once again tend to be subtle. More work is thus needed to evaluate these relationships.

3 Case study – use of respiration-based measurements to evaluate conservation management practices

In some areas, including the Mid-Atlantic region of the United States, farmers are being incentivized to include cover crops in their rotations as a way to build soil health. Farmers have a number of options for cover crops that can be readily incorporated into production systems (Schonbeck and Morse, 2006). Grasses and legumes are the most commonly used cover crop species, although brassicas are sometimes used if hard pans are present or pest management characteristics are desired (Reeves, 1994). Factors such as the carbon-to-nitrogen ratio of the plant material is very important in determining how much of the cover crop becomes mineralized by soil microorganisms and incorporated into soil organic matter, with legumes and legume-grass mixtures associated with greater soil carbon increases than grasses alone (Jian et al., 2020a). Nonetheless, total soil C can take multiple years to meaningfully increase after adding cover crops to rotations. There is a need for tools that can evaluate how microbiological functions, along with the type and lability of carbon pools, respond to shifts in management practices on much shorter time-scales (Morrow et al., 2016). Respiration measurements hold great potential to fill this requirement, and in the process reveal many of the direct and indirect effects that cover crops can have on soil microbial function (Nivelle et al., 2016, Fierer, 2017).

We have applied and interpreted respiration-based measurements as part of ongoing work to better understand how cover crops, along with tillage practices, affect soil health. Starting in September 2015, we installed replicated treatments at sites across the state of Virginia, U.S., to evaluate interactions between tillage and cover cropping practices on different soil health-related parameters. The data presented here were collected in two of the sites, both of which were located in silt loam soils. Site 1 was located near Blacksburg, VA, U.S., in a soil classified as a Duffield-Ernest-Purdy undifferentiated group: Duffield – *Fine-loamy, mixed, active, mesic Ultic Hapludalfs*; Ernest – *Fine-loamy, mixed, superactive, mesic Aquic Fragiudults*; Purdy – *Fine, mixed, active, mesic Typic Endoaquults* (NRCS, 2017). Site 2 was located near Harrisonburg, VA, U.S., in a soil classified as a Frederick and Lodi undifferentiated group: Frederick – *Fine, mixed, semiactive, mesic Typic Paleudults*; Lodi – *Fine, mixed, subactive, mesic Typic Hapludults*. Both fields had been managed in no-till grain production with winter cover crops and regular manure applications prior to the start of the study.

We used a split plot design with tillage as the main plot (no-till versus conventional tillage twice a year with a disc furrow to ~15 cm) and cover cropping as the sub-plot (a three-species winter cover crop mixture versus no-cover). The experiment started in summer of 2015. The cover crops were a mixture of winter barley ([*Hordeum vulgare*]; 1.4×10^6 seeds per hectare), crimson clover ([*Trifolium incarnatum*]; 1.9×10^6 seeds per hectare), and tillage radish ([*Raphanus raphanistrum* subsp. *sativus*]; 2.1×10^5 seeds per hectare). Fields at both sites were planted in maize [*Zea mays* subsp. *mays*] in April 2016 and April 2017. The maize was harvested each September prior to tillage and planting of cover crops. All fields were fertilized with urea at planting using agronomic rates for maize.

In September 2015, April 2016, September 2016, and April 2017, we collected soil samples from 0–10 and 10–20 cm depths from each subplot (n = 4 samples per treatment) and also measured in-field R_s using a NDIR CO_2 analyzer (Li-COR 8100; Li-COR, Nebraska, U.S.) with a static chamber (n = 4 measurements per treatment). September samples were taken within 2 weeks after completing the maize harvest, fall tillage, and cover crop planting activities. April samples were collected prior to cover crop termination and spring tillage each year. We analyzed the soil samples for (1) SIR using a 4-h incubation; (2) mineralizable carbon (Cmin) using a 60-day incubation; and (3) catabolic response profile using water plus five other substrates (glucose, glycine, oxalic acid, cellulose, and chitin). The catabolic response profile measurements were used to calculate normalized microbial functional evenness (nMFE) with the Simpson-Yule index: $nMFE = 1/\Sigma(p_i/NP)^2$ where p_i is the respiration from each substrate, N is the number of substrates ($N = 6$), and P is the total respiration from all substrates. More details on the experimental design and measurement protocols can be found in Gyawali et al. (2022).

Our repeated measurements show how historic and current agricultural management practices interact with seasonal variables to affect soil microbial biomass and function. For example, the introduction of disk tillage into a row-cropping system that had been managed using no-till systems caused immediate changes in respiration, with the disk-tilled plots in Site 1 having greater R_s than the no-till plots when measured in September 2015 (two-way ANOVA; $P = 0.0043$; Fig. 3a). The laboratory-based respiration measurements for samples collected on that date showed the opposite response, with greater SIR values in the no-till plots compared to the recently tilled plots ($P = 0.025$; Fig. 3b). The two parameters had a possible negative correlation (adjusted $R^2 = 0.20$; $P = 0.054$; data not shown). One potential explanation is that the disruption caused by tillage temporarily reduced the active microbial biomass, but the surviving microbes were able to respire more as they processed extracellular carbon made available by the disturbance. Alternatively, tillage may have caused a stress response in the soil microbes that resulted in greater respiration per unit biomass (i.e. mass-specific respiration). In other words, the microbes needed to allocate more resources toward cellular repair and maintenance as opposed to growth (Schimel et al., 2007).

The SIR and Cmin values both tended to be numerically higher in cover crop treatments versus no cover crop controls, with significant differences detected in both sites and depths in April 2017 ($P < 0.05$). These two respiration-based measures were significantly correlated to each other ($P < 0.0001$; Fig. 4). The active microbial biomass in these sites therefore appears to have responded to the amount of bioavailable carbon in the soil, regardless of the specific management practices being applied or the time of year that the soils were sampled. We also detected significant correlations between SIR and the quantity of water-stable aggregates <0.25 mm in size ($P = 0.022$ for Site 1 and $P = 0.020$ for Site 2; data not shown), showing the importance of active microbial biomass in sustaining soil aggregation.

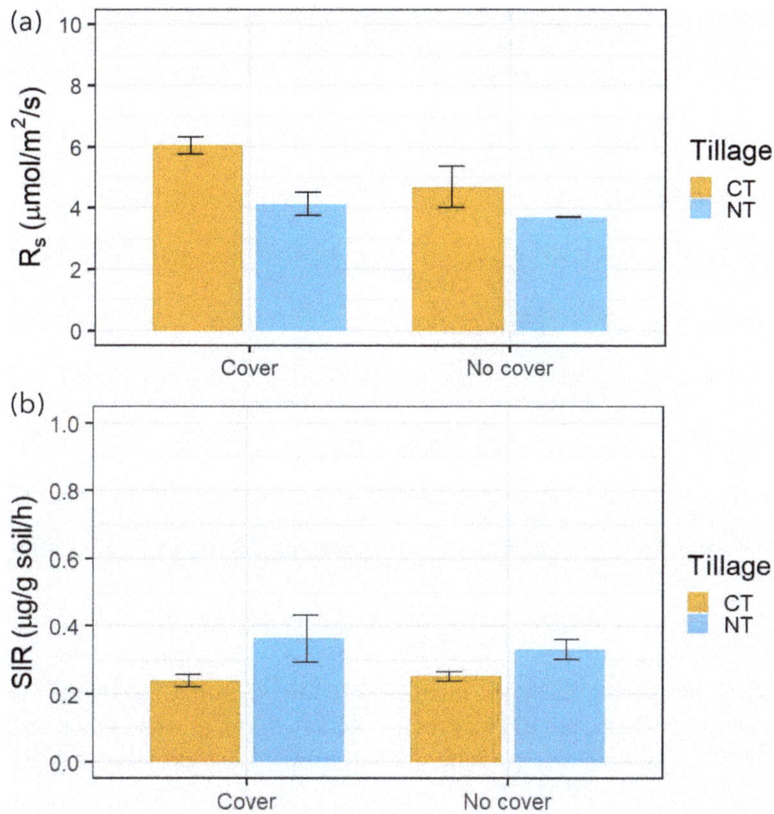

Figure 3 Measurements of (a) soil respiration (R_s) at the soil surface and (b) substrate-induced respiration (SIR) for samples collected at 10–20 cm depth from a replicated field experiment located in Site 1 (near Blacksburg, Virginia, U.S.) in September 2015, near the beginning of the experiment. Responses are separated by treatments (n = 4): CT = conventional tillage via a disk harrow; NT = no-till; Cover = three-way mixture of wintertime cover crops; No Cover = soil was left fallow after harvesting the maize cash crop each year.

Our repeated measurements revealed that nMFE significantly increased through time at both sites (Fig. 5). On the first sampling date (September 2015), the no-till plots had numerically higher nMFE values in the upper 0–10 cm samples compared to the disk-tilled sites ($P = 0.17$ for Site 1 and $P = 0.11$ for Site 2). In April 2016, nMFE was possibly different between tillage ($P = 0.057$) and cover crop ($P = 0.072$) plots at Site 1. Samples from that site also indicated a possible tillage effect in September 2016 (10–20 cm sample depth; $P = 0.08$; data not shown), and a cover crop effect in April 2017 ($P = 0.05$). We hypothesize that the general increase in functional evenness at these sites may have been in response to a change in fertility management from repeated

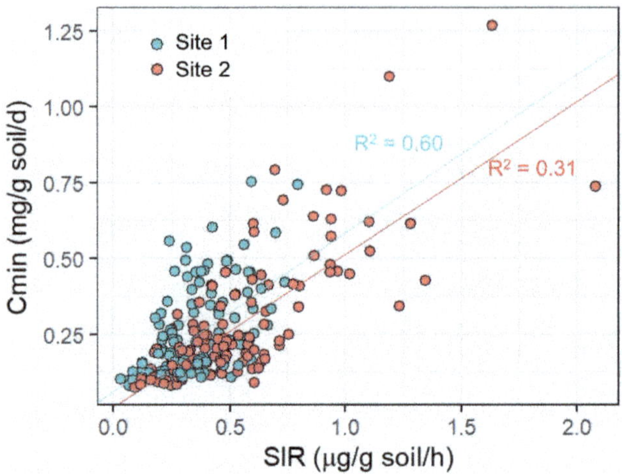

Figure 4 Correlation between substrate induced respiration (SIR) and mineralizable carbon (Cmin) for samples collected at four different times between September 2015 and April 2017. Samples came from two sites in Virginia, U.S., with replicated tillage and cover cropping treatments.

animal manure applications prior to the experiment to urea applications during the experiment. Specifically, the microbial communities may have shifted as the microorganisms adapted to changes in substrate type, as well as to a possible decrease of antibiotic compounds that are often found in manures (Wepking et al., 2017, Hilaire et al., 2022).

In summary, respiration-based measurements form the basis of many field- and laboratory-based indicators of soil microbial functionality. Our replicated field experiment displayed the potential of using respiration-based measures to assess effects of cover cropping and different tillage practices, while also revealing how these indicators can be highly variable within the same fields both in space and through time. We also note that the results presented here, which indicate that microbial functionality may be at least partially responsive to changing agricultural management practices, are not a universal response. Other work in similar agricultural systems, as well as restored forest areas, has found that belowground microbial community composition can resist changes in above-ground management practices (Strickland et al., 2017, 2019). Additional research is clearly needed to link meaningful and easily measured indicators of soil health with specific soil health outcomes. In the next section, we provide more information on how the measurements and interpretations of respiration-based indicators can be improved to help fill this gap.

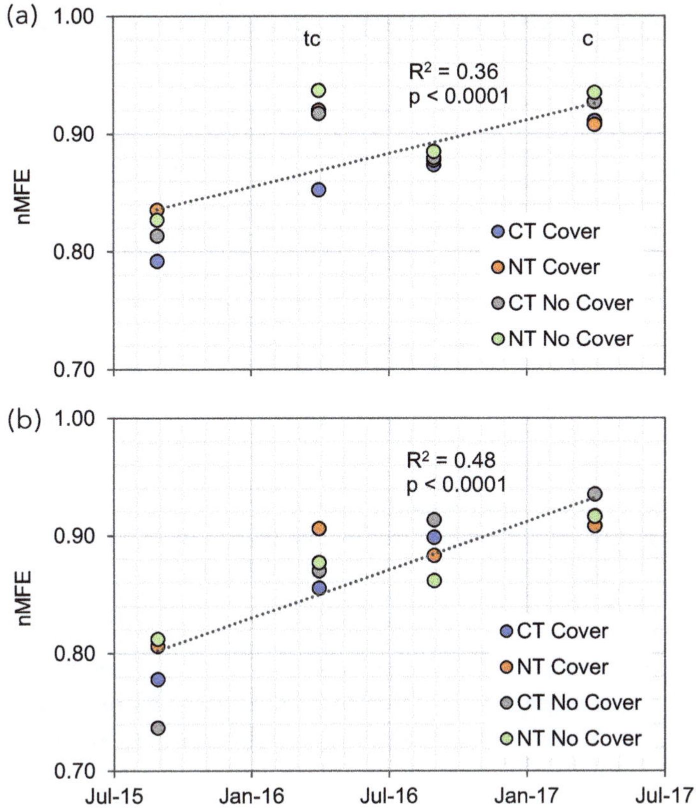

Figure 5 Changes in normalized microbial functional evenness (nMFE) through time for (a) Site 1 and (b) Site 2, both row-cropped fields in Virginia, U.S., with responses separated by treatments (n = 4): CT = conventional tillage via a disk harrow; NT = no-till; Cover = three-way mixture of wintertime cover crops; No Cover = soil was left fallow after harvesting the maize cash crop each year. *t* indicates a potential tillage effect and *c* indicates a potential cover crop effect ($P < 0.1$).

4 Future trends in research

Having a healthy microbiome is a key component of sustainable soil management; however, there are many barriers and uncertainties when applying respiration-based measurements to agroecosystems and other landscapes. Future research should address challenges associated with the cost, technology, and training required to perform these measurements. There also needs to be a concurrent focus on better developing, applying, understanding, and interpreting such analyses. Here we summarize a few of the many possible future trends in research.

4.1 Determining type and timing of measurements to fine-tune management

Respiration-based measurements have the potential to inform agricultural management decisions and even guide cost-share programs that pay for practices such as cover cropping. To be fully realized, however, much still needs to be learned about the types and timing of measurements that best capture the role of soil microbiological functions in sustaining crop productivity. Microbial community composition, resource use efficiency, and energy storage all shift with environmental conditions and stressors. These variations translate to continued changes in the amount of necromass versus active biomass in soils (Kastner et al., 2021), which can ultimately affect carbon stocks and fertility of different agricultural soils. Successful application of respiration-based measurements requires being able to perform measurements with sufficient temporal resolution to untangle these dynamics; otherwise, they must be applied in a manner that best captures the integrated effects of management on the soil microbiome. As illustrated in the case study (Section 3), indicators like SIR and Cmin fluctuate over seasonal to annual timescales. Better understanding the temporal variability of these indicators within and between production cycles should provide sampling recommendations that will most effectively track the overall health of a given soil. There is a concurrent need to interpret how indicators such as nMFE, Cmin, and SIR are associated with microbial carbon use efficiency and carbon cycling processes (Fulton-Smith et al., 2024, Zhou et al., 2025).

Considerable interest, along with much uncertainty, exists for applying respiration-based tests to assess nitrogen availability in soil. Some studies have pointed to the ability to predict bioavailable nitrogen using STCM measurements (Franzluebbers et al., 2018), with moderate-to-strong correlations ($R^2 > 0.6$) between the 3-day STCM test and the net amount of nitrogen mineralized during aerobic incubations (Farmaha et al., 2022, McDonald et al., 2014). While the relationships are less strong, STCM values also have correlation with optimum nitrogen rates for grain production (Franzluebbers, 2020). However, more work is required to ascertain if and how these measurements link to anaerobic measurements that are commonly performed to measure potentially mineralizable nitrogen (Clark et al., 2019). There is also need for field validations to ensure that any nitrogen recommendations given based on respiration measurements do not incur yield penalties, as individual observations can have wide variations in the crop response based on actual nitrogen added versus that which is predicted based on STCM tests.

4.2 Integration with soil carbon sequestration and conservation management efforts

Tremendous resources are being devoted to efforts to sequester excess atmospheric carbon in soils via better management (Minasny et al., 2017). While forested lands have the most potential to store extra carbon, working agricultural lands are also of great interest due to their large global extent and potential to derive other agronomic- and soil-related benefits from greater carbon content (Jian et al., 2020a). Soil carbon stocks can take years to decades to measurably increase and often have year-to-year variations when repeatedly measured in the same field (Nicholakos et al., 2025). Both of these factors make it difficult to incentivize carbon sequestration efforts based on direct measurements. Respiration-based approaches can help to pinpoint how soil microbiological communities respond to management, and thereby provide a predictor of how soil carbon stocks may change over medium-to-long periods of management. Mineralizable carbon may be particularly useful for this purpose, as it has linkages to eventual formation of mineral-associated organic carbon, which is typically the more stable and longer-persisting form of carbon in soils (Fulton-Smith et al., 2024).

4.3 Technology-based innovations

Integrating low-cost infrared analyzers into incubation chambers can increase the ease of measurements and provide better temporal resolution of soil respiration. These sensors can be integrated into static chambers used for in-situ field assessments and laboratory measurements such as SIR and Cmin (Gyawali et al., 2019). While lower-cost options tend to have much poorer resolution than commercial units, they can still obtain similar results in terms of measured concentrations and fluxes (Fig. 6). They also can be integrated into autonomous chambers (Forbes et al., 2023), allowing for increases in the spatial and temporal coverage of R_s measurements by orders of magnitude. Opportunities exist to combine these types of sensors into multi-parameter assessments, such as an integrated quantification of R_s and soil water potential (Gan et al., 2023). Other options, such as mid-infrared spectroscopy, can enable rapid estimates of R_s and Q_{10} relationships (Meyer et al., 2018a), so development and dissemination of these methods should continue. Altogether, our collective ability to measure and interpret soil microbiological functioning via respiration measurements will continue to improve as sensor performance-to-cost ratios increase, data-logger capacities expand, new analysis packages are developed (e.g. the fluxfinder R package; Wilson et al., 2024), and approaches such as machine learning and Internet of Things monitoring become more appropriately applied to these types of analyses.

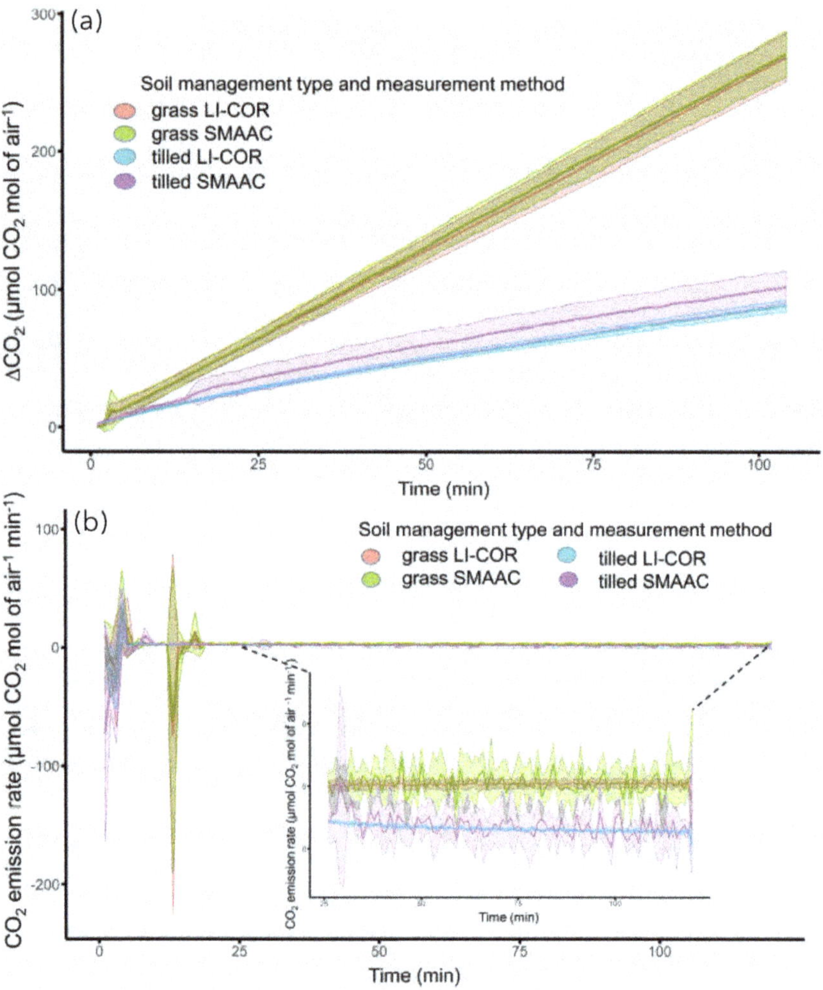

Figure 6 (a) Laboratory measurement of change in CO_2 through time (min) for grass-covered and tilled soils, measured using a commercial nondispersive infrared gas analyzer (LI-COR 8100) versus an Arduino-powered low-cost unit called the soil microbial activity assessment contraption, or SMAAC. Solid lines represent mean values, and shaded areas represent standard deviations from the means. (b) CO_2 emission rates for grass-covered and tilled soils, measured using the LI-COR and the SMAAC. The inset shows the noise profile through time for the two instruments. Figure from Gyawali et al. (2019).

5 Conclusions

Respiration-based methods are used to infer a variety of soil microbiological properties and responses. Over the past century these measurements have helped to better understand carbon mineralization and stability, microbial diversity and functioning (including the effects of different management practices), biological responses to environmental perturbations, and resource allocations within soil. The protocols used - and the questions posed - continue to evolve as technologies emerge, leading to new knowledge and insight. Nonetheless, many uncertainties exist regarding the exact mechanisms that underlie many of these assessments. It is critical to resolve the biophysical underpinnings of soil respiration, particularly in the context of responses to stress, before these methods can be appropriately and accurately used to assess soil health and guide fertility management decisions.

Most respiration measurements are done by experts using specialized, and often expensive, equipment. This type of approach will continue to be important for advancing scientific understanding and testing hypotheses. However, there is also an opportunity to broaden the application of respiration as a tool for characterizing soil microbial activity. Tests such as SIR and STCM could be part of routine soil testing offered by non-profit and commercial laboratories, particularly as CO2 sensor costs decrease. Assays can be incorporated into more educational and training efforts, for example by comparing basal respiration from soils representing different tillage practices or land covers, while at the same time teaching students how to perform and interpret such tests. It is clear that respiration will continue to be an important aspect of soil assessment, so it is necessary to continue to refine these procedures and guide their interpretation for maximum utility.

Most respiration measurements are done by experts using specialized, and often expensive, equipment. This type of approach will continue to be important for advancing scientific understanding and testing hypotheses. However, there is also an opportunity to broaden the application of respiration as a tool for characterizing soil microbial activity.Tests such as SIR and STCM could become part of routine soil testing offered by non-profit and commercial laboratories, particularly as CO_2 sensor costs decrease. Assays can be incorporated into more educational and training efforts, for example by comparing basal respiration from soils representing different tillage practices or land covers, while at the same time teaching students how to perform and interpret such tests. It is clear that respiration will continue to be an important aspect of soil assessment, so it is necessary to continue to refine these procedures and guide their interpretation for maximum utility.

6 Where to look for further information

Numerous resources are available in print and on the internet related to soil respiration and its use. Some pertinent resources include:

- The history of many respiration-based measurements are extensively detailed in several classic reviews, including Anderson and Domsch (1978), Martens (1995), Jenkinson et al. (2004), and Anderson and Domsch (2010).
- *Building Soils for Better Crops*, by Magdoff and Van Es (2000), is available both as a free PDF and softcover book. The text covers many basics, including the types of soil organisms and their importance to health soils. It is available at https://www.sare.org/resources/building-soils-for-better-crops/.
- Standardized procedures for SIR and basal respiration can be found in ISO 16072:2002, Soil quality - Laboratory methods for determination of microbial soil respiration. This document can be found at: https://www.iso.org/standard/32096.html.
- Many good references exist for best practices when performing field measurements of soil respiration and other greenhouse gases, including De Klein et al. (2020), Maier et al. (2022), and Mumu et al. (2024).
- A larger, more holistic set of methods related to soil microbiological assessment is published by the Soil Science Society of America (SSSA), titled *Methods of Soil Analysis: Part 2. Microbiological and Biochemical Properties*. This text covers soil sampling procedures, statistical treatments, and biogeochemical processes and includes many of the respiration-based measurements described in this chapter. The book can be purchased in print or found in digital form at: https://acsess.onlinelibrary.wiley.com/doi/book/10.2136/sssabookser5.2.
- The USDA Natural Resources Conservation Service (NRCS) has prepared several documents detailing methods to determine soil quality and health. The first focuses on in-field assessments using the Soil Quality Test Kit, and can be found at: https://efotg.sc.egov.usda.gov/references/Public/WI/Soil_Quality_Test_Kit_Guide.pdf. The second is aimed at laboratory assessments and can be found at: https://efotg.sc.egov.usda.gov/references/public/CO/Tech_Note_450_06_Recommended_Soil_Health_Indicators_and_Lab_procedures.pdf.
- Active scientific communities that focus on understanding the role of soil microbiology in agriculture management include the Soil Science Society of America (SSSA), the American Society of Agronomy (ASA), the Ecological Society of America (ESA), the American Geophysical Union (AGU), the European Geophysical Union (EGU), and the Soil and Water Conservation Society (SWCS) of the United States. These societies hold annual meetings and solicit contributions to sessions that align with this topic.
- Two nationwide research networks based in the United States provide soil respiration-related measurements. The first, the National Ecological Observatory Network (NEON), has in-field measurements of CO_2

concentrations at multiple soil depths, which can be combined with a gas diffusivity model to estimate soil respiration rates. The NEON data can be accessed at: https://www.neonscience.org/ The second, the USDA Agricultural Research Service Long-Term Agroecosystem Research (LTAR) network, provides agriculture-related data from different systems across the nation. Many of the sites include soil respiration measurements, both measured in the field and under laboratory conditions. Those data can be found at: https://ltar.ars.usda.gov/ Similar networks exist in other parts of the world, such as the Terrestrial Environmental Observatories (TERENO) network of Germany: https://www.tereno.net/.

- The Soil Health Database (SoilHealthDB) has compiled the results of many studies related to soil health and includes numerous results for respiration-based measurements (Jian et al., 2020b). The database can be accessed at https://github.com/jinshijian/SoilHealthDB.
- The Deep Soil Ecotron is a U.S. National Science Foundation-funded facility that probes interactions between soils, ecology, and biogeochemistry, and has collaborators across the United States. More information can be found at: https://deepsoilecotron.org/.

7 References

Al-Kaisi, M. M. & Yin, X. 2005. Tillage and crop residue effects on soil carbon and carbon dioxide emission in corn–soybean rotations. *Journal of Environmental Quality*, 34, 437–445.

Anderson, J. P. & Domsch, K. H. 1978. A physiological method for the quantitative measurement of microbial biomass in soils. *Soil Biology and Biochemistry*, 10, 215–221.

Anderson, T.-H. & Domsch, K. 1985. Maintenance carbon requirements of actively-metabolizing microbial populations under in situ conditions. *Soil Biology and Biochemistry*, 17, 197–203.

Anderson, T. H., and Domsch, K. H. (1990). Application of eco-physiological quotients (qCO2 and qD) on microbial biomasses from soils of different cropping histories. *Soil Biology and Biochemistry*, 22(2), 251-255.

Anderson, T.-H. & Domsch, K. H. 2010. Soil microbial biomass: the eco-physiological approach. *Soil Biology and Biochemistry*, 42, 2039–2043.

Barnard, R. L., et al. 2020. Rewetting of soil: revisiting the origin of soil CO_2 emissions. *Soil Biology and Biochemistry*, 147, 107819.

Bierer, A. M., et al. 2020. Evaluating effects of dairy manure application method on soil health and nitrate. *Journal of Soil and Water Conservation*, 75(5), 533–544.

Birch, H. & Friend, M. 1956. Humus decomposition in East African soils. *Nature*, 178, 500–501.

Bossio, D., et al. 2005. Soil microbial community response to land use change in an agricultural landscape of western Kenya. *Microbial Ecology*, 49, 50–62.

Bouma, T. J., et al. 1997. Estimating respiration of roots in soil: Interactions with soil CO_2, soil temperature and soil water content. *Plant and Soil*, 195, 221–232.

Bradford, M. A., et al. 2013. Empirical evidence that soil carbon formation from plant inputs is positively related to microbial growth. *Biogeochemistry*, 113, 271–281.

Brown, R. & Markewitz, D. 2018. Soil heterotrophic respiration: Measuring and modeling seasonal variation and silvicultural impacts. *Forest Ecology and Management*, 430, 594–608.

Bryla, D., et al. 2001. Influence of temperature and soil drying on respiration of individual roots in citrus: integrating greenhouse observations into a predictive model for the field. *Plant, Cell & Environment*, 24, 781–790.

Buyanovsky, G. & Wagner, G. 1983. Annual cycles of carbon dioxide level in soil air. *Soil Science Society of America Journal*, 47, 1139–1145.

Chen, D., et al. 2005. Corn root influence on automated measurement of soil carbon dioxide concentrations. *Soil Science*, 170, 779–787.

Clark, J. D., et al. 2019. Predicting economic optimal nitrogen rate with the anaerobic potentially mineralizable nitrogen test. *Agronomy Journal*, 111, 3329–3338.

Clough, T. J., et al. 2020. Global research alliance N2 O chamber methodology guidelines: design considerations. *Journal of Environmental Quality*, 49, 1081–1091.

Collier, S. M., et al. 2014. Measurement of greenhouse gas flux from agricultural soils using static chambers. *Journal of Visualized Experiments: JoVE*, 90, 52110.

Congreves, K. & Wu, Q. 2024. Using soil classification to improve interpretation of biological soil health indicators. *Geoderma*, 451, 117085.

Darbishire, F. V. & Russell, E. J. 1907. Oxidation in soils, and its relation to productiveness. *The Journal of Agricultural Science*, 2, 305–326.

De Klein, C. A. M., et al. 2020. Global research alliance N2 O chamber methodology guidelines: Introduction, with health and safety considerations. *Journal of Environmental Quality*, 49, 1073–1080.

Degens, B. & Harris, J. 1997. Development of a physiological approach to measuring the catabolic diversity of soil microbial communities. *Soil Biology and Biochemistry*, 29, 1309–1320.

Dehérain, M. P.-P. & Demoussy, E. 1896. Sur l'oxydation de la matière organique du sol. *Annales Agronomiques*, 22, 305–337.

Drobnik, J. 1960. Primary oxidation of organic matter in the soil: I. The form of respiration curves with glucose as the substrate. *Plant and Soil*, 12, 199–211.

Escalas, A., et al. 2019. Microbial functional diversity: from concepts to applications. *Ecology and Evolution*, 9, 12000–12016.

Farmaha, B. S., et al. 2022. Cover cropping and conservation tillage improve soil health in the southeastern United States. *Agronomy Journal*, 114, 296–316.

Fierer, N. 2017. Embracing the unknown: disentangling the complexities of the soil microbiome. *Nature Reviews Microbiology*, 15, 579.

Fierer, N. & Schimel, J. P. 2003. A proposed mechanism for the pulse in carbon dioxide production commonly observed following the rapid rewetting of a dry soil. *Soil Science Society of America Journal*, 67, 798.

Fierer, N., et al. 2003. Controls on microbial CO_2 production: a comparison of surface and subsurface soil horizons. *Global Change Biology*, 9, 1322–1332.

Forbes, E., et al. 2023. Fluxbots: a method for building, deploying, collecting and analyzing data from an array of inexpensive, autonomous soil carbon flux chambers. *Journal of Geophysical Research: Biogeosciences*, 128, e2023JG007451.

Franzluebbers, A. J. 2020. Soil-test biological activity with the flush of CO_2: V. Validation of nitrogen prediction for corn production. *Agronomy Journal*, 112, 2188–2204.

Franzluebbers, A. J., et al. 1995. Tillage and crop effects on seasonal dynamics of soil CO_2 evolution, water content, temperature, and bulk density. *Applied Soil Ecology*, 2, 95–109.

Franzluebbers, A. J. & Veum, K. S. 2020. Comparison of two alkali trap methods for measuring the flush of CO_2. *Agronomy Journal*, 112(2), 965–971.

Franzluebbers, A. J., et al. 2018. Soil-test biological activity with the flush of CO_2: I. C and N characteristics of soils in corn production. *Soil Science Society of America Journal*, 82, 685–695.

Fulton-Smith, S., et al. 2024. Depth impacts on the aggregate-mediated mechanisms of root carbon stabilization in soil: trade-off between MAOM and POM pathways. *Geoderma*, 452, 117078.

Gan, H., et al. 2023. Automated sensor-based quantification of soil water retention and microbial respiration across drying conditions. *Soil Biology and Biochemistry*, 180, 108987.

Gyawali, A. J., et al. 2019. Talking SMAAC: a new tool to measure soil respiration and microbial activity. *Frontiers in Earth Science*, 7, 138.

Gyawali, A. J., et al. 2022. Quantifying short-term responsiveness and consistency of soil health parameters in row crop systems. Part 1: developing a multivariate approach. *Soil and Tillage Research*, 219, 105354.

Haney, R., et al. 2008. Soil CO2 respiration: comparison of chemical titration, CO_2 IRGA analysis and the Solvita gel system. *Renewable Agriculture and Food Systems*, 23, 171–176.

Hendrix, P., et al. 1988. Soil respiration in conventional and no-tillage agroecosystems under different winter cover crop rotations. *Soil and Tillage Research*, 12, 135–148.

Hilaire, S. S., et al. 2022. Culturable antibiotic-resistant fecal coliform bacteria in soil and surface runoff after liquid dairy manure surface application and subsurface injection. *Journal of Environment Quality*, 41, 288–300.

Jenkinson, D. 1966. Studies on the decomposition of plant material in soil: II. Partial sterilization of soil and the soil biomass. *Journal of Soil Science*, 17, 280–302.

Jenkinson, D. S., et al. 2004. Measuring soil microbial biomass. *Soil Biology and Biochemistry*, 36, 5–7.

Jenkinson, D. S. & Powlson, D. S. 1976. The effects of biocidal treatments on metabolism in soil–V: a method for measuring soil biomass. *Soil Biology and Biochemistry*, 8, 209–213.

Jian, J., et al. 2018. Measurement strategies to account for soil respiration temporal heterogeneity across diverse regions. *Soil Biology and Biochemistry*, 125, 167–177.

Jian, J., et al. 2020a. A meta-analysis of global cropland soil carbon changes due to cover cropping. *Soil Biology and Biochemistry*, 143, 107735.

Jian, J., et al. 2020b. A database for global soil health assessment. *Scientific Data*, 7, 16.

Kästner, et al (2021). Microbial necromass in soils–linking microbes to soil processes and carbon turnover. *Frontiers in Environmental Science*, 9, 756378.

Knoepp, J. D. & Vose, J. M. 2002. *Quantitative comparison of in situ soil CO_2 flux measurement methods*. United States Department of Agriculture, Forest Service, Southern Region Station.

Kuzyakov, Y. & Larionova, A. A. 2005. Root and rhizomicrobial respiration: a review of approaches to estimate respiration by autotrophic and heterotrophic organisms in soil. *Journal of Plant Nutrition and Soil Science*, 168, 503–520.

Le Roux, X., et al. 2008. Effects of aboveground grazing on coupling among nitrifier activity, abundance and community structure. *The ISME Journal*, 2, 221–232.

Macdonald, J. D., et al. 2010. Plowing a poorly drained grassland reduced soil respiration. *Soil Science Society of America Journal*, 74, 2067–2076.

Magdoff, F. & Van Es, H. 2000. *Building soils for better crops.* Sustainable Agriculture Network Beltsville.

Magurran, A. E. 1988. *Ecological diversity and its measurement.* Springer Science & Business Media.

Maier, M., et al. 2022. Introduction of a guideline for measurements of greenhouse gas fluxes from soils using non-steady-state chambers. *Journal of Plant Nutrition and Soil Science*, 185, 447–461.

Martens, R. 1995. Current methods for measuring microbial biomass C in soil: potentials and limitations. *Biology and Fertility of Soils*, 19, 87–99.

McBride, S. G., et al. 2020. Volatile organic compounds from leaf litter decomposition alter soil microbial communities and carbon dynamics. *Ecology*, 101, e03130.

McCloskey, C. S., et al. 2020. A field system for measuring plant and soil carbon fluxes using stable isotope methods. *European Journal of Soil Science*, 72, 2330–2342.

McCourty, M. A., et al. 2018. Of macropores and tillage: influence of biomass incorporation on cover crop decomposition and soil respiration. *Soil Use and Management*, 34, 101–110.

McDonald, N. T., et al. 2014. Evaluation of soil tests for predicting nitrogen mineralization in temperate grassland soils. *Soil Science Society of America Journal*, 78, 1051–1064.

McGowen, E. B., et al. 2018. An automated laboratory method for measuring CO_2 emissions from soils. *Agricultural & Environmental Letters*, 3, 1–5.

Meyer, N., et al. 2018a. Soil respiration and its temperature sensitivity (Q_{10}): rapid acquisition using mid-infrared spectroscopy. *Geoderma*, 323, 31–40.

Meyer, N., et al. 2018b. The temperature sensitivity (Q_{10}) of soil respiration: controlling factors and spatial prediction at regional scale based on environmental soil classes. *Global Biogeochemical Cycles*, 32, 306–323.

Minasny, B., et al. 2017. Soil carbon 4 per mille. *Geoderma*, 292, 59–86.

Moldrup, P., et al. 2005. Predictive-descriptive models for gas and solute diffusion coefficients in variably saturated porous media coupled to pore-size distribution. II. Gas diffusivity in undisturbed soil. *Soil Science*, 170, 854–866.

Morrow, J. G., et al. 2016. Evaluating measures to assess soil health in long-term agroecosystem trials. *Soil Science Society of America Journal*, 80, 450–462.

Mumu, N. J., et al. 2024. Methodological progress in the measurement of agricultural greenhouse gases. *Carbon Management*, 15, 2366527.

Nicholakos, S. A., et al. 2025. Multi-year soil response to conservation management in the Virginia Coastal Plain. *Soil and Tillage Research*, 245, 106303.

Nielsen, U. N., et al. 2015. Soil biodiversity and the environment. *Annual Review of Environment and Resources*, 40, 63–90.

Nivelle, E., et al. 2016. Functional response of soil microbial communities to tillage, cover crops and nitrogen fertilization. *Applied Soil Ecology*, 108, 147–155.

Nogueira, M. A., et al. 2006. Promising indicators for assessment of agroecosystems alteration among natural, reforested and agricultural land use in southern Brazil. *Agriculture, Ecosystems & Environment*, 115, 237–247.

NRCS. 2017. *Web soil survey* [Online]. Available: http://websoilsurvey.nrcs.usda.gov [Accessed 10 January 2025].

NRCS. 2019. *Recommended soil health indicators and associated laboratory procedures*. U.S. Department of Agriculture, N. R. C. S.

Possinger, A. R., et al. 2021. Climate effects on subsoil carbon loss mediated by soil chemistry. *Environmental Science & Technology*, 55, 16224–16235.

Reeves, D. 1994. Cover crops and rotations. *Advances in Soil Science: Crops Residue Management*, 125–172.

Reicosky, D. 1997. Tillage-induced CO_2 emission from soil. *Nutrient Cycling in Agroecosystems*, 49, 273–285.

Reicosky, D., et al. 1997. Tillage-induced soil carbon dioxide loss from different cropping systems. *Soil and Tillage Research*, 41, 105–118.

Savage, K., et al. 2013. Diel patterns of autotrophic and heterotrophic respiration among phenological stages. *Global Change Biology*, 19, 1151–1159.

Schimel, J., et al. 2007. Microbial stress-response physiology and its implications for ecosystem function. *Ecology*, 88, 1386–1394.

Schindlbacher, A., et al. 2012. Soil respiration under climate change: prolonged summer drought offsets soil warming effects. *Global Change Biology*, 18, 2270–2279.

Schipper, L., et al. 2001. Changes in microbial heterotrophic diversity along five plant successional sequences. *Soil Biology and Biochemistry*, 33, 2093–2103.

Schonbeck, M. & Morse, R. 2006. *Cover crops for all seasons*. Virginia Association for Biological Farming, pp. 3–6.

Slessarev, E. W. & Schimel, J. P. 2020. Partitioning sources of CO_2 emission after soil wetting using high-resolution observations and minimal models. *Soil Biology and Biochemistry*, 143, 107753.

Stewart, R. D., et al. 2018. What we talk about when we talk about soil health. *Agricultural & Environmental Letters*, 3, 1–5.

Störmer, K. 1908. Über die Wirkung des Schwefelkohlenstoffs und ähnlicher Stoffe auf den Boden. *Zentbl. Bakt. II*, 20, 282–286.

Strickland, M. S., et al. 2010. Rates of in situ carbon mineralization in relation to land-use, microbial community and edaphic characteristics. *Soil Biology and Biochemistry*, 42, 260–269.

Strickland, M. S., et al. 2013. Trophic cascade alters ecosystem carbon exchange. *Proceedings of the National Academy of Sciences*, 110, 11035–11038.

Strickland, M. S., et al. 2015. Compositional differences in simulated root exudates elicit a limited functional and compositional response in soil microbial communities. *Frontiers in Microbiology*, 6, 817.

Strickland, M. S., et al. 2017. Response of soil microbial community composition and function to a bottomland forest restoration intensity gradient. *Applied Soil Ecology*, 119, 317–326.

Strickland, M. S., et al. 2019. Short-term effects of cover crops on soil microbial characteristics and biogeochemical processes across actively managed farms. *Agrosystems, Geosciences & Environment*, 2, 1–9.

Tang, J., et al. 2003. Assessing soil CO_2 efflux using continuous measurements of CO_2 profiles in soils with small solid-state sensors. *Agricultural and Forest Meteorology*, 118, 207–220.

Thompson, T. K., et al. 2023. Assessing landscape and seasonal controls on soil CO_2 fluxes in a Karst Sinkhole. *Environmental & Engineering Geoscience*, 29, 217–229.

Trevors, J. 2010. One gram of soil: a microbial biochemical gene library. *Antonie Van Leeuwenhoek*, 97, 99–106.

Turcu, V. E., et al. 2005. Continuous soil carbon dioxide and oxygen measurements and estimation of gradient-based gaseous flux. *Vadose Zone Journal*, 4, 1161–1169.

Unger, S., et al. 2010. The influence of precipitation pulses on soil respiration–assessing the "Birch effect" by stable carbon isotopes. *Soil Biology and Biochemistry*, 42, 1800–1810.

USDA. 1999. *Soil quality test kit guide. USDA agricultural research service*. National Consevation Service. Soil Quality Institute.

Van Bodegom, P. 2007. Microbial maintenance: a critical review on its quantification. *Microbial Ecology*, 53, 513–523.

Vance, E. D., et al. 1987. An extraction method for measuring soil microbial biomass C. *Soil Biology and Biochemistry*, 19, 703–707.

Wade, J., et al. 2018. Sources of variability that compromise mineralizable carbon as a soil health indicator. *Soil Science Society of America Journal*, 82.

Warner, D., et al. 2019. Spatial predictions and associated uncertainty of annual soil respiration at the global scale. *Global Biogeochemical Cycles*, 33, 1733–1745.

Wepking, C., et al. 2017. Exposure to dairy manure leads to greater antibiotic resistance and increased mass-specific respiration in soil microbial communities. *Proceedings of the Royal Society B: Biological Sciences*, 284, 20162233.

Wepking, C., et al. 2019. Prolonged exposure to manure from livestock-administered antibiotics decreases ecosystem carbon-use efficiency and alters nitrogen cycling. *Ecology Letters*, 22, 2067–2076.

Wilson, S. J., et al. 2024. fluxfinder: an R package for reproducible calculation and initial processing of greenhouse gas fluxes from static chamber measurements. *Journal of Geophysical Research: Biogeosciences*, 129, e2024JG008208.

Wittebolle, L., et al. 2009. Initial community evenness favours functionality under selective stress. *Nature*, 458, 623–626.

Wright, R. R. & Hobbie, J. E. 1966. Use of glucose and acetate by bacteria and algae in aquatic ecosystems. *Ecology*, 47, 447–464.

Wu, Q., et al. 2024. Soil health indicator responses to three years of cover crop and crop rotation in a northern semi-arid region, the Canadian prairies. *Agriculture, Ecosystems & Environment*, 359, 108755.

Yonemura, S., et al. 2014. Soil respiration, N2O, and CH4 emissions from an Andisol under conventional-tillage and no-tillage cultivation for 4 years. *Biology and Fertility of Soils*, 50, 63–74.

Zhou, J., et al. 2025. Dilemmas in linking microbial carbon use efficiency with soil organic carbon dynamics. *Global Change Biology*, 31, e70047.

Chapter 5

Enzyme assays for measuring soil microbial activity

Paolo Nannipieri, Università degli Studi di Firenze, Italy; and Carmen Trasar-Cepeda, MBG Sede Santiago-CSIC, Spain

1 Introduction

Soil enzymology has a long history. The earliest report of soil enzymes was by Woods (1899), who determined the activity of oxidizing enzymes and hypothesized that these enzymes occurred in the extracellular soil environment. This report was followed by a series of studies mainly dealing with catalase activity, which is relatively easy to detect (see review by Skujins, 1978). In the first half of the twentieth century, soil solutions were compared with animal blood, and Quastel (1946) suggested that soil could be considered a living tissue. This period, termed "the long childhood" by Skujins (1978), can be characterized by the development of several assays for measurement of hydrolase and oxidase activities in soil. However, scientific advancements in the field of soil enzymology were made in the 1920s with the establishment of accurate analytical assays and discussion of (1) the meaning of enzyme measurements in soil, (2) the different locations of enzymes in the soil matrix and (3) the ecological importance of extracellularly stabilized enzymes, which are protected from microbial degradation by adsorption onto surface-reactive particles or entrapment by soil particles. Skujins (1978) coined the term "abiontic" to represent all extracellular enzymes, i.e. those stabilized in soil and also those released by active cells or after cell lysis and those present in

http://dx.doi.org/10.19103/AS.2025.0159.06

dead cells or cell debris. Importantly, the activity of abiontic enzymes is not controlled by the soil microbiome, and these enzymes can thus contribute to soil functions even when the conditions do not favour microbial activity. Burns (1982) hypothesized that the presence of stabilized extracellular enzymes could favour the detection and use of exogenous substrates by the soil microbiome. In the 1920s, measurement of enzyme activities was combined with that of other biochemical and microbiological properties to obtain insights into soil fertility and crop yields. The enzyme activities were considered indicators of soil fertility and of the status of polluted and degraded/restored soils and prediction of the effects of climate change on soil metabolic activity. Between the 1950s and 1970s, both Western and Eastern European research groups contributed extensively to soil enzymology (Skujins, 1978). Although some of the most productive laboratories published their studies in Russian, the main findings are available in English, thanks to the excellent reviews by Skujins (1978) and Kuprevich and Shcherbakova (1971). The fact that important reviews on soil enzymology were published in books and journal issues not accessible to electronic searches led to some shortcomings in soil enzyme measurements and the complexity of the measurement interpretations being ignored around the turn of the century. In particular, it was (and still is) generally assumed that enzyme assays determine extracellular activity and that the activities of one or two enzymes can be used as proxies for C-, N- or P-acquiring processes. These assumptions led to the spread of misconceptions, as already discussed by Nannipieri et al. (2018). The early 2000s were characterized by the use of molecular techniques to characterize the diversity and activity of the soil microbiome, which provided insights into the source of enzymes contributing to the measured activity, a long-standing knowledge gap in soil enzymology.

The scope of this review is to discuss the advantages and limits of enzyme assays used as proxies for nutrient availability in soil and as indicators of soil health. Correct sampling and handling of soil samples and the use of accurate methods are essential to prevent both over- and underestimation of enzyme activities. The fact that only a small number of enzyme activities can be measured in soil is often overlooked, and this hampers the understanding of metabolic processes related to nutrient dynamics, which depend on a large number of different enzyme activities that usually occur as cascade reactions. We will also discuss the complexity of interpreting soil enzyme activities. As mentioned above, the use of molecular techniques combined with enzyme assays has provided important insights into the origin of enzymes in soil; the present review also underlines research shortcomings and proposes future research lines that will address some of the current knowledge gaps. Finally, we propose further studies to improve the use of enzyme activities as indicators of soil health. The literature on soil enzymology is extensive, including several books (Burns, 1978a; Kiss et al., 1998; Burns and Dick, 2002; García et al., 2003;

Dick, 2011a), various reviews and a huge number of articles, which cannot be completely covered here. We therefore apologize to the authors of the great majority of these publications not mentioned here.

2 Use of accurate enzyme assays and relative standardization

2.1 Bench-scale enzyme assays

Burns (1978b) highlighted the use of (in)accurate methods and the lack of standardization among the primary obstacles to advances in soil enzymology. However, although both problems have been widely debated (Malcom, 1983; Saratchandra et al., 1984; Gianfreda and Bollag, 1996), the importance of the availability of accurate enzyme assays is often overlooked (Nannipieri et al., 2018), and no consensus has been reached regarding the standardization of soil enzyme measurements. One of the problems is that soil enzyme assays have been adapted from methods developed for other biological systems, often without prior verification of the factors involved, including the type and concentration of buffer used, optimal pH of the enzyme reaction and appropriate concentration of substrate. Another common problem is that different authors modify established methods without first verifying how the changes will affect the measurement of the enzyme activity in question. The results obtained in different laboratories are therefore not comparable. This is exacerbated by the fact that the modifications are often not described. Indeed, information relevant to the enzyme activity assay (pH at which the assay is performed, type and concentration of buffer, substrate concentration, etc.) is often not provided. The use of commercial kits, which are easy to use but are not tested with different soil types, and techniques that collect massive amounts of data rapidly and inexpensively using small amounts of soil, has further exacerbated the situation. The use of well-established assays is important to enable comparison of the same enzyme activity measured in different soils and in different laboratories. The data thus obtained would allow definition of the range of the enzyme activity measured in different soil types, under different types of use and management and in different climatic zones.

Soil enzyme activity assays are usually performed under conditions that ensure optimal rates of catalysis, i.e. at optimum pH and temperature, with excess substrate and a reaction volume and test conditions that allow free diffusion of the substrate and facilitate its interaction with the enzyme (Burns, 1978a; Alef and Nannipieri, 1995; Taylor et al., 2002). The measured activity therefore represents the maximum potential activity of the enzyme and not the actual activity of the enzyme under field conditions. Thus, development of soil enzyme assays should follow the criteria proposed by Tabatabai and co-workers, including (1) use of an adequate number of soils (usually no less than five or six) differing in physico-chemical properties; (2) accurate determination of the product (although disappearance of the substrate could also be measured, this approach is less sensitive and therefore the former is usually preferred); (3) selection of optimal pH and temperature (usually between 25°C and 37°C); (4) use of an adequate amount of soil sample, considering that decreasing the quantity decreases the sensitivity and the representativeness of the soil being assayed; (5) maintenance of zero-order kinetics during the assay (favoured by use of an adequate concentration of substrate); (6) prevention of microbial growth and enzyme production during the assay, achieved by short incubation assays lasting no more than a few hours; (7) use of a proper control by adding the substrate at the end of the incubation period or by using the same procedure with soil treated to completely eliminate all enzyme activity; and (8) construction of standard calibration curves verifying that the soil does not retain (adsorb) the product to be measured and, where appropriate, establishing standard calibration curves for soil under the same conditions of the enzyme assay. These recommendations are not always followed. For example, enzyme assays lasting 24 hours are still used, despite the availability of shorter assays, as in the case of urease activity (Dick, 2011b). Enzyme assays should only be prolonged up to 24 hours in cases where a sufficient amount of product must be released for accurate measurement of difference relative to a control, as in the CM-cellulase activity assay (Schinner and von Mersi, 1990) (i.e. when this is the only way to measure the activity of the enzyme in question). It is also important that pH is maintained constant during the assay. Enzyme activity assays are usually performed at the optimum pH of the enzyme, using a buffer to maintain the pH of the reaction mixture. However, some authors recommend that the enzyme assay should be performed at the soil pH, without adding any buffer (Duxbury and Tate, 1981; Kandeler and Gerber, 1988; von Mersi and Schinner, 1991; Taylor et al., 2002) or adding a buffer at the soil pH (Harrison and Pearce, 1979; Sinsabaugh et al., 2000). Whether the activity is measured at the optimum pH of the enzyme (the most widely used and recommended option) or at the soil pH, it is important that the pH does not change during the assay (Malcom, 1983; Vaughan and Malcom, 1984). The capacity of a buffer to maintain the pH depends on several factors (e.g. soil pH, buffer pH and p*Ka*, buffering capacity of the soil itself), and the pH of the reaction mixture does not always coincide with the pH of the buffer. The pH of both should be measured, and monitoring should be carried out throughout the incubation period to ensure that the pH of the reaction mixture period remains stable; however, this is rarely done. Although the enzyme activity measured at similar pH may not vary greatly, significant variations can also occur, with activities 28% lower and a pH difference of only half a pH unit (Trasar-Cepeda and Gil-Sotres, 1988). The variability in the buffering capacity of the buffers used is also often overlooked. Thus, for example, polybasic buffer systems, such as

Modified Universal Buffer (MUB), have a wide range of p*Ka* values and work within a very wide range of pH (from pH 3 to pH 11). MUB is therefore the reference buffer used in for many enzyme assays (especially those developed by Tabatabai and collaborators over the years). However, MUB does not have a strong buffering capacity, and bringing the reaction mixture to the desired pH may not be effective (depending on the soil properties). For example, in a study of acid soils rich in organic matter, Trasar-Cepeda et al. (1985) found that MUB was unable to maintain the pH of the soil-buffer-substrate reaction mixture at the intended pH. The deviation increased with the buffer pH values; for example, for MUB of pH 11, the pH of the reaction mixture did not exceed 6.2. Additionally, in soils with carbonate, MUB also fails to maintain the reaction mixture at an acidic pH (unpublished data). The strong adsorption capacity many soils have for the reaction product being measured is yet another overlooked aspect. This has been demonstrated, for example, with *p*-nitrophenol (PNP), which is measured as the product in many enzyme assays (Gerritse and van Dijk, 1978; Sarathchandra and Perrott, 1981; Malcom, 1983; Vuorinen, 1993), and with iodonitrotetrazolium formazan (INTF), determined in dehydrogenase activity assays (Camiña et al., 1998). In both cases, adsorption of the product is strongly correlated with the soil organic matter content (Gerritse and van Dijk, 1978; Vuorinen, 1993; Camiña et al., 1998) and varies depending on the concentration of PNP (Sarathchandra and Perrott, 1981) or INTF (Camiña et al., 1998). The amount of product adsorbed by the soil must therefore be corrected to prevent underestimation of the enzyme activity, and constructing calibration curves for the soil under consideration is recommended (Sarathchandra and Perrott, 1981; Vuorinen, 1993; Camiña et al., 1998).

2.2 Fluorimetric microplate assays

Fluorimetric microplate assays for soil enzyme activities were proposed and developed by, respectively, Freeman et al. (1995) and Marx et al. (2001), and they have since become very popular and are widely used nowadays. These methods of measuring enzyme activity are based on substrates which, after enzymatic cleavage, release 4-methylumbelliferone (MUF) or other products (e.g. 7-amino-4-methyl coumarin, 7-AMC) that emit fluorescence, thus allowing the amount of product to be measured. Microplate assays have expanded rapidly and been adopted in numerous laboratories, because they are less time-consuming than traditional bench-scale assays, different enzyme activities can be determined using a small quantity of soil and the same soil suspension, and fluorescence detection is highly sensitive. Moreover, microplate readers can measure the fluorescence of numerous samples simultaneously (e.g. in 96 wells) in microliter volumes, enabling substantial reductions in reagent costs relative to conventional bench-scale assays (Deng et al., 2011). However, fluorescence-based assays have several shortcomings, including (1) the small amounts of soil used; (2) pipetting errors caused by preparation of soil suspensions in microliter volumes; (3) standard calibration curves must be prepared for each soil sample to account for fluorescence quenching (Deng et al., 2011), which varies temporally and spatially (Freeman et al., 1995); (4) fluorescence is affected by pH and temperature (Lakowicz, 1983, cited in Deng et al., 2011); and (5) the additional precautions required (treating standards, blanks and samples in exactly the same way, preparing all solutions using the same reagents and preparation techniques, and measuring at the same temperature and for the same time) add considerable work, expense and sources of error to the assay (Deng et al., 2011). Moreover, many researchers ignore the rigorous assay conditions required in microplate fluorometric assays (Marx et al., 2001; Deng et al., 2011; German et al., 2011), as discussed above for colorimetric bench-scale enzyme assays. These conditions include, for example, optimization of the amount of NaOH added to increase the pH before measuring fluorescence, and determination of the optimal pH of the enzyme being measured. Thus, the same suspension is often used at the same pH to determine the activity of different enzymes, even though the optimal pH may be different for different enzymes. Similarly, different buffers are often used, although this change might affect the activities being measured. Other common issues include the use of sub-saturating substrate concentrations, not checking the assay duration to ensure that the reaction remains linear and not optimizing the buffer-to-soil/litter sample ratio (Marx et al., 2001; German et al., 2011).

2.3 Bench-scale vs. microplate and chromogenic vs. fluorogenic methods

Popova and Deng (2010) described a high-throughput assay for the simultaneous quantification of multiple soil enzyme activities based on the use of *p*-nitrophenyl substrates, which enabled the use of small amounts of soil and determination of multiple enzyme activities in the same soil suspension. Although the repeatability of the assay was lower than that of the bench-scale assay, it had the advantage of being a cost-effective, high-throughput assay. Deng et al. (2013) compared bench-scale and microplate assays of soil enzyme activities using MUF- and PNP-labelled substrates to determine β-glucosaminidase, β-glucosidase and acid phosphomonoesterase in 16 different soils. Although different values were obtained, the enzyme activities were correlated and ranked in the same order, suggesting that the methods are comparable. Subsequently, Deng et al. (2017) published an article summarizing the most common enzyme assays, pointing out the need for standardization of methods and also the most common problems and some solutions or recommendations to overcome these. These authors reported that assay conditions must be carefully controlled in both microplate assays with MUF-based substrates and in benchtop assays with PNP substrates, in order to obtain comparable measurements. They also pointed out that one of the problems with most MUF-based microplate assays is that they do not include test replicates of soil suspensions. Thus, as soil is very heterogeneous, the amount of soil included in each well will not be representative and can be considered a potentially important source of analytical error. Dick et al. (2018) compared β-glucosidase and acid phosphomonoesterase activities determined in four different soils by fluorimetric microplate and colorimetric bench-scale methods, in five different laboratories. The enzyme activities determined by the two methods were strongly correlated, and the same rankings were obtained. One of the objectives of the study was to compare whether MUF-based microplate assays provided comparable results to the bench-scale PNP methods for both enzymes. It was concluded that the MUF microplate methods could be used as a standard protocol for measuring the activity of both enzymes if the buffer pH was adjusted to the optimal values of the enzyme of interest. In conclusion, both types of methods are suitable for measuring soil enzyme activity. Each has advantages and disadvantages, but the procedures must be followed rigorously and precautions must be taken to address all the issues associated with each method.

The use of accurate and standardized methods is obviously important to enable advances in soil enzymology. Burns (1978b) stressed that the use of different methods was the most serious problem regarding comparison of the same enzyme activity in different soils determined in different laboratories, but this remains a problem today. Although some progress has been made in developing accurate and standardized approaches to measuring enzyme activity, further work is still needed. Standardization of methods is the first step towards valid comparison and meaningful interpretation of data to relate enzyme activities to ecosystem functions (Deng et al., 2017). Generally, broad ranges of values of hydrolases and oxidase have been reported by Nannipieri et al. (2002) due to the use of different soil types and different enzyme assays. It is important to build databases that include the range of values for each of the enzyme activity determined by bench-scale assays or by MUF-microplate format assays. Tscherko and Kandeler (1999) proposed differentiating classes of activity by applying the 30th and 70th percentiles to each of the specific variables in a dataset and thus classifying the enzyme activities as low, normal and high. It must be taken into account that enzyme activity assays will always produce data, whether the method is applied correctly or not. Thus, and because of the many possible sources of error in enzyme assays, particular care should be taken to avoid publishing and perpetuating erroneous databases of enzyme activity values.

3 Meaning of enzyme measurements and their use

The literature on soil enzymology is extensive, because soil enzyme activities can provide information about the ability of soils to perform biogeochemical reactions, they can be used to detect impacts of anthropogenic management (agriculture and forestry) or pollution on soils, and they are generally simple, rapid, accurate and inexpensive to perform. However, the meaning of the measurements is rarely considered in interpreting the data, and studies are often biased by conceptual mistakes, as discussed below. The scope of this section is to address both misconceptions about these measurements and misinterpretation of the results.

3.1 Current enzyme assays determine potential enzyme activities

Burns (1978b) first pointed out that the standardized protocols measure potential rather than actual enzyme activities because of the conditions used in assays, including the use of synthetic substrates that do not occur *in situ*. This has been repeated in several reviews (Gianfreda and Ruggiero, 2006; Nannipieri et al., 2002, 2012; Tabatabai, 1994) and is generally taken into consideration. However, new methods of determining enzyme activities have been developed since Burns' comments, including high-throughput fluorimetric and colorimetric assays. These also use synthetic substrates and are performed under optimal conditions, and therefore they also measure potential activities (Dick, 2011b). Although it is generally accepted that measurement of potential activity is ecologically relevant and, as indicated above, recommended, the activities should ideally be measured under conditions that reflect the enzyme activity that exists *in situ* (Dick, 2011b). The use of water or buffer at the soil pH, simulating *in situ* conditions, has been proposed to make enzyme activities more realistic (Kandeler and Gerber, 1988; Sinsabaugh et al., 2000; Taylor et al., 2002). However, as a basic principle of biochemistry, it is important to conduct enzyme assays at optimal pH values, to yield the maximum potential soil enzyme activity, and also under reproducible conditions, to enable comparison of enzyme activities determined in different studies (Burns, 1978b; Dick, 2011b). Conducting enzyme assays at both optimal pH and the soil pH may be useful to enable evaluation of the effect of soil pH on the tested enzyme activity and also to obtain values that are comparable with previously reported values. On the other hand, even when measurements are conducted at soil pH, some of the other conditions under which the test is carried out will still be unrealistic; for example, it is not always possible to use natural substrates, and it is sometimes only possible to measure the reaction product reliably by using artificial substrates.

Although the concept of potential enzyme activity, measured by currently available enzyme assays, is well established, the meaning of these enzyme activities is often overlooked, despite the extensive literature published in the 1970s, 1980 and 1990s.

3.2 Misuse of the term extracellular enzyme activity

Burns (1978b) discussed the problems associated with assessing the contribution of extracellular enzyme activity to the overall measured activity in soil. By definition, extracellular enzyme activity is the activity caused by an enzyme acting in the extracellular environment, but spatially and temporally separated from its cellular source (Burns, 1978b). Many researchers assume that extracellular enzyme activity can be determined with available enzyme assays because the short incubation time (maximum of a few hours) is not sufficient to determine the intracellular enzyme activity (i.e. for the uptake of the substrate and its transformation by the intracellular enzyme activity in the product, which is finally released into the extracellular medium). This assumption holds true for high molecular weight substrates, such as MUF-bound substrates. However, it is not valid for assays using small molecular weight substrates, such as urea. These hypotheses must be tested scientifically. With currently available methods, it is not possible to differentiate the type of enzyme activity being measured. The distinction between truly extracellular enzyme activity and intracellular enzyme activity is complicated by the following: (1) microorganisms possess enzymes in the periplasmic space, attached to the external wall surface and which thus act in the extracellular environment and (2) microorganisms release viscous polymers that remain attached to the cell and include active enzymes that operate in the extracellular environment (Burns, 1978b). In addition, dead cells contain active enzymes related to the activity of the original cell (Renella et al., 2002). Finally, enzymes released into the soil by microorganisms or plants can be quickly denatured or degraded by soil proteases (Burns, 1982). It is therefore difficult to detect free extracellular enzymes in the soil, as they are very ephemeral. However, a portion of the enzymes released can be adsorbed onto humic compounds and clay colloids in the soil: these are referred to as immobilized enzymes. Binding to soil colloids often confers the enzymes some degree of protection and can allow them to remain active for long periods of time (Burns, 1982, and other references). It is therefore generally assumed that the enzyme activities measured represent both the extracellular enzyme activity (free and/or immobilized) and also the state of the activity of microbial origin at the time of soil sampling. In conclusion, as it is not possible to differentiate the type of activity being measured, the term "extracellular" should be not used for the activity measured by the available assays, unless the action can be scientifically demonstrated. New methods are required that can reliably differentiate which type of activity is being measured.

3.3 It is conceptually wrong to assume that a few enzyme activities can be used as proxies for complex pathways represented by C-, N- or P-acquiring processes

Skujins (1978) suggested that individual soil enzyme measurements do not reflect either the total nutrient status or answer questions regarding specific decomposition processes in the soil or questions about specific nutrient cycles, because they are substrate specific. Nonetheless, the concept that some enzyme activities can be considered proxies for C-, N- or P-acquiring processes was later introduced. Thus, it was hypothesized that extracellular enzymes release assimilable C, N and P and that the relative abundance of each of these enzymes reflects the stoichiometry of microbial biomass (Sinsabaugh et al., 2009). The term "Ecoenzyme" was also introduced to include all enzymes located outside the cell membranes (Sinsabaugh and Follstad Shah, 2012). The term "acquiring" when applied to any nutrient is a process that involves many factors. First, the process involves a huge number of enzymes, most of which cannot be determined in soil. For example, leucine aminopeptidase (LAP) is one of the most commonly determined enzyme activities, together with *N*-acetyl-β-glucosaminidase (NAG), as a proxy for N-acquiring processes. This enzyme catalyses the hydrolysis of leucine residues at the N-terminus of peptides and proteins, and it is therefore one of the several exopeptidases present in microbial cells. If we assume that protein-N is the main organic N in soil, mineralization of this N will involve both exopeptidases and endopeptidases, and when the amino acid is produced, amino acid oxidases and amino acid dehydrogenases will catalyse reactions releasing ammonium, which is taken up by cells (Nannipieri et al., 2012). If all of the organic N is protein-N and peptide-N, a huge number of enzymes will be involved in the N mineralization process. There are no accurate assays for measuring all soil enzymes that release ammonium. Another shortcoming relative to the use of the term "acquiring" is that the uptake of any nutrient by cells is not mediated by enzymes, but by transporters or through ion channels, both of which are regulated by mechanisms other than those involved in enzyme synthesis and activity. In enzyme reactions, the substrates are transformed into products, whereas in the uptake processes, the target molecule enters the cell and is not transformed into another molecule. The assumption that one or two enzyme activities can be considered proxies for C-, N- or P-acquiring processes is therefore a misconception.

4 Enzyme assays used with other assays for assessing soil health

4.1 What is soil health?

Although discussed in several chapters of this book, we think that it is important to summarize the evolution of the term soil health, which was coined in the 1970s, but gained greater importance in the 1990s (Bünemann et al., 2018) and replaced the use of the term soil quality (Powlson, 2020). There was then an exponential increase in publications defining the term and proposing indicators or evaluation methods. Perhaps one of the most widely accepted definitions is that proposed by Karlen et al. (1997): "Soil quality is the ability of a specific soil type to function, within natural or managed ecosystem boundaries, to sustain plant and animal productivity, maintain or improve water and air quality, and support human health and habitation." Other important definitions include those suggested by Lehmann et al. (2020), i.e. "Soil health is the continued capacity of soil to function as a vital living ecosystem that sustains plants, animals and humans, and connects agricultural and soil science to policy, stakeholder needs and sustainable supply-chain management", and by Kopittke et al. (2023), who stated that "a 'healthy soil' is one that provides the multifunctionality that underpins planetary and human health". Some authors differentiate soil quality and soil health by pointing out that *quality* refers to the suitability of the soil to perform a specific function, while *health* is more holistic and refers to the general state of the soil, encompassing biological aspects (Doran and Safley, 1997; Doran, 2002; Powlson, 2020). However, both terms are generally used as synonyms and interchangeably. The distinction between soil quality and soil health has shifted from a matter of principle to a matter of preference, and the two terms are therefore considered equivalent (Doran and Parkin, 1994; Bünemann et al., 2018; Powlson, 2020). However, using the term soil health is valuable in communication, as the word "health" seems to resonate more strongly than "quality" among non-specialists (including policymakers) and with the unspoken idea that soil may have health attributes that parallel those of human health (Powlson 2020). In this chapter, we will therefore refer to soil health and soil quality as synonymous and interchangeably.

4.2 Enzyme as indicators of soil health

Despite the abundant literature on soil quality, studies attempting to determine reliable methods of assessing soil quality are relatively scarce. This is at least partly due to the difficulty in developing quantitative indices that reflect changes in soil quality (Nortcliff, 2002) as many changes occur over the long term, so that changes in soil quality can only be perceived when the effects are combined over a period of time. Physical, chemical, biochemical and biological properties are all important for the correct functioning of soil (Arshad and Coen, 1992; Parr et al., 1992; Trasar-Cepeda et al., 1998, 2008), but characterization of soil quality should focus on the properties most sensitive to environmental stress, i.e. mainly biological and biochemical properties (Dick and Gupta, 1994; Vanhala and Ahtiainen, 1994). These properties are usually considered the most appropriate for the purpose of assessing soil quality (Fig. 1; Pankhurst et al., 1995; Yakovchenko et al., 1996; Elliott, 1994; Dawson et al., 2007). In particular, enzyme activities provide information on the potential capacity of the soil to carry out specific reactions and are fundamental in numerous biochemical processes, such as the decomposition of organic remains, nutrient cycling and many processes related to microbial activity (Frankenberger and Dick, 1983; García and Hernández, 2000; Ros et al., 2002; Leirós et al., 2000; Trasar-Cepeda et al., 1999). Furthermore, Kandeler et al. (1996) indicated that the composition of the microbial community determines the potential of that community for enzyme synthesis and, therefore, any modification of microbial diversity caused by environmental factors should be reflected in the level of enzyme activities in the soil. The dehydrogenase enzyme is perhaps one of the most commonly considered, as it is an intracellular enzyme that reflects the overall microbial activity and is related to the oxidation of organic compounds (Bastida et al., 2008; Camiña et al., 1998), being very sensitive to the toxic effects of many compounds (Bello et al., 2013; Dawson et al., 2007). Additionally, the activity of various hydrolase enzymes (phosphomonoesterase, β-glucosidase, urease and arylsulphatase) is often determined because these enzymes participate in nutrient cycling (C, N, P and S) and catalyse reactions occurring during the latter stages of decomposition of the organic residues (Trasar-Cepeda et al., 2007).

In a review paper, Gil-Sotres et al. (2005) considered the different approaches that have been used to assess how land use and/or management, pollution and other anthropogenic activities affect soil quality, by measuring biological and biochemical properties, mainly enzymatic activity, used individually or in combination as indicators. These researchers concluded that

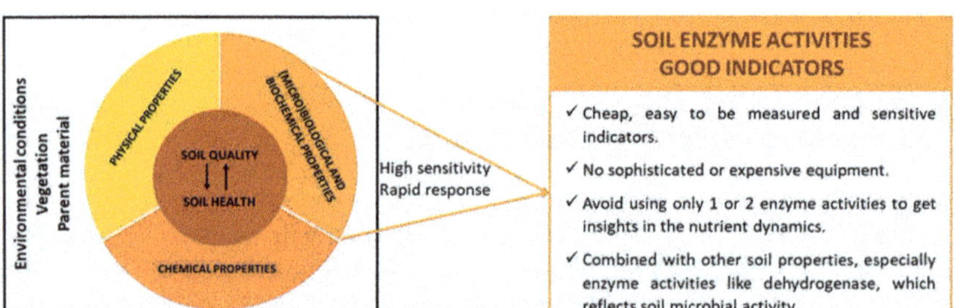

Figure 1 Schematic representation of key factors determining soil health and its best indicators.

(1) individual enzyme activities are of limited use as indicators of degradation: enzymes catalyse specific reactions in the soil and cannot be used as indicators of more complex functions such as total microbial activity, soil fertility or soil quality, which depend on other reactions and factors (Skujins, 1978; Nannipieri et al., 1990) and (2) enzyme activities are highly variable owing to different factors (climate, season, geographical location, etc.), and they are difficult to compare as they are often determined following very different protocols and are subjected to different types of pre-treatment (storage, preservation mode, pre-incubation, etc.) prior to the assay. However, the use of simple indexes, proposed by Dalal (1998), as a way of overcoming the limitations of individual soil enzymes as indicators of soil health, is also affected by the same limitations as assays of individual biochemical properties (Gil-Sotres et al., 2005; Bastida et al., 2008; Paz-Ferreiro et al., 2013). Potentially promising indices include the ratio of enzyme activity to C content or microbial biomass, the ratio of dehydrogenase to soluble carbon (metabolic ratio, Masciandaro et al., 1998) and another two indices not involving enzyme activities but closely related to soil microbial activity, i.e. the metabolic quotient (respiration to microbial biomass C ratio) and the microbial quotient (microbial biomass C to soil C ratio). However, none of these reflect changes in soil biochemical and biological properties (affecting soil functions) as a basis for estimating soil quality. It is also difficult to determine the meaning of these determinations, depending on several variables as in the case of the metabolic quotient, which depends on microbial stress and also on the ratio between bacterial and fungal biomass (Wardle and Ghani, 1995). Therefore, although those simple indices are easy to apply, they are not recommended as indicators of soil quality.

Soil quality can also be evaluated by using complex indices calculated by combinations of different soil biochemical properties, as first carried out in the 1980s with the indices proposed by Stefanic et al. (1984) and Beck (1984), although these were designed for soil fertility. Gil-Sotres et al. (2005) discussed a series of complex indices developed by different authors by using different approaches. These included the lignocellulosic index (Sinsabaugh et al., 1994), the decomposition factor, (Monreal and Bergstrom, 2000), an index in which only two enzyme activities involved in litter decomposition were considered (Fioretto et al., 2001), and an index developed after verification that the biochemical equilibrium is altered in soils developed under climax vegetation and subjected to some type of disturbance (Trasar-Cepeda et al., 1998). Other authors have also evaluated the use of complex indices of soil quality involving the use of enzyme activities (Bastida et al., 2008; Paz-Ferreiro and Fu, 2013). These authors concluded that determining enzyme activities and other biochemical properties reflects the complexity of the edaphic system, and complex expressions seem appropriate for evaluating soil quality, at least for the conditions under which they have been designed. However, they also point out that the main drawback of these complex indices is that they have not been tested in locations or under conditions other than those for which they were designed, and there is no way to verify whether or not they are universally valid. In this regard, Paz-Ferrerio and Fu (2013) pointed out that it is unrealistic to expect the same soil quality index to be applicable in areas affected by highly divergent climatic parameters, such as precipitation. For example, they highlight that small rainfall events in arid and semi-arid regions can trigger enzymatic and microbiological processes of great magnitude (as observed by Miralles et al., 2013). They also point out that climatic conditions, differences in the composition of the vegetation cover, or in soil type can limit the applicability of complex indices developed for natural ecosystems under varying conditions. They emphasize that for several soil orders (such as Gelisols, Vertisols and Spodosols), biochemical equilibria have not yet been studied, and complex indices have not been developed, despite the fact that the existence of some of these soils (Gelisols) may be threatened by climate change. Enzyme activities are known to be very sensitive to changes due to soil management practices, climatic variations, contamination, etc., and they are theoretically highly suitable for establishing quality indices for both agricultural and non-agricultural soils affected by various human activities (indeed, many soil quality indices include some enzymatic activities). However, the currently available information does not enable establishment of widely applicable soil quality indicators based on soil enzyme activities.

In conclusion, despite the efforts of many researchers and the fact that different studies and different complex indexes using different enzymatic activities have been developed and published in the last 20 years (Hinojosa et al., 2004; Bastida et al., 2006; Armas et al., 2007; Paz-Ferreiro et al., 2007; Zornoza et al., 2007; Puglisi et al., 2008; Aponte et al., 2020, 2024), enzyme assays cannot yet be considered effective diagnostic tools. This is because neither the individual properties nor the indexes obtained can be considered of general use, as those properties or indexes that appear to be valid in some situations are not at all effective in others. Further research is required to advance in the development of a general expression for the evaluation of soil quality and imaginative, coordinated work is required at the international level.

5 Perspectives

Despite the extensive literature on bibliography, there remain several issues, which should be addressed in future research by (1) establishing methods for the many enzymes whose activities cannot be determined in soil but are important for gaining insights into nutrient cycling; (2) avoiding conceptual mistakes and applying methods correctly, which especially important for assessment of soil quality; (3) applying imaginative approaches, which also visualize the location of the target enzyme activity in the soil matrix; and (4) combining measurement of enzyme activities with molecular techniques especially for assessing the origin of the enzyme activity.

5.1 Methods for determining important enzyme activities which cannot be determined in soil

As mentioned, accurate assays are only available for a few soil enzyme activities. Thus, Dick (2011a) reported only 35 bench-scale assays established according to the classical protocol proposed by Tabatabai and co-workers. In many studies, only one or two enzyme activities are measured, and this is not sufficient to reach firm conclusions about the type of variation experienced by the relative metabolic process, as the behaviour of a single enzyme does not necessarily reflect the responses of the other enzymes involved in the target process. For example, the mineralization of organic N to ammonium is considered important to increase N availability to soil microbes, and it can be assumed that organic N is mainly present in soil as protein-N and peptide-N. However, although casein hydrolysis and the activity of a few exopeptidases can be determined accurately, new methods must be developed to measure the activity of the various exopeptidase and endopeptidases involved in the hydrolysis of protein-N and amino acid-N and amino oxidase and amino dehydrogenase activities releasing ammonium from amino acids. Monitoring the sequence of enzyme activities involved in a process such as organic N mineralization can provide insights into the reaction limiting the rate of the process. On the other hand, the activities of phosphodiesterase and phosphomonoesterase, which act sequentially in soil, with the former hydrolysing phosphodiesters and the latter hydrolysing phosphomonoesters, can also be accurately determined. The activity of both enzymes can be measured in order to determine organic P mineralization, because phosphodiesters and phosphomonoesters are important components of organic P in soil (Nannipieri et al., 2011). In biochemistry, phosphodiesterases and phosphomonoesterases are included in the phosphatase class, whereas in soil enzymology the term phosphatase is commonly used to refer to phosphomonoesterases (Tabatabai, 1994).

In conclusion, the scientific community should devote further efforts to developing new enzyme assays to determine the many enzyme activities involved in important soil processes. Rather than focusing on whether the assays are bench-scale or microplate-scale, the most important point is to follow the protocol of Tabatabai and co-workers, i.e. to (1) test an adequate number of soils; (2) choose whether to determine formation of the product or substrate disappearance; (3) select the optimal pH and temperature; (4) use an appropriate size of soil sample; (5) maintain zero-order kinetics;

(6) avoid microbial growth by using short incubation periods; and (7) construct appropriate standard calibration curves.

5.2 Enzyme activities as soil health indicators

Future research should aim to develop soil quality indicators by developing indices (including enzyme activity) that can describe soil complexity more accurately than at present, while complying with several requirements. According to Dalal (1998), a good indicator should (1) measure one or more soil functions; (2) be sufficiently sensitive to reflect changes due to alteration, restoration or management; (3) provide reference values, critical levels or thresholds; (4) be easily interpretable; and (5) be easy and cost-effective to obtain. Some preconceived notions that are often applied in developing indices of the quality of agricultural soils should also be avoided. For example, it is often generally assumed that the productive function of soil and modification of this as a consequence of land management (e.g. tilling vs. no-tilling, organic vs. inorganic fertilization) are the most important factors to be taken into account. This has led to the identification of indicators that simply confirm what researchers had already decided *a priori*, i.e. classification of different soils already considered of good/poor quality. In other words, the aim is to find an indicator (often based on just one or two enzymatic activities) that demonstrates what "we already know" *a priori*. They, therefore, aim to seek evidence that, for example, conventional practices are worse than ecological/organic farming, that tillage is worse than reduced tillage or no tillage, or that organic fertilization is better than inorganic fertilization. This is largely due to the use of inaccurate reference levels or even not considering reference levels. When comparing two management practices, an additional, external reference must be used (rather than one of the management types). In such cases, any possible decision will almost certainly be biased by our "*a priori* knowledge". We agree with Kopittke et al. (2023), who suggest that when determining soil health, the reduction in multifunctional capacity of a soil impacted by human activities must be compared with that of a similar, natural soil, which will represent the reference state (citing Román Dobarco et al., 2021). They note that although the multifunctional capacity may be inherently lower in some soils than in others (e.g. a sandy soil in an arid climate compared to a clay soil in a humid environment), this does not imply that certain natural soils are inherently healthier than others. This perspective is consistent with the idea of Gil-Sotres et al. (2005), who proposed that soils developed under climax vegetation, representing the optimal vegetation for a given area (based on specific climatic conditions and parent material) and therefore exhibiting the highest level of multifunctional capacity achievable under the specific conditions, should be used as reference soils. However, one of the problems highlighted regarding the use of climax soils or natural soils not affected by anthropogenic influences is that many of these soils have disappeared.

Another problem related to the use of soil quality indicators is that researchers often assume that "more is better". When enzymatic activity is used to assess soil quality (typically focusing on a single enzyme or several enzymes considered individually), it is often assumed that higher activity indicates better soil quality. Agricultural practices or methods of rehabilitating degraded soils that increase enzyme activity are thus assumed to be the most appropriate and to improve soil quality or health. However, this notion is conceptually erroneous. Furthermore, as already discussed, the lack of standardized analytical methods applied in all laboratories is a fundamental problem when interpreting enzyme activity. Differences in sample collection, storage and pre-treatment protocols for determination of enzyme activity (where temperature, substrate concentration, incubation time, etc. are crucial) make it very difficult to compare the data obtained in different laboratories. The high degree of variability in enzyme activity, due to both seasonal and edaphic factors, as well as the lack of reference values or comprehensive databases for high quality soils that can be used for comparison, must also be taken into account. The following must also be considered possible explanations for the lack of standard patterns of enzyme activities in response to changes in land use and management or contamination (among many other factors that modify soil quality): (1) the small number of soils usually analysed in each study, which may not be sufficient to overcome statistical uncertainty (given the high degree of spatial and temporal variability) and (2) the small number of enzymes usually considered. Thus, different researchers considering the effects of anthropogenic activity (pollution, intensive land use, land management systems, forest fires, climate change, etc.) on soil quality often obtain contradictory results.

5.3 Visualization of enzyme activity in the soil matrix

Visualization of enzyme activity in the soil matrix is important. This approach can localize the activities associated with abiontic and intracellular enzymes and can also provide some insights into their contribution to soil organic matter transformation, their response to nutrient deficiencies and changes in climatic conditions and their relation to the presence of roots, microbiome and fauna in the soil matrix. Phosphomonoesterase activity associated with a small (7 × 20 nm) microbial membrane fragment was observed in soil by scanning electron microscopy (SEM)-based ultra cytochemical observation of soil preparations (Ladd et al., 1996). Unfortunately, extracellular enzymes immobilized or entrapped by organic-mineral complexes in soil were not detected because this method can only visualize active enzymes in electron transparent materials and thus cannot be used to examine electron-dense (mineral components of soil) or OsO_4 reactive materials (humified organic matter) (Ladd et al., 1996).

The application of soil zymography has enabled localization of the activities of some hydrolases (β-glucosidase, cellobiohydrolase, xylanase, cellulase, N-acetyl-glucosaminidase, leucine aminopeptidase, acid and alkaline phosphomonoesterases and arylsulfatase) and oxidases (phenol oxidase and peroxidase) in soil. This non-destructive technique visualizes the involvement of enzyme activity in substrate conversion by fluorescence or colorimetry, with the former being more commonly used than the latter (see extensive bibliography in the review by Bylera and Kuzyakov, 2024). Among the fluorescent substrates, 4-methylumbelliferone (MUF) is usually preferred to 7-amino-methylcoumarin (AMC) for saturating membranes, which are attached to the soil surface, to enable reaction of the enzyme with the substrate. After removal of the membrane, the fluorescence is usually photographed under UV light, and the light intensity in grey values is calibrated to the enzyme activity by MUF or AMC standards. During the incubation, the substrate diffuses from the membrane to soil and the product makes the opposite movement. This non-destructive technique enables the temporal dynamics of the enzyme activity to be monitored, different enzyme activities to be determined and images of enzyme activities to be combined with those of other soil parameters, such as soil respiration and C and N soil dynamics. The different enzyme activities should be determined by starting with the lowest levels of activity, because higher enzyme activities can leave MUF residues and lead to overestimation of the lower enzyme activities (Bylera and Kuzyakov, 2024). However, problems associated with zymography include quenching and the lack of contact between the membrane and soil particles. Different soil compounds (organic matter components, salts, heavy metals, etc.) can induce quenching, and it has been suggested that the amount of MUF should be increased to ensure saturation of the substrate (Bylera and Kuzyakov, 2024) and that MUF calibration curves should be prepared for each soil (Deng et al., 2017). In addition, zymography does not distinguish abiontic and intracellular enzyme activities, and the resolution is low and dependent on the method used, with resolution of direct zymography in the range 15–38 μm and that of in-gel zymography in the range 65–200 μm (Bylera and Kuzyakov, 2024).

Visualization of the different locations of the enzymes contributing to the overall activity measured is an important knowledge gap. Addressing this gap could yield insights into the quantitative contribution, and thus the importance of the activity associated with enzyme in dead cells, resting cells, cell remains and active cells, or with free extracellular enzyme or extracellular enzyme stabilized by surface-reactive particles. The same samples can be analysed by zymography at different times to identify when peak enzyme activity occurs;

however, the resolution of this method is lower than that of high-resolution imaging mass spectrometry (NanoSIMS), transmission electron microscopy (TEM) and SEM. However, the latter methods require destruction of the samples without time-resolved analysis, and the related microscale observations may not be representative of larger scales processes (Vedere et al., 2022). Nonetheless, combining observation methods at different scales can overcome this problem, as shown by Schlüter et al. (2019), who described the presence of bacteria in pores of diameter less than 10 μm and in macropores close to SOM, by the sequential application of X-ray tomography (at the centimetre scale), optical microscopy and fluorescence (at the millimetre scale), electron microscopy (at the scale of hundreds of micrometres) and NanoSIMS (at the scale of tens of micrometres).

5.4 Combining measurements of enzyme activity with detection of genes encoding the related enzyme: an approach to obtain insights into the origin of the measured enzyme activity

The use of molecular techniques has enabled the question of the origin of enzymes contributing to the measured enzyme activity to be addressed. However, study of the relationships between the overall diversity of soil microbiome and the target enzyme activity is complicated, because many of the microbial species bearing the enzyme encoding gene are rare (Wei et al., 2019). The relationship between the enzyme-encoding gene and the relative enzyme activity can provide some insights into the enzyme origin. Indeed, the use of molecular techniques represents an advancement in soil enzymology as both the abundance and diversity of the microbiome contributing to the enzyme activity can be measured. The milestone paper by Metcalfe et al. (2002) showed that chitinase activity in a brown forest soil increased on addition of sludge but that the diversity of the actinobacterium-like chitinase sequences contributing to the determined enzyme activity decreased. Several primers have been used to detect enzyme-encoding genes and to relate their abundance to the relative soil enzyme activity (Table 1). The number of sub and npr (encoding neutral metalloproteases) genes was positively correlated with casein-hydrolysing activities in a sandy soil but not in a clay soil, probably because the enzyme activity in the latter soil mainly depended on extracellular proteases adsorbed by clays (Mrkonjic Fuka et al., 2008). The N-benzyloxy-carbonyl-L-phenylalanyl L-leucine (ZFL)-hydrolysing activities in soil were found to depend on the composition of both npr and apr (encoding neutral metalloproteases) bacterial communities in rhizosphere and bulk soil fertilized with organic or inorganic fertilizers (Sakurai et al. 2007). The activity of both N-benzoyl-L-argininamide (BAA) hydrolase (a

Table 1 Some enzyme-encoding genes and the relative enzyme activity

Gene	Enzyme activity	Reference
*ure*C	Urease activity	Singh et al. (2009)
*pho*D	Alkaline phosphomonoesterase activity	Tan et al. (2013)
npr (neutral metallo-peptidases)	Casein-hydrolysing activity	Bach et al. (2001)
apr (alkaline metallo-peptidases)	Casein-hydrolysing activity	Bach et al. (2001)
Sub (serine peptidases)	N-benzoyl-L-argininamide (BAA)-hydrolysing activity	Bach et al. (2001)
*bgl*A, *bgl*B, and Abg	ß-glucosidase activity	Cañizares et al. (2011)
cbhI	Cellulase activity	Baldrian et al. (2012)
*chi*A	Chitinase activity	Johnson et al. (2014)

trypsin-like protease) and casein-hydrolase (a less specific protease activity) was higher in the rhizosphere of two inbred maize lines with high than low nitrogen use efficiency; the same was found for the abundance and diversity of bacterial communities bearing alkaline metallopeptidase (apr) and neutral metallopeptidases (npr) genes (Baraniya et al., 2016). A single primer is usually used, as in the case of *pho*D, which is ubiquitous through the bacterial kingdom (Gómez and Ingram, 1995), and predominates over *pho*A and *pho*X for detecting alkaline phosphomonoesterase-encoding genes in the soil bacterial community (Wei et al., 2019). However, more than one primer may be required for some enzymes, as in the case of acid phosphomonoesterase, with three classes (A, B and C) lacking specificity and therefore indicated as non-specific acid phosphomonoesterase (NSAP) (Neal et al., 2018). However, correct interpretation of the relationship between the measured enzyme activity and the relative enzyme-encoding gene should consider the following: (1) The primer used may not represent all microbes bearing the relative enzyme encoding genes. It is reasonable to hypothesize that some of the microbes and probably unknown species may synthesize enzyme molecules different from that universally known, whose encoding gene is known at least for a part of the enzyme. The primer usually covers a gene encoding a part of the enzyme. For example, *ure*C, commonly used to detect ureolytic microorganisms, encodes the alpha subunit of urease (Reed, 2001); (2) detection of the enzyme-encoding gene does not necessarily indicate that the gene is active, and studying the gene expression is therefore important. The best approach is to detect the expressed enzyme, as reported by Johnson-Rollings et al. (2014), who showed that chitinolytic activity was probably due to *Nocardiopsis chi*A-like chitinase in K_2SO_4 soil extracts. However, considering the present difficulties in soil proteomics, such as the inability to extract expressed proteins (Nannipieri et al., 2025), determination of the expressed mRNA is the best approach to determine whether the enzyme-encoding gene is active.

6 Conclusion

The literature on soil enzymology is extensive, as enzyme activities are easy to measure and can potentially indicate the capacity of soil to carry out specific functions. However, although widely discussed, the shortcomings of these methods are often overlooked. First, it is frequently assumed that currently available enzyme assays determine extracellular enzyme activity, although this has never actually been proved. Second, one or two enzyme activities are used as proxies for C-, N- and P-acquiring processes, even though different enzymes catalyse specific reactions, the metabolic processes involving nutrient dynamics involve many different enzymes (most of which cannot be determined in soil because of the lack of accurate assays) and nutrient take up by microbial cells involves transporters or ion channels and not enzyme activities. Enzyme activity can be included, along with other soil properties, in indices reflecting soil quality/soil health (Fig. 1), although the indices should be tested in locations or under conditions other than those for which they were designed so that they can be generalized. Accurate assays are required to determine the activities of many important enzymes, such as nuclease. Future research should also consider both enzyme activity and expression of the relative encoding genes, in order to gain insights into the origin of the enzyme activity measured thereby improving interpretation of the role of the target enzyme in the soil microbial functions.

7 References

Alef, K. and Nannipieri, P., 1995. *Methods in Applied Soil Microbiology and Biochemistry*, Academic Press, London. https://doi.org/10.1016/B978-0-12-513840-6.X5014-9

Aponte, H., et al. (2020), 'Soil quality indices for metal(loid) contamination: an enzymatic perspective', *Land Degr Develop*, 31 (17), 2700–2719. https://doi.org/10.1002/ldr.3630

Aponte, H., et al. (2024), 'Soil quality indices based on biological properties to assess chemical soil degradation by metal(loid)s', *Soil Sediment Contam*, 33 (1), 1–23. https://doi.org/10.1080/15320383.2024.2358426

Armas, C.M., et al. (2007), 'A biological quality index for volcanic Andisols and Aridisols (Canary Islands, Spain): variations related to the ecosystem development', *Sci Total Environ*, 378 (1–2), 238–244. https://doi.org/10.1016/j.scitotenv.2007.01.053

Arshad, M.A. and Coen, G.M. (1992), 'Characterization of soil quality: physical and chemical criteria', *Am J Altern Agric*, 7 (1–2), 25–31. https://doi.org/10.1017/S0889189300004410

Bach, H.J., et al. (2001), 'PCR primers and functional probes for amplification and detection of bacterial genes for extracellular peptidases in single strains and in soil', *J Microbiol Meth*, 44 (2), 173–182. https://doi.org/10.1016/S0167-7012(00)00239-6

Baldrian, P., et al. (2012), 'Response of the extracellular enzyme activity in hardwood forest to soil temperature and seasonality and the potential effects of climate change', *Soil Biol Biochem*, 56, 60–68. https://doi.org/10.1016/j.soilbio.2012.01.020

Baraniya, D., et al. (2016), 'Protease encoding microbial communities and protease activity of the rhizosphere and bulk soils of two maize lines with different N uptake efficiency', *Soil Biol Biochem*, 96, 176–179. https://doi.org/10.1016/j.soilbio.2016.02.001

Bastida, F., et al. (2006), 'Microbiological degradation index of soils in a semiarid climate', *Soil Biol Biochem*, 38 (12), 3463–3473. https://doi.org/10.1016/j.soilbio.2006.06.001

Bastida, F., et al. (2008), 'Past, present and future of soil quality indices: a biological perspective', *Geoderma*, 147 (3–4), 159–171. https://doi.org/10.1016/j.geoderma.2008.08.007

Beck, T. (1984), *Methods and applications of soil microbiological analyses at the Landsensanstalt für Bodenkultur und Pfanzenban (LBB) in Munich for the determination of some aspects of soil fertility,* in: Nemes, M.P., Kiss, S., Papacostea, P., Stefanic, G., Rusan, M. (Eds.), Fifth Symposium on Soil Biology. Roman National Society of Soil Science, Bucharest, pp. 13–20

Bello, D., et al. (2013), 'Modification of enzymatic activity in soils of contrasting pH contaminated with 2,4-dichlorophenol and 2,4,5-trichlorophenol', *Soil Biol Biochem*, 56, 80–86. https://doi.org/10.1016/j.soilbio.2012.02.011

Burns, R.G. (1978a), *Soil Enzymes*, Academic Press, New York.

Burns, R.G. (1978b), 'Enzyme activity in soil: some theoretical and practical considerations'. In: R.G. Burns (Ed.), *Soil Enzymes*, Academic Press, New York, pp. 295–340.

Burns, R.G. (1982), 'Enzyme activity in soil: location and a possible role in microbial ecology', *Soil Biol Biochem*, 14 (5), 423–427. https://doi.org/10.1016/0038-0717(82)90099-2

Burns, R.G. and Dick, R.P. (2002), *Enzymes in the Environment. Activity, Ecology and Applications*, Marcel Dekker, New York.

Bünemann, E.K., et al. (2018), 'Soil quality–a critical review', *Soil Biol Biochem*, 120, 105–125. https://doi.org/10.1016/j.soilbio.2018.01.030

Bylera, N. and Kuzyakov, Y. (2024), 'Soil zymography: a decade of rapid development of microbial hotspot imaging', *Soil Biol Biochem*, 189, 109264. https://doi.org/10.1016/j.soilbio.2023.109264

Camiña, F., et al. (1998), 'Measurement of dehydrogenase activity in acid soils rich in organic matter', *Soil Biol Biochem*, 30 (8–9), 1005–1011. https://doi.org/10.1016/S0038-0717(98)00010-8

Cañizares, R., et al. (2011), 'Molecular analyses of β-glucosidase diversity and function in soil', *Eur J Soil Biol*, 47, 1–8. https://doi.org/10.1016/j.ejsobi.2010.11.002

Dalal, R.C. (1998), 'Soil microbial biomass what do the numbers really mean?', *Aust J Exp Agric*, 38 (7), 649–665. https://doi.org/10.1071/EA97142

Dawson, J.J.C., et al. (2007), 'Application of biological indicators to assess recovery of hydrocarbon impacted soils', *Soil Biol Biochem*, 39 (1), 164–177. https://doi.org/10.1016/j.soilbio.2006.06.020

Deng, S., et al. (2011), 'Microplate fluorimetric assay of soil enzymes'. In: R.P. Dick (Ed.), *Methods of Soil Enzymology, vol 9*. Soil Science Society of America, Madison, WI, pp. 311–318.

Deng, S., et al. (2013), 'Bench scale and microplate format assay of soil enzyme activities using spectroscopic and fluorometric approaches', *Appl Soil Ecol*, 64, 84–90. https://doi.org/10.1016/j.apsoil.2012.11.002

Deng, S., et al. (2017), 'Comparison and standardization of soil enzyme assay for meaningful data interpretation', *J Microbiol Methods*, 133, 32–34 https://doi.org/10.1016/j.mimet.2016.12.013

Dick, R.P. (2011a), *Methods of Soil Enzymology*, Soil Science Society of America, Madison, WI.

Dick, W.A. (2011b), 'Development of a soil enzyme reaction assay'. In: R.P. Dick (Ed.), *Methods of Soil Enzymology*, Soil Science Society of America, Madison, WI, USA, pp. 71–84.

Dick, R.P. and Gupta, V.V.S.R. (1994), 'A conceptual model for the role of abiotic soil enzymes in microbial ecology: a potential analogue for soil quality'. In: C.E. Pankhurst, et al. (Eds.), *Soil Biota: Management in Sustainable Farming Systems*, CSIRO, Melbourne, pp. 167–168.

Dick, R.P., et al. (2018), 'Cross-laboratory comparison of fluorimetric microplate and colorimetric bench-scale soil enzyme assays', *Soil Biol Biochem*, 121, 240–248. https://doi.org/10.1016/j.soilbio.2017.12.020

Doran, J.W. (2002), 'Soils health and global sustainability: translating science into practice', *Agric Ecosyst Environ*, 88 (2), 119–127. https://doi.org/10.1016/S0167-8809(01)00246-8

Doran, J.W. and Parkin, T.B. (1994), 'Defining and assessing soil quality'. In: J.W. Doran, et al. (Eds.), *Defining Soil Quality for a Sustainable Environment*, Soil Science Society of America, Special Publication 35. SSSA-ASA, Madison, WI, pp. 3–21. http://dx.doi.org/10.2136/sssaspecpub35.c1

Doran, J.W. and Safley, M. (1997). 'Defining and assessing soils health and sustainable productivity'. In: C.E. Pankhurst, et al. (Eds.), *Biological Indicators of Soil Health*, CAB International, Wallingford, pp. 1–28.

Duxbury J.M. and Tate, R.L. III. (1981), 'The effect of soil depth and crop cover on enzymatic activities in Pahokee Muck', *Soil Sci Soc Am Proc*, 45 (2), 322–328. https://doi.org/10.2136/sssaj1981.03615995004500020019x

Elliott, E.T. (1994), 'The potential use of soil biotic activity as an indicator of productivity'. In: C.E. Pankhurst, et al. (Eds.), *Soil Biota: Management in Sustainable Farming Systems*, CSIRO, Melbourne, pp. 250–256.

Fioretto, A., et al. (2001), Decomposition of *Cistus incanus* leaf litter in a Mediterranean maquis ecosystem: mass loss, microbial enzyme activities and nutrient changes. *Soil Biol Biochem* 33 (3), 311–321. https://doi.org/10.1016/S0038-0717(00)00142-5

Frankenberger, W.T. and Dick, W.A. (1983), 'Relationship between enzyme activities and microbial growth and activity indices in soil', *Soil Sci Soc Am J*, 47 (5), 945–951. https://doi.org/10.2136/sssaj1983.03615995004700050021x

Freeman, C., et al. (1995),' The use of fluorogenic substrates for measuring enzyme activity in peatland', *Plant Soil*, 175 (1), 147–152. https://doi.org/10.1007/BF02413020

García, C., et al. (2003), *Técnicas de Análisis de Propiedades Bioquímicas es Suelos: Medida de Actividades Enzimáticas y Biomasa Microbiana*, MundiPrensa Editorial, Madrid, Spain.

García, C. and Hernández, T. (1997), Biological and biochemical indicators in derelict soils subject to erosion. *Soil Biol Biochem* 29 (2), 171–177. https://doi.org/10.1016/S0038-0717(96)00294-5

German, D.P., et al. (2011), 'Optimization of oxidative and hydrolytic enzyme methods for ecosystem studies', *Soil Biol Biochem*, 43 (7), 1387–1397. https://doi.org/10.1016/j.soilbio.2011.03.017

Gerritse, R.G. and van Dijk, H. (1978), 'Determination of phosphatase activities of soils and animal wastes', *Soil Biol Biochem*, 10 (6), 545–551. https://doi.org/10.1016/0038-0717(78)90051-2

Gianfreda, L. and Bollag, J.-M. (1996), 'Influence of natural and anthropogenic factors on enzyme activity in soil'. In: G. Stotzky and J.-M. Bollag (Eds.), *Soil Biochemistry, vol. 9.* Marcel Dekker, New York, pp. 123–193.

Gianfreda, L. and Ruggiero, P. (2006), 'Enzyme activities in soil'. In: P. Nannipieri and K. Smalla (Eds.), *Nucleic Acids and Proteins in Soil, vol 8*, Springer, Berlin, Germany, pp. 257–311.

Gil-Sotres, F., et al. (2005), 'Different approaches to evaluate soil quality using biochemical properties', *Soil Biol Biochem*, 37 (5), 877–887. https://doi.org/10.1016/j.soilbio.2004.10.003

Gómez, P.F. and Ingram, L.O. (1995), 'Cloning, sequencing and characterization of gene (phoD) from *Zymomonas mobilis*', *FEMS Microbiol Lett*, 125 (2–3), 237–245. https://doi.org/10.1111/j.1574-6968.1995.tb07364.x

Harrison, A.F. and Pearce, T. (1979), 'Seasonal variation of phosphatase activity in woodland soils', *Soil Biol Biochem*, 11 (4), 405–410. https://doi.org/10.1016/0038-0717(79)90054-3

Hinojosa, M.B., et al. (2004), 'Microbiological rates and enzyme activities as indicators of functionality in soils affected by the Aznalcóllar toxic spill'. *Soil Biol Biochem*, 36 (10), 1637–1644. https://doi.org/10.1016/j.soilbio.2004.07.006

Johnson-Rollings, A.S., et al. (2014), 'Exploring the functional soil-microbe interface and exoenzymes through soil metaexoproteomics', *ISME J*, 8, 2148–2150 https://doi.org/10.1038/ismej.2014.130

Kandeler E. and Gerber, H. (1988), 'Short-term assay of soil urease activity using colorimetric determination of ammonium', *Biol Fert Soils*, 6, 68–72. https://doi.org/10.1007/BF00257924

Kandeler, E., et al. (1996), 'Influence of heavy metals on the functional diversity of soil microbial communities', *Biol Fert Soils*, 23, 299–306. https://doi.org/10.1007/BF00335958

Karlen, D.L., et al. (1997), 'Soil quality: a concept, definition, and framework for evaluation', *Soil Sci Soc Am J*, 61 (1), 4–10. https://doi.org/10.2136/sssaj1997.03615995006100010001x

Kiss, S., et al. (1998), *Enzymology of Disturbed Soils*, Elsevier Science, Amsterdam.

Kopittke, P.M., et al. (2023), 'Healthy soil for healthy humans and a healthy planet', *Crit Rev Environ Sci Technol*, 54 (3), 210–221. https://doi.org/10.1080/10643389.2023.2228651

Kuprevich, V.F. and Shcherbakova, T.A. (1971), 'Comparative enzymatic activity in diverse types of soil'. In: A.D. McLaren and J. Skujins (Eds.), *Soil Biochemistry, vol 2*, Marcel Dekker, New York, pp. 167–201.

Ladd, J.N., et al. (1996), 'Soil structure and biological activity'. In: G. Stotzky and J.M. Bollag (Eds.), *Soil Biochemistry, vol 9*, Marcel Dekker, New York, pp. 23–78.

Lakowicz, J.R. (1983), *Principles of fluorescence spectroscopy*, Plenum Press, New York.

Leirós, M.C., et al. (2000), 'Biochemical properties of acid soils under oak vegetation (Atlantic oakwood) in an area of the European temperate-humid zone (Galicia, NW

Spain): general parameters', *Soil Biol Biochem*, 32 (6), 733–745. https://doi.org/10.1016/S0038-0717(99)00195-9

Lehmann, J., et al. (2020), 'The concept and future prospects of soil health', *Nat Rev Earth Environ*, 1, 544–553. https://doi.org/10.1038/s43017-020-0080-8

Malcom, R.E. (1983), 'Assessment of phosphatase activity in soils', *Soil Biol Biochem*, 15 (4), 403–408. https://doi.org/10.1016/0038-0717(83)90003-2

Marx, M.C., et al. (2001), 'A microplate fluorimetric assay for the study of enzyme diversity in soils', *Soil Biol Biochem*, 33 (12–13), 1633–1640. https://doi.org/10.1016/S0038-0717(01)00079-7

Masciandaro, G., et al. (1998), 'Organic matter properties in cultivated versus set-aside arable soils', *Agric Ecosyst Environ*, 67 (2–3), 267–274. https://doi.org/10.1016/S0167-8809(97)00124-2

Metcalfe, A.C., et al. (2002),'Molecular analysis of a bacterial chitinolytic community in an upland pasture', *Appl Environ Microbiol*, 68 (10), 5042–5050. https://doi.org/10.1128/AEM.68.10.5042-5050.2002

Miralles I., et al. (2014), 'Capacity of biological soil crusts colonized by the lichen *Diploschistes* to metabolize simple phenols', *Plant Soil*, 385, 229–240. https://doi.org/10.1007/s11104-014-2220-8

Monreal, C.M. and Bergstrom, D.W. (2000), Soil enzymatic factors expressing the influence of land use, tillage system and texture on soil biochemical quality. *Can J Soil Sci* 80 (3), 419–428. https://doi.org/10.4141/S99-088

Mrkonjic Fuka, M., et al. (2008), 'Factors influencing variability of proteolytic genes and activities in arable soils', *Soil Biol Biochem*, 40 (7), 1646–1653. https://doi.org/10.1016/j.soilbio.2008.01.028

Nannipieri, P., et al. (1990), 'Ecological significance of biological activity in soil'. In: G.M. Bollag and G. Stotzky (Eds.), *Soil Biochemistry, vol. 6*, Marcel Dekker, New York, pp. 293–355.

Nannipieri, P., et al. (2002). 'Enzyme activities and microbiological and biochemical processes in soil'. In: R.G. Burns and R.P. Dick (Eds.), *Enzymes in the Environment, Activity, Ecology, and Applications*, Marcel Dekker, New York, pp. 1–36. https://doi.org/10.1201/9780203904039

Nannipieri, P., et al. (2011), 'Role of phosphatase enzymes in soil'. In: E.K. Bünemann, et al. (Eds), *Phosphorus in Action, Biological Processes in Soil Phosphorus Cycling*, Springer, Heidelberg, Germany, pp. 215–253. https://doi.org/10.1007/978-3-642-15271-9

Nannipieri, P,. et al. (2012), 'Soil enzymology: classical and molecular approaches', *Biol Fert Soils*, 48, 743–762. https://doi.org/10.1007/s00374-012-0723-0

Nannipieri, P., et al. (2018), 'Soil enzyme activity: a brief history and biochemistry as a basis for appropriate interpretations and meta-analysis', *Biol Fert Soils*, 54, 11–19. https://doi.org/10.1007/s00374-017-1245-6

Nannipieri, P., et al. (2025), 'Advances in metaproteomics for analyzing soil microorganisms'. In K. Dunfield (Ed.), *Understanding and Utilising Soil Microbes for a More Sustainable Agriculture*, Burleigh Dodds Science Publishing, Cambridge, UK, under publication.

Neal, A.L., et al. (2018), 'Phylogenetic distribution, biogeography and the effects of land management upon bacterial non-specific acid phosphatase gene diversity and abundance', *Plant Soil*, 427, 175–189. https://doi.org/10.1007/s11104-017-3301-2

Nortcliff, S., (2002), Standardisation of soil quality attributes. *Agricult Ecosys Environ,* 88 (2), 161–168. https://doi.org/10.1016/S0167-8809(01)00253-5

Pankhurst, C.E., et al. (1995), 'Evaluation of soil biological properties as potential bioindicators of soil health', *Aust J Exp Agric*, 35, 1015–1028.

Parr, J.F., et al. (1992), 'Soil quality: attributes and relationship to alternative and sustainable agriculture', *Am J Altern Agric*, 7 (1–2), 5–11. https://doi.org/10.1017/S0889189300004367

Paz-Ferreiro J. and Fu S. (2013), 'Biological índices for soil quality evaluation: perspectives and limitations', *Land Degrad Dev*, 27 (1), 14–25. https://doi.org/10.1002/ldr.2262

Paz-Ferreiro J., et al. (2007), 'Biochemical properties of acid soils under native grassland in a temperate humid zone', *N Z J Agric Res*, 50 (4), 537–548. https://doi.org/10.1080/00288230709510321

Popova, I.E. and Deng, S. (2010), 'A high-throughput microplate assay for simultaneous colorimetric quantification of multiple enzyme activities in soil'. *Appl Soil Ecol*, 45 (3), 315–318. https://doi.org/10.1016/j.apsoil.2010.04.004

Powlson, D.S. (2020), 'Soil health–useful terminology for communication or meaningless concept? Or both?' *Front Agric Sci Eng*, 7 (3), 246–250 https://doi.org/10.15302/J-FASE-2020326

Puglisi, E., et al. (2006), 'Development and validation of numerical indexes integrating enzyme activities of soils', *Soil Biol Biochem*, 38 (7), 1673–1681. https://doi.org/10.1016/j.soilbio.2005.11.021

Quastel, J.H. (1946). *Soil Metabolism*. The Royal Institute of Chemistry of Great Britain and Ireland, London.

Reed, K.E. (2001), 'Restriction enzyme mapping of bacterial urease genes: Using degenerate primers to expand experimental outcomes', *Biochem Mol Biol Educ*, 29 (6), 239–244. https://doi.org/10.1111/j.1539-3429.2001.tb00131.x

Renella, G., et al. (2002), 'Hydrolase activities during and after chloroform fumigation of soil as affected by protease activity', *Soil Biol Biochem* 34 (1), 51–60. https://doi.org/10.1016/S0038-0717(01)00152-3

Román Dobarco, M., et al. (2021), 'A framework to assess changes in soil condition and capability over large areas', *Soil Secur*, 4, 100011. https://doi.org/10.1016/j.soisec.2021.100011

Ros, M., et al. (2003), Soil microbial activity after restoration of a semiarid soil by organic amendments. *Soil Biolo Biochem* 35(3), 463–469. https://doi.org/10.1016/S0038-0717(02)00298-5

Sakurai, M., et al. (2007), 'Analysis of bacterial communities in soil by PCR-DGGE targeting protease genes', *Soil Biol Biochem*, 39 (11), 2777–2784. https://doi.org/10.1016/j.soilbio.2007.05.026

Sarathchandra, S.U. and Perrott, K.W. (1981), 'Determination of phosphatase and arylsulphatase activities in soil', *Soil Biol Biochem*, 13 (6), 543–545. https://doi.org/10.1016/0038-0717(81)90050-X

Saratchandra, S.U., et al. (1984), 'Microbiological and biochemical characteristics of a range of New Zealand soils under stabilized pasture', *Soil Biol Biochem*, 16 (2), 177–183. https://doi.org/10.1016/0038-0717(84)90109-3

Saviozzi, A., et al. (2001), 'A comparison of soil quality in adjacent cultivated, forest and native grassland soils', *Plant Soil*, 233, 251–259. https://doi.org/10.1023/A:1010526209076

Schinner, F. and von Mersi, W. (1990), 'Xylanase-, CM-cellulase- and invertase activity in soil: an improved method', *Soil Biol Biochem*, 22 (4), 511–515. https://doi.org/10.1016/0038-0717(90)90187-5

Schlüter, S., et al. (2019), 'Correlative imaging reveals holistic view of soil microenvironments', *Environ Sci Technol*, 53 (2), 829–837. https://doi.org/10.1021/acs.est.8b05245

Singh, B.K., et al. (2009), 'Response of fungal, bacterial and ureolytic communities to synthetic sheep urine deposition in a grassland soil', *FEMS Microbiol Ecol*, 70 (1), 109–117. https://doi.org/10.1111/j.1574-6941.2009.00731.x

Sinsabaugh, R.L. and Follstad Shah, J.J. (2012), 'Ecoenzymatic stoichiometry and ecological theory', *Annu Rev Ecol Evol Syst*, 43, 313–343. https://doi.org/10.1146/annurev-ecolsys-071112-124414

Sinsabaugh, R.L., et al. (2000), 'Rapid assay for amidohydrolase (urease) activity in environmental samples', *Soil Biol Biochem*, 32 (14), 2095–2097. https://doi.org/10.1016/S0038-0717(00)00102-4

Sinsabaugh, R.L., et al. (2009), 'Ecoenzymatic stoichiometry of microbial organic nutrient acquisition in soil and sediment', *Nature* 462, 795–798. https://doi.org/10.1038/nature08632

Sinsabaugh, R.L., et al. (1994), The enzymatic basis of plant litter decomposition: emergence of an ecological process. *Appl Soil Ecol* 1 (2), 97–111. https://doi.org/10.1016/0929-1393(94)90030-2

Skujins, J, (1978), 'History of abiotic soil enzyme research'. In R.G. Burns (Ed.), *Soil Enzymes*, Academic Press, London, pp. 1–49.

Stefanic, G., et al. (1984), *Researches concerning a biological index of soil fertility*, in: Nemes, M.P., Kiss, S., Papacostea, P., Stefanic, G., Rusan, M. (Eds.), Fifth Symposium on Soil Biology. Roman National Society of Soil Science, Bucharest, pp. 35–45.

Tabatabai, M.A. (1994), 'Soil enzymes'. In R.W. Weave, et al. (Eds.), *Methods in Soil Analysis. Part 2. Microbiological and Biochemical Properties*, Soil Science Society of America, Madison, pp. 775–833.

Tan, H., et al. (2013), 'Long-term phosphorus fertilization increased the diversity of the total bacterial community and the phoD phosphorus mineralizer group in pasture soil', *Biol Fertil Soils*, 49 (6), 661–672. https://doi.org/10.1007/s00374-012-0755-5

Taylor, J.P., et al. (2002), 'Comparison of microbial numbers and enzymatic activities in surface soils and subsoils using various techniques', *Soil Biol Biochem*, 34 (3), 387–401. https://doi.org/10.1016/S0038-0717(01)00199-7

Trasar-Cepeda, C. and Gil-Sotres, F. (1988), 'Kinetics of acid phosphatase activity in various soils of Galicia (NW Spain)', *Soil Biol Biochem*, 20 (3), 275–280. https://doi.org/10.1016/0038-0717(88)90003-X

Trasar-Cepeda, C., et al. (1985), 'Determinación de la actividad fosfatasa en suelos gallegos. Precisiones al método de Saratchandra y Perrott?', *An Edafol Agrobiol*, 44, 987–991.

Trasar-Cepeda, C., et al. (1998), 'Towards a biochemical quality index for soils: an expression relating several biological and biochemical properties', *Biol Fertil Soils*, 26, 100–106. https://doi.org/10.1007/s003740050350

Trasar-Cepeda, C., et al. (1999), 'Biochemical properties of acid soils under climax vegetation (Atlantic oakwood) in an area of the European temperate–humid zone (Galicia, NW Spain): specific parameters', *Soil Biol Biochem*, 32 (6), 747–755. https://doi.org/10.1016/S0038-0717(99)00196-0

Trasar-Cepeda C., et al. (2007), 'Thermodynamic parameters of enzymes in grassland soils from Galicia, NW Spain', *Soil Biol Biochem*, 39 (1), 311–319. https://doi.org/10.1016/j.soilbio.2006.08.002

Trasar-Cepeda, C., et al. (2008), 'Hydrolytic enzyme activities in agricultural and forest soils. Some implications for their use as indicators of soil quality', *Soil Biol Biochem*, 40 (9), 2146–2155. https://doi.org/10.1016/j.soilbio.2008.03.015

Tscherko, D. and Kandeler, E. (1999), 'Classification and monitoring of soil microbial biomass, N-mineralization and enzyme activities to indicate environmental changes', *Bodenkultur*, 50 (4), 215–226. https://diebodenkultur.boku.ac.at/volltexte/band-50/heft-4/tscherko.pdf

Vaughan, D. and Malcom, R.E. (1984), 'Influence of assay conditions on invertase activity in different Scottish soils', *Plant Soil*, 80, 285–292. https://doi.org/10.1007/BF02161184

Vanhala, P.T. and Ahtiainen, J. (1994), 'Soil respiration, ATP content and photobacterium toxicity test as indicators of metal pollution in soil', *Environ Toxicol Water Qual*, 9 (2), 115–121. https://doi.org/10.1002/tox.2530090207

Vedere, C., et al. (2022), 'Opportunities and limits in imaging microorganisms and their activities in soil microhabitats', *Soil Biol Biochem*, 174, 108807. https://doi.org/10.1016/j.soilbio.2022.108807

Vidal, A., et al. (2019), 'Earthworm cast formation and development: a shift from plant litter to mineral associated organic matter', *Front Environ Sci*, 7, 55. https://doi.org/10.3389/fenvs.2019.00055

Visser, S. and Parkinson, D. (1992), 'Soil biological criteria as indicators of soil quality: soil microorganisms', *Am J Altern Agric*, 7 (1–2), 33–37. https://www.jstor.org/stable/44479648

Von Mersi, W. and Schinner, F. (1991), 'An improved and accurate method for determining the dehydrogenase activity of soils with iodonitrotetrazolium chloride'. *Biol Fertil Soils*, 11, 216–220. https://doi.org/10.1007/BF00335770

Vuorinen, A.H. (1993), 'Requirement of *p*-nitrophenol standard for each soil', *Soil Biol Biochem*, 25 (2), 295–296. https://doi.org/10.1016/0038-0717(93)90042-A

Wardle, D.A. and Ghani, A. (1995), 'A critique of the microbial metabolic quotient (qCO_2) as a bioindicator of disturbance and ecosystem development', *Soil Biol Biochem*, 27 (12), 1601–1610. https://doi.org/10.1016/0038-0717(95)00093-T

Wei, X., et al. (2019), 'Rare taxa of alkaline phosphomonoesterase-harboring microorganisms mediate soil phosphorus mineralization', *Soil Biol Biochem*, 131, 62–70. https://doi.org/10.1016/j.soilbio.2018.12.025

Witzgall, K., et al. (2021), 'Particulate organic matter as a functional soil component for persistent soil organic carbon', *Nat Commun*, 12, 4115. https://doi.org/10.1038/s41467-021-24192-8

Woods, A.F. (1899), 'The destruction of chlorophyll by oxidizing enzymes', *Zentralbl Bakteriol Parasitenkd Abt*, 5, 745–754.

Yakovchenko, V.I., et al. (1996). 'A biologically based indicator of soil quality', *Biol Fertil Soils*, 21, 245–251. https://doi.org/10.1007/BF00334899

Zornoza, R., et al. (2007), 'Evaluation of soil quality using multiple lineal regression based on physical, chemical and biochemical properties', *Science Tot Environ*, 378 (1–2), 233–237. https://doi.org/10.1016/j.scitotenv.2007.01.052

Chapter 6

Advances in stable isotope profiling (SIP) techniques to track plant-microbial interactions

Jennifer L. Kane and Kinsey M. Reed, West Virginia University, USA; Paul Dijkstra, Egbert Schwartz, Victor O. Leshyk and Bruce A. Hungate, Northern Arizona University, USA; Kirsten Hofmockel, Pacific Northwest National Laboratory, USA; and Debjani Sihi, North Carolina State University, USA

http://dx.doi.org/10.19103/AS.2025.0159.07

1 Introduction

Soil microorganisms play a central role in agriculture. They interact with plants, affecting productivity and crop yield. They defend against disease, protect against nutrient loss, and build soil carbon. Soil microbes also cause disease, promote nutrient loss, degrade soil carbon, and alter gas emissions, with aggregated effects large enough to influence climate.

We know microbes do all these things, yet we know far less about which specific microbial taxa do them and how fast, and how those taxa are impacted by changing environmental conditions. Given the high diversity of soil microorganisms, it makes sense that microbes vary in their likely impacts on agroecosystems. Quantifying microbial processes with taxonomic resolution is a pressing research frontier in agriculture with many potential applications toward sustainable agriculture, like developing effective biofertilizers and other agricultural outcomes,including quantitative metrics of ecosystem services.

By quantifying the incorporation of heavy isotopes into microbial DNA coupled with taxonomic identifications from DNA sequencing, quantitative stable isotope probing (qSIP) allows for taxon-specific activity measurements. This method greatly increases the resolution with which we can study microbial growth rates and soil biogeochemical cycles, critical needs in understanding the productivity and climate impacts of agricultural systems.

2 Quantitative stable isotope probing (qSIP)

Stable isotope probing (SIP) refers to tracing isotopically labeled atoms into molecules like nucleic acids, lipids, and proteins. As the name indicates, SIP techniques use stable isotopes as element tracers, which are powerful for measuring element flow in ecosystems. Unlike unstable isotopes which can be detected by the radiation emitted as they decay to another isotope of the element (or to another element), stable isotopes have no known decay constant, so detection methods rely on the difference in mass caused by having one or more extra neutrons in the atom's nucleus. Analysis of stable isotopes is possible by measuring differences in mass, or density, of the analyte. Mass spectrometers do so by stripping an electron from the analyte and passing it by a magnet in a curved flight tube. The charge of the magnet deflects the charged analyte in a radius proportional to its mass (mass-to-charge ratio). Detectors at the end of the flight tube are positioned for those deflection radii, registering the quantities of masses of analytes with and without the isotope label, thereby quantifying tracer abundance in the analyte.

In DNA stable isotope probing (Fig. 1), the DNA molecules are separated by density when placed in a test tube filled with a solution of cesium chloride and spun in an ultracentrifuge. The DNA molecules in the density gradient are then

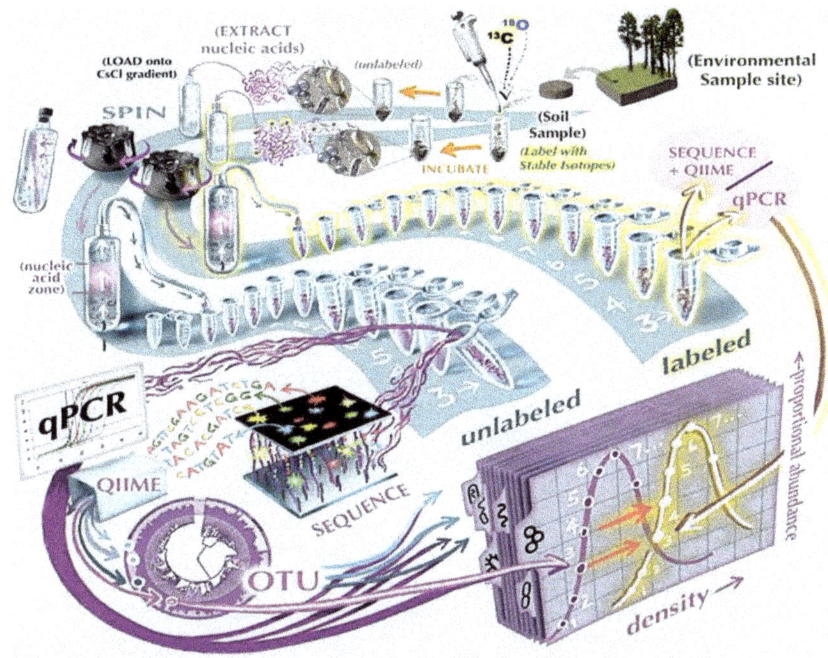

Figure 1 Conceptual model of the quantitative stable isotope probing technique, from sample collection to determining the density of 16S rRNA gene fractions for individual taxa. The density shift is proportional to the excess atom fraction stable isotope composition. Note that, except for the addition of the stable isotope tracer at the beginning of the incubation, all steps are identical for both labeled and unlabeled samples. Source: Artwork by Victor Leshyk. Reprinted from Hungate et al. (2015).

retrieved by fraction collection, after which the individual fractions are analyzed for density and sequenced. The more of the rare, heavier, stable isotope an organism uses to grow and metabolize, the denser its DNA becomes (i.e. its weighted average density increases, see Table 1), as newly synthesized DNA using the labeled substrate incorporates the heavier stable isotopes from the labeled substrate into its molecular structure. With ^{18}O-H_2O SIP, the oxygen from water becomes the oxygen in DNA – same with the 13-carbon from labeled glucose (Papp et al., 2020), methane (Cébron et al., 2007), or pesticides like atrazine which can serve as a carbon source for microbial growth (Monard et al., 2011). Compared to measuring isotope composition by mass spectrometry, the centrifuge approach used in SIP is much less precise. For example, qSIP (quantitative SIP) can detect an EAF (excess atom fraction) of 0.05 (Sieradzki et al., 2020). In contrast, dual-inlet IRMS can detect a difference of 0.05 ‰ (per mil), which corresponds to 0.0000002 EAF ^{13}C, a difference in precision of 5 orders of magnitude.

Table 1 A glossary of key terms related to DNA stable isotope probing of microbial communities

Term	Definition
Stable isotopes	Atoms of the same element that differ in mass with no known decay constants. Mass differences occur because of differences in the number of neutrons in the atoms' nuclei. Isotopes used in tracer experiments typically have extra neutrons compared to the more common stable isotope of a given element. For example, ^{13}C is the rare, stable isotope of carbon and has seven neutrons instead of the six found in the more common isotope of carbon, ^{12}C.
Isotopic labeling	The process of adding a compound that contains a stable isotope to a sample. In SIP, isotopically labeled compounds (e.g. ^{13}C-glucose or ^{15}N-ammonium) are introduced to track microbial uptake of these elements.
Density-gradient centrifugation	A lab technique that separates DNA based on density. This separation allows researchers to identify microbes that have assimilated the isotopic label, and, with qSIP, quantify the rate of assimilation.
Microbial assimilation	The process by which microbes take up and incorporate a compound into their cellular biomass, often specifically as DNA or protein, primarily to support growth and maintenance.
Microbial immobilization	Refers to the broader process of nutrient uptake by microbes from the environment, effectively "locking up" these nutrients within microbial biomass. Unlike assimilation, immobilization encompasses all nutrient uptake by microbes, regardless of whether the nutrients are used directly for growth or are temporarily stored within the cell. All assimilated nutrients are immobilized, but not all immobilized nutrients are necessarily assimilated.
Microbial Taxa	Groups or categories of microbes, classified by genetic similarity. With qSIP it is possible to quantify isotope composition of any desired level of taxonomic resolution (e.g. from Domain to genus to ASV), supported by the depth of the sequencing and the sequecing target (e.g. amplicon or meta-genome).
Operational taxonomic unit (OTU)	A microbial taxonomic designation used to categorize microbes based on DNA sequence similarity, commonly at a 97% sequence similarity threshold. OTUs represent species-like groups in microbial studies.
Amplicon sequence variant (ASV)	A microbial taxonomic designation that represents exact DNA sequences in microbial communities, offering higher resolution than OTUs. ASVs are identified through error-corrected sequence clustering.
Microbial clade	A group of microbes with a common evolutionary ancestor, often defined by phylogenetic lineage.
Natural abundance isotope atom fraction	The average abundance of stable isotopes of an element that occur naturally. Natural abundance isotope composition are often expressed in delta values, with units of per mil (‰) relative to a standard (Coplen, 2011). Natural abundance isotope composition serves as a baseline for measuring the increase in labeled isotope in the sample.

(Continued)

Table 1 (*Continued*)

Term	Definition
GC content	The percentage of nucleotides in DNA that are either guanine (G) or cytosine (C). Higher GC content often reflects genomic stability and can vary across microbial taxa. In qSIP, GC content can impact DNA density during gradient centrifugation, influencing separation of labeled versus unlabeled DNA.
Excess atom fraction (EAF)	The molar proportion of an element that is a heavy isotope (e.g. ^13C or ^15N) in a sample above the natural abundance level of that isotope. EAF values above zero indicate isotope enrichment. qSIP accounts for natural abundance background by focusing on the change in weighted average density of DNA compared to unlabeled DNA which is proportional to EAF. EAF is preferred over atom percent excess (Coplen et al., 2011).
Weighted average density (WAD)	The calculated density (g/cm^3) of the 16S rRNA gene copies for a taxon, community, or any group as inferred from sequencing multiple fractions in qSIP. The density is estimated as the weighted average density (weighted by abundance) of a given taxonomic group across all density fractions. The density of DNA in cesium chloride depends on both isotopic composition and GC content. qSIP accounts for the natural differences in organisms' DNA density by measuring the fractions in a spin of unlabeled DNA controls. With the use of isotope standards of known composition and density, it may be possible to drop the unlabeled control. An assumption of the qSIP method is that the shift in weighted average density after isotope labeling is proportional to isotope assimilation (Hungate et al., 2015).

Interpreting qSIP estimates of EAF requires assumptions typical of isotope tracer experiments, including:

- abiotic processes contribute little to assimilation of the isotope label. In other words, the isotopic label is incorporated only by active microbes. Any detected incorporation is taken as evidence of microbial growth or substrate use, and inactive microbes are not expected to incorporate the isotope. qSIP assumes further that isotope incorporation into microbial DNA is quantitatively proportional to microbial activity or growth. This assumption underpins the quantification of microbial activity based on isotope incorporation.
- The isotope tracer is distributed uniformly in the environment, ensuring that microbes have equal access to the labeled substrate and that observed differences in isotope uptake are due to metabolic differences rather than access to the label.
- Some mathematical models used to interpret isotope tracer data assume minimal label recycling (cross-feeding) to estimate process rates (Koch et al., 2018). Cross-feeding occurs when one labeled microbe releases

labeled metabolites or dies, and the metabolites or cellular remains are then consumed by another microbe. Other mathematical models allow relaxing this assumption (Braun et al., 2018). SIP experiments on populations with generation times much faster than the incubation period will likely underestimate incorporation.

- High concentrations of the rare, heavier, isotopes have negligible effects on metabolism. The mass difference in isotopes is known to influence metabolism. Replacing ~20% of body water with deuterated water is lethal in mammals because of these mass effects (Jones and Leatherdale, 1991; Thomson, 1960), while some algae and bacteria can grow in nearly pure D_2O (Kushner et al., 1999).

Scientists pursing SIP studies should be aware of these possible effects, as SIP designs advocate adding substrates with high EAF values (often >0.9) (Simpson et al., 2024) (Table 2).

Even with low precision, there is value in measuring the isotope values of specific DNA sequences because of the wealth of biological information in DNA and the power of combining this with quantitative biogeochemistry through isotope tracing. Plant and animal ecology advanced by combining species identity with isotope tracer measurements of element fluxes (Michener and Latjtha, 2008; Schimel, 1993). Techniques like qSIP make it possible to combine 'omics and a quantitative approach to isotope tracers for microbial ecology.

EAF can express tracer distribution among element pools and is commonly used in isotope tracer experiments to model isotope distributions and infer rates of element fluxes (Schimel, 1993). Isotope tracers can be applied in many kinds of environments, including soil in agriculture (Hauck and Bremner, 1976). Combining isotope tracers and DNA sequencing in qSIP makes it possible to estimate rates of element fluxes by individual microbial taxa in agroecosystems.

^{18}O-H_2O is a useful tracer, because all organisms that have been tested use oxygen from water to synthesize DNA. Given the central role of DNA in cell division and population growth, qSIP with ^{18}O-labeled water has been used to estimate growth rates of microbial populations in soil (Foley et al., 2024b), in other words not only to find out which organisms are growing, but also how fast, because oxygen from water is incorporated into DNA (and RNA) during synthesis. Chip-SIP is another technique that allows for monitoring of growth rate as well as simultaneous tracing of multiple elements (e.g. ^{18}O, ^{15}N, and ^{13}C together). Chip-SIP requires a Nano-SIMS and a Phylochip with DNA probes to bind DNA from target taxa, which must be selected *a priori* (Mayali et al., 2012).

Table 2 Further information on selected stable isotope probing methods

Method	Description	Advantages	Disadvantages	Key references
Stable isotope probing (SIP)	Uses stable isotope tracers (e.g. ^{13}C, ^{15}N) to label microbes based on their metabolic activity in incorporating specific substrates into biomolecules.	Identifies active microbial populations without needing genetic markers. Broadly applicable across different environments.	Does not quantify isotope assimilation, so activity inferences are qualitative.	Radajewski et al., 2000 (introduced SIP with ^{13}C-labeled compounds in soil); Dumont & Murrell, 2005
DNA-SIP	A variation of SIP that isolates only DNA to identify active microbial populations, linking their genetic identity to substrate assimilation.	Directly links microbial identity to metabolic activity, making it easier to study specific taxa.	Less suitable for quantitative measurements. Provides a snapshot of activity without quantifying metabolic rates.	Neufeld et al., 2007 (critical review on DNA-SIP limitations).
Quantitative SIP (qSIP)	An advanced version of DNA-SIP that quantitatively measures isotope incorporation, linking microbial activity with specific rates of substrate assimilation.	Provides quantitative insights into microbial activity levels, allowing for comparisons between taxa. Can reveal metabolic rates in microbial communities.	More complex data analysis and sample processing. Requires precise isotope measurements and access to density-gradient centrifugation equipment.	Hungate et al., 2015 introduced qSIP for quantifying EAF; Koch et al. 2018 present a model for qSIP with ^{18}O to estimate growth rates)

3 Using quantitative stable isotope probing in agriculture

Many microbial process rates affect ecosystem services. In agriculture, qSIP can help quantify the contributions of different microbial taxa to these ecosystem services, because the rate of growth of a population integrates many of its impacts on the environment (Foley et al., 2024b). Cells are made of molecules composed of elements like carbon, nitrogen, oxygen, etc. To grow, organisms must assimilate those elements from the environment. Growth therefore reflects element assimilation according to the ratios of elements (stoichiometry) required for microbial growth, and these can be measured using qSIP for different elements independently (Fig. 2). Measurements of microbial growth rates using ^{18}O qSIP are therefore likely to correlate with assimilation rates of other elements as well. For example, rates of nitrogen assimilation measured using qSIP in an agricultural soil were also correlated with rates of carbon assimilation from ^{13}C-labeled straw (Dong et al., 2022).

As our growing population faces a changing climate, optimizing agricultural practices for productivity and sustainability (e.g. minimizing negative environmental impacts) is a pressing need. Given the influence of microbial processes in governing the flow of carbon and nutrients throughout ecosystems, understanding the function of soil microbiome in agricultural contexts is key to educate management practices that consider both productivity and sustainability. Soil microbes can aid plants in nutrient acquisition and pathogen resistance, major factors that contribute to maximizing yields. Additionally, microbial metabolism strongly influences the fate of carbon in soil as microbially processed carbon can be retained in soil microbial biomass in the soil organic matter (SOM) or lost as CO_2 through respiration. Thus, leveraging microbial metabolism to enhance plant growth and retain carbon in soil is a promising pathway to not only productive but also sustainable agricultural systems. Still, many questions remain about what management tools can be employed to achieve these goals. This knowledge gap is driven in part by the incredible taxonomic and functional diversity of the soil microbiome. qSIP provides us with a tool to quantify elemental transformations at the taxon-level, data which could be used to inform management decisions with more precision than community-wide functional measurements can provide.

Because of the many roles of microorganisms in agriculture, research questions like these are possible for many areas, including areas around sustainability (Fig. 3). It is possible to imagine experiments using qSIP to test, in the field, how individual taxa contribute to ecosystem processes important in agriculture. Additionally, SIP is useful in that it can identify organisms involved with functions important to agroecosystems such as nitrogen fixation (Buckley et al., 2007), antibiotic resistance (Hernández et al., 2023), and carbon cycling

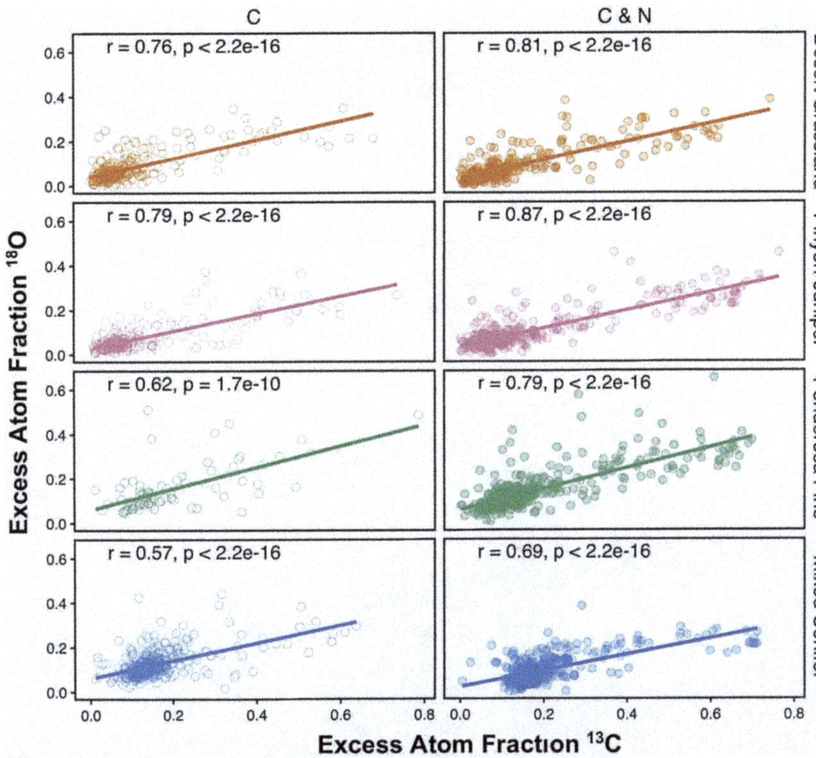

Figure 2 Consistent and positive correlation between excess atom fraction ^{13}C and ^{18}O across four soils. Panels on the left show soils that received a glucose amendment of 1000 μgC/g soil. Panels in the right show soils receiving the same glucose amendment plus 100 μgN/g soil ammonium sulfate. Each data point represents a bacterial operational taxonomic unit (based on 16S sequencing) measured in parallel incubations, one with an ^{18}O-H_2O tracer, and the other with a ^{13}C-glucose tracer. The positive relationship indicates that rates of growth (^{18}O) reflect rates of carbon assimilation (^{13}C). Source: Data from: Li et al. (2019).

(Barnett et al., 2021). The advantage of qSIP is that it provides a foundation for quantitative scaling and modeling.

4 Using qSIP: increasing soil carbon decreasing carbon emissions

As primary producers, plants fix CO_2 during photosynthesis and translocate some of this carbon belowground as exudates in exchange for critical nutrients made available by their microbial counterparts. Carbon also enters soil as leaves and root tissues turn over, an important source of nutrients which can be recycled by microbial processing. As these organic matter sources are metabolized by microbes, carbon can be retained in microbial biomass or

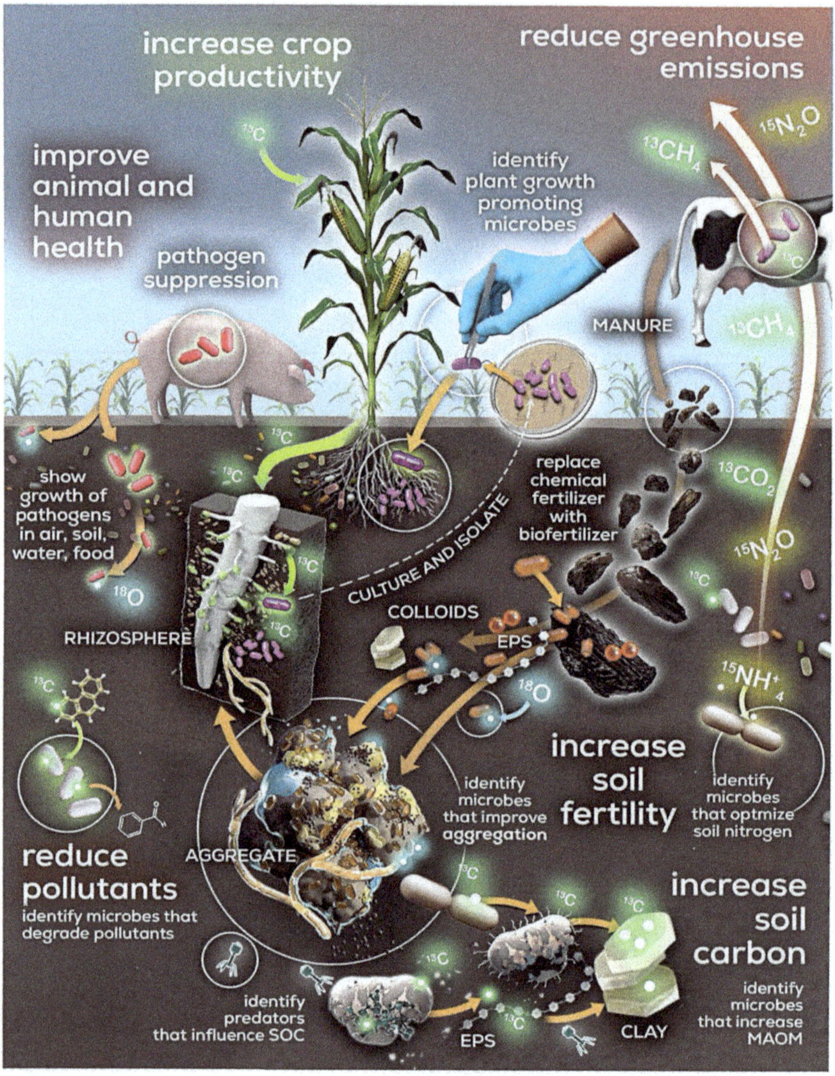

Figure 3 Ecosystem services in agriculture that depend on microorganisms, and where insights from quantitative stable isotope probing may lead to applications.

lost as CO_2 during respiration. Carbon in microbial biomass may be further stabilized on mineral surfaces during microbial turnover. As such, the carbon metabolism of the soil microbiome is of key interest when considering the propensity of agricultural systems to sequester carbon. Measurements from qSIP can aid in this understanding in several ways. For example, the amount of microbially derived carbon that may form stable mineral-associated organic matter (MAOM) can be estimated by tracking isotopic signatures from labeled substrates into extracellular DNA and amino sugars (Liang et al., 2019). Another source of carbon emissions from agricultural systems is methane (CH_4). A large source of agricultural methane emissions is ruminant digestion, a process dependent upon methanogen activity. The ecology of the rumen and activity of methanogens could be explored productively with qSIP experiments.

Soil carbon is a microbial graveyard – composed of the remains and excretions of former microbial cells (Angst et al., 2021; Buckeridge et al., 2020b, 2022; Rempfert et al., 2024). A simple prediction is that the rate of C flux through mortality predicts SOC accumulation. SOC accumulation may be adequately modeled as proportional to the rate of C flow through mortality. On the other hand, if microbial biodiversity is a strong influence, then the rate of C flux in mortality that ends up as SOC may depend more on the source species and the conditions of its death.

Dead microbial cells decompose at different rates, depending on species, and the rate is higher for taxa with high GC content and low genome size but with little influence of cell wall chemistry (Dong et al., 2021). Differences in susceptibility to predation are also likely to affect SOC, also their inclination to form biofilms on minerals. Different taxa are likely susceptible to different sources of mortality: viruses are highly taxon-specific, and protists and nematodes are selective as well. In addition to cellular chemistry (Buckeridge et al., 2020a; Malik et al., 2015; Rempfert et al., 2024), the agent of mortality influences how different taxa contribute to SOM formation (Sokol et al., 2022). qSIP can help explore how different microbes that form detritus influence soil carbon accumulation.

Because mortality is an important source of persistent carbon in soil, understanding the modes and rates of mortality are also important, and their links to qSIP shows promise for identifying pathways of energy flow through microbial food webs and identifying agents of microbial mortality including predatory bacteria (Hungate et al., 2021), as well as protists and nematodes (Greenlon et al., 2022; Starr et al., 2018). Characterizing trophic interactions in soil, and developing models that represent interactions such as predation, may improve our understanding of soil carbon cycling (Morrissey et al., 2023).

Rates of growth and mortality may strongly inform our understanding of which microbes most contribute to soil carbon accumulation in agriculture. A first step might be to compare rates of growth and mortality of individual taxa with metrics of necromass formation like bacterial and fungal amino sugars. The carbon balance implications of growth depend critically on CUE, which should also be measured in such experiments. Unknowns from a single line of evidence may be substantially reduced when many relevant data streams are considered, such as meta-metabolomic profiles for inputs to models of central metabolism (Dijkstra et al., 2011) or PLFA markers of groups of bacteria and fungi.

5 Using qSIP: reducing dependence on synthetic agrochemicals, increasing nitrogen retention and decreasing nitrogen losses

Synthetic agrochemicals (e.g. fertilizers, pesticides, herbicides) have revolutionized agricultural production systems by alleviating challenges of nutrient limitation, insect damage, and weed pressure. These advantages have not come without environmental consequences, e.g. the widespread eutrophication in the Gulf of Mexico caused by fertilizer leaching into the Mississippi river, and the pollution of watersheds by pesticides and herbicides that threaten human health. Decreasing reliance on fertilizer, pesticides, and herbicides will improve environmental and human health outcomes. Monitoring the mortality and growth rate of soil organisms could allow more effective management of these organisms, increasing beneficial organisms and reducing pathogens. This knowledge could also lead to identification of taxa that are specifically critical to plant success which could then be included in biofertilizers, which are microbial inocula aimed at increasing plant productivity. For example, qSIP with $^{15}N_2$ could be used to quantify active nitrogen fixers (Buckley et al., 2007, 2008).

Nitrogen cycling is a key area of interest in agricultural systems since nitrogen is often a limiting nutrient for crop growth. For this reason, nitrogen is often added in the form of fertilizers. Nitrogen from livestock excrement and synthetic fertilizers is rapidly transformed by microbial metabolic processes, namely nitrification and denitrification. Nitrification occurs as ammonium is transformed to nitrate, a much more mobile form, often resulting in leaching of nitrate into surface or groundwaters. Nitrous oxide, a byproduct of denitrification, is of major concern as it has significantly higher warming potential than CO_2 and contributes to loss of stratospheric ozone.

The growth and activity of soil organisms could be traced with qSIP estimate taxon-specific contributions of methane and nitrous oxide fluxes. Additional analyses like metabolomics and metagenomics could be utilized alongside qSIP measurements to better understand the ecology of denitrifiers and cause microbially driven pathways to decrease emissions. For example, Hiis et al. (2024) detailed a pathway to enrich bacteria that consume nitrous oxide with the potential to decrease emissions by 5–20%. Xu et al. (2021) observed that straw additions to soil decreased total losses of fertilizer N and lessened nitrous oxide emissions overall. Taxon-specific measurements of microbial carbon and nitrogen cycling dynamics under such conditions would surely increase our understanding of how nitrous oxide emissions could best be mitigated. For example, Dong et al. (2022) utilized co-addition of ^{15}N-labeled fertilizer and ^{13}C-labeled crop residues and performed qSIP, observing that the taxa most responsible for fertilizer N immobilization also assimilated large amounts of crop residue carbon. These microbial measurements could be considered alongside ecosystem scale carbon and nitrogen flux measurements, improving the resolution of ecosystem scale isotope tracer studies.

6 Using qSIP: increasing plant resistance to environmental stress

Plant growth-promoting rhizobacteria can increase plant resistance to water stress (Rubin et al., 2018), as isotope labeling experiments have demonstrated (Dijkstra et al., 2015a). qSIP experiments show interactions within the plant microbiome that modulate plant responses to environmental stressors such as drought: specifically, soil fungi altered bacterial growth potential, growth efficiency, and the diversity of the actively growing bacterial community with water stress (Hestrin et al., 2022). qSIP metagenomics also illustrate whole-microbiome effects related to plant response to water availability (Greenlon et al., 2022). There is tremendous potential for techniques like qSIP, especially coupled with other 'omics, to leverage microorganisms to optimize plant performance or agroecosystem sustainability (Fig. 2). Root and rhizosphere microbes in agricultural systems are highly diverse, but only a subset of the organisms may be interacting with the plant. qSIP can quantify interactions between microorganisms, focusing on highly active taxa that may have a disproportionately large influence on the rhizosphere environment. Additionally, growing plants in a $^{13}CO_2$-enriched atmosphere can be used to track the uptake of plant-derived carbon into soil microorganisms in the rhizosphere, facilitating the identification of microorganisms subsisting on plant exudates.

Experiments using $^{13}CO_2$ SIP have great potential to advance our understanding of plant–microbe interactions. For instance, labeling plants with $^{13}CO_2$ can lead to arbuscular mycorrhizal fungi becoming highly enriched in ^{13}C, demonstrating that qSIP can be used to quantify carbon assimilation by plant symbionts (Nuccio et al., 2013). Analysis of bacterial taxa enriched with ^{13}C in the rhizosphere may facilitate the identification of novel plant growth promoting microorganisms. Further, SIP in combination with metagenomic analysis can provide information into the metabolic capabilities of microorganisms acquiring carbon from plant roots (Nuccio et al., 2022). For instance, bacteria enriched following $^{13}CO_2$ SIP with an annual grass (*Avena barbata*) system identified many taxa capable of polysaccharide degradation (Nuccio et al., 2022) suggesting the grass may be allocating carbon below ground to prime the decomposition of SOM for nutrient release. If plants are grown in specialized rhizoboxes that have distinct rhizosphere and hyphosphere compartments, $^{13}CO_2$ SIP can be used to quantify hyphal transfer of recently fixed carbon to soil microorganisms (Kakouridis et al., 2024; Nuccio et al., 2022; Wang et al., 2016). Hyphae can facilitate bacterial dispersal and fuel bacterial activity, and interactions between mycorrhiza and bacteria are likely to have strong influences on how much plant-derived C is transported below ground and to which microorganisms. $^{13}CO_2$ qSIP can assess when and how plants rely on their microbial partners and the consequences of rhizodeposition for SOM stabilization processes.

7 Using qSIP: suppressing pathogens and reducing antibiotic use in livestock

Microbiomes can suppress crop pathogens, and those microbiomes can be transferred from one soil to another, conferring the suppressive effect (Durán et al, 2021). SIP demonstrated how rhizosphere bacteria of the genus *Burkholderia* relied on fixed carbon from the host plant, and simultaneously suppressed the growth of the pathogenic fungus, *Alternaria alstromemeriae* (Lu et al., 2023). Specific organisms are thought to be responsible, such as nematode-trapping fungi, organisms that could be targeted for development as bioinoculants. ISP could quantify how the organisms survive, what substrates are used for growth, and how the bioinoculants affect other members of the soil microbial assemblage. qSIP could document the effect of an organism thought to suppress the growth of plant pathogens, quantifying the nature of the interaction and possibly providing data useful to documenting the efficacy of bio-inoculation for pathogen suppression.

Using SIP, it is possible to test for the effects of antibiotics on the growth rates of soil microorganisms, in situ, a way to identify sensitive and resistant organisms (Hernández et al., 2023). SIP using ^{13}C-labeled antibiotics makes it possible to discover which organisms can degrade the antibiotic compound and use it as a carbon source (Zhang et al., 2024; Ouyang et al., 2019).

Human pathogens live in soil (Stone et al., 2022). SIP measurements could document whether and how fast these organisms grow in soil or confirm that they exist there primarily in dormant phases. If these organisms do indeed grow in soil, qSIP measurements would be useful to understand what drives population dynamics and potentially infectious events. Patterns of substrate use in soil through qSIP would help quantify their ecology in soil and of the ecosystem of food animal production (Koch et al., 2018), including infections in humans from strains derived from agriculture, now well documented; 8% of extraintestinal *E. coli* infections were caused by zoonotic strains, as identified by the unique portfolio of genetic elements indicating a livestock host (Liu et al., 2023).

8 Using qSIP: reducing greenhouse gas emissions

Soil microbes emit and consume greenhouse gases, influencing their concentration in the atmosphere and thus the radiation balance of the planet.

Respiration releases carbon dioxide, the major flux of carbon out of soil. Microbial phototrophs, like algae in biocrusts, bring carbon into the soil through photosynthesis, as do chemolithotrophs, though in soil these carbon fixation pathways are typically much smaller than CO_2 release through respiration. Denitrification and nitrification both release nitrous oxide, and the terminal step in denitrification consumes it. Similarly, methane is produced in soil by methanogens and consumed by methane oxidizers. With qSIP, its possible to detect the growth rates of taxa whose genomes match these physiological pathways. Such results are scalable to flux estimates, even though the fluxes themselves are dissimilatory with respect to the greenhouse gas emitted, and thus to the element composition of the substrate. With dissimilatory processes like nitrification and denitrification, qSIP is not a direct measure of the rate of that dissimilatory flux. One simple way to estimate dissimilatory fluxes from qSIP data is to through stoichiometry: for example, lab studies estimate that archaeal nitrifiers consume 10 moles of NH_3 to generate 1 moles of biomass (Tijhuis et al., 1993). Assuming all nitrifiers grow solely through nitrification, the known stoichiometry (from culture) means an absolute growth rate (e.g. 15 nmol C g soil^{-1} day^{-1} for taxon i) scales by a factor of 10 to the nitrification rate of taxon (e.g. 150 nmol N g soil^{-1} day^{-1} for taxon i). Using qSIP it is possible to demonstrate this very tight connection between EAF ^{18}O of mRNA of the amoA gene and EAF ^{18}O of the 16S rRNA genomic DNA (Papp et al., 2020). Alternatively, interpreting the meta-transcriptome to estimate nitrifier gene expression could help partition growth into that supported by nitrification versus other pathways, because nitrifiers can be mixotrophic.

Climate mitigation schemes include the idea of adding to agricultural soils microorganisms that have been engineered to consume greenhouse gases. *Cloacibacterium* sp. CB-01 is a N_2O respiring bacterium that showed promise in field experiments for reducing N_2O fluxes from agricultural soils (Hiis et al., 2024). Techniques like qSIP could be applied in such cases to document growth (fitness) associated with changes in the N_2O flux, along with changes in the expression of the NosZ gene, the enzyme responsible for N_2O reduction to N_2. Another approach is to modify the soil conditions to stimulate the desired change in greenhouse gas fluxes. For example, by stimulating nitrogen assimilation into soil microorganisms, straw addition in agriculture can reduce nitrous oxide emissions soils (Xu et al., 2021). Combining such experiments with taxon-specific qSIP estimates of ^{15}N assimilation could help identify organisms contributing to the reduction in greenhouse gas emissions.

9 Linking qSIP to other 'omics' technologies

DNA replication occurs at the scale of the genome: if cells are dividing, they need a new copy of the entire genome, not just the particular genes directly supporting growth. For example, soil carbon depends on inputs through microbial growth and death, as well as respiratory carbon losses. The efficiency of microbial biomass production is also thought to influence SOC formation (Prommer et al., 2020), where efficiency is the rate of biomass formation per unit substrate consumed. While qSIP can estimate growth, estimating respiration with taxonomic resolution in soil will require methods that can measure dissimilatory processes. Progress with metabolic flux analysis shows promise (Dijkstra et al., 2011), suggesting coupling with isotope-informed meta-metabolomics.

On a cellular basis, CUE has been described as the result from the balance between biosynthesis and energy production. However, as CUE is measured in practice as an aggregated parameter of the microbial assemblage, it often includes processes beyond biosynthesis and energy generation that affect CUE, such as turnover of biomass by death or into higher trophic levels. When measuring the efficiency of biomass formation at the cellular level using metabolic flux analysis with ^{13}C-isotopomers of metabolic intermediates (Dijkstra et al., 2015b), efficiency values range from 0.65 to 0.70, biochemical efficiencies that are higher and less variant that community-scale estimates of carbon-use efficiency (Geyer et al., 2016). The differences may be explained by trophic dynamics: transfer of C across two trophic levels (e.g. the primary consumers and their predator) may each have a high efficiency (0.7 in this example), yielding a much lower combined efficiency ($0.7 \times 0.7 = 0.49$).

It seems obvious that a high CUE is beneficial for SOM formation – high biomass efficiency is thought to promote necromass and soil C formation (Tao et al., 2023). Experimental results have been mixed, in part because SOC formation is not just a function of the efficiency of biomass formation but also of the absolute C flow through the system. More detailed studies on the relationship between microbial biomass formation, necromass formation, trophic energy dissipation, and SOM formation are urgently needed. In response to a labile C addition, we recently observed very dynamic changes in gene expression, indicating that some microbes go from highly active exponential growth phase to much less active stationary growth phase. Predatory bacteria also respond rapidly, within 24 hours of glucose addition. These observations suggest a new model for SOM formation in response to labile C additions: (1) the rates are high and (2) there is very fast succession between primary consumers (those consuming the glucose) and secondary consumers (predators, viruses, etc.). From the perspective of CUE, separating primary and secondary consumers may have beneficial effects on SOC formation: specifically changing the balance of activity between primary consumers and predators. This can in theory be done in multiple ways: increase the abundance of primary consumers by inoculation, decrease the abundance of predators through chemical or physical means, (3) interrupt the succession through applying growth rate suppressing measure after primary consumers enter stationary phase (experience starvation) and the start of the growth of the predator populations.

To study these very fast transitions between growth and inactivity of various clades of microorganisms requires the use of a method of taxon-specific growth rates at a temporal resolution that, we believe only qSIP can provide. Other techniques in complement may be required to resolve impacts of longer-term processes, e.g. slowly growing fungi that respond slowly to pulsed resource events but nevertheless have large impacts on soil C accumulation. Combining qSIP with multi-omic and other signatures of dissimilatory processes is poised to advance this science.

Other dissimilatory processes face the same challenge. For example, facultative denitrifiers often grow using oxygen as a terminal electron acceptor, not relying on denitrification at all. With data from qSIP using $^{18}O\text{-}H_2O$ as a tracer to measure growth, it would not be possible to determine whether denitrifier growth occurred through aerobic respiration (using O_2 as an electron acceptor) or denitrification (using NO_3 as an electron acceptor). During cellular replication, the entire genome is copied, so all the genes acquire the same $^{18}O\text{-}H_2O$ label at the same time, depending on the rate of DNA replication, not on the rate of gene expression. However, with an accompanying meta-transcriptomic dataset, one could test how activities and abundances of specific genes and proteins were associated with differences in growth rates or changes in growth in response to external forcings.

The underlying processes of growth can be categorized as biosynthesis of all cell materials and energy production. New techniques are becoming available that allow us to study these processes in detail as never before (Foley et al., 2024a). This allows us to identify the ecophysiological processes that are associated with fast and slow growth for a particular taxon. For example, using known biochemical principles, we can identify using transcripts, proteins, and metabolite concentrations when carbohydrates are consumed or when small organic molecules are consumed resulting in gluconeogenesis. Similarly, we can identify when growth takes place using transcriptional regulators of the exponential and stationary growth phase and dormancy. Since each living cell has the potential to divide, the resulting exponential growth will quickly deplete any natural substrate pulse, after which starvation is experienced, and growth rates decline. Similarly, using transcripts, it is possible to identify limiting nutrient conditions, e.g. whether N or P, or other elements are growth-limiting. With this information, we can optimize the growth rates of the natural community to benefit a specific ecosystem goal. For example, we can stimulate those organisms that produce SOM or slow down pathogens by reducing the conditions for their growth.

Biotic interactions can influence population growth, including predation, viruses, and grazing. We have begun identifying food web structure in soil communities using qSIP (Hungate, 2021). Pulses of growth of primary consumers are likely to be quickly followed by pulses of growth of their predators and viruses, thereby limiting their growth rate. Understanding the temporal dynamics of these trophic levels and their controls is crucial to understand and manipulate microbial food webs in agriculture. This provides us with new opportunities that design management to benefit ecosystem goals. For example, identifying and introducing or stimulating natural enemies of pathogenic microbial organisms may be a way to control their population sizes; or by manipulating soil food web structure, we may be able to stimulate SOM formation. Each of the possibilities requires that we quantify microbial growth rates with high precision and over short periods of time. With qSIP, this is definitely possible.

10 Scaling and modeling: from taxon-specific information to ecosystem functions and services

Taxon-specific information on different groups of microorganisms allows us to create more accurate models to predict how changes in farming practices could impact nutrient availability and cycling efficiency. We can leverage qSIP to identify key species crucial for maintaining soil fertility. By incorporating information on functional traits into biogeochemical models, we can develop targeted strategies to optimize nutrient use, improve crop yields, and promote sustainable agricultural practices, while minimizing environmental impacts. An improved understanding of the complex interactions between soil microbes and nutrient cycles can enhance our ability to effectively scale up microbial contributions in carbon/nutrient storage (or loss) and greenhouse gas fluxes in agricultural ecosystems.

Growth and mortality rates are often included as parameters in microbial models (e.g. Sihi et al., 2021) (see equations 2-4 for a generalized framework). Some models partition the flux of carbon associated with microbial mortality into C entering a rapidly cycling SOC pool versus C becoming complexed with minerals and cycling more slowly (e.g. Fan et al., 2021 see equations 2-5). In addition to growth rate, mortality can be estimated from qSIP data (Koch et al., 2018) by modeling the rate of disappearance of unlabeled DNA from a population. Thus, there is potential to conduct data-model comparisons using qSIP estimates of growth and mortality along with ^{13}C-tracer distribution to mineral-bound versus soluble soil pools, e.g. providing data to test which microbial formulations are most useful in soil C simulations.

Most quantitative models of soil carbon cycle simulate carbon fluxes in soil with little-to-no representation of the biodiversity of microorganisms, which represents most of the genetic biodiversity over the entire tree of life.

Macroscopic species affect ecosystem processes (Vitousek, 1990), and the bulk of biodiversity is microbial, so the influence of microbial species on ecosystem processes could be substantial. Recently, so-called "microbial models" of soil biogeochemistry have been developed. They represent microbial biodiversity by modeling as many as two microbial groups (Wieder et al., 2015), and thus still do not represent the immensity of the microbial biodiversity obvious from nature.

With qSIP, it is possible to estimate taxon-specific rates of element fluxes (Stone et al., 2021) with units compatible with biogeochemical models. Doing so makes it possible to estimate how growth of individual microbial species affects biogeochemical fluxes, or whether growth rates of genes are better predictors, or functional groups based on life strategies. Such data also provides a platform for discovery without *a priori* postulates, e.g. through machine learning.

Using qSIP, we can measure taxon-specific functional traits, but scaling up these traits to understand ecosystem function remains challenging. In plant communities, functional traits are often combined with biomass estimates and used to calculate community-weighted mean trait values (Enquist et al., 2015). In this approach, the traits of more abundant organisms are more heavily weighted as dominant organisms have a disproportionate impact on their ecosystems. This is challenging in microbial ecosystem because although we can now quantify taxon-specific microbial traits, measuring taxon-specific microbial biomass remains challenging. Many soil microbial data sets yield information on a taxon's relative abundance within the 16S rRNA gene pool which may not be proportional to its biomass, given that soil bacteria vary dramatically with regard to cell size (Leichty et al., 2021; Portillo et al., 2013). However, using relative abundance estimates in combination with estimates of total microbial biomass may help facilitate the estimation of community-weighted microbial traits (as in Walkup and Morrissey, 2025). Alternatively, taxon-specific microbial biomass can be estimated using the relationship between genome size and cellular mass (West and Brown, 2005) as described in Stone et al. (2021). Such approaches may help connect taxon-specific traits (measured via qSIP) with soil biogeochemical process rates (Walkup and Morrissey, 2025).

A major development in the field would be techniques that isolate extracellular, relic, DNA and distinguish it from DNA enclosed within still-viable cells. The propensity for DNA to remain in soil after mortality could be a powerful signal of taxon-specific post-mortem legacies of microbial cells in soil, yet distinguishing this fraction remains difficult. DNA that persists in the environment could identify the recently deceased, possibly a foundation for inference about who matters for SOM formation and loss.

11 Case study: identifying microbial allies in nitrogen retention

Nitrogen retention in the soil microbial biomass is a mechanism to reduce leaching and greenhouse gas emissions from soils. Inorganic nitrogen flows into microbial cells and then becomes the proteins and nucleic acids that fuel those cells' growth and survival. In the past, we have measured this primarily as total nitrogen retained in the microbial biomass. Discovering which specific microbes immobilize effectively could help us manage this critical ecosystem process of nitrogen retention. In this case study, we assessed nitrogen mineralization and immobilization relative to plant N demand and N losses in an agroecosystem.

Nitrogen management remains a challenge in both conventional and organically managed agricultural systems. Too little nitrogen limits plant productivity, while too much leads to greenhouse gas production (N_2O), nitrogen leaching and subsequent pollution of neighboring aquatic ecosystems (Kuypers et al., 2018). Microbial assimilation of fertilizer nitrogen may help retain N in agricultural soils until it is needed by plants (King and Hofmockel, 2017).

Nitrogen immobilized by the microbial biomass is assimilated into proteins and nucleic acids, unavailable for denitrification and leaching, the two major pathways of soil nitrogen loss. While microbes are often thought to compete with plants for N, fertilizer is often added early in the growing season, when plant demand is low (Crews and Peoples, 2005). Microbial assimilation of this N may benefit plants as the N may be slowly released as microbial biomass turns over during the growing season (Grandy et al., 2022). However, during peak plant N demand microbial assimilation may impair plant productivity because the N trapped in microbial biomass is not immediately available for plant uptake. Optimizing microbial immobilization when plant demand is low, then switching to nitrogen release (mineralization) when the plant demand is high may allow us to reduce reliance on synthetic fertilizers. By regulating the size and flow through of the major inorganic forms of nitrogen abundant in agriculture (ammonium, nitrate, and nitrite), microbes that immobilize nitrogen have a major ecosystem impact. Agricultural practices already manage to optimize nitrogen availability to plants while simultaneously minimizing nitrogen losses from the system (Matson et al., 1998).

Using qSIP with ^{15}N, we can measure taxon-specific N immobilization (Morrissey et al., 2018), and this may allow us to identify key taxa involved in N assimilation at various points in the growing season. In a recent study (Reed et al., 2025a) nitrogen assimilation was measured in rhizosphere soils of two maize systems in the field and in the laboratory (Fig. 4). The experiment was designed with parallel measurements in the field and laboratory to determine whether

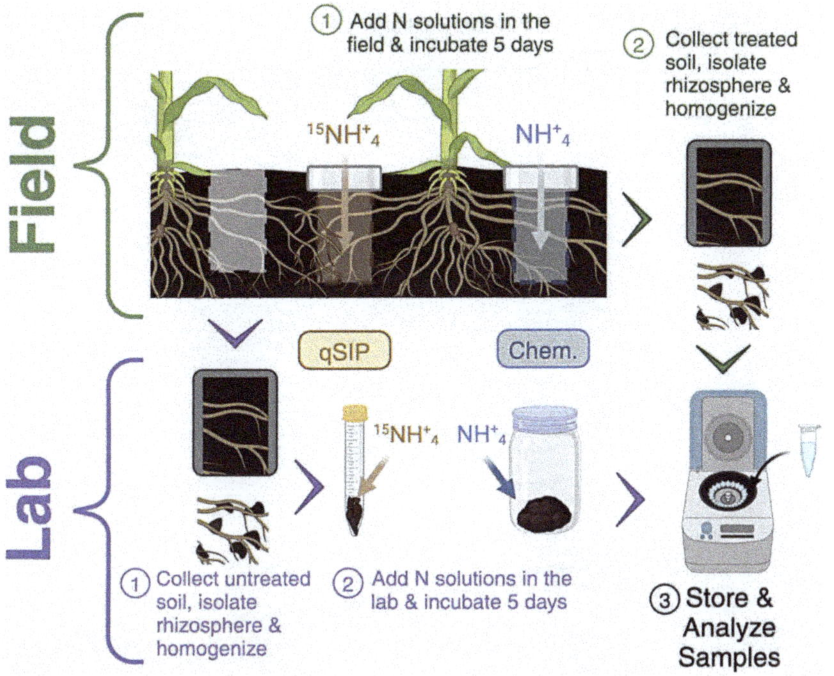

Figure 4 Experimental design comparing lab and field qSIP measurements in conventional and organic agriculture. Source: From Reed et al. (in review).

microbial function measured in the lab reflects activities in situ. Highly enriched (~0.97 EAF) $^{15}NH_4^+$ was added for the qSIP measurements, and NH_4^+ with a natural abundance level of ^{15}N was added (Fig. 4) in an identical manner to permit tandem measurements of biogeochemical process rates (N transformations, microbial biomass C, and CO_2 flux). These incubations are often conducted in parallel, because high amounts of ^{15}N are required in SIP experiments in order to detect assimilation via qSIP, whereas ^{15}N tracer and pool dilution measurements (Wessel and Tietema, 1992)) can use much smaller ("trace") amounts to avoid artifacts caused by altering the size of the substrate pool.

To determine if agreement between field and lab measurements depends upon soil characteristics, this experiment was repeated at two topographically distinct sites with differing soil properties. A "Ridge" site with comparatively sandy well-draining soil and a "Valley" site where the soil had more organic matter, greater bulk density, and higher soil moisture (Reed et al., 2025b).

This experiment was able to quantify the rates of relative N assimilation by individual taxa and the results demonstrate that some taxa assimilate N much more rapidly than others (Fig. 5, Fig. 6). For instance, *Chitinophaga*, *Oxalobacteraceae*, and *Luteolibacter* had high rates of N assimilation at both

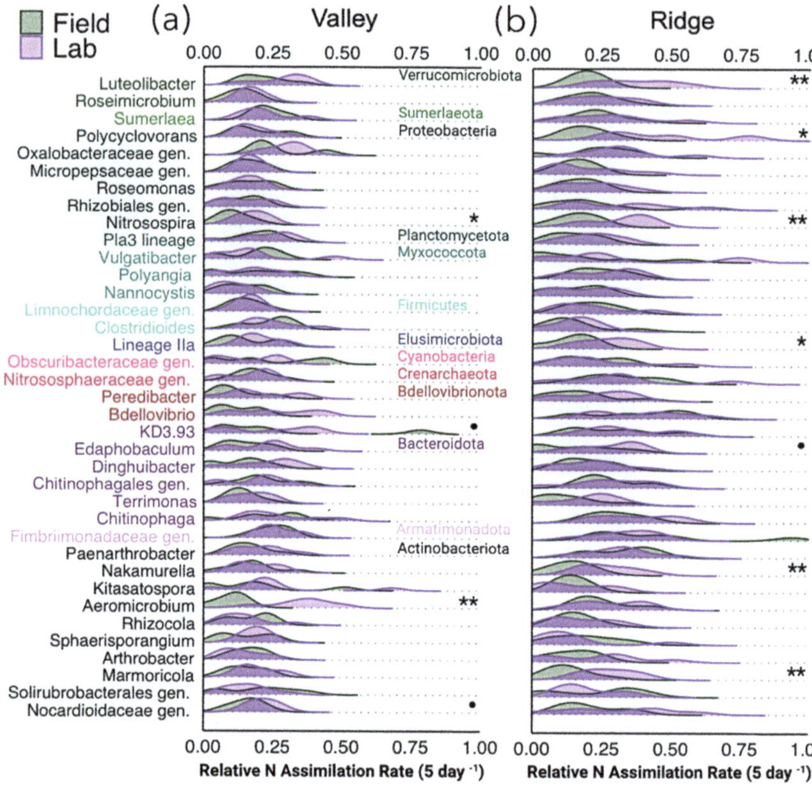

Figure 5 Taxon-specific nitrogen assimilation rates by soil bacteria at two agricultural sites as estimated by qSIP. Genera shown have the greatest median values for each method (Field and Lab) averaged across sites.

sites suggesting the physiology of these organisms may make them especially competitive for NH_4^+ in various soil conditions.

qSIP can be used to estimate the proportion of N assimilated by individual microbial populations. The total amount of N assimilated by a microbial population is dependent not only on the N assimilation rate but also on the population size. Consequently, to determine which taxa are more important for N assimilation within a given community, the proportion of N assimilated by each genus "% N assimilated" was calculated based as the product of each taxon's relative abundance (expressed as a proportion) and ^{15}N EAF divided by the sum of the product of relative abundance and ^{15}N EAF for all genera as in Morrissey et al. (2018).

This analysis suggests that just three genera, *Bacillus*, *Gaiella*, and *Vicinmibacterales*, were responsible for over 10% of the N assimilated within the microbial community, primarily because they are dominant members of the

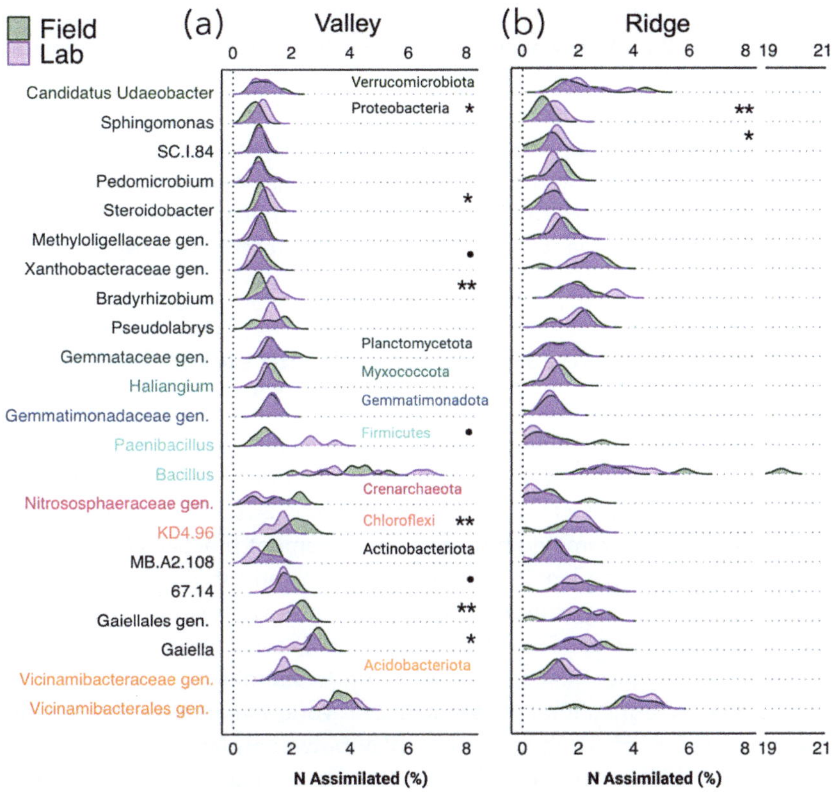

Figure 6 Taxon-specific nitrogen assimilation, shown here as the proportion of N assimilated by each genus "% N assimilated," by soil bacteria at two agricultural sites as estimated by qSIP. Genera shown have the greatest median values for each method (Field and Lab) averaged across sites.

community as they did not have especially high relative N assimilation rates. Most of these microbes are known for their abundance in the crop rhizospheres and their responsiveness to N applications. In this example, results were quantified in the field, with other root-associated taxa present and, importantly, interacting with live plants.

This information – how much and how fast certain microbes can assimilate N – can be used to further optimize precision fertilization techniques (timing, fertilizer type, incorporation method, etc.) that reduce runoff and N_2O emissions while maximizing short- and long-term crop-N availability. This data can be used to connect the composition and traits of microbes with rates of N transformations in the soil (Walkup and Morrissey, 2025). Balancing the need for plant-available N with microbial assimilation and overall loss from the soil is challenging and system-specific. Traditional 16S rRNA sequencing methods have identified broadly which microbes respond to N fertilization (Ramirez et al., 2012). However, those methods are not able to demonstrate who is assimilating the added N (though you can *assume* they are to some degree) or at what rate. Quantifying taxon-specific rates of N assimilation may allow us to identify agronomic management strategies that select for certain microbes more adept at immobilizing N and perhaps less prone to produce a net loss of N from the system. This case study illustrates how ^{15}N-qSIP can be applied in the field to identify microbial allies in N retention, a key ecosystem service in agriculture.

12 Case study: Miscanthus, manure and soil carbon in agriculture

Soil carbon sequestration in agricultural systems could aid in climate change mitigation, offsetting greenhouse gas emissions from agricultural practices, which constituted ~22% of total global emissions in 2022 (Pörtner et al., 2022). Still, many questions remain about how management practices (e.g. nutrient management) impact plant–microbe interactions and thus, the cycling and storage of carbon in agricultural soils. Quantitatively tracking carbon uptake into microbial DNA with qSIP allows for analysis of carbon cycling at the taxon level, bringing higher resolution to our understanding of carbon cycling and storage in agricultural systems under varied management regimes. SOM dynamics depend upon the interplay between plant carbon inputs to soil and microbial processing of those inputs. As such, SOM pools can be conceptualized into particulate and mineral-associated pools which relate to carbon origin and vulnerability to decomposition (Lavallee et al., 2020). MAOM is made up of low molecular weight organic compounds interacting with mineral surfaces and is thought to largely be composed of microbially derived compounds released during turnover (Kallenbach et al., 2015, 2016). MAOM is thought to persist in soil for decades or even centuries (Kögel-Knabner et al., 2008; Lavallee et al., 2020), and thus bolstering its formation is of great interest to help mitigate climate change by sequestering carbon in soil. On the other hand, particulate organic matter (POM) is composed of undecomposed plant residues and is generally more vulnerable to decomposition than MAOM, although it can become physically protected from decomposition within aggregates (aggregate-occluded POM). POM from root and leaf litter, alongside direct carbon inputs into the rhizosphere as exudates, serves as an important source of carbon and nutrients in soil, eventually contributing to the MAOM carbon pool following microbial processing (Cotrufo et al., 2013; Lehmann, 2009). Agricultural management practices, including the addition of organic and inorganic fertilizers, can alter microbial C cycling processes. Inorganic fertilizer additions can reduce the diversity and abundance of bacterial decomposers (Jangid et al., 2008; Zhou et al., 2016) and suppress decomposition (Li et al., 2017), while organic fertilizers often stimulate microbial growth and carbon cycling functions (Chakraborty et al., 2011; Ma et al., 2020). As such, relating microbial activity measurements to POM and MAOM dynamics may provide insights into the stability of soil carbon agricultural systems under varied management regimes. The resolution provided by qSIP allows for an understanding of these dynamics at the taxon-level, e.g. by allowing for identification of the specific microbial taxa that may be especially active in cycling carbon and forming MAOM. In this way, qSIP applied to agricultural field experiments can help us understand how fertilization-driven changes in microbiome composition and function drive soil carbon responses.

A recent study by Kane et al. (in review) utilized a field experiment aimed at understanding the response of Miscanthus x giganteus cropping systems to either organic (manure), conventional (N-P-K), or no nutrient amendments on marginal land in Appalachia. This study used qSIP with ^{13}C to understand microbial carbon cycling and relate this to carbon storage in distinct SOM pools (POM and MAOM). The results revealed that taxon-level responses to nutrient additions were more driven by manure addition than by N-P-K addition (Fig. 7), similar to another experiment comparing manure with inorganic fertilizer (Xu et al., 2024). Several key nitrogen cycling taxa showed increased activity under manure addition (Burkhoderiales, Nitrosomonadaceae, *Rhodoplanes*, *Reyranella sp.* (Lee et al., 2017; Prosser et al., 2014; Zhang et al., 2022). These taxa may be directly increasing plant access to nitrogen, contributing to the higher plant yield that was observed under manure addition. The community weighted average uptake of ^{13}C was higher as well under manure addition (Fig. 8) suggesting overall increases in bacterial activity. This increase in carbon uptake under manure addition was accompanied by increases in MAOM (Fig. 9).

The accumulation of carbon in the mineral-associated pool accounted for about 43% of the total increase in soil carbon stocks. This suggests a relatively large contribution of microbially-derived, chemically stabilized organic matter to total carbon stocks under manure addition. These results suggest that manure addition increases microbial carbon assimilation which may in turn contribute to increases in mineral-associated organic matter accumulation. In this way, qSIP with ^{13}C can help reveal how increases in C assimilation by individual taxa relate to ecosystem level carbon cycling dynamics.

This study also suggests that qSIP may aid in the understanding of aggregate formation, a key process in the physical protection of carbon in soil. We observed a trending increase in aggregate-occluded POM carbon under manure addition (Fig. 9) which seemed to be associated with bacterial carbon cycling activity. This was evidenced by a relationship between community-wide

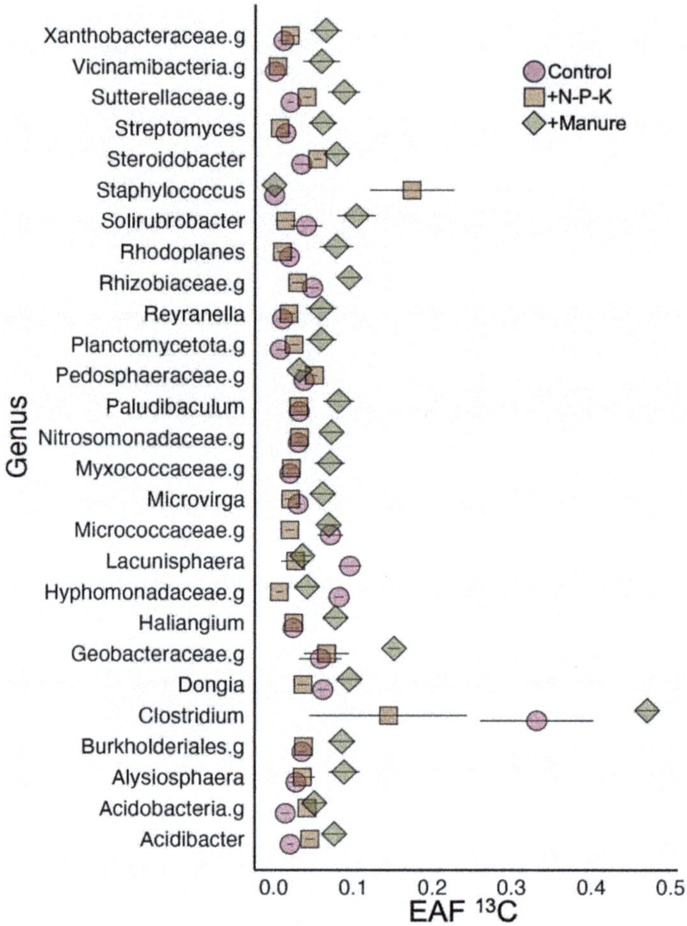

Figure 7 a) Uptake of added ^{13}C substrate by treatment-responsive genera represented as average (±S.E.) excess atom fraction (EAF) ^{13}C. These genera were considered treatment responsive since their EAF ^{13}C was significantly explained ($P < 0.05$) by nutrient treatment as determined by one-way ANOVA with treatment as the main factor and block as a random factor. When genus-level classification was not successful, the finest taxonomy available was listed followed by ".g".

^{13}C uptake and aggregate-occluded (heavy) POM-C (Fig. 10). This could be related to microbially-driven aggregate formation, for instance through extracellular polysaccharide production and hyphal exploration (Costa et al., 2018; Lehmann et al., 2020). More C assimilation into microbial biomass (as measured via qSIP) could be, at least in part, responsible for the accumulation of such compounds possibly driving aggregate formation and contributing to carbon accumulation in soil. All told, our qSIP results showing increased taxon-level activity under manure addition, coupled with higher community wide

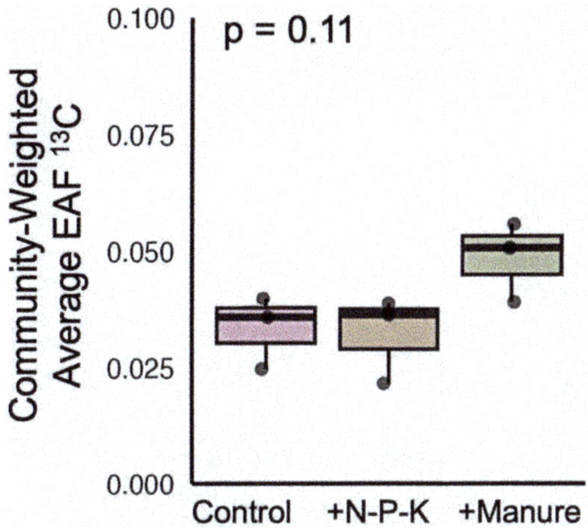

Figure 8 Community-weighted average excess atom fraction (EAF) ^{13}C of the active bacterial community. The active community is defined as those taxa, which assimilated the labeled carbon substrate at a greater-than-average rate. The community-weighted average EAF ^{13}C is calculated as each taxon's EAF ^{13}C multiplied by their 16S gene abundance summed for all taxa in the sample. Statistical significance was tested using one-way analysis of variance with treatment as the main factor and block as a random factor.

measurements of activity (e.g. respiration), support that increases in the carbon stored in this pool may be driven by higher microbial activity with manure addition.

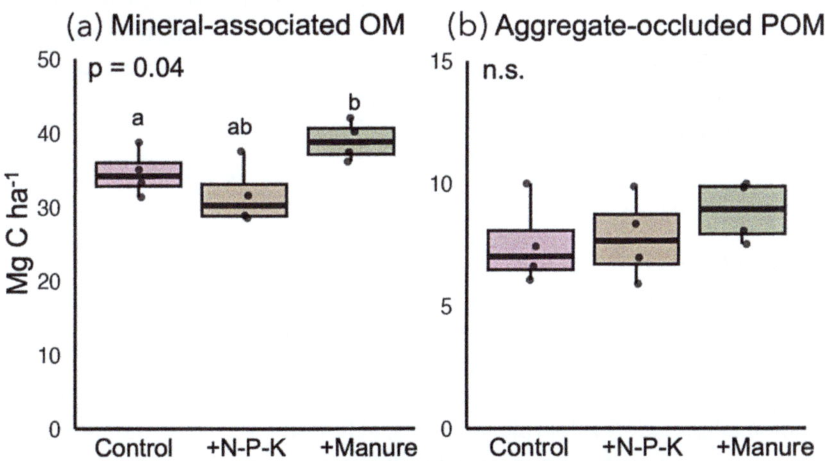

Figure 9 Soil carbon stocks in the mineral-associated OM (a) and aggregate-occluded POM (b) pools. Statistical significance was tested using one-way ANOVA with fertilization treatment as the factor and block as a random factor. Pairwise differences were determined using Tukey's HSD test and are represented by letters where appropriate.

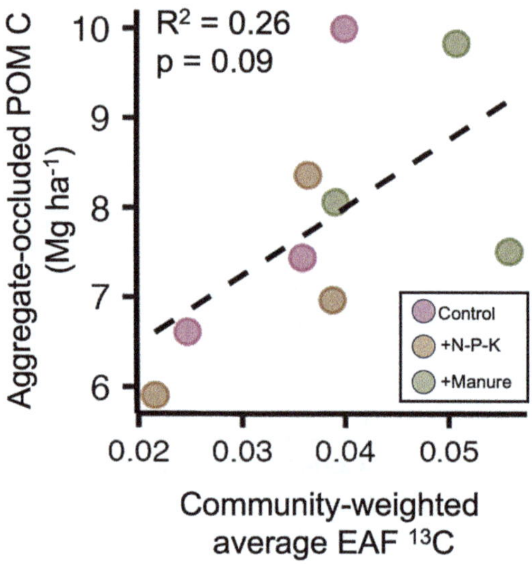

Figure 10 Relationship between ^{13}C tracer uptake into bacterial DNA weighted by taxon abundance (community-weighted average EAF ^{13}C) and carbon stocks in the aggregate-occluded soil carbon pool.

13 qSIP for agriculture: technical challenges and future trends

qSIP offers unique opportunities for linking microbial identity to ecosystem processes, but several technical hurdles remain. Addressing these is critical to expanding qSIP's utility in agricultural research. Continued methodological development – especially to improve precision, reduce costs, and extend qSIP to a wider range of soil organisms–will enable broader application of qSIP in agricultural research and management.

13.1 Cost, throughput, and access

qSIP experiments remain expensive and labor-intensive due to the need for ultracentrifugation, density fractionation, and deep sequencing. Reducing costs and increasing throughput are priorities for broader adoption. Advances such as automated density fractionation (Nuccio et al., 2022) and potential miniaturized workflows (e.g. microfluidics) show promise. Currently, no commercial providers offer qSIP analyses, although the Joint Genome Institute (JGI) supports projects on a competitive basis (Institute, 2024).

13.2 Standards and precision

Researchers are currently developing standards to increase the accuracy, precision, and inter-lab comparability of qSIP (Vyshenska et al., 2023). New efforts to use internal DNA standards (Schwartz et al., in prep) aim to improve both precision and inter-lab comparability. Standards could allow small deviations to be corrected, enhancing qSIP's reliability across studies. Additional strides are being made as well to standardize SIP terminology with the goal of making SIP data reproducible and reusable to the scientific community (Simpson et al., 2024). Automated fraction collectors should also help improve precision (Nuccio et al., 2022).

13.3 Sequencing resolution

qSIP is most reliable for abundant taxa. Low-abundance but highly active taxa may remain undetected without high sequencing coverage. Metagenomic approaches, internal standards, and carefully tailored experimental designs can help overcome this limitation.

13.4 Cross-kingdom qSIP

A rich body of work has used SIP technology to qualitatively identify active members of the soil microbiome including animals, fungi, archaea, and bacteria and gain insights into the flow of nutrients between soil trophic levels (Crotty et al., 2012). Most qSIP studies have focused on bacteria and archaea, but expanding the approach to fungi, protists, and other soil organisms is essential for a full view of ecosystem function (Greenlon et al., 2022; Hayer et al., 2021; Mau et al., 2024; Ochs et al., 2025; Raczka et al., 2021; Trejos-Espeleta et al., 2024). Technical optimizations tailored to physiological differences among taxa are needed.

13.5 Environmental heterogeneity and isotope recycling

Field soils are spatially heterogeneous, introducing variance into qSIP measurements. Mathematical modeling and fine-scale sampling designs could improve inference. Dual-isotope labeling (e.g. ^{18}O and ^{13}C together) can help distinguish direct assimilation from cross-feeding.

13.6 Stakeholder engagement

For qSIP to meaningfully impact agricultural sustainability, results must be co-developed with stakeholders and validated in production settings.

Experiments with clear agricultural goals and subsequent field validation will be critical.

13.7 Balancing precision and cost

A key challenge is balancing precision against sequencing costs. Sieradzki et al. (2020) recommend sequencing nine to twelve fractions to maintain sufficient precision while minimizing expenses, compared to older approaches using up to 40 fractions. Metagenomic qSIP, while less quantitative, can yield ecologically meaningful insights with as few as three fractions (Starr et al., 2018), offering a path toward lower-cost applications. Ongoing development of high-throughput workflows and standardized protocols will continue to improve precision and reduce costs.

14 Conclusion: can agriculture be more sustainable through quantitative stable isotope probing-informed microbial engineering?

qSIP provides taxon-specific insights into microbial activity in the wild, opening new possibilities for precision microbiome management in agriculture. By informing targeted alterations to microbial communities – including changes in composition, gene expression, or even genome editing (e.g. CRISPR) – qSIP could improve the efficiency of agricultural inputs and mitigate environmental impact. Yet, biological solutions also carry risks, including species invasions (Jack et al., 2021) and unintended gene flow across farms (Lopes et al., 2024; Novaes et al., 2024). These concerns must be carefully considered along with scalability.

One promising application is using qSIP is to track microbial responses to bioinoculants, assessing their growth rates, persistence, and contributions to carbon and nitrogen cycling (Kaminsky et al., 2019). Precise evaluations of microbial functions within managed soils are feasible using isotope-labeled substrates such as $^{15}N_2$ for nitrogen fixers, $^{15}NH_4^+$ for taxa that optimize ammonia uptake, and ^{13}C or ^{15}N for nutrient tracking (Fig. 2).

In the face of environmental change, qSIP is a key tool to bridge microbial ecology and agricultural practice, improving plant productivity, soil health, and sustainability. The next steps are to refine qSIP methods, validate them through field trials, and translate microbial data into practical, scalable management strategies for agriculture.

15 Where to look for further information

15.1 Relevant papers, protocols, and further reading

- Hungate, B. A., et al. 2015. Quantitative microbial ecology through stable isotope probing. *Applied and Environmental Microbiology* 81, 7570–7581.
- Foley, M. M., et al. 2024. Growth rate as a link between microbial diversity and soil biogeochemistry. *Nature Ecology & Evolution* 8, 2018–2026.
- Koch, B. J., et al. 2018. Estimating taxon-specific population dynamics in diverse microbial communities. *Ecosphere* 9, e02290. https://doi.org/10.1002/ecs2.2090.
- Morrissey, E.M., et al. 2018. Taxonomic patterns in the nitrogen assimilation of soil prokaryotes. *Environmental Microbiology* 20, 3103–3115. https://doi.org/10.1111/1462-2920.14051.
- Purcell, A.M., et al. 2019. Quantitative stable isotope probing with $H_2^{18}O$ to measure taxon-specific microbial growth. *Methods of Soil Analysis* 4, 3043-3045. https://doi.org/10.2136/msa2018.0083.
- Vyshenska, D., et al. 2023. A standardized quantitative analysis strategy for stable isotope probing metagenomics. *mSystems* 8, e01280-22. https://doi.org/10.1128/msystems.01280-22.
- Walkup, J. G. and Morrissey, E. M. 2024. Scaling up taxon-specific microbial traits to predict community-level microbial activity in agricultural systems. *Soil Biology and Biochemistry*, 198, 109622.

15.2 Pipelines and packages

- https://github.com/bramstone/qsip "qsip" R package by Dr. Stone.
- https://github.com/ZielsLab/SIPmg "SIPmg" package for SIP metagenomic data Zeils Lab.
- QIIME qSIP plug-in.

15.3 Workshops and trainings

- (Vyshenska et al., 2023).doe.gov/.

Considerable background information exists on methodological advancements and ecological applications of stable isotope probing techniques that are relevant to agriculture. Radajewski et al. (2000) cover SIP as a method to track microbial activity through isotope incorporation, a broadly applicable technique for identifying active microbial populations in soil without relying on genetic markers. For a critical discussion of DNA-SIP's advantages and limitations, including its suitability for soil research, Neufeld et al. (2007) provided an essential review. For those specifically interested in quantitative approaches, Hungate et al. (2015) introduced qSIP, a technique that estimates isotope incorporation in units that are scalable to biogeochemical processes. Koch et al. (2018) developed one example, converting ^{18}O isotope incorporation from water into relative growth rates of populations in units of organisms per day, normalized by population size (i.e. dN/Ndt). Such metrics have the advantage of being scalable to biogeochemical processes such as nitrogen cycling and carbon sequestration. For methodological guidance on improving qSIP's precision and scaling it for field applications, Sieradzki et al. (2020) provide a detailed examination of measurement error, resolution, and best practices for experimental designNuccio et al. (2022) provide a detailed description of the state-of-the-art in automated fraction collection.

Additionally, researchers interested in microbial growth rates in the context of soil biogeochemistry will find Foley et al. (2024a) useful in linking microbial diversity to ecosystem functions, while Sokol et al. (2022) explore the broader role of microbial metabolism and necromass in soil carbon storage, an essential consideration for sustainable agriculture. Collectively, these sources offer a comprehensive foundation for applying SIP tools to the study of microbial contributions to agricultural sustainability and soil health, providing valuable insights for researchers looking to integrate isotope-based methods into their work on ecosystem services.

16 References

Angst, G., et al. (2021). Plant- or microbial-derived? A review on the molecular composition of stabilized soil organic matter. *Soil Biology and Biochemistry* 156, 108189.

Barnett, S. E., et al. (2021). Multisubstrate DNA stable isotope probing reveals guild structure of bacteria that mediate soil carbon cycling. *Proceedings of the National Academy of Sciences* 118, e2115292118.

Braun, J., et al. (2018). Full 15N tracer accounting to revisit major assumptions of 15N isotope pool dilution approaches for gross nitrogen mineralization. *Soil Biology and Biochemistry* 117, 16–26.

Buckley, D. H. et al. 2007. Stable isotope probing with $^{15}N_2$ reveals novel noncultivated diazotrophs in soil. Appl. *Environ. Microb.* 73(10): 3196-3204.

Buckeridge, K. M., et al. (2022). Deconstructing the microbial necromass continuum to inform soil carbon sequestration. *Functional Ecology* 36, 1396–1410.

Buckeridge, K. M., et al. (2020a). Sticky dead microbes: rapid abiotic retention of microbial necromass in soil. *Soil Biology and Biochemistry* 149, 107929.

Buckeridge, K. M., et al. (2020b). Environmental and microbial controls on microbial necromass recycling, an important precursor for soil carbon stabilization. *Communications Earth & Environment* 1, 36.

Cébron, A. et al. 2007. Nutrient amendments in soil DNA stable isotope probing experiments reduce observed methanotroph diversity. Appl. *Environ. Microb.* 73(3): 798-807.

Chakraborty, A., et al. (2011). Effect of long-term fertilizers and manure application on microbial biomass and microbial activity of a tropical agricultural soil. *Biology and Fertility of Soils* 47, 227–233.

Coplen, T. B. (2011). Guidelines and recommended terms for expression of stable-isotope-ratio and gas-ratio measurement results. *Rapid Communications in Mass Spectrometry* 25, (17), 2538–2560.

Costa, O. Y. A., et al. (2018). Microbial extracellular polymeric substances: ecological function and impact on soil aggregation. *Frontiers in Microbiology* 9, 1636.

Cotrufo, M. F., et al. (2013). The Microbial Efficiency-Matrix Stabilization (MEMS) framework integrates plant litter decomposition with soil organic matter stabilization: do labile plant inputs form stable soil organic matter? *Global Change Biology* 19, 988–995.

Crews, T. E. and Peoples, M. B. (2005). Can the synchrony of nitrogen supply and crop demand be improved in legume and fertilizer-based agroecosystems? A review. *Nutrient Cycling in Agroecosystems* 72, 101–120.

Crotty, F. V., et al. (2012). Using stable isotopes to differentiate trophic feeding channels within soil food webs. *Journal of Eukaryotic MicrobiologyJournal of Eukaryotic MicrobiologyJournal of Eukaryotic Microbiology* Journal of Eukaryotic MicrobiologyJournal of Eukaryotic Microbiology*Journal of Eukaryotic Microbiology*59, 520–526.

Dijkstra, P., et al. (2011). Modeling soil metabolic processes using isotopologue pairs of position-specific (13)C-labeled glucose and pyruvate. *Soil Biology & Biochemistry* 43, 1848–1857.

Dijkstra, F. A., et al. (2015a). Plant and microbial uptake of nitrogen and phosphorus affected by drought using 15N and 32P tracers. *Soil Biology and Biochemistry* 82, 135–142.

Dijkstra, P., et al. (2015b). High carbon use efficiency in soil microbial communities is related to balanced growth, not storage compound synthesis. *Soil Biology & Biochemistry* 89, 35–43.

Dong, W., et al. (2021). Decomposition of microbial necromass is divergent at the individual taxonomic level in soil. *Frontiers in Microbiology* 12.

Dong, W., et al. (2022). Investigating bacterial coupled assimilation of fertilizer-nitrogen and crop residue-carbon in upland soils by DNA-qSIP. *Science of The Total Environment* 845, 157279.

Durán, P. et al. 2021. Engineering Multigenerational Host-Modulated Microbiota against Soilborne Pathogens in Response to Global Climate Change J Biology 10: article 865

Enquist, B. J., et al. (2015). Chapter Nine - Scaling from Traits to Ecosystems: Developing a General Trait Driver Theory via Integrating Trait-Based and Metabolic Scaling Theories. In *Advances in Ecological Research* (S. Pawar, et al., eds.), Vol. 52, pp. 249–318. Academic Press.

Fan, X., et al. (2021). Improved model simulation of soil carbon cycling by representing the microbially derived organic carbon pool. *Isme j* 152248–2263.

Foley, M. M., et al. (2024a). Microbial growth in soil. *EcoEvoRxiv*. https://doi.org/10.32942/osf.io/

Foley, M. M., et al. (2024b). Growth rate as a link between microbial diversity and soil biogeochemistry. *Nature Ecology & Evolution* 8, 2018–2026.

Geyer, K. M., et al. (2016). Microbial carbon use efficiency: accounting for population, community, and ecosystem-scale controls over the fate of metabolized organic matter. *Biogeochemistry* 127, 173–188.

Grandy, A. S., et al. (2022). The nitrogen gap in soil health concepts and fertility measurements. *Soil Biology and Biochemistry* 175, 108856.

Greenlon, A., et al. (2022). Quantitative Stable-Isotope Probing (qSIP) with Metagenomics Links Microbial Physiology and Activity to Soil Moisture in Mediterranean-Climate Grassland Ecosystems. *mSystems* 7, e00417-22.

Hauck, R. and Bremner, J. (1976). Use of tracers for soil and fertilizer nitrogen research. *Advances in Agronomy* 28, 219–266.

Hayer, M., et al. (2021). Microbes on decomposing litter in streams: entering on the leaf or colonizing in the water? *The ISME Journal* 16, 717–725.

Hernández, M. et al. 2023. Identification of diverse antibiotic resistant bacteria in agricultural soil with H218O stable isotope probing combined with high-throughput sequencing. *Environmental Microbiome* 18(34): 2524-6372.

Hestrin, R., et al. (2022). Plant-associated fungi support bacterial resilience following water limitation. *The ISME Journal* 16, 2752–2762.

Hiis, E., et al. (2024). Unlocking bacterial potential to reduce farmland N_2O emissions. *Nature* 630, 421–428.

Hungate, B. A., et al. (2021). The functional significance of bacterial predators. *mBio* 12, e00466-21.

Hungate, B. A., et al. (2015). Quantitative microbial ecology through stable isotope probing. *Applied and Environmental Microbiology* 81, 7570–7581.

Institute, J. G. (2024). Community sequencing program.

Jack, C. N., et al. (2021). Microbial inoculants: silver bullet or microbial jurassic park? *Trends in Microbiology* 29, 299–308.

Jangid, K., et al. (2008). Relative impacts of land-use, management intensity and fertilization upon soil microbial community structure in agricultural systems. *Soil Biology and Biochemistry* 40, 2843–2853.

Jones, P. J. H. and Leatherdale, S. T. (1991). Stable isotopes in clinical research: safety reaffirmed. *Clinical Science* 80, 277–280.

Kakouridis, A., et al. (2024). Arbuscular mycorrhiza convey significant plant carbon to a diverse hyphosphere microbial food web and mineral-associated organic matter. *New Phytologist* 242, 1661–1675.

Kallenbach, C. M., et al. (2016). Direct evidence for microbial-derived soil organic matter formation and its ecophysiological controls. *Nature Communications* 7, 13630.

Kallenbach, C. M., et al. (2015). Microbial physiology and necromass regulate agricultural soil carbon accumulation. *Soil Biology and Biochemistry* 91, 279–290.

Kaminsky, L. M., et al. (2019). The inherent conflicts in developing soil microbial inoculants. *Trends Biotechnol* 37, 140–151.

King, A. E. and Hofmockel, K. S. (2017). Diversified cropping systems support greater microbial cycling and retention of carbon and nitrogen. *Agriculture, Ecosystems & Environment* 240, 66–76.

Koch, B. J., et al. (2018). Estimating taxon-specific bacterial growth rates in intact soil communities. *Ecosphere* 9, e02090.

Kögel-Knabner, I., et al. (2008). Organo-mineral associations in temperate soils: Integrating biology, mineralogy, and organic matter chemistry. *Journal of Plant Nutrition and Soil Science* 171, 61–82.

Krause K. et al. 2020. The Role of Metabolites in the Link between DNA Replication and Central Carbon Metabolism in *Escherichia coli*. Genes (Basel) 11(4): article 447.

Kushner, D. J., et al. (1999). Pharmacological uses and perspectives of heavy water and deuterated compounds. *Canadian Journal of Physiology and Pharmacology* 77, 79–88.

Kuypers, M. M. M., et al. (2018). The microbial nitrogen-cycling network. *Nature Reviews Microbiology* 16, 263–276.

Lavallee, J. M., et al. (2020). Conceptualizing soil organic matter into particulate and mineral-associated forms to address global change in the 21st century. *Global Change Biology* 26, 261–273.

Lee, H., et al. (2017). Reyranella terrae sp. nov., isolated from an agricultural soil, and emended description of the genus Reyranella. *International Journal of Systematic and Evolutionary Microbiology* 67, 2031–2035.

Lehmann, A., et al. (2020). Fungal traits important for soil aggregation. *Frontiers in Microbiology* 10, 1333.

Lehmann, J. (2009). Biological carbon sequestration must and can be a win-win approach. *Climatic Change* 97, 459–463.

Leichty, S. I., et al. (2021). Site and bioenergy cropping system similarly affect distinct live and total soil microbial communities. *Frontiers in Microbiology* 12, 701094.

Li, J., et al. (2019). Predictive genomic traits for bacterial growth in culture versus actual growth in soil. *The ISME Journal* 13, 2551–2564

Li, X. G., et al. (2017). Nitrogen fertilization decreases the decomposition of soil organic matter and plant residues in planted soils. *Soil Biology and Biochemistry* 112, 47–55.

Liang, C., et al. (2019). Quantitative assessment of microbial necromass contribution to soil organic matter. *Global Change Biology* 25, 3578–3590.

Liu, C. M., et al. (2023). Using source-associated mobile genetic elements to identify zoonotic extraintestinal E. coli infections. *One Health* 16, 100518.

Lopes, E. D. S., et al. (2024). Cross-environmental cycling of antimicrobial resistance in agricultural areas fertilized with poultry litter: a one health approach. *Environmental Pollution* 363, 125177.

Lu, J., et al. (2023). Orientation-driven photosynthesized carbon belowground mediates intercropped peanut microbiota changes for pathogen resistance. *Plant and Soil.* 506(1): 209-226

Ma, Q., et al. (2020). Farmyard manure applications stimulate soil carbon and nitrogen cycling by boosting microbial biomass rather than changing its community composition. *Soil Biology and Biochemistry* 144, 107760.

Malik, A. A., et al. (2015). Rhizosphere bacterial carbon turnover is higher in nucleic acids than membrane lipids: implications for understanding soil carbon cycling. *Frontiers in Microbiology* 6, 656.

Matson, P. A., et al. (1998). Integration of environmental, agronomic, and economic aspects of fertilizer management. *Science* 280, 112–115.

Mau, R. L., et al. (2024). Measurements of soil protist richness and community composition are influenced by primer pair, annealing temperature, and bioinformatics choices. *Applied and Environmental Microbiology* 90, e00800-24.

Mayali, X., et al. (2012). High-throughput isotopic analysis of RNA microarrays to quantify microbial resource use. *ISME Journal* 6, 1210–1221.

Michener, R. and Latjtha, K., eds. (2008). *Stable Isotopes in Ecology and Environmental Science*, pp. 1–594. Blackwell Publishing Ltd.

Monard, C. et al. 2011. Relationship between bacterial diversity and function under biotic control: the soil pesticide degraders as a case st udy. *The ISME Journal* 5: 1048-1056.

Morrissey, E. M., et al. (2023). Carbon acquisition ecological strategies to connect soil microbial biodiversity and carbon cycling. *Soil Biology and Biochemistry* 177, 108893.

Morrissey, E. M., et al. (2018). Taxonomic patterns in the nitrogen assimilation of soil prokaryotes. *Environmental Microbiology* 20, 1112–1119.

Neufeld J. D et al. 2007. Methodological considerations for the use of stable isotope probing in microbial ecology. *Microb Ecol*. 53(3): 435-42.

Novaes, D. R., et al. (2024). Natural habitat connectivity and organic management modulate pest dispersal, gene flow, and natural enemy communities. *Ecollogical Applications* 34, e2938.

Nuccio, E. E., et al. (2022). HT-SIP: a semi-automated stable isotope probing pipeline identifies cross-kingdom interactions in the hyphosphere of arbuscular mycorrhizal fungi. *Microbiome* 10, 199.

Nuccio, E. E., et al. (2013). An arbuscular mycorrhizal fungus significantly modifies the soil bacterial community and nitrogen cycling during litter decomposition. *Environmental Microbiology* 15, 1870–1881.

Ochs, C., et al. (2025). From treetops to river bottoms: Exploring the role of phyllosphere fungi in aquatic fungal communities. *Ecology* 106, e4513.

Ouyang, W.-Y., et al. (2019). Identification of dominant sulfamethoxazole-degraders in pig farm-impacted soil by DNA and protein stable isotope probing. *Environment International* 126, 118–126.

Papp, K et al. 2020. Glucose triggers strong taxon-specific responses in microbial growth and activity: insights from DNA and RNA qSIP. *Ecology* 101(1): e02887.

Portillo, M. C., et al. (2013). Cell size distributions of soil bacterial and archaeal taxa. *Applied and Environmental Microbiology* 79, 7610–7617.

Pörtner, H.-O., et al. (2022). *Climate Change 2022: Impacts, Adaptation, and Vulnerability. Contribution of Working Group II to the Sixth Assessment Report of the Intergovernmental Panel on Climate Change*. Cambridge, UK and New York, NY, USA.

Prommer, J., et al. (2020). Increased microbial growth, biomass, and turnover drive soil organic carbon accumulation at higher plant diversity. *Global Change Biology* 26, 669–681.

Prosser, J. I., et al. (2014). The Family Nitrosomonadaceae. In *The Prokaryotes: Alphaproteobacteria and Betaproteobacteria* (E. Rosenberg, E. F. DeLong, S. Lory, E. Stackebrandt and F. Thompson, eds.), pp. 901–918. Springer Berlin Heidelberg.

Raczka, N. C., et al. (2021). Interactions between microbial diversity and substrate chemistry determine the fate of carbon in soil. *Scientific Reports* 11, 19320.

Ramirez, K. S., et al. (2012). Consistent effects of nitrogen amendments on soil microbial communities and processes across biomes. *Global Change Biology* 18, 1918–1927.

Radajewski S. et al. 2000. Stable-isotope probing as a tool in microbial ecology. *Nature* 403(6770): 646-9.

Reed, K., et al. (2025a). Comparing field and lab quantitative stable isotope probing for nitrogen assimilation in soil microbes. *Applied and Environmental Microbiology* 0, e01849-24.

Reed, K., et al. (2025b). Comparing field and lab quantitative stable isotope probing for nitrogen assimilation in soil microbes. *Applied and Environmental Microbiology* 91, e01849-24.

Rempfert, K. R., et al. (2024). Biomolecular budget of persistent, microbial-derived soil organic carbon: the importance of underexplored pools. *Science of The Total Environment* 932, 172916.

Rubin, R. L., et al. (2018). Developing climate-smart restoration: can plant microbiomes be hardened against heat waves? *Ecological Applications* 28, 1594-1605.

Schimel, D. S. (1993). *Theory and Application of Tracers*. Academic Press.

Sihi, D., et al. (2021). Representing methane emissions from wet tropical forest soils using microbial functional groups constrained by soil diffusivity. *Biogeosciences* 18, 1769–1786.

Sieradzki E. T. et al. 2020. Measurement Error and Resolution in Quantitative Stable Isotope Probing: Implications for Experimental Design. *mSystems* 5(4): e00151-20.

Simpson, A., et al. (2024). MISIP: a data standard for the reuse and reproducibility of any stable isotope probing-derived nucleic acid sequence and experiment. *GigaScience* 13, giad045.

Sokol, N. W., et al. (2022). Life and death in the soil microbiome: how ecological processes influence biogeochemistry. *Nature Reviews MicrobiologyNature Reviews MicrobiologyNature Reviews MicrobiologyNature Reviews Microbiology* 20, 415–430.

Starr, E. P., et al. (2018). Stable isotope informed genome-resolved metagenomics reveals that Saccharibacteria utilize microbially-processed plant-derived carbon. *Microbiome* 6, 122.

Stone, B. W., et al. (2021). Nutrients cause consolidation of soil carbon flux to small proportion of bacterial community. *Nature Communications* 12, 3381.

Stone, N. E., et al. (2022). Diverse lineages of pathogenic Leptospira species are widespread in the environment in Puerto Rico, USA. *PLOS Neglected Tropical Diseases* 16, e0009959.

Tao, F., et al (2023). Microbial carbon use efficiency promotes global soil carbon storage. *Nature* 618, 981–985.

Thomson, J. F. (1960). Physiological effects of D2O in mammals. *Annals of the New York Academy of Sciences* 84, 736–744.

Tijhuis, L., et al. (1993). A thermodynamically based correlation for maintenance gibbs energy requirements in aerobic and anaerobic chemotrophic growth. *Biotechnology and Bioengineering* 42, 509–519.

Trejos-Espeleta, J. C., et al. (2024). Principal role of fungi in soil carbon stabilization during early pedogenesis in the high Arctic. *Proceedings of the National Academy of Sciences* 121, e2402689121.

Vitousek, P. M. (1990). Biological invasions and ecosystem processes - towards an integration of population and ecosystem studies. *Oikos* 57, 7–13.

Vyshenska, D., et al. (2023). A standardized quantitative analysis strategy for stable isotope probing metagenomics. *mSystems* 8, e01280-22.

Walkup, J. G. and Morrissey, E. M. (2025). Scaling up taxon-specific microbial traits to predict community-level microbial activity in agricultural systems. *Soil Biology and Biochemistry* 200, 109622.

Wang, F., et al. (2016). In situ stable isotope probing of phosphate-solubilizing bacteria in the hyphosphere. *Journal of Experimental Botany* 67, 1689–1701.

Wessel, W. W. and Tietema, A. (1992). Calculating gross N transformation rates of 15N pool dilution experiments with acid forest litter: Analytical and numerical approaches. *Soil Biology and Biochemistry* 24, 931–942.

West, G. B. and Brown, J. H. (2005). The origin of allometric scaling laws in biology from genomes to ecosystems: towards a quantitative unifying theory of biological structure and organization. *The Journal of Experimental Biology* 208, 1575–1592.

Wieder, W. R., et al. (2015). Representing life in the Earth system with soil microbial functional traits in the MIMICS model. *Geoscientific Model Development* 8, 1789–1808.

Xu, C., et al. (2021). Crop straw incorporation alleviates overall fertilizer-N losses and mitigates N2O emissions per unit applied N from intensively farmed soils: an in situ 15N tracing study. *Science of The Total Environment* 764, 142884.

Xu, Q., et al. (2024). Active microbial population dynamics and life strategies drive the enhanced carbon use efficiency in high-organic matter soils. *mBio* 15, e00177-24.

Zhang, H., et al. (2022). Subsoiling and conversion to conservation tillage enriched nitrogen cycling bacterial communities in sandy soils under long-term maize monoculture. *Soil and Tillage Research* 215, 105197.

Zhang, S., et al. (2024). Biodegradation of penicillin G sodium by Sphingobacterium sp. SQW1: performance, degradation mechanism, and key enzymes. *Journal of Hazardous Materials* 468, 133485.

Zhou, J., et al. (2016). Thirty four years of nitrogen fertilization decreases fungal diversity and alters fungal community composition in black soil in northeast China. *Soil Biology and Biochemistry* 95, 135–143.

Chapter 7

Applications and advances in using phospholipid fatty acids (PLFAs) as biomarkers in analyzing soil microbial communities

Sarina Claassens, Curtin University, Australia

1 Introduction

Microbial communities are an integral part of healthy soil ecosystems and plant production, and microbial community function and structure are responsive to a myriad of environmental factors and management practices. The cell membranes of bacteria and fungi contain phospholipids as a key component, and when extracted from soil and analyzed, they provide a snapshot of the microbial community structure in that particular soil environment. Microbial ecologists have extensively used phospholipid fatty acid (PLFA) analysis for more than 40 years in environmental studies. This biomarker-based method is a robust tool for understanding microbial dynamics and soil health, and its versatility across a range of ecosystems has been well established.

A unique feature of PLFA analysis is that microbial lipids can be quantitatively extracted from almost any sample matrix (Peacock and White, 2016; Quideau et al., 2016). Results obtained from the analysis of PLFAs with gas chromatography (GC) and mass spectrometry (MS) provide (1) an estimate

http://dx.doi.org/10.19103/AS.2025.0159.08

of viable microbial biomass, (2) an outline of community structure based on signature PLFA biomarkers, and (3) an indicator of microbial physiological status (Peacock and White, 2016). The latter is possible because the lipid composition of microbial cell membranes can change due to changes in the environment induced by climatic variation, disturbance, or contamination, and analyzing certain signature fatty acid biomarkers then provides information about physiological status (Willers et al., 2015a, 2015b). In studies of agricultural soils, the method has often been used to screen for the effects of management inputs on bacterial versus fungal biomass.

This chapter covers the principles and techniques of PLFA analysis and explains the functional and structural aspects of microbial communities that can be described by PLFA analysis. It also gives an overview of the interpretation and associated caveats of the data, how it compares to other methods, and how it is applied in agriculture. Lastly, the chapter provides an overview of recent advances and possible future trends.

1.1 Historical development and key applications

Quideau et al. (2016) provides an informative introduction to the origins and early developments of the contemporary PLFA analytical method, including original modifications to the Bligh and Dyer (1959) method that is still the basis of most isolation protocols used today. Phospholipid fatty acid analysis has been used in soil studies since the 1980s, becoming popular during the time before molecular methods as a rapid and inexpensive technique to sensitively determine biomass and community structure without the bias associated with cultivation (Zelles, 1999). Today, it is still used to quantify living biomass and to describe microbial community structure according to broad phylogenetic groups. The method itself is still being modified to facilitate more comprehensive and standardized analyses (Camenzind et al., 2024; Gorka et al., 2023).

Early studies were primarily performed to establish the foundational use of PLFA profiles as reliable indicators of microbial community structure linked to environmental effects (Frostegård et al., 1993; Frostegård et al., 2011). Over time, its effectiveness across various ecosystems, including agricultural soils and contaminated sites, was also demonstrated (Frostegård et al., 2011; Zelles, 1999) and later extended beyond soil to aquatic environments, sediments (Boon et al., 1996; Li, et al., 2007), and even extreme habitats like deserts (Ben-David et al., 2011) and permafrost (Wagner et al., 2005).

Developments in GC-MS and high-resolution applications in the 1990s, combined with improved spectral libraries and software solutions, significantly enhanced the robustness and reproducibility of PLFA profiling. Integration of PLFA data with data from other techniques and multivariate statistical tools, such as principal component analysis (PCA), redundancy analysis, and correspondence analysis, enabled researchers to interpret complex datasets and infer relationships between environmental factors and microbial community attributes (Ben-David et al., 2011; Frostegård et al., 1993; Grayston et al., 2004, Velasco et al., 2010).

Despite the increased accessibility to molecular methods, PLFA analysis is still widely applied in studies of soil environments, including agricultural soils, often in combination with molecular or other methods. This enables researchers to examine shifts in microbial community structure under varying environmental conditions or agricultural management practices more comprehensively, leading to a better understanding of soil health and a more holistic view of ecosystem responses.

1.2 Assessing and monitoring soil health in agriculture

Phospholipid fatty acid analysis is widely used in agricultural soil health monitoring to study microbial responses to agricultural practices such as tillage, crop rotation, and organic amendments (Norris et al., 2023). The characterization of viable biomass is a valuable measure to understand temporal plant-microbe interactions (Oates et al., 2017) and microbial community structure can be studied with insight in long-term field trials under different management systems (Daquido et al., 2013). Since certain PLFA biomarkers are sensitive to soil disturbances, application of pesticides or herbicides, and changes in temperature and moisture content, they can act as early warning signals of soil stress before significant declines in soil function occur (Zhang et al., 2019). The high turnover rate of these biomarkers also means that PLFA gives an accurate representation of active microbial communities that allows for monitoring microbial shifts in real time, rather than relying on the legacy effects of soil (Buyer et al., 2019).

2 Principles and techniques of PLFA analysis

2.1 What are PLFAs and what insights do they provide?

Phospholipids are polar lipids that occur in intact cells and are the main structural elements of bacterial and fungal cell membranes, where PLFAs of different chain lengths and branching are important to maintain the integrity and function of the cell membrane (Peacock and White, 2016). Soil PLFAs with a chain length between 14 and 20 carbon atoms are typically considered of microbial origin (Joergensen, 2022). The phospholipids are arranged in the membrane in a bilayer with a polar (hydrophilic) head group and two fatty acid tails (hydrophobic) arranged toward one another inside the bilayer (Fig. 1). These fatty acid tails have different chemical structures based on which

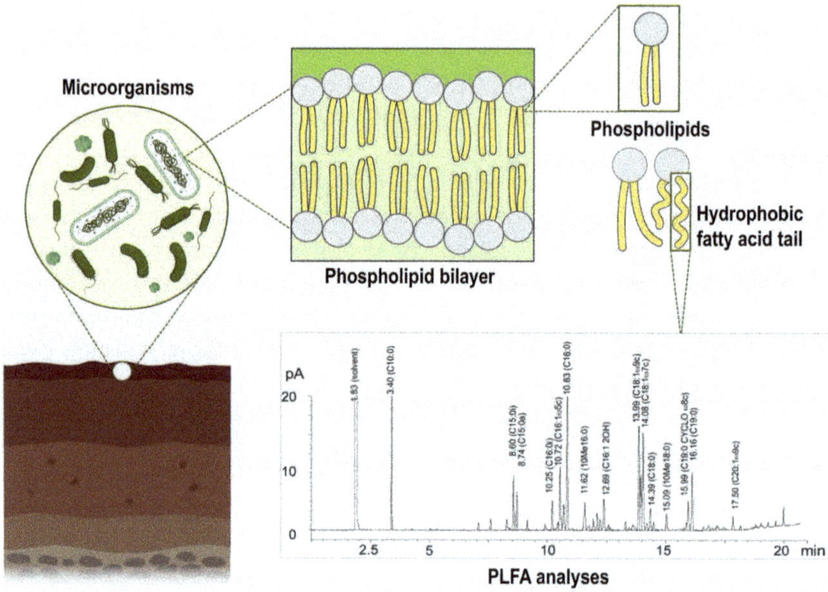

Figure 1 Diagram illustrating the structure of the phospholipid bilayer, highlighting the arrangement of polar head groups and hydrophobic fatty acid tails and showing that fatty acid tails are analyzed based on differences in their structures. PLFA: Phospholipid fatty acid .

microorganisms they come from and therefore can be used as group-specific biomarkers, offering insights into the composition of the microbial community. Importantly, phospholipids only occur in intact cells and rapidly degrade when cells die; they also do not occur in storage components of cells (Zelles, 1999) and therefore provide a reflection of the viable microbial biomass *in situ* at a specific time (White et al., 1979; Zhang et al., 2019).

Phospholipid fatty acid analysis relies on extracting and identifying the unique fatty acids associated with microbial membranes to quantitatively infer viable biomass, community structure, physiological states, and environmental interactions. First, the basis of microbial group-specific biomarkers is important, meaning that specific PLFAs are linked to distinct microbial groups. For example, branched-chain fatty acids such as *iso*- and *anteiso*-branched PLFAs indicate Gram-positive bacteria (Peacock and White, 2016). These group-specific biomarkers provide a "snapshot" of the entire microbial community present in a soil sample at a specific time, enabling assessments of diversity in terms of broad functional groups, community structure, and shifts due to agricultural practices or environmental changes.

Fatty acid nomenclature follows the convention where the fatty acid is identified by X:YωZ and where X represents the number of carbon atoms, Y represents the number of double bonds, and Z indicates the position of the first double bond from the aliphatic (ω) end of the molecule. The suffixes "c" and "t" indicate *cis* and *trans* geometric isomers. The prefixes and suffixes "a" and "i" refer to *anteiso* and *iso* branching, and Me and OH indicate methyl groups and hydroxyl groups, respectively (Quideau et al., 2016).

Some bacteria respond to environmental conditions by changing the specific fatty acids in their cell membrane, which can subsequently be used as biomarkers of physiological stress or changes (e.g. cyclopropyl fatty acids). This maintains cellular activity under adverse or shifting environmental conditions because it enables adaptation or regulation of cellular fluidity and permeability. Ratios of specific fatty acids, including the ratio of saturated to unsaturated PLFA, serve as indicators of ecosystem functionality and soil health (Willers et al., 2015b) and are discussed in more detail in Section 3.

2.2 Extraction and detection of PLFAs

The PLFA method consists of four distinct parts (Fig. 2), namely:

1. extraction of the total lipids from cells;
2. fractionation of the lipid extract into various lipid classes;
3. methylation of polar lipids into fatty acid methyl esters (FAMEs); and
4. the separation, quantification, and identification of the FAMEs using GC-MS.

Peacock and White (2016) and Quideau et al. (2016), are good examples of easy-to-follow protocols based on these four parts. Both also include informative notes about the development of the method, caveats and interpretation of results, and Quideau et al. (2016) includes a URL link to a step-by-step video of the method described in the article (see Section 7).

1) Extraction of the total lipids from cells

Various modifications of the original Bligh and Dyer (1959) procedure are used to quantitatively extract the lipid soluble components from viable cells, resulting in a total or representative extraction of fatty acids from soil samples. The procedure uses a single-phase chloroform:methanol:water extraction system (White et al., 1979) that is optimized for high recovery rates of phospholipids from microbial membranes.

2) Fractionation of the lipid extract

The whole lipid extracts obtained from soil samples consist of different lipid classes that are categorized into neutral-, glyco-, and polar lipids (phospholipids) based on increased polarity. Fractionation is achieved by using silicic acid column chromatography and organic solvents (chloroform, methanol, and water) to sequentially separate these fractions into liquid mobile

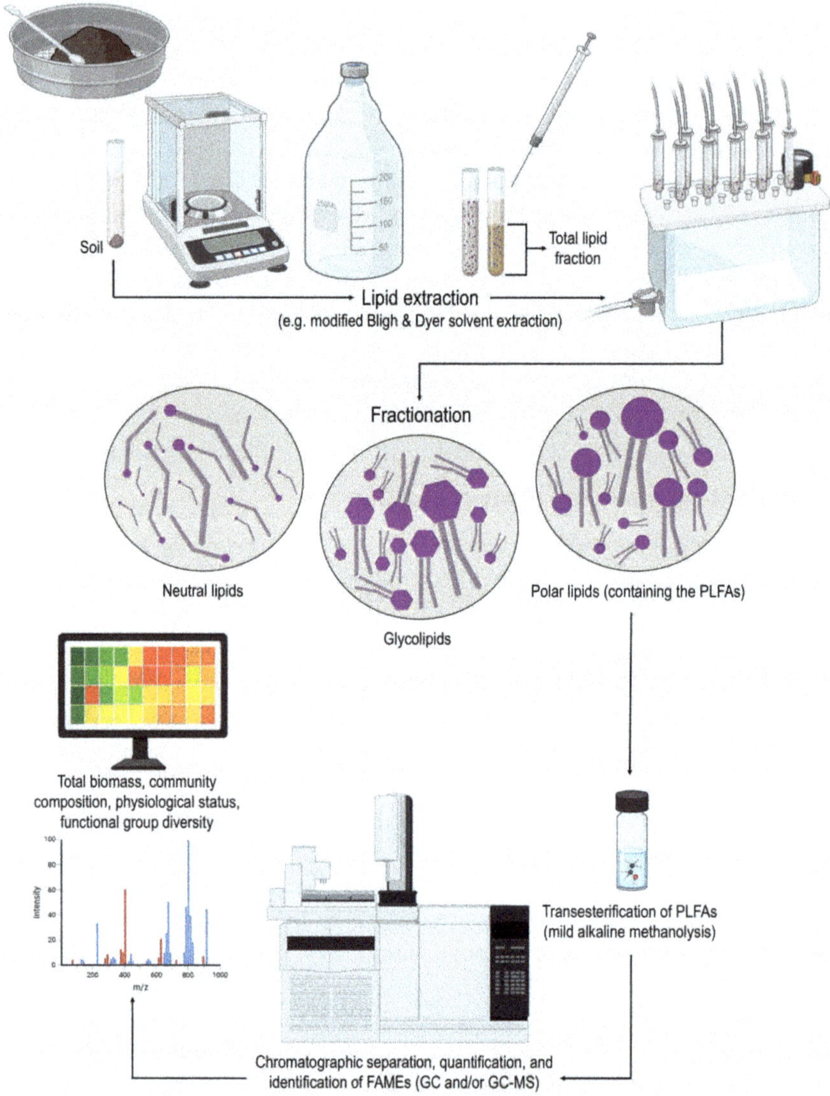

Figure 2 Flow diagram of the extraction, fractionation, methylation, analysis and data processing of phospholipid fatty acids. PLFA: phospholipid fatty acids, FAME: fatty acid methyl ester, GC: gas chromatograph with a flame ionization detector; MS: mass spectrometer/spectrometry (created with BioRender.com).

phases (Peacock and White, 2016). If the analyte (the lipid fraction) has a higher affinity for the organic solvent than for the stationary phase (the silicic acid), it will be eluted into the mobile phase (Willers et al., 2015a).

3) Methylation of phospholipids to FAMEs

Methylation is an example of derivatization, the objective of which is to increase the volatility of analytes for better separation, improve thermal stability, and decrease the interaction of the sample with the GC column (Halket, 2003). The fraction of the lipid extract containing the phospholipids is subjected to mild alkaline methanolysis to convert fatty acids into FAMEs (transesterification). The FAMEs are volatile and suitable for GC analysis.

4) Analysis of FAMEs

The FAMEs are separated using GC and detected with flame ionization detection (FID) or MS. Retention times and mass spectral libraries are used to identify specific fatty acids. Peak areas in chromatograms are used to quantify the relative abundance of each fatty acid (Peacock and White, 2016).

2.3 Standardization of procedures

Due to the sensitivity of PLFA biomarkers, it is important to ensure the minimization of potential error and misinterpretation of data due to inconsistent procedures and to consider the factors that could influence extraction and analysis of PLFAs.

Sample handling and storage conditions can lead to significant losses of PLFA biomarkers and will have a disproportionate effect on microbial groups (Veum et al., 2019). It is recommended that samples be handled as little as possible, with sterile sampling tools, and that samples are immediately frozen and lyophilized with storage at −80°C before extraction (Veum et al., 2019; White et al., 1979). Optimal sample handling with immediate lyophilization is often impractical, particularly during field sampling in remote locations far from a laboratory. In such cases, especially when multiple sampling events are conducted and results are compared over time, it is crucial to follow a consistent approach to sample handling and storage. This ensures that variations in data are not due to inconsistent methodology but to actual changes in the microbial profile. Extraction parameters, including the choice of solvent and buffer (Papadopoulou et al., 2011; Willers et al., 2015a), and soil properties, such as soil moisture and hydrophobicity, have a notable effect on the quantity and profile of PLFAs extracted from soils (Joergensen, 2022; Papadopoulou et al. 2011).

When comparing results, it is important to ensure that the unit in which the lipid data is expressed is considered. Lipid data can be analyzed in various ways but is generally expressed in terms of abundance (nmol.g^{-1} soil), mole fraction (nmol individual lipid.nmol^{-1} total lipid), and mole percent (mole fraction × 100) (Oates et al., 2017). Abundance is the amount of a given lipid extracted per a given amount of soil and gives a reliable indication of the total abundance of lipids in a sample, while signature lipid biomarker abundance indicates the abundance of the corresponding microbial functional group

(Willers et al., 2015a). Mole percent and mole fraction are normalized by the total lipids in a sample to give the relative abundance of specific biomarkers. Arcsine square root transformation will render data that is appropriate for use in multivariate ordinations (Oates et al., 2017).

Standardization and reproducibility of PLFA analysis across studies remain challenging due to variations in extraction techniques and chromatographic conditions. It is therefore important to: (i) include the use of internal standards to ensure consistency in quantification, (ii) validate the method, and (iii) correctly assign indicator PLFAs during interpretation. For studies that will monitor the same soils over time, it is important to use the same standardized protocols or to send samples to the same laboratory for analysis.

3 Interpretation of results and comparison with other methods

Comparing microbial community profiles based on PLFA patterns across different studies is often challenging because interpretations of PLFA data are inconsistent. These discrepancies, along with their implications for environmental studies, are discussed in Willers et al. (2015b). Frostegård et al. (2011) discussed the "use and misuse" of PLFA in soil studies and includes the flawed use of diversity indices and the extent to which PLFA turnover in soil can affect interpretation.

The reasons for using PLFA analysis in agricultural studies can vary, potentially providing insight into microbial biomass, community structure, functional status, or a combination of these and generally connected to management practices and/or environmental variables. Each of these applications requires interpretation of the data obtained from the analysis, and this is often where inconsistencies arise that limit subsequent comparisons with similar studies (Watzinger, 2015; Willers et al., 2015b).

3.1 Interpretation of PLFA data

One of the most important aspects in interpretation is to ensure correct and consistent assignment of indicator PLFA to microbial groups. While certain PLFAs are commonly associated with specific microbial groups (e.g. 16:1ω7c for Gram-negative bacteria, 18:2ω6, 9 for fungi), many PLFAs are present in multiple taxa. This can lead to ambiguity in interpretation (Ruess and Chamberlain, 2010; Zelles, 1999). Joergensen (2022) states that the assignment of indicator PLFA to microbial groups is often not compatible with basic microbiological knowledge, e.g. when failing to assign Actinobacterial PLFA to the group of Gram-positive bacterial PLFA or arbuscular mycorrhizal fungi (AMF) PLFA to the group of fungal PLFA. Recommended assignment of PLFA to different main microbial groups is presented in Table 1. The number

Table 1 Assignment of PLFA to different main microbial groups (Joergensen 2022 and references therein)

Bacterial PLFA	Gram-positive + Gram-negative
G-positive	Firmicutes + Actinobacteria
Firmicutes	i14:0, **i15:0**[a], i16:0[a], **i17:0**, i18, **a15:0**, a16:0, a17:0, a18:0, a19:0
Actinobacteria	10Me16:0, **10Me17:0**, **10Me18:0**
G-negative	**cy17:0**, cy19:0, 16:1ω7, 16:1ω9, **17:1ω8**, **18:1ω7**
Fungal PLFA	AMF + Zygomycota + Ascomycota + Basidiomycota + unspecific fungal
AMF	**16:1ω5c**[b]
Zygomycota	**18:1ω9c**[c]
Ascomycota and Basidiomycota	**18:2ω6c**[d]
Unspecific fungal PLFA	18:3ω6,9,12[d]
Unspecific microbial PLFA	**14:0**, **15:0**, **16:0**[d], **17:0**, 18:0, 20:0, **20:4ω6,9,12,15**
Total microbial PLFA	Bacterial + fungal, unspecific microbial

Bold > 1 mol% contribution to total PLFA in soil[a] Also some cultured G- bacteria contained higher concentrations of PLFA i15:0 and i16:0[b] Also some cultured G- bacteria contained minor concentrations of PLFA 16:1ω5[c] Also some cultured G- bacteria and plants contained higher concentrations of PLFA 18:1ω9[d] Also plants contained higher concentrations of PLFA 18:2ω6,9, 18:3ω6,9,12, and 16:0

AMF, arbuscular mycorrhizal fungi; G-, Gram-negative; PLFA, phospholipid fatty acid.

of PLFA used as a group-specific indicator can also vary and total PLFA is not consistent across studies (Joergensen, 2022). For this reason, it is very important to list all PLFA calculations clearly in the materials and methods section of a paper.

3.2 Differences in FAME profiling methods

A popular alternative to the extraction of phospholipid-derived fatty acids is the whole-cell FAME extraction analysis using the Microbial Identification Inc. (MIDI) method. Lipids are extracted directly from pure cultures or soil/sediment samples through saponification (Fernandes et al., 2013). This rapid method has lower lipid loss due to having none of the concentration or purification steps of the PLFA method and uses a smaller amount of soil. However, it was originally designed for identifying organisms in pure culture, so there are no initial extraction or purification steps. The main disadvantage for soil studies is that the extracts can include lipid-like compounds co-extracted from soil organic matter that distort biomass measures and microbial community signatures.

It has therefore typically been used to broadly and qualitatively describe soil lipids (Fernandes et al., 2013; Oates et al., 2017).

Another method to characterize FAMEs is to use ester-linked FAMEs (EL-FAMEs), a lower cost and simpler method that makes it more practical. Li et al. (2020) found that both the PLFA and EL-FAME methods produced similar profile abundances for important microbial markers, but that PLFA was more sensitive to soil chemical properties such as pH, clay content, soil organic matter, and active carbon. The study also showed that bacterial biomarkers were key to distinguishing between soil samples with PLFA, but that fungal biomarkers were more important in EL-FAME analysis, especially the high abundance of the AMF biomarker.

3.3 PLFA versus other methods

Studies comparing PLFA analysis to community level substrate utilization, for example denaturing gradient gel electrophoresis (DGGE) and DNA-based methods, have generally found good differentiation by all the methods. When comparing DNA-based methods with PLFA analysis, some found that in many cases the PLFA allowed detection of microbial community responses with greater sensitivity (Duncan et al., 2016; Liang et al., 2016; Ramsey et al., 2006). However, the results obtained by different methods, although comparable, give different perspectives on microbial communities, and each method has strengths and weaknesses.

Monitoring of microbial diversity and community structure through analysis of DNA metabarcoding is increasingly popular and accessible. This approach characterizes complex microbial communities in detail and can be utilized for a range of applications and in various matrices, such as soil, sediments, and water but is not without bias (Lear et al., 2018; Orwin et al., 2018). Lewe et al. (2021) found that metabarcoding and PLFA analysis provided different views of bacterial composition and that these differences varied among substrates, but that relative biomass and relative sequence abundances aligned reasonably well. It was recommended that using PLFA in parallel with metabarcoding (as complementary methods) can strengthen comparisons between samples, substrates, and studies. Importantly, the results demonstrate that PLFA correction adds bacterial biomass as an ecologically meaningful dimension to the data and can mitigate problems arising from compositional sequencing results (Lewe et al., 2021). In a comparison of EL-FAME, PLFA, and quantitative PCR methods to detect changes in the abundance of soil bacteria and fungi, Siles et al. (2024) found that PLFA analysis performed better than EL-FAME in forest soil and in detecting small antibiotic-induced decreases in microbial abundances. Both the PLFA and

EL-FAME methods showed stronger relationships with soil basal respiration and induced soil changes, indicating that these methods relate better to the living microbial community than qPCR.

4 Applications of PLFA analysis in agricultural soils

Considerable work has been done to understand the dynamics of microbial communities in agricultural soils and how management practices, such as tillage, crop rotation, organic amendments, and pesticides impact the function and structure of these communities. Their ability to deliver the processes that uphold healthy soil function such as nutrient cycling, formation of organic matter, carbon sequestration, and soil structure is of the utmost importance to agriculture and the environment. Phospholipid fatty acid analysis has proven popular and useful for providing insight into the relative abundance of microbial groups in response to agricultural practices and for tracking trends over time. It is especially useful when combined with other soil health indicators such as organic carbon and enzymatic activity. Besides the application of microbial biomass and the abundance of functional groups of microorganisms, ratios of certain PLFAs are also used as indicators of certain soil health parameters. Although PLFA indicators show general trends and are used based on a number of accepted assumptions, it is important to consider any ratios separately and to interpret values based on context (Fierer et al., 2021; Willers et al., 2015b).

4.1 The ratio of fungal to bacterial PLFA

The ratio of fungal to bacterial PLFA gives an indication of the abundance of fungi versus bacteria in the soil and is expressed as a ratio of biomass. It is considered an important ecological indicator, with higher fungal to bacterial ratios being indicative of a more sustainable soil system (Camenzind et al., 2024; Fierer et al., 2021). Relative abundances of fungi and bacteria correlate to soil parameters relevant for carbon cycling, and high fungal to bacterial ratios have been shown to correlate with carbon to nitrogen ratios and organic carbon stocks. Bacteria tend to dominate in systems with lower organic matter and lower carbon to nitrogen ratios (Agnihotri et al., 2021; Camenzind et al., 2024; Strickland and Rousk, 2010). This ratio has also been shown to correlate well with soil pH (Bååth and Anderson, 2003; Rousk et al., 2010). The caveat of using the fungal to bacterial ratio is that fungi and bacteria often have overlapping niches and functions in soil and that ratios may vary for many reasons, making interpretation difficult (Fierer et al., 2021).

4.2 Gram-positive to Gram-negative bacterial PLFA

It is generally understood that Gram-positive and Gram-negative bacteria use different carbon sources, and the ratio of Gram-positive to Gram-negative biomass is used as an indicator of the relative carbon availability for soil bacterial communities in organic soils and of energy limitation in natural ecosystems (Fanin et al., 2019). Gram-positive bacteria survive better under certain environmental conditions such as drought or extreme temperatures, due to their ability to form spores, and they use soil organic matter-derived carbon sources that are more recalcitrant, while Gram-negative bacteria use more plant-derived carbon sources, which are labile.

4.3 The influence of environmental conditions on stress and community activity ratios

Bacteria alter their membranes under various environmental conditions to maintain optimal fluidity for the transport of nutrients and waste products. By looking at the ratios of certain signature fatty acids, we can determine whether microbial communities are stressed (Norris et al., 2023). The following ratios are considered indicative of stress due to environmental conditions such as drought, temperature, pH, or lack of nutrients (Guckert et al., 1986; Willers et al., 2015b):

1. **sat:unsat** – the ratio of saturated to unsaturated fatty acids, where stress will lead to a higher proportion of unsaturated fatty acids;
2. **mono:poly** – a lower ratio of monounsaturated to polyunsaturated fatty acids is used to indicate stress;
3. ***trans:cis*** – *trans-* to *cis* monoenoic fatty acids, where most bacterial cultures and environmental samples will have a ratio of less than 0.1 and a ratio significantly greater is interpreted as a starvation or stress index; and
4. **16:1ω7c:cy17:0** and **18:1ω7c:cy19:0** – the ratio of cyclopropyl fatty acids and their monoenoic precursors indicates stress conditions that affect growth and the fluidity of bacterial membranes. Cyclopropyl fatty acids are more prominent during stationary phases of growth, and a higher proportion of the precursors (16:1ω7c and 18:1ω7c) indicates an actively growing community.

4.4 The influence of management practices on microbial communities

Phosholipid fatty acid analysis has been widely used in studies of agricultural soil health, conservation agriculture, and to compare more traditional practices with those employing fewer inputs and more organic approaches. The success of the method in improving our understanding of shifts in microbial communities in this context is well established. It shows the relationship between microbial community structure and environmental conditions (e.g. Chen et al., 2019; Mann et al., 2019; Rousk et al., 2010; Yang et al., 2022). It can distinguish different agricultural management practices, such as the influence of soil types, crop species, crop rotation, tillage practices, irrigation, and application of pesticides (e.g. Drenovsky et al., 2010; Du Preez et al., 2024; Francisco et al., 2022; Habig et al., 2018; Norris et al., 2023).

4.5 The North American Project to Evaluate Soil Health Measurements dataset

Norris et al. (2023) set out to address challenges in processing of data and interpretation of results of PLFA analysis through the North American Project to Evaluate Soil Health Measurements dataset. This dataset consisted of the results of more than 1800 agricultural soil samples collected from more than 100 sites and representing large variations in climate, parent material, and management practices (crop rotation, tillage, cover crops) (Norris et al., 2020). The study also included other laboratory testing methods such as pH and organic carbon, but the focus was on using PLFA-derived biomass and ratios as phenotypic biomarkers for soil health and the influence of management practices (Norris et al., 2023).

The first objective of the study (Norris et al., 2023) was to support standardization of PLFA data processing by evaluating different methods for calculating total PLFA biomass. The authors calculated total PLFA biomass of all samples from the study by summing the biomarkers in four different ways:

1 All biomarkers between C14:0 and C20:0 (Quideau et al., 2016);
2 C9:0 to C20:0 (Francisco et al., 2016);
3 A selected subset of 27 biomarkers according to Frostegård et al. (1993);
4 A selected subset of biomarkers according to the method of Bradford et al. (2019), using 18 of the 27 biomarkers identified in Frostegård et al. (1993).

The sum of PLFAs from (2) had a linear correspondence to the sum from (1), while (3) and (4) were also linear but less than (1), with (4) being the lowest.

The second objective was to compare PLFA biomass, common genotypic biomarker ratios and novel PLFA biomarker ratios for correlation to environmental conditions (Norris et al., 2023). Specific biomarkers used in calculating different PLFA composition ratios were:

1 ((a15:0 + a17:0)/(i15:0 + i17:0)) = ARR: Adaptation response ratio;
2 (10Me16,0/16:0) = BRR: Branch response ratio; and
3 (cy17:0/16:0) = MRR: Metabolic response ratio.

It was found that the ARR was significantly correlated with environmental conditions across different land management systems and broad geographic regions. The ARR also detected significant differences in the microbial community as a result of tillage. The MRR, on the other hand, detected significant differences with soil conservation practices, including reduced tillage, organic nutrient application, and use of cover crops and rotation diversity.

This study (Norris et al., 2023) showed that the PLFA method as an indicator of soil health has several strengths and that although challenges in PLFA data interpretation remain, the data can be well correlated to environmental conditions and agricultural management practices.

5 Recent advances in PLFA analysis

Despite its long history, PLFA analysis continues to advance, enhancing its application and value in soil health assessment. The interpretation of PLFA data is still frequently reviewed and refined, including assignment of indicator PLFA to microbial groups (Joergensen, 2022; Quideau et al., 2016) and conversion factors for microbial biomass (Willers et al., 2015b) and fungal biomarkers (Agnihotri et al., 2021; Camenzind et al., 2024). In a study assessing the relationship between field management, soil health, and microbial community composition, Mann et al. (2019) found that there were correlations between PLFA and the Cornell Soil Health Assessment (CSHA). The PCA successfully related shifts in the PLFA profile to CSHA, and the authors concluded that the integration of these two approaches could provide powerful new insights into soil health.

Over the past decade, methodological refinements have improved sample processing and analysis rates, while the integration of PLFA profiling with complementary techniques, such as metagenomics, metabolomics, and stable isotope probing (SIP), has broadened its scope. Standardization, automation, and optimization of high-throughput and miniaturized analysis are recent themes.

5.1 Advances in methodology and instrumentation

Contemporary PLFA analytical methods have evolved over the past six decades through key advancements by multiple research groups (Quideau et al., 2016). These developments were reviewed by Willers et al. (2015a), providing an overview of the optimization of lipid extraction based on the original Bligh and Dyer (1959) method, fractionation using silicic acid solid phase extraction, and methylation followed by analysis via GC-FID or GC-MS. More recently, Saini et al. (2021), reviewed the advances in lipid extraction methods – a critical step in PLFA analysis. The review includes plant and animal lipids but gives a useful overview of classical methods and modern modifications such as high-throughput extraction methodologies and the emerging green solvents-based extraction methods. A recent study by Gorka et al. (2023), introduced a high-throughput method for the simultaneous extraction and fractionation of neutral and glycolipid fatty acids (NLFA and GLFA) along with PLFA that yielded more information than PLFA alone. The combined analysis provided sensitive information about fungal and bacterial biomass and physiology, such as carbon storage and stress response.

Automation and high-throughput analysis reduce time, costs, and solvent waste. However, when applied to soil samples, the reduction in sample size may introduce analytical bias, particularly depending on the soil particle size and sample homogeneity.

High-throughput analysis of soil microbial communities was achieved by developing a hybrid method from the more indiscriminate but rapid total lipid FAME analysis (associated with MIDI) and the precise but time-consuming PLFA analysis (Balser et al., 2019). The authors concluded that this protocol provides a balance between time and accuracy and may be helpful for large-scale ecosystem studies. In another study, a unified protocol for high-throughput measurement of PLFA, NLFA, GLFA, and sterols from soil was developed by Gorka et al. (2023), demonstrating improved recovery and efficiency by optimizing solid phase extraction. Buyer et al. (2019) modified the Bligh and Dyer extraction method to enhance soil metabolite detection while maintaining compatibility with PLFA analysis. This approach allows simultaneous metabolomic and PLFA analysis for comprehensive microbial profiling.

Detection of signature biomarkers from FAMEs with GC-FID remains the preferred and most robust technique for analyzing PLFAs (e.g. Camenzind et al., 2024; Chen et al., 2019; Norris et al., 2023; Siles et al., 2024). It is cost-effective to own and operate, simplistic in operation, has a broad detection range, and high quantitative accuracy (Quideau et al., 2016). One of the criticisms against FID is that it relies solely on retention times. However, approaches such as retention time locking (David et al., 2002) and retention time prediction based on equivalent chain lengths (Mjøs, 2003; Torres et al., 2002) overcomes this limitation.

The use of GC with MS instead of FID offers structural identification with a higher specificity and greater ability to resolve isomeric and branched-chain fatty acids, which eliminates ambiguity in compound identification (Zeng et al., 2013). Quantification of PLFAs using isotopically labeled internal standards can further be used to improve quantitative accuracy compared to FID and enables detailed tracking of microbial lipid transformations. It improves detection of complex fatty acid metabolic pathways, such as carbon and nitrogen cycling, and allows for the identification of microbial-derived fatty acids that may be overlooked by FID (Dippold and Kuzyakov, 2016; Evershed et al., 2006). An example of this application is a study combining SIP with PLFA analysis (^{13}C-PLFA) in three cropping systems (Kong et al., 2011). The results described the role of microbial community composition in the short-term processing of root and residue-C and long-term C sequestration, illustrating that microaggregates are important microenvironments for fungi and that more C was processed here than in silt-and-clay particles.

5.2 Integration with other methods

The integration of PLFA analysis with "omics" technologies such as genomics, transcriptomics, proteomics, and metabolomics can provide a more comprehensive understanding of metabolic processes, enabling researchers to link community structure to functional roles in ecosystems more effectively (Chakraborty et al., 2023; Nkongolo and Narendrula-Kotha, 2020). Buyer et al., (2019) refined the Bligh and Dyer extraction method for soil metabolite detection and effectively combined it with PLFA analysis to study specific crop production factors on the soil metabolome.

Combining PLFA with high-throughput DNA sequencing analyses, the short-term shifts in soil microbial community composition after spring precipitation were investigated and highlighted the advantage of using more than one approach to expand the utility of diversity analysis (Chen et al., 2019). The results from both approaches showed that spring watering increased Gram-negative bacteria and fungi, with the DNA biomarkers discriminating these as Proteobacteria, Bacteroides, and Ascomycota. While the PLFA method was more sensitive to short-term shifts in bacterial and fungal abundance, the DNA-based method provided finer taxonomic resolution.

Camenzind et al. (2024) compared interspecific variability in PLFA, ergosterol, and rDNA copy numbers for fungal biomarkers and conversion factors. Lewe et al. (2021) investigated the effect of substrate type on the relative performance of DNA metabarcoding in parallel with PLFA analysis. Both studies found that ecological interpretation and coverage of biomarker data will benefit from the utilization of a combination of different biomarkers.

5.3 Advances in data analysis and interpretation

By leveraging advanced statistical approaches and software tools, researchers can gain deeper insights from PLFA profiles, further enhancing the application value of PLFA-based characterization of microbial communities in soil ecosystems (Docherty et al., 2015; Nadler, 2019). The study by Buyer et al.,

(2019) which used soil metabolite detection combined with PLFA analysis to study specific crop production factors on the soil metabolome, effectively employed multivariate statistical analysis to show how soil type and crop species influence microbial lipid composition. Statistical modelling and machine learning have been shown to strengthen the interpretation of complex datasets by integrating PLFA and 16S sequencing data with a range of environmental factors over multiple biomes (Docherty et al., 2015), elucidating the impact of soil factors on microbial communities.

6 Conclusion and future trends

6.1 Conclusion

Assessing and monitoring soil microbial parameters are similar to soil physical and chemical properties in that the interpretation of what is a "good" or "healthy" level will be context-specific (Fierer et al., 2021). The PLFA method has both strengths and weaknesses (as do all methods) but overall provides an easy-to-use and robust measure of changes in soil microbial communities that can be related to environmental conditions and agricultural management practices. The method is especially valuable when combined with other analyses of soil microbial communities, including DNA-based sequencing and omics approaches to give a more comprehensive view of ecosystem dynamics. Leveraging PLFA as part of soil health indices and making it applicable and accessible to stakeholders in agriculture is paramount in policymaking, conservation, and restoration projects.

In summary, the strengths of the PLFA method include:

- Being a reliable indicator of viable microbial biomass and an accurate representation of active microbial communities;
- Enabling stakeholders to monitor microbial shifts in real time rather than relying on the legacy effects of soil;
- Yielding results that inform on the structure and abundance of microbial groups;
- Showing more sensitivity for detecting shifts in microbial community composition compared to DNA-based methods;
- Providing a way to calculate the ratio of fungal to bacterial biomass; and
- Giving information on the physiological state of microbial communities by means of stress ratios.

The weaknesses of the PLFA method include:

- Poor taxonomic resolution, especially for fungi;

- The interpretation of PLFA signatures is not straightforward and needs to be context-specific; and
- Data from PLFA analysis cannot be used to calculate diversity indices.

6.2 Future trends

The PLFA method will remain a popular method when assessing microbial community composition and will still be of interest in the future, especially in combination with other methods. Improvements are likely to come from greater automation and accessibility through the development of fully automated workflows for sample handling, extraction, and data analysis. This will save time, enhance reproducibility, and minimize human error. Researchers are also calling for the elimination of more toxic organic solvents used for extraction, the reduction of solvents and chemical reagents through integration of analytical steps or miniaturization of sample preparation, and the reduction in energy consumption (Arteaga-Clemente et al., 2024).

By leveraging advanced statistical approaches, software tools, and machine learning frameworks, researchers can gain deeper insights from PLFA profiles, further enhancing the application value of PLFA-based characterization of microbial communities in soil ecosystems. Statistical modelling and machine learning frameworks in R and Python have been demonstrated to effectively predict microbial community changes using environmental data (Kibbey et al., 2020; Tajik et al., 2020), making complex datasets more accessible and demonstrating relevance for PLFA-based analyses. Expansion of PLFA biomarker libraries to enhance taxonomic and functional resolution, developments that can link specific PLFA profiles to metabolic pathways such as nitrogen fixation, and predictive modeling for climate impact soil carbon studies are areas likely to receive attention in the future.

7 Acknowledgements

I would like to thank Peet Jansen van Rensburg (North-West University, Potchefstroom, South Africa) for discussions on content and technical aspects of this chapter.

Figures were produced by Leonie Venter (Auckland University of Technology, Auckland, New Zealand).

8 Where to look for further information

The fourth edition of 'Soil Microbiology, Ecology and Biochemistry' by Eldor A. Paul (2015) offers a comprehensive overview of soil microbial processes, including biomarker techniques like PLFA analysis, in the context of soil biochemistry and ecosystem functions. To develop a better understanding of the importance of microbial biomass, the book 'Microbial Biomass: A Paradigm Shift in Terrestrial Biogeochemistry' by Kevin R. Tate (2017) is useful and discusses PLFA and complimentary approaches and their implications for ecosystem modelling and nutrient cycling.

Online resources that are useful for a better understanding of PLFA analysis include:

- The AOCS Lipid Library® provides comprehensive, curated information on lipid science and technology: https://www.aocs.org/publications-resources/resource-library/lipid-library/
- Ward Laboratories, Inc. provides helpful resources for understanding PLFA analysis in the context of agricultural testing: https://www.wardlab.com/resources/ward-guide
- The video accompanying the article by Quideau et al. (2016), 'Extraction and analysis of microbial phospholipid fatty acids in soils', is available here: https://app.jove.com/v/54360/extraction-and-analysis-of-microbial-phospholipid-fatty-acids-in-soils
- The video accompanying the article by Oates et al. (2017), 'A Lipid Extraction and Analysis Method for Characterizing Soil Microbes in Experiments with Many Samples', is available here: https://pmc.ncbi.nlm.nih.gov/articles/PMC5553326/

9 References

Agnihotri, R., et al. (2021), 'Glomalin related protein and C16:1ω5 PLFA associated with AM fungi as potential signatures for assessing the soil C sequestration under contrasting soil management practices', *Eur. J. Soil Biol.*, 103, 103286. https://doi.org/10.1016/j.ejsobi.2021.103286

Arteaga-Clemente, G., et al. (2024), 'Soil lipid analysis by chromatography: a critical review of the current state in sample preparation', *J. Chromatogr. Open*, 6, 100173. https://doi.org/10.1016/j.jcoa.2024.10017

Bååth, E. and Anderson, T.-H. (2003), 'Comparison of soil fungal/bacterial ratios in a pH gradient using physiological and PLFA-based techniques', *Soil Biol. Biochem.*, 35, 955–963. https://doi.org/10.1016/S0038-0717(03)00154-8

Balser, T.C., et al. (2019), 'Linking microbial community analysis and ecosystem studies: a rapid lipid analysis protocol for high throughput' *Soil Ecol. Lett.*, 1(1–2), 22–32. https://doi.org/10.1007/s42832-019-0003-0

Ben-David, E.A., et al. (2011), 'Assessment of the spatial distribution of soil microbial communities in patchy arid and semi-arid landscapes of the Negev Desert using combined PLFA and DGGE analyses', *FEMS Microbiol. Ecol.*, 76(3), 492–503. https://doi.org/10.1111/j.1574-6941.2011.01075.x

Bligh, E.G. and Dyer, W.J. (1959), 'A rapid method of total lipid extraction and purification', *Can. J. Biochem. Physiol.*, 37, 911–917. https://doi.org/10.1139/o59-099

Boon, P.I., et al. (1996), 'Microbial consortia in wetland sediments: a biomarker analysis of the effect of hydrological regime, vegetation and season on benthic microbes', *Mar. Freshwater Res.*, 47, 27–41. https://doi.org/10.1071/MF9960027

Bradford, M.A., et al. (2019), 'Cross-biome patterns in soil microbial respiration predictable from evolutionary theory on thermal adaptation', *Nat. Ecol, Evol.*, 3, 223–231. https://doi.org/10.1038/s41559-018-0771-4

Buyer, J.S., et al. (2019), 'Combined extraction method for metabolomic and PLFA analysis of soil', *Appl. Soil Ecol.*, 135, 129–136. https://doi.org/10.1016/j.apsoil.2018.11.012

Camenzind, T., et al. (2024), 'Revisiting soil fungal biomarkers and conversion factors: interspecific variability in phospholipid fatty acids, ergosterol and rDNA copy numbers', *Soil Ecol. Lett.*, 6(4), 240–243. https://doi.org/10.1007/s42832-024-0243-5

Chakraborty, A., et al. (2023), 'Impact of 30 years precipitation regime differences on forest soil physiology and microbial assemblages', *Front. For. Global Change*, 6, 1142979. https://doi.org/10.3389/ffgc.2023.1142979

Chen, H., et al. (2019), 'Using a combination of PLFA and DNA-based sequencing analyses to detect shifts in the soil microbial community composition after a simulated spring precipitation in a semi-arid grassland in China', *Sci. Total Environ.*, 657, 1237–1245. https://doi.org/10.1016/j.scitotenv.2018.12.126

Daquido, A.R., et al. (2013), 'Microbial community structure of paddy soil under long-term fertilizer treatment using phospholipid fatty acid (PLFA) analysis', *Korean J. Soil Sci. Fert.*, 46(6), 474–481. https://doi.org/10.7745/KJSSF.2013.46.6.474

David, F., et al. (2002) *Improving Analysis of Fatty Acid Methyl Esters Using Retention Time Locked Methods and Retention Time Databases*. Agilent Technologies Application Note 5988-5871EN, Palo Alto, CA.

Dippold, M.A. and Kuzyakov, Y. (2016), 'Direct incorporation of fatty acids into microbial phospholipids in soils: Position-specific labeling tells the story', *Geochim. Cosmochim. Acta*, 174, 211–221. https://doi.org/10.1016/j.gca.2015.10.032

Docherty, K.M., et al. (2015), 'Key edaphic properties largely explain temporal and geographic variation in soil microbial communities across four biomes', *PLoS One*, 10(11), e0135352. https://doi.org/10.1371/journal.pone.0135352

Drenovsky, R.E., et al. (2010), 'Land use and climatic factors structure regional patterns in soil microbial communities', *Global Ecol. Biogeogr.*, 19(1), 27–39. https://doi.org/10.1111/j.1466-8238.2009.00486.x

Du Preez, G.C., et al. (2024), 'Time matters: a short-term longitudinal analysis of conservation agriculture and its impact on soil health', *J. Soil Sci. Plant Nutr.*, 24, 1319–1334. https://doi.org/10.1007/s42729-024-01636-w

Duncan, D.S., et al. (2016), 'Detection of short-term cropping system-induced changes to soil bacterial communities differs among four molecular characterization methods', *Soil Biol. Biochem.*, 96, 160–168. https://doi.org/10.1016/j.soilbio.2016.02.002

Evershed, R.P., et al. (2006), '13C-labelling of lipids to investigate microbial communities in the environment', *Curr. Opin. Biotechnol.*, 17(1), 72–82. https://doi.org/10.1016/j.copbio.2006.01.003

Fanin, N., et al. (2019), 'The ratio of gram-positive to gram-negative bacterial PLFA markers as an indicator of carbon availability in organic soils', *Soil Biol. Biochem.*, 128, 111–114. https://doi.org/10.1016/j.soilbio.2018.10.010

Fernandes, M.F., et al. (2013), 'Comparison of whole-cell fatty acid (MIDI) or phospholipid fatty acid (PLFA) extractants as biomarkers to profile soil microbial communities', *Microb. Ecol.*, 66(1), 145–157. https://doi.org/10.1007/s00248-013-0195-2

Fierer, N., et al. (2021), 'How microbes can, and cannot, be used to assess soil health', *Soil Biol. Biochem.*, 153, 108111. https://doi.org/10.1016/j.soilbio.2020.108111

Francisco, R., et al. (2016), 'European scale analysis of phospholipid fatty acid composition of soils to establish operating ranges', *Appl. Soil Ecol.* 97, 49–60. https://doi.org/10.1016/j.apsoil.2015.09.001

Francisco, R., et al. (2022), 'Effects of a bionematicide 1,4-naphthoquinone solution on soil microbial community assessed by PLFA: Tracing toxicity indicators', *Appl. Soil Ecol.*, 174, 104417. https://doi.org/10.1016/j.apsoil.2022.104417

Frostegård, A., et al. (1993), 'Shifts in the structure of soil microbial communities in limed forests as revealed by phospholipid fatty acid analysis', *Soil Biol. Biochem.*, 25(6), 723–730. https://doi.org/10.1016/0038-0717(93)90113-P

Frostegård, Å., et al. (2011), 'Use and misuse of PLFA measurements in soils', *Soil Biol. Biochem.*, 43(8), 1621–1625. https://doi.org/10.1016/j.soilbio.2010.11.021

Gorka, S., et al. (2023), 'Beyond PLFA: concurrent extraction of neutral and glycolipid fatty acids provides new insights into soil microbial communities', *Soil Biol. Biochem.*, 187, 109205. https://doi.org/10.1016/j.soilbio.2023.109205

Grayston, S.J., et al. (2004), 'Assessing shifts in microbial community structure across a range of grasslands of differing management intensity using CLPP, PLFA and community DNA techniques', *Appl. Soil Ecol.*, 25(1), 63–84. https://doi.org/10.1016/S0929-1393(03)00098-2

Guckert, J.B., et al. (1986), 'Phospholipid ester-linked fatty acid profile changes during nutrient deprivation of Vibrio cholerae: increases in the trans/cis ratio and proportions of cyclopropyl fatty acids', *Appl. Environ. Microbiol.*, 52(4), 794–801. https://doi.org/10.1128/aem.52.4.794-801.1986

Habig, J., et al. (2018), 'The effect of a medic-wheat rotational system and contrasting degrees of soil disturbance on nematode functional groups and soil microbial communities', *Agric. Ecosyst. Environ.*, 268, 103–114. https://doi.org/10.1016/j.agee.2018.09.013

Halket, J.M. and Zaikin, V.G. (2003), 'Derivatization in mass spectrometry–1. Silylation', *Eur. J. Mass Spectrom.*, 9, 1–21. https://doi.org/10.1255/ejms.527

Joergensen, R.G. (2022), 'Phospholipid fatty acids in soil–drawbacks and future prospects', *Biol. Fertil. Soils*, 58, 633–648. https://doi.org/10.1007/s00374-021-01613-w

Kibbey, T.C.G., et al. (2020), 'Supervised machine learning for source allocation of per- and polyfluoroalkyl substances (PFAS) in environmental samples', *Chemosphere*, 252, 126593. https://doi.org/10.1016/j.chemosphere.2020.126593

Kong, A.Y.Y., et al. (2011), 'Microbial community composition and carbon cycling within soil microenvironments of conventional, low-input, and organic cropping systems' *Soil Biol. Biochem.*, 43(1), 20–30. https://doi.org/10.1016/j.soilbio.2010.09.005

Lear, G., et al. (2018), 'Methods for the extraction, storage, amplification and sequencing of DNA from environmental samples', *N. Z. J. Ecol.*, 42(1), 10–50. https://doi.org/10.20417/nzjecol.42.9

Lewe, N., et al. (2021), 'Phospholipid fatty acid (PLFA) analysis as a tool to estimate absolute abundances from compositional 16S rRNA bacterial metabarcoding data', *J. Microbiol. Meth.*, 188, 106271. https://doi.org/10.1016/j.mimet.2021.106271

Li, C., et al. (2020), 'A comparison between fatty acid methyl ester profiling methods (PLFA and EL-FAME) as soil health indicators', *Soil Sci. Soc. Am. J.*, 84, 1153–1169. https://doi.org/10.1002/saj2.20118

Li, Y.-L., et al. (2007), 'Spatial patterns of bacterial signature biomarkers in marine sediments of the Gulf of Mexico', *Chem. Geol.* 238, 168–179. https://doi.org/10.1016/j.chemgeo.2006.11.007

Liang, C., et al. (2016), 'Switchgrass rhizospheres stimulate microbial biomass but deplete microbial necromass in agricultural soils of the upper Midwest, USA', *Soil Biol. Biochem.*, 94, 173–180. https://doi.org/10.1016/j.soilbio.2015.11.020

Mann, C., et al. (2019), 'Relationships between field management, soil health, and microbial community composition', *Appl. Soil Ecol.*, 144, 12–21. https://doi.org/10.1016/j.apsoil.2019.06.012

Mjøs, S.A. (2003), 'Identification of fatty acids in gas chromatography by application of different temperature programs on a single capillary column', *J. Chromatogr. A*, 1015, 151–161. https://doi.org/10.1016/s0021-9673(03)01240-8

Nadler, D.W. (2019), 'Decision support: using machine learning through MATLAB to analyze environmental data', *J. Environ. Stud. Sci.*, 9, 419–428. https://doi.org/10.1007/s13412-019-00558-9

Nkongolo, K.K. and Narendrula-Kotha, R. (2020), 'Advances in monitoring soil microbial community dynamics and function', *J. Appl. Genet.*, 61, 249–263. https://doi.org/10.1007/s13353-020-00549-5

Norris, C.E., et al. (2020), 'Introducing the North American project to evaluate soil health measurements', *Agron. J.*, 112(4), 3195–3215. https://doi.org/10.1002/agj2.20234

Norris, C.E., et al. (2023), 'Use of phospholipid fatty acid analysis as phenotypic biomarkers for soil health and the influence of management practices', *Appl. Soil Ecol.*, 185, 104793. https://doi.org/10.1016/j.apsoil.2022.104793

Oates, L.G., et al. (2017), 'A lipid extraction and analysis method for characterizing soil microbes in experiments with many samples', *J. Visualized Exp.*, 125, e55310. https://doi.org/10.3791/55310

Orwin, K.H., et al. (2018), 'A comparison of the ability of PLFA and 16S rRNA gene metabarcoding to resolve soil community change and predict ecosystem functions', *Soil Biol. Biochem.*, 117 27–35. https://doi.org/10.1016/j.soilbio.2017.10.036

Papadopoulou, E.S., et al. (2011), 'Extraction parameters significantly influence the quantity and the profile of PLFAs extracted from soils', *Microb. Ecol.*, 62(4), 704–714. https://doi.org/10.1007/s00248-011-9863-2

Peacock, A.D. and White, D.C. (2016), 'Microbial biomass and community composition analysis using phospholipid fatty acids'. In T. McGenity, et al. (Eds), *Hydrocarbon and Lipid Microbiology Protocols*, Springer Protocols Handbooks, Springer, Berlin, Heidelberg. https://doi.org/10.1007/8623_2016_213

Paul, E.A. (Ed.) (2015), Soil Microbiology, Ecology and Biochemistry, 4th edition, *Academic Press*. https://doi.org/10.1016/B978-0-12-415955-6.09995-3.

Quideau, S.A., et al. (2016), 'Extraction and analysis of microbial phospholipid fatty acids in soils', *J. Visualized Exp.*, 114, e54360. https://doi.org/10.3791/54360

Ramsey, P.W., et al. (2006), 'Choice of methods for soil microbial community analysis: PLFA maximizes power compared to CLPP and PCR-based approaches', *Pedobiologia*, 50(3), 275–280. https://doi.org/10.1016/j.pedobi.2006.03.003

Rousk, J., et al. (2010), 'The microbial PLFA composition as affected by pH in an arable soil', *Soil Biol. Biochem.*, 42, 516–520. https://doi.org/10.1016/j.soilbio.2009.11.026

Ruess, L. and Chamberlain, P.M. (2010), 'The fat that matters: Soil food web analysis using fatty acids and their carbon stable isotope signature', *Soil Biol. Biochem.*, 42(11), 1898–1910. https://doi.org/10.1016/j.soilbio.2010.07.020

Saini, R.K., et al. (2021), 'Advances in lipid extraction methods–a review'. *Int. J. Mol. Sci.*, 22, 13643. https://doi.org/10.3390/ijms222413643

Siles, J.A., et al. (2024), 'A comparison among EL-FAME, PLFA, and quantitative PCR methods to detect changes in the abundance of soil bacteria and fungi', *Soil Biol. Biochem.*, 198, 109557. https://doi.org/10.1016/j.soilbio.2024.109557

Strickland, M.S. and Rousk, J. (2010), 'Considering fungal: bacterial dominance in soils - Methods, controls, and ecosystem implications', *Soil Biol. Biochem.* 42, 1385–1395. https://doi.org/10.1016/j.soilbio.2010.05.007

Tajik, S., et al. (2020), 'Soil microbial communities affected by vegetation, topography and soil properties in a forest ecosystem', *Appl. Soil Ecol.*, 149, 103514. https://doi.org/10.1016/j.apsoil.2020.103514

Tate, K.R. (Ed.) (2017), Microbial Biomass: A Paradigm Shift in Terrestrial Biogeochemistry. *World Scientific Publishing*, Great Britain, 346p.

Torres, A.G., et al. (2002), 'Mathematical method for the prediction of retention times of fatty acid methyl esters in temperature-programmed capillary gas chromatography', *J. Agric. Food Chem.*, 50, 4156–4163. https://doi.org/10.1021/jf011259j

Velasco, A.G.V., et al. (2010), 'Characterization of the rhizosphere microbial community from different Arabidopsis thaliana genotypes using phospholipid fatty acids (PLFA) analysis', *Plant Soil*, 329(1), 315–325. https://doi.org/10.1007/s11104-009-0160-5

Veum, K.S., et al. (2019), 'Phospholipid fatty acid profiles of soils under variable handling and storage conditions', *Agron. J.*, 111(3), 1090–1096. https://doi.org/10.2134/agronj2018.09.0628

Wagner, D., et al. (2005), 'Methane fluxes in permafrost habitats of the Lena Delta: effects of microbial community structure and organic matter quality', *Environ. Microbiol.*, 7(10), 1582–1592. https://doi.org/10.1111/j.1462-2920.2005.00849.x

Watzinger, A. (2015), 'Microbial phospholipid biomarkers and stable isotope methods help reveal soil functions', *Soil Biol. Biochem.*, 86, 98–107. https://doi.org/10.1016/j.soilbio.2015.03.019

White, D.C., et al. (1979), 'Biochemical measurements of microbial mass and activity from environmental samples'. In J.W. Costerton and R.R. Colwell (Eds), *Native Aquatic Bacteria: Enumeration, Activity and Ecology*, vol 695, ASTM STP. American Society for Testing Materials, Philadelphia, pp. 69–81.

Willers, C., et al. (2015a), 'Microbial signature lipid biomarker analysis – an approach that is still preferred, even amid various method modifications', *J. Appl. Microbiol.*, 118(6), 1251–1263. https://doi.org/10.1111/jam.12798

Willers, C., et al. (2015b), 'Phospholipid fatty acid profiling of microbial communities–a review of interpretations and recent applications', *J. Appl. Microbiol.*, 119(5), 1207–1218. https://doi.org/10.1111/jam.12902

Yang, Y., et al. (2022), 'Deciphering factors driving soil microbial life-history strategies in restored grasslands', *iMeta*, 2, e66. https://doi.org/10.1002/imt2.66

Zelles, L. (1999), 'Fatty acid patterns of phospholipids and lipopolysaccharides in the characterisation of microbial communities in soil: a review', *Biol. Fertil. Soils*, 29, 111–129. https://doi.org/10.1007/s003740050533.

Zeng, A.X., et al. (2013), 'Profiling of soil fatty acids using comprehensive two-dimensional gas chromatography with mass spectrometry detection'. *J. Chromatogr. A*, 1317, 239–245. https://doi.org/10.1016/j.chroma.2013.08.095

Zhang, Y., et al. (2019), 'High turnover rate of free phospholipids in soil confirms the classic hypothesis of PLFA methodology', *Soil Biol. Biochem.*, 135, 323–330. https://doi.org/10.1016/j.soilbio.2019.05.023

Part 2

Fauna as indicators of soil health

Chapter 8

Macrofauna as indicators of soil health: earthworms and enchytraeids

Alix Vidal, Wageningen University, The Netherlands; Céline Pelosi†, UMR EMMAH, France; George G. Brown‡,§, Universidade Federal do Paraná and Embrapa Florestas, Brazil; Nicole Schon¶, AgResearch Limited, New Zealand; Tullia Calogiuri and Jan Willem Van Groenigen*, Wageningen University, The Netherlands*

1 Introduction

Reports highlighting the degradation of soils in Europe and worldwide are flourishing (Akça et al., 2024; Pennock et al., 2015; Smith et al., 2024). These reports emphasize that 33% of soils are moderately to highly degraded and face significant threats, including biodiversity loss (Pennock et al., 2015). This is particularly concerning as soils support 59% of all life on Earth (Anthony et al., 2023). In this context, developing biological indicators for soil health is essential for identifying strategies for soil monitoring and improvement. Soil

* Soil Biology Group, Department of Environmental Sciences, Wageningen University, Wageningen, The Netherlands.
† INRAE, Avignon Université, UMR EMMAH, Avignon, France.
‡ Universidade Federal do Paraná, Departamento de Solos e Engenharia Agrícola, R. dos Funcionários, Curitiba, Brazil.
§ Embrapa Florestas, Estrada da Ribeira, Colombo, PR, Brazil.
¶ AgResearch Limited, Private Bag 4749, Christchurch 8140, New Zealand.

http://dx.doi.org/10.19103/AS.2025.0159.09

fauna, almost exclusively consisting of invertebrates, represent a substantial portion of life in soils, i.e. around 23% of total species number (Decaëns et al., 2006), and play a critical role in maintaining and enhancing soil quality and providing ecosystem services (Lavelle et al., 2006). This chapter focuses on two key soil faunal groups, earthworms and enchytraeids, as potential soil health indicators.

Earthworms and enchytraeids both belong to the phylum Annelida and the class Oligochaeta, and both are considered as ecosystem engineers that play a crucial role in soil functioning and provide numerous ecosystem services (Blouin et al., 2013; Conti & Mulder, 2022). Earthworm species are often classified according to their feeding behavior and habitat in three main ecological categories: *epigeics* that feed on litter at the soil surface, *endogeics* that feed on soil organic matter, and form non-permanent burrows, and *anecics* that feed on litter at the soil surface and live in permanent vertical burrows (Bouché, 1972; Lee, 1985). Enchytraeids, also known as 'potworms,' are often classified into two groups according to their reproduction strategy: *r strategists* that have a high reproduction rate but high juvenile mortality, and *K strategists* that have slower reproduction rates but longer life span (Graefe & Schmelz, 1999). The main difference between earthworms and enchytraeids is their size; earthworms are classified as macrofauna (diameter > 2 mm) while enchytraeids belong to the mesofauna (diameter: 200 μm to 2 mm) (Potapov et al., 2022). Earthworms and enchytraeids are typically considered to have similar functions in soils, albeit at different orders of magnitude and scale given their difference in size, making it logical to treat both groups within the same chapter. While the role of earthworms on soil functions has been extensively studied and multiple high-quality reviews are available (Blouin et al., 2013; Lavelle et al., 2006; Vidal et al., 2023), enchytraeids are particularly understudied (Pelosi & Römbke, 2016).

For earthworms and enchytraeids to impact soil functions, they must be present in the soil in sufficient amounts. Overall, earthworms and enchytraeids are widely distributed in soils with abundances often reaching 500 and 100 000 ind. m^{-2}, respectively (Potapov et al., 2022). Yet, many agricultural management practices and climatic conditions adversely affect earthworms and enchytraeids, with populations often varying dramatically even in similar soil types and climate conditions. The ubiquitous presence of earthworms and enchytraeids in soils in combination with their sensitivity to anthropogenic and climatic factors make them good candidates for soil health indicators (Brussaard et al., 2004; Creamer et al., 2022; Turbé et al., 2010), the topic of this chapter. We chose to focus primarily on agricultural soils, where concerns for soil degradation and biodiversity loss are particularly pertinent, also given their role in ensuring food security (Kopittke et al., 2019). However, we do not

Published by Burleigh Dodds Science Publishing Limited, 2025.

exclude instructive examples from forest and urban soils, where earthworms and enchytraeids also play important roles.

Figure 1 illustrates the structure of the chapter. It focuses on the impact of land use, agricultural management practices and climate on earthworms (Section 2) and enchytraeids (Section 3), as well as the interactions between these two faunal groups (Section 4). Our ultimate goal is to discuss their potential as indicators of soil health (Section 7). However, this is conditional on the capacity to sample and quantity them efficiently (Section 5). It is also important to emphasize that soil biota presence does not equate soil biota being very active. Thus, we also present some methods to assess the activity of earthworms and enchytraeids in soils (Section 6).

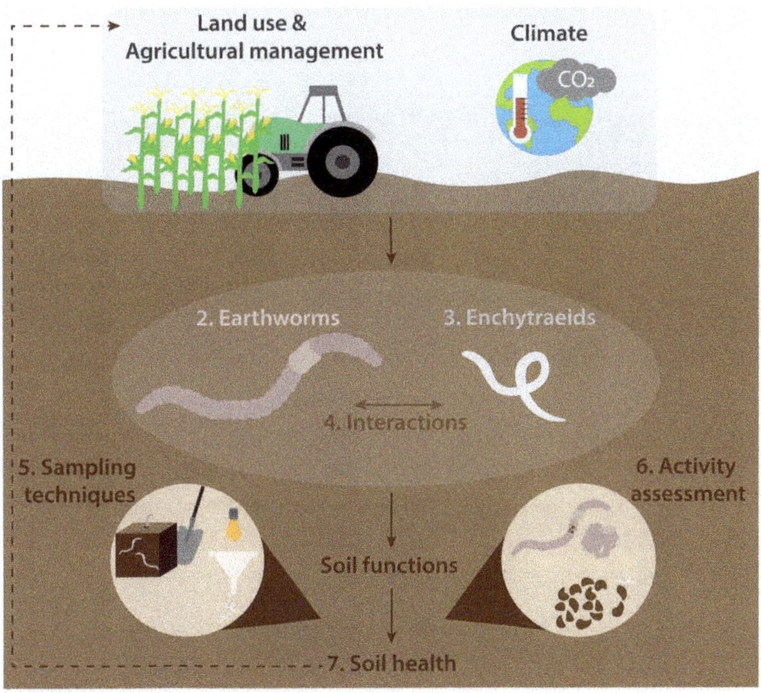

Figure 1 Structure of this chapter, which focuses on the role of earthworms and enchytraeids as indicators of soil health. It examines the impact of land use, agricultural management practices and climate on earthworms (Section 2) and enchytraeids (Section 3), as well as the potential interactions between these two soil organisms (Section 4). For their effective use as soil indicators, it is essential to sample and quantify earthworms and enchytraeids using efficient techniques (Section 5) and evaluate their activity in the soil, as their mere presence does not guarantee activity (Section 6). Finally, the chapter highlights the opportunities and challenges of utilizing earthworms and enchytraeids as reliable indicators of soil health (Section 7).

Published by Burleigh Dodds Science Publishing Limited, 2025.

2 Effect of land use, agricultural management, and climate on earthworm communities

Earthworms serve as valuable indicators of soil health due to their sensitivity to environmental changes, with land use, management practices, and climatic conditions all being important. In terms of land use, this section primarily compares grasslands and croplands. We subsequently explore management practices, in particular those that increase or decrease earthworm density and diversity by modifying the physical and chemical soil properties, as well as the organic matter availability as food for earthworms (Fig. 2). The impact of these management practices on earthworm populations may be exacerbated by climate change, particularly through increases in droughts and flooding events (Fig. 3), which are also examined in this section. The increase in atmospheric CO_2

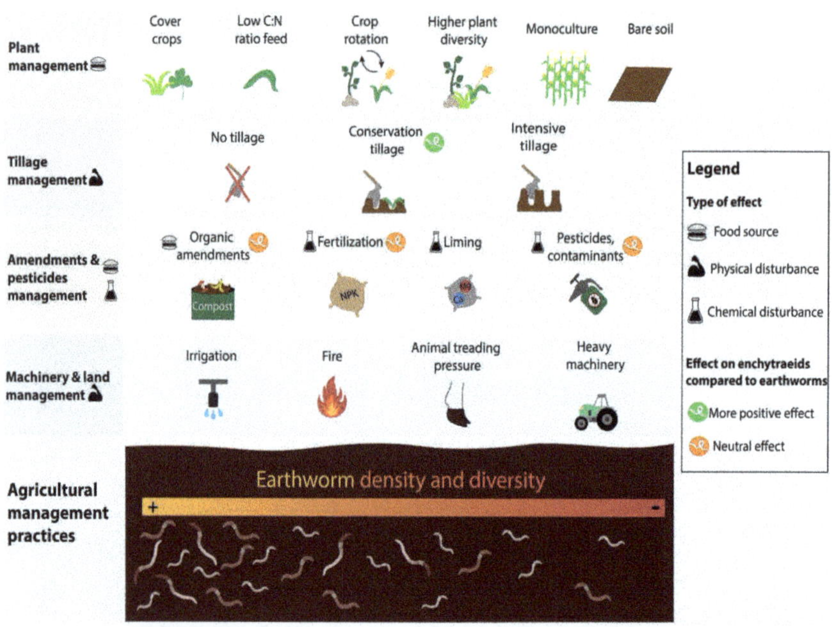

Figure 2 Potential impact of various agricultural management practices on earthworm density and diversity is categorized based on management type, with a primary focus on croplands and grasslands. These effects are further classified into three categories: chemical disturbance, physical disturbance, and the availability of food source. The impact of some practices is also highlighted for enchytraeids, with the color of the circle indicating the relative response compared with earthworms. However, studies on enchytraeids are limited and less abundant compared to research on earthworms. It is important to note that this figure is conceptual, designed to represent average effects of practices, while substantial variability in responses has been observed depending on the specific context and species studied. The initial idea of this figure is inspired from Brown et al. (2006).

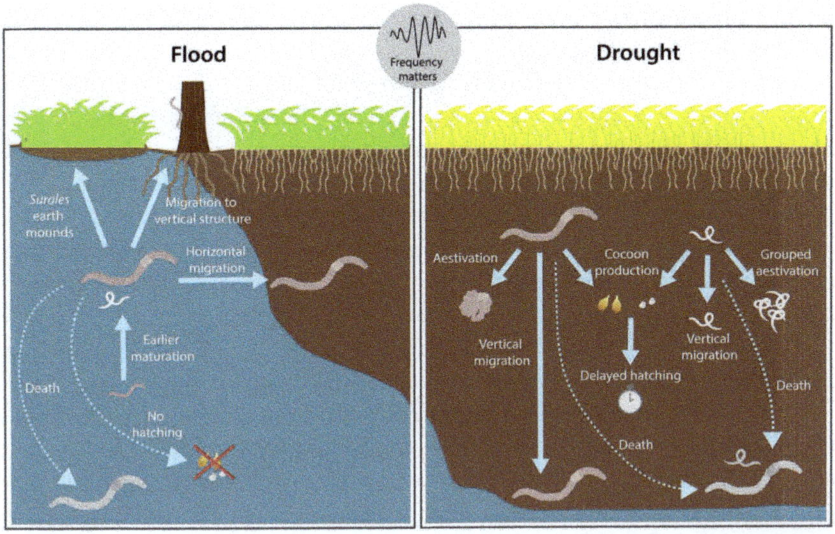

Figure 3 Effects of flooding and drought conditions on earthworms and enchytraeids. Frequent flood and drought events can lead to mortality in earthworms and enchytraeids and halt cocoon hatching (dashed arrows), ultimately causing a decline in their abundance and diversity. However, these organisms can also exhibit adaptive responses (solid arrows) to extreme climatic conditions, such as horizontal or vertical migration, aestivation, or alterations in reproductive dynamics. It is important to note that this figure is conceptual, intended to illustrate average effects of floods and droughts, while considerable variability in responses has been observed depending on the specific context and species studied.

concentrations is not expected to directly affect earthworms and enchytraeid populations, as these organisms are adapted to significantly higher CO_2 levels within the soil (Blankinship et al., 2011; Goncharov et al., 2023). Potential indirect effects, such as those influencing plant primary production and soil microbial communities, and consequently habitat quality for earthworms and enchytraeids, are not addressed in this chapter.

2.1 Temperate and continental grasslands and croplands

Earthworms are widely distributed across temperate and continental areas and often dominate invertebrate biomass (Curry, 2004; Turbé et al., 2010). The most abundant earthworm populations are typically found in managed grasslands, with lower abundances found in croplands, forests, vineyards and native pastures (Curry, 1993; King & Hutchinson, 2007; Lavelle, 1983; Rutgers et al., 2016). Earthworm abundance and ecological diversity are both impacted by land use and management. Even when earthworms are present in significant quantities, one or two of the three ecological categories can be absent or lacking, as widely observed with anecic species across grasslands and croplands (Stroud, 2019; van de Logt et al., 2023b).

Agricultural soils in temperate and continental regions across the world are dominated by earthworm species originating from Europe. Earthworms are common across Europe and include Lumbricidae species, such as *Aporrectodea caliginosa*, *Allolobophora chlorotica* and *Lumbricus rubellus* (Ashwood et al., 2024; Rutgers et al., 2016). These lumbricid species have inadvertently or purposely spread to many other temperate and continental regions. In northern forests of North America, where indigenous earthworms are lacking (Hendrix et al., 1986), the spread of invasive lumbricids (e.g. *Lumbricus terrestris*) and Asian Megascolecids (e.g. *Amynthas* spp.) is impacting the structure and function of temperate forest ecosystems through the removal of litter on the forest floor (Bohlen et al., 2004; Chang et al., 2021). In other temperate regions (e.g., Australia and New Zealand), indigenous earthworms such as *Anisochaeta gigantea* are found in native forests but have not adapted to managed grasslands and croplands (Dalby et al., 1998; Lee, 1961). In these agricultural systems, the introduction of *A. caliginosa* improved pasture productivity by 28–113% (Stockdill, 1982).

Earthworm populations of over 1200 ind. m^{-2} have been observed in managed temperate grasslands although lower abundances are more common (Curry, 1993; Rutgers et al., 2016). A positive relationship between pasture production and earthworm abundance has been observed, with earthworm numbers increasing in response to both the quantity and quality of the plant biomass available to soil fauna. The increased earthworm activity in turn improves soil quality and pasture productivity. Increases in earthworm populations as a result of increased pasture productivity have also been reported in systems amended with synthetic fertilisers (Curry et al., 2008; Schon et al., 2008). In both grasslands and croplands, larger increases in earthworm populations are often observed with the combined application of synthetic fertiliser and organic amendments (Curry, 1987; Leroy et al., 2008). Improvements in pasture quality through the addition of legumes can also benefit earthworm populations (Van Eekeren et al., 2009) (Fig. 2).

Conversely, the application of some fertilisers can contribute to soil acidification, which is detrimental to earthworm populations (Pelosi et al., 2020) as they prefer soils of neutral pH. The long-term Park Grass Experiment at Rothamsted, running since 1856, showed that earthworms were absent in soils with a pH < 4 (Edwards & Lofty, 1974), resulting in an accumulation of undecomposed organic matter forming a layer on the soil surface (Silvertown et al., 2006). However, soil acidification can be remediated with the application of lime, and liming is known to increase earthworm populations (Holland et al., 2018; Springett & Syers, 1984) (Fig. 2).

Typically, increased pasture production also leads to increases in livestock numbers, liveweight loading and treading pressure. Increased treading pressure can lead to soil compaction, especially on vulnerable and/or heavier soil types when wet. These conditions make the soil physical environment more challenging for earthworm presence (Schon et al., 2012). Surface-dwelling epigeic earthworms are most sensitive to the treading pressures of livestock (Cluzeau et al., 1992; de Bruyn & Kingston, 1997; Schon et al., 2011, 2012). With cattle treading pressure greater than that of sheep (Drewry, 2006), earthworm populations are more abundant under the latter (Schon et al., 2011). Rotational grazing practices appear to benefit earthworm populations under pastures compared to continuous grazing (Alvarez et al., 2024; Voisin, 1960).

The disruption of the soil physical environment and food supply through cultivation can have major detrimental impacts on earthworms. Hence, earthworm populations are generally less abundant in annual cropping systems in comparison to grasslands (Fraser et al., 1996; van Eekeren et al., 2008) (Fig. 4). The intensity of cultivation (e.g., frequency of interventions, length of rotations) is an important factor. Practices such as no-till, strip-till, conservation and minimum tillage reduce the level of soil disturbance (Fig. 2). No-tillage and conservation tillage systems have been observed to sustain higher earthworm abundances (+137%) than conventional ploughing (Briones & Schmidt, 2017). However, even a one tillage event during pasture renewal can have a

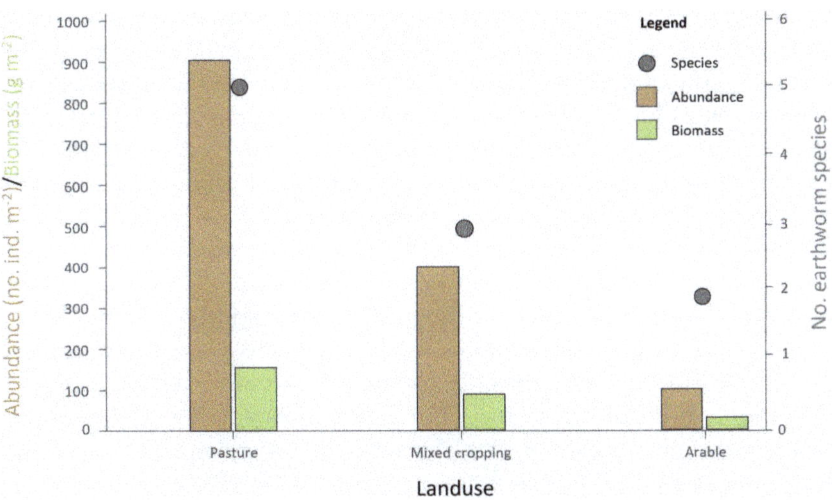

Figure 4 Changes in earthworm abundance, biomass, and species diversity under improved pasture, mixed cropping, and continuous arable use in the Canterbury Plains of New Zealand, which has a temperate climate. Abundance is expressed as number of individuals per meter square and biomass as gram per meter square. Source: Adapted from: Fraser et al. (1996).

detrimental impact on earthworms, taking more than a year to recover to pre-cultivation levels (Bell et al., 2009).

Decreases in earthworm abundance in cropping systems often coincide with reduced species diversity (Briones & Schmidt, 2017; Fraser et al., 1996) (Fig. 4). In particular, anecic earthworms, with their (semi-)permanent burrows that are vulnerable to cultivation, are often found at low densities or are completely absent from arable fields (Edwards, 1983; van Eekeren et al., 2008). A global meta-analysis reports also epigeic earthworms to be sensitive to cultivation (Briones & Schmidt, 2017). The reduction in two of the three earthworm ecological groups under cropping systems has obvious implications for soil functions mediated by earthworms, including decomposition of organic material and maintenance of soil structure (Blouin et al., 2013).

It is not only the physical disturbance of cultivation in cropping systems that has a negative impact on earthworm populations. Edwards and Lofty (1982) reported a continuous decline in earthworm populations under no-till treatments during 8 years of continuous cereal cropping. Similar declines in earthworm populations under conventional tillage were reported where corn was continually cropped, although the decline was slowed by the addition of manure (Whalen et al., 1998). Other practices that can mitigate the impacts of cropping systems on earthworm communities include the management of crop residues, the use of cover crops and the addition of organic amendments (Briones & Schmidt, 2017). D'Hose et al. (2018) reported increases in earthworm abundance when applying either farmyard manure (+151%), slurry (+103%) or compost (+63%), in comparison to synthetic fertiliser addition. Often, tillage treatments and crop residue inputs change simultaneously between systems, making it difficult to determine the role of either factor. For example, Blanco-Canqui (2022) found that the combination of cover crops with low C/N ratios (e.g. *Pisum sativum* L.) promote earthworm abundance under no-till or reduced-till systems, as compared to a combination of lower quality cover crops (e.g. *Sinapis alba* L.) under conventionally tilled systems.

Pesticides have been reported to alter the growth, reproduction, behavior, essential enzymes and DNA of earthworms (Yatoo et al., 2022). In a review by Pelosi et al. (2014), insecticides and fungicides used in agricultural systems were found to have the greatest detrimental impact on survival and reproduction of earthworms. No significant effect of glyphosate was observed on earthworm populations in cropping systems (Briones & Schmidt, 2017). However, a recent comprehensive review on the subject highlighted that glyphosate and glyphosate-based herbicides can have negative effects on earthworms at the recommended application rate. These effects may not always manifest themselves after a single application or when focusing solely on adult mortality. However, when evaluating sensitive endpoints such as reproduction and growth, and considering field-relevant species subjected to multiple applications per year, negative effects were observed at sub-individual, individual, population and community levels (de Lima E Silva & Pelosi, 2024).

Since earthworms breathe through their skin, they need to always keep it moist, and the availability of water in the soil plays a fundamental role in their activity, life cycle, reproduction and abundance. Earthworms are therefore most abundant in the upper soil profile when soil moisture is not limiting. When irrigation is used to overcome seasonal soil moisture deficits, higher earthworm populations during the drier summer period are expected (Fraser et al., 2012). Drought or flooding conditions can affect earthworm population negatively for several years (Curry, 1993; Kiss et al., 2021; Plum & Filser, 2005). They have developed mechanisms to survive dry conditions, moving deeper into the soil profile and entering a state of aestivation (Fig. 3). For instance, *A. caliginosa* is known to be particularly sensitive to drought, responding by entering aestivation, drastically decreasing its metabolic rates (Bayley et al., 2010). For species such as *L. rubellus*, laying cocoons that hatch when conditions are more favorable is another adaptive response (Edwards & Bohlen, 1996). When soil moisture is not limiting, increasing temperatures in cooler temperate regions may benefit earthworm activity, with earthworm burrowing increasing from 15°C to 20°C. During flooding events, earthworms may migrate horizontally to find unflooded areas or vertical structures (e.g. trees or poles) (Singh et al., 2019). The success of these adaptations depends on the scale and frequency of the weather events. As the frequency and intensity of extreme climatic conditions will increase as a result of climate change, the ability of earthworm populations to recover may be reduced, especially if the time interval between two extreme events is less than the development time from cocoon to the adult stage of the earthworm life cycle (i.e. about 6 months for many lumbricid earthworm species) (Plum & Filser, 2005). For instance, at sites subject to frequent flooding, earthworms matured at a lower weight and at a younger age, and cocoons were unable to hatch (Klok et al., 2006; Thonon & Klok, 2007). Managed grasslands which have a higher earthworm abundance and diversity may be better placed to buffer adverse environmental conditions than croplands (Kiss et al., 2021; Liu et al., 2025). These results emphasize the importance of examining the interactive effects of climate factors, land use and management practices on earthworms to improve system resilience, an area that remains largely understudied (Singh et al., 2019).

2.2 Tropical grasslands and croplands

In tropical regions, native species have high endemism levels with limited geographical distributions, mainly due to habitat variability and restricted dispersal ability (Lavelle & Lapied, 2003). Preferences for particular soil conditions are commonplace in earthworms, and some have very restrictive distributions, such as *Glossoscolex ibira*, a large species found only in a low pH sandy Acrisol, near the town of Ibirá, in São Paulo state, Brazil (Caballero, 1976). On the other hand, very successful pantropical species like *Pontoscolex corethrurus* display a very wide tolerance for soil types, as well as chemical and physical conditions (Taheri et al., 2018), attributes also found in several other widespread earthworm species, such as those in the *Amynthas* genus (Steffen, 2012). Very few of the 2000 or more tropical earthworm species have been the object of detailed studies. Nevertheless, earthworms have been proposed as bioindicators of land use disturbance and/or change in several tropical countries including Ivory Coast, India, Mexico, Peru and Brazil (Fragoso et al., 1997; Guéi & Tondoh, 2012; Tondoh et al., 2007). Furthermore, they have been used as indicators of soil management practices, such as no-tillage systems, particularly in Brazil (Bartz et al., 2014).

Earthworms are particularly affected by land use and management practices. As an example of the potential differences between earthworm populations in various agricultural and natural habitats in the tropics, Fig. 5

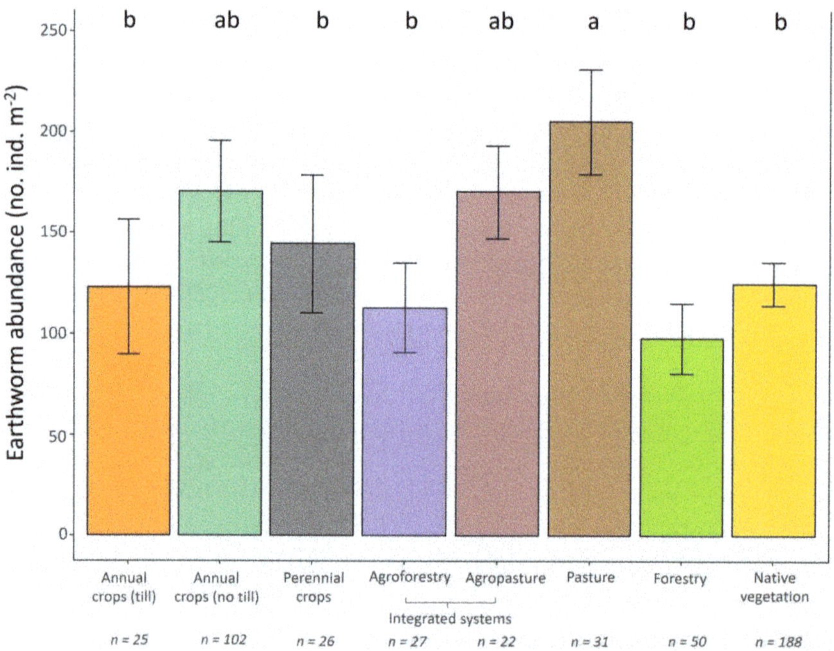

Figure 5 Earthworm abundance was evaluated across 468 sites in Brazil under various land-use systems. Abundance is reported as the number of individuals per meter square. Significant differences between land uses are indicated by different letters. Details of species composition are provided in Fig. 6. Significant differences between land uses are indicated by different letters as a result of a Kruskal-Wallis test followed by a post-hoc Dunn test. Values represent the mean and bars the standard deviation.

summarizes the content of an unpublished database of one of the authors (G. Brown) on the total earthworm abundance and species richness across 468 sites in Brazil. Land use included annual and perennial crops, pastures, and integrated production (agroforestry and agropasture) systems, as well as native vegetation and forestry plantations. Earthworm abundance was the highest in pastures (Fig. 5). The species most commonly encountered in these Brazilian agroecosystems were in the genus *Dichogaster*, representing small, exotic earthworms native to Africa; and *P. corethrurus*, the most abundant species encountered (Fig. 6). Only one of the ubiquitous species may be native to Latin America: a small and thin earthworm in the Ocnerodrilidae family (*Nematogenia panamaensis*), commonly associated with annual cropping systems.

There are only a few publications describing potential impacts of pesticides on tropical earthworms in the laboratory, and even fewer reporting field studies. Most studies focused on vermicomposting species like *Perionyx excavatus*, which are not the aim of this chapter. Some common ubiquitous endogeic species, such as *P. corethrurus* and *Amynthas gracilis*, were evaluated

Land use systems	Earthworm species	No. sites (%)	Abundance (ind. m^{-2})	Range (min - max ind. m^{-2})
Annual crops including tillage and no-tillage systems (n = 127)	*Dichogaster saliens*	37 (30)	16 ± 3	3 - 225
	Dichgaster gracilis	33 (26)	13 ± 2	5 - 100
	Nematogenia panamaensis?	32 (26)	15 ± 3	5 - 220
	Pontoscolex corethurus	25 (20)	60 ± 12	5 - 320
	Dichogaster affinis	24 (19)	11 ± 2	3 - 65
	Dichogaster bolaui	22 (18)	15 ± 3	5 - 208
Integrated systems including agroforestry and agropastoral systems (n = 49)	*Pontoscolex corethurus*	29 (59)	170 ± 12	1 - 300
Perennial crops (n = 26)	*Pontoscolex corethurus*	21 (81)	144 ± 30	4 - 765
Pastures (n = 31)	*Pontoscolex corethurus*	25 (80)	155 ± 29	4 - 419

Figure 6 Mean abundance (individuals m^{-2}) of the most frequently encountered earthworm species across various land-use systems (i.e. species present at more than 15% of all sampled sites; percentage occurrence is indicated in parentheses), along with the range of abundance (minimum and maximum values) observed across 468 sites in Brazil.

in the lab (Awaknavar & Karabhantanal, 2004, 2005; Buch et al., 2013; Cantelli, 2011; Chang & Bruno, 1970; Khalil, 2013, 2015; Morowati, 2000; Mostert, 2001; Mostert et al., 2002; Parelho et al., 2018). Unfortunately, very few studies have addressed impacts of pesticides on native earthworm species. Observed effects ranged from almost absent (Brazil; insecticide lambda-cyhalothrin and the fungicide carbendazim (Garcia, 2004)) to slightly negative (India; insecticide malthion; Panda & Sahu, 2000; Senapati et al., 1992, 1994) or strongly negative (Mexico, Glyfos®, Classical Faena®, and Faena Fuerte®) (García-Pérez et al., 2014). Garcia-Perez et al. (2020) emphasized the growing evidence that epigeic, endogeic and anecic earthworms are vulnerable to repeated applications of glyphosate. More detailed and, at the same time, broader field tests, which would help us understand the potential short- and long-term effects of pesticides on tropical earthworm communities are still largely missing. There is also a need for studies comparing different species under identical test conditions and a range of contaminants. Finally, very little is known about the effects of mixtures of pesticides (Pelosi et al., 2021b).

The impact of other agricultural management practices such as soil tillage have been reviewed globally, but always with few results from tropical countries (Briones & Schmidt, 2017). Two recent reviews (Bartz et al., 2024; Demetrio et al., 2020) clearly showed higher populations in no-tillage sites although the type and frequency of the soil management employed is critical: periodic tillage can be very detrimental, and the lack of crop rotation was also highlighted as a limiting factor (Fig. 2). Nonetheless, few experiments have addressed crop rotation effects, particularly the inclusion of legumes in cropping systems. Oluko et al. (2014) and Guéi et al. (2019) highlighted the importance of external N inputs into the system for increasing earthworm populations and optimizing their roles in crop production. Long-term no-tillage can cause major soil compaction when there is too much traffic or no adequate cover cropping, and both have been associated with lower earthworm populations in Southern and Southeastern Brazil (Dudas et al., 2020).

Organic matter additions via manure deposition on the surface or incorporated into the soil have helped increase earthworm populations in several locations in the tropics (Oliveira Filho et al., 2018; Senapati et al., 1999) (Fig. 2). However, the long-term impacts of high heavy metal concentrations (particularly Cu, Ni or Cd) brought with these wastes is a concern and must be monitored to avoid a decrease in earthworm populations (Cotton DCF & Curry JP, 1980). Similar concerns are associated with antibiotics in some wastes (such as ivermectin), as high concentrations of this acaricide can lower reproduction rates (Ma et al., 2023), though other studies have shown little to no non-target effects (Kaneda et al., 2006; Scheffczyk et al., 2016). Further, many highly weathered tropical soils, particularly in the wetter regions, are highly acidic and require liming in order to be useful and productive for agricultural purposes.

Published by Burleigh Dodds Science Publishing Limited, 2025.

Similar to temperate soils, acidification is expected to negatively impact earthworm populations and liming likely remediates this effect (Fig. 2), but very few studies on tropical soils exist.

As noted earlier (Section 2.1), earthworms are particularly sensitive to climatic factors, and especially moisture levels. Among the 40 species collected in the Manaus region of the Amazon (database G. Brown), 33 lived strictly in areas close to watercourses, while four inhabited soils with large moisture variations. The remainder lived in rotting tree trunks (Ayres & Guerra, 1981). In environments with seasonal rainfall, such as tropical Savannas, many species enter aestivation and/or seek deeper soil layers with higher moisture contents (Fig. 3). For instance, *Rhinodrilus alatus* enters aestivation for more than 6 months, at an average depth of 30 cm (Drumond et al., 2015), while *Glossoscolex paulistus* is normally active in 30 cm deep U-shaped burrows but aestivates in chambers at 50 cm depth (Abe & Buck, 1985). An Amazonian species *Chibui bari* undergoes diapause for around 6 months at approximately 1 m depth (Guerra, 1988).

Earthworms also have evolved responses to flooding events. When the tropical grassland plains of Colombia (Orinoco Llanos) are flooded, earthworms, primarily *Andiorrhinus* species, construct *surales*. These are large mounds composed mainly of casts, built by earthworms searching for aeration (Zangerlé et al., 2016b). Interestingly, similar structures have also been observed in a solar park located on temperate flooded grasslands in the Netherlands. These mounds appear to be primarily constructed by *Aporrectodea caliginosa* (Ron de Goede, personal communication, 01-09-2024). Given the increasing frequency of rainfall events due to climate change, these mounds could play a growing role in both temperate and tropical regions.

The potential impact of climate change on tropical earthworms has not yet been adequately determined. Two recent literature reviews, for instance, did not contain any data on tropical climates and earthworm species (Ghosh, 2021; Singh et al., 2019). Only one study was found, which evaluated the impact of predicted climate change on the distribution of a widely used fish-bait species, *R. alatus* (Hughes et al., 2018). The authors concluded that, under a realistic scenario of climate change for 2070, the distribution of this earthworm species would become restricted, and its habitat fragmented. This would lead to a decline in earthworm populations, the collapse of harvesting activities and severe impacts on the local community that relies on this traditional practice for subsistence. The global analysis by Phillips et al. (2019), though based on limited data from tropical regions, revealed that climate is one of the most significant factors influencing earthworm abundance, biomass and species richness worldwide. It may also further contribute to earthworm invasion. This finding suggests that any potential changes in global climate could have substantial impacts on these parameters.

3 Effect of land use, agricultural management, and climate on enchytraeid communities

Enchytraeids are regarded as valuable indicators of land use and management practices, particularly in agriculture (Pelosi & Römbke, 2016, 2018) due to their abundance across various ecosystems, their essential role in ecosystem functioning and their sensitivity to a wide range of environmental stresses. The first section focuses on the effect of different land uses on enchytraeid communities, comparing forests, grasslands and arable soils. A next section is dedicated to agricultural practices, highlighting how population size and species composition are affected by ploughing, pesticides, and other factors (Fig. 2). In a final section, we review how climate parameters influence enchytraeids, which, in turn, significantly affects ecosystem processes.

3.1 Land use and enchytraeid communities

Enchytraeids are often considered typical forest-dwelling organisms, but they are also commonly found in agroecosystems. Their abundance – often dominated by *Cognettia sphagnetorum* – is generally higher in forest than in grasslands and croplands, likely due to the low pH levels of forest soils, which enchytraeids tend to prefer (Beylich & Graefe, 2009; Pelosi & Römbke, 2018). However, this pattern does not always hold. For instance, Springett et al. (1996) found higher enchytraeid abundance in kiwifruit orchards and hill pastures than in a dairy pasture and native forest. As microbivorous and saprovorous organisms (Didden W A M, 1993), enchytraeids often find greater food availability and encounter fewer disturbances in grasslands, which generally results in higher populations compared to arable fields (Barbercheck et al., 2009). Yet, some studies suggested that land use tends to affect species composition more strongly than total abundance (Beylich & Graefe, 2010; Nowak, 2004).

3.2 Agricultural management practices and enchytraeid communities

In grasslands, enchytraeids are influenced by livestock grazing and mowing, with grazing having a more negative effect than mowing, likely due to soil compaction induced by animal trampling (Maraldo et al., 2015). In arable fields, most studies have focused on the effects of soil tillage, particularly ploughing, on enchytraeid communities. Ploughing alters the vertical distribution and composition of enchytraeid community, but the overall impact is unclear due to contrasting findings. Some positive effects of ploughing on enchytraeids may offset the mechanical damage caused by conventional tillage. For instance, the incorporation of crop residues within the soil profile can provide enchytraeids with food (Lagerlöf et al., 1989; Manetti et al., 2010; Severon et al., 2010);

Published by Burleigh Dodds Science Publishing Limited, 2025.

promote bacterial growth, which favors bacterial-feeding organisms with high metabolic activity like enchytraeids (Hendrix et al., 1986); and increase pore space, which facilitates enchytraeid movement through the soil (Lagerlöf et al., 1989). As a result intermediate disturbance, such as non-inversion tillage, has been reported to favor enchytraeids (Chan & Heenan, 1995) (Fig. 2).

As with earthworms, many laboratory studies report harmful effects of pesticides on enchytraeids (Römbke et al., 2017). However, these controlled tests often lack ecological relevance due to factors like test duration, use of artificial soils, model species and single-species tests. The response of enchytraeids to pesticides at the community level can be studied using semi-field methods, such as Terrestrial Model Ecosystems (Burrows & Edwards, 2004; Scholz-Starke et al., 2013) or by directly sampling communities in the field. Harmful effects of pesticides on enchytraeids were reported in both field and semi-field studies, though rarely at the recommended application rate (Pelosi & Römbke, 2016). Results vary due to numerous interacting factors, and more studies are needed on the long-term effects of pesticide mixtures applied at agronomically recommended doses.

Other practices, such as inorganic fertilization and organic matter inputs, also affect enchytraeid community abundance and composition, though data on these effects remain limited. Generally, organic matter inputs appear beneficial to enchytraeid communities (Andrén & Lagerlöf, 1983; Mitchell et al., 1978). The effects of inorganic fertilization are more variable and may be negative (Birkhofer et al., 2008) possibly due to changes in soil pH and electrical conductivity following fertilizer applications (Lagerlöf et al., 1989) (Fig. 2). Contrasting results when assessing effects of land use and management on enchytraeids may also arise from annual climatic variations in enchytraeid communities (Maraldo et al., 2015).

3.3 Climate and enchytraeid communities

Enchytraeids are soft-bodied organisms highly dependent of soil moisture for survival, reproduction and activity. Vulnerable to drought, they are typically confined to moist habitats (Didden, 1991). Severe droughts can dramatically alter enchytraeid water content, reproduction and survival, affecting their abundance, diversity and composition (Beylich & Achazi, 1999; Coyle et al., 2017; Maraldo & Holmstrup, 2010). Like earthworms, enchytraeids have different strategies to cope with drought (Fig. 3). Studies from over 50 years ago (Dózsa-Farkas, 1973, 1977; Willard, 1972) reported that enchytraeids, such as *Stercutus niveus*, enter aestivation during dry summer periods. Dózsa-Farkas (1973) observed in laboratory conditions that enchytraeids cluster in small, dense, inactive groups in response to dry soil conditions, a phenomenon that remains largely overlooked in the literature. When conditions are unfavorable, enchytraeids can also survive in a cocoon stage (Christensen, 1956; Lagerlöf & Strandh, 1997) or migrate to more humid microhabitats (Didden, 1991; Maraldo, 2009; Springett et al., 1970). However, due to their small size, enchytraeids cannot migrate as deep as earthworms. Maraldo and Holmstrup (2009) identified *Cognettia sphagnetorum* in 5–10 cm deep layers below the surface, during dry periods, whereas *Lumbricus terrestris* can reach depths of up to 2 m under similar conditions (van de Logt et al., 2023a).

Optimal conditions of 20% gravimetric water content (−2.57 kPa) were found for *E. albidus*, supporting high survival, feeding activity and reproductive rates. As for earthworms, overly moist soil conditions may reduce enchytraeid reproduction by preventing cocoons laying (Cécile Serbource, personal communication, 01-09-2024) (Fig. 3). Temperature also influences enchytraeid activity, productivity and reproduction (Briones et al., 2007; Dial et al., 2016; Maraldo et al., 2010; Patricio Silva et al., 2014). For instance, freeze tolerance in the enchytraeid *Fridericia ratzeli* relies on glucose accumulation, a mechanism similar to that of freeze-tolerant earthworm species (Holmstrup & Sjursen, 2001). Optimal temperatures for enchytraeids, such as model species *Enchytraeus* sp., are around 18°C. In a meta-analysis, Briones et al. (2007) found that a mean annual temperature above 16°C can limit the presence of *C. sphagnetorum*. Some enchytraeid species can live in areas with very dry summers and severe winters, indicating distinct physiological or behavioral adaptations (Grongroft & Miehlich, 1983; Pedersen & Holmstrup, 2003; Willard, 1974). Other species, such as *Fridericia galba* and *Enchytraeus buchholzi*, can tolerate dry conditions as long as temperatures remain moderate (Abrahamsen, 1971; Dózsa-Farkas, 1977).

In conclusion, enchytraeids in agricultural fields are sensitive to both management practices and climatic variability. Climate change is expected to increase the frequency and intensity of heavy rainfall events and extend drought periods, posing challenges for enchytraeid populations (Holmstrup et al., 2012). These effects may be subtle at the level of total abundance, making genus- or species-level indicators more informative.

4 The interaction between earthworms and enchytraeids

4.1 An unclear relationship

Enchytraeids and earthworms can provide similar benefits to the soil, but at different scales. However, there is still limited data on interactions between these two groups under natural conditions. Some studies revealed a competitive relationship, as both inhabit the same soil horizon and consume similar food sources (Gajda et al., 2017; Larsen et al., 2016). Beylich and Graefe

(2010) highlighted that enchytraeid populations tend to increase in areas with low earthworm activity, suggesting a potential exclusion relationship. The authors also demonstrated that smaller earthworm species or juveniles of medium-sized species (belonging to endogeic or epigeic groups) exhibit more antagonistic relationships with enchytraeids compared to larger anecic species. The presence of invasive earthworm species has also been shown to decrease the enchytraeid densities (Schlaghamerský et al., 2014). However, earthworms and enchytraeids may also display positive relationships. For instance, Schlaghamerský et al. (2014) and Nuutinen et al. (2017) demonstrated that the presence of *Lumbricus terrestris* may facilitate the presence of enchytraeids by increasing macroporosity. Schrader and Seibel (2001) and Beylich and Graefe (2009) suggested that the casting of anecic earthworms near the soil surface may locally improve conditions for enchytraeids and other soil organisms. Karaban and Uvarov (2014) observed a positive influence of *Aporrectodea caliginosa* on enchytraeid abundance, attributing this to its non-trophic activities such as burrowing, cast deposition and mucus release. Several authors underlined that, in the absence of resource limitation, earthworms and enchytraeids can coexist without competitive exclusion (Beylich & Graefe, 2010; Makulec & Pilipiuk, 2000; Ricci et al., 2015).

4.2 Diverging responses to external factors

Very little is known about the factors driving the interactions between earthworms and enchytraeids, including climate, food resources and agroecological practices. Few studies have documented differing responses between earthworms and enchytraeids to external factors. For instance, Bart et al. (2017) and Kobetičová et al. (2010) demonstrated that earthworms are more sensitive to contaminated soils and pesticides exposure than enchytraeids, using avoidance tests. However, Serbource et al. (2024) compared the sensitivity of earthworms and enchytraeids to different stressors under similar test conditions and found them to be equally sensitive to chemical stressors regardless of the studied endpoint (mortality or reproduction). This study also underlined the lack of studies (1) on environmentally relevant (i.e. representative of natural soils) enchytraeid and earthworm species and (2) on other driving factors than chemicals such as agricultural practices and climate change. Topoliantz et al. (2000) found that after tillage, the activity of enchytraeids increased when earthworm activity declined (Fig. 2). Further, while this has not been tested, enchytraeids' preference for acidic soils suggests they may be less affected by fertilizer-induced soil acidification than earthworms. These few studies point toward the need to quantify both earthworms and enchytraeids when assessing soil health, as enchytraeids could partially maintain the soil functions performed by earthworms.

Published by Burleigh Dodds Science Publishing Limited, 2025.

5 How to sample and monitor earthworm and enchytraeid populations

Sampling earthworms and enchytraeids in an efficient, reliable and reproducible way is a crucial step in assessing the effects of land use, management practices and climate on these soil organisms and for developing reliable soil health indicators. Below, we summarize the most common methods for sampling earthworms and enchytraeids. We also direct readers to key references for further information.

5.1 How to sample earthworms

There is a wide diversity of earthworm sampling methods, which often makes comparing studies challenging. This variation is partly due to the fact that not every sampling method is suitable for all soil conditions or earthworm species. We can broadly distinguish two main groups of methods to systematically sample earthworms: the *ethological methods* that mainly rely on the earthworms being stimulated to come out of the soil themselves, and the *excavation methods* followed by hand sorting approaches with or without wet sieving (Bouche, 1969; Satchell, 1969; Singh et al., 2016) (Fig. 7). Ethological approaches include the use of electricity, i.e., the octet method (Eisenhauer et al., 2008; Schmidt, 2001a; Weyers et al., 2008) and the use of chemicals such as formaldehyde (Raw, 1959), hot mustard (Gronstol et al., 2000; Lawrence & Bowers, 2002), onion solution (Steffen et al., 2013) and allyl isothiocyanate (AITC) (Pelosi et al., 2009; Zaborski, 2003). Another less well-known approach involves actively searching for earthworms within specific habitats, such as beneath dead logs in forests, within and under dung pats in pastures, or in wetter environments like stream margins, alongside or inside marshes, in bromeliads and in suspended soils (Bartz et al., 2014; Decaens et al., 2016; James, 1996). If needed, these specific niches should be considered in addition to soil sampling.

The effectiveness of each method varies between ecological categories of earthworms. Hand sorting is more effective in detecting small individuals, including endogeics and juvenile anecics, which are often underestimated with ethological methods due to their lower mobility (Bartlett et al., 2006; Callaham Jr & Hendrix, 1997; Chan & Munro, 2001). Larger anecics with deeper burrowing strategies are often missed with hand sorting. Among ethological methods, the octet approach has the advantage of not disrupting the soil biogeochemistry, which can be a significant advantage in urban areas where the use of chemicals and excavation may not be viable options (Pelosi et al., 2021a) or in manipulative experiments where imposed treatments should not be disturbed. The downside of this method is that it requires specialized equipment (Schmidt, 2001a), and it is highly dependent on adequate soil water content to transmit the electric signal. When a chemical approach is used, Pelosi

Published by Burleigh Dodds Science Publishing Limited, 2025.

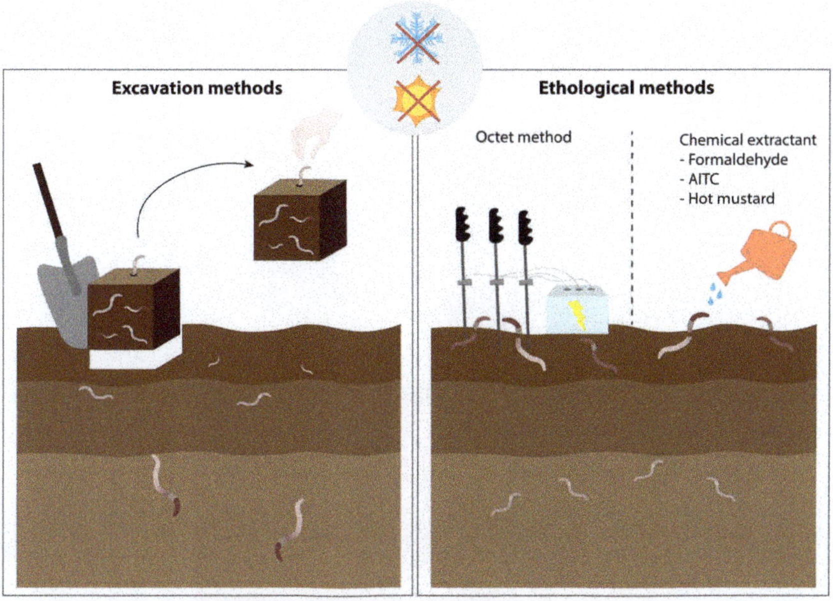

Figure 7 Overview of earthworm sampling methods, divided into two main categories: excavation methods (left) and ethological methods (right). Excavation methods involve removing a block of soil and hand-sorting earthworms. Ethological methods utilize external stimuli, such as electricity (e.g. the octet method) or chemical extractants applied to the soil surface, including formaldehyde, allyl isothiocyanate (AITC) and hot mustard. Excavation methods tend to favor the collection of smaller individuals, such as juveniles and endogeic earthworms but often miss species that burrow deeper. Conversely, ethological methods are more effective for collecting epigeic and anecic species, while underestimating smaller species due to their lower mobility. Field sampling is generally discouraged during very cold or very dry weather conditions. We do not aim to provide an exhaustive list of sampling methods; instead, we direct readers to key references in the text for further information.

et al. (2009) advocate the use of AITC, which is not as toxic as formaldehyde and is easier to standardize than hot mustard. Although considered to be the most reliable approach, hand sorting is very labor intensive. To overcome the challenges related to hand-sorting, Schmidt (2001b) suggests a time-limited (40 minutes per 25 cm × 25 cm × 25 cm block) hand-sorting approach for large and long-term monitoring campaigns. In Fig. 8, we provide a brief description of the main methods used for earthworm sampling, together with their pros and cons. Providing a detailed analysis of these approaches is beyond the scope of this review. We refer readers to Singh et al. (2016) who performed a thorough review of sampling methods.

Given all the pros and cons of each methods, the most reliable approach might be to combine hand sorting with chemical extraction. In this line, the latest ISO norm 23611-1:2018(E) (ISO, 2018) recognizes the human and

		Brief description	Soil disturbance	Time	Efficiency				Toxicity	Cocoon extraction	Earthworm injuries
					Adult anecics	Juvenile anecics	Endogeics	Epigeics			
Excavation methods	Hand-sorting	A block of soil is excavated and earthworms are hand sorted	Soil extraction		Migrate deeper						Some earthworms cut in two with spade
	Hand-sorting with wet-sieving	A block of soil is hand sorted for earthworms and washed with mesh sieve	Soil extraction		Migrate deeper						Some earthworms cut in two with spade
Ethological methods	Formaldehyd	A formaline solution is applied to a determined soil surface	Pollution			Low mobility	Low mobility	Lateral migration	Carcinogenic		Die if not washed under water
	AITC	A allyl isothiocyanate solution is applied to a determined soil surface				Low mobility	Low mobility	Lateral migration			
	Mustard	A hot mustard solution is applied to a determined soil surface				Low mobility	Low mobility	Lateral migration			
	Octet	Electroshocks are sent to the soil				Low mobility	Low mobility	Lateral migration			

Figure 8 A brief description of the main methods used for earthworm sampling, along with their pros and cons. The color scale provides a relative comparison of methods based on key factors to consider during sampling, ranging from yellow (positive) to red (negative). A detailed analysis of these approaches is beyond the scope of this review. Readers are directed to key references in the text for further information.

ecosystem toxicity of formaldehyde and proposes hand sorting followed by AITC extraction when deep burrowing earthworm species are present.

Regardless of the sampling method chosen, climatic conditions should be considered when sampling earthworms, with a special focus on soil moisture and temperature (Eggleton et al., 2009). Periods when earthworms are inactive or suffering stressful conditions, i.e. winter and summer, should be avoided (Section 2). In temperate regions, the activity of earthworms drastically decreases when the soil temperature is below 5°C in winter and when the soil moisture content is below the wilting point in summer (Curry, 2004; Daugbjerg, 1988). In tropical regions, sampling is advised during the wet season, when temperatures are relatively high without drought stress (Fragoso et al., 1999; Fragoso & Lavelle, 1992; Singh et al., 2019).

5.2 How to sample enchytraeids

Enchytraeids are typically sampled using a soil corer with a diameter between 3 cm and 6 cm (Fig. 9). In agricultural soils, it is common to collect samples for enchytraeid extraction up to 30 cm where most of the enchytraeids are thriving.

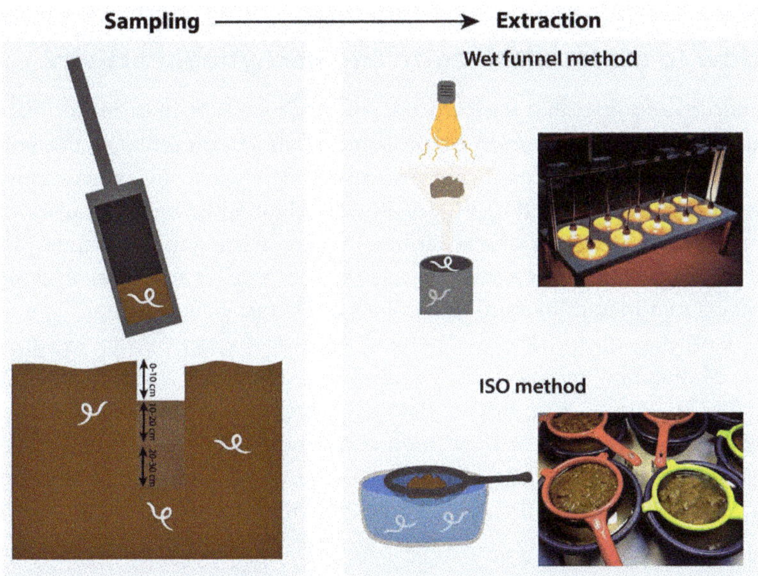

Figure 9 Different stages of enchytraeid sampling, from field collection. Sampling is typically performed at a depth of 0–10 cm but can also include other depths where enchytraeids are found, such as 10–20 cm and 20–30 cm. Two main methods are available for extraction: the wet funnel method, which requires a heat source, and the ISO method, which involves submerging samples in water and does not necessarily require heat. Counting and identification are usually carried out using a petri dish placed under the microscope when the enchytraeids are still alive.

Published by Burleigh Dodds Science Publishing Limited, 2025.

Enchytraeids are more abundant in the upper layers (0–10 cm) in plots under reduced or no-tillage while under conventional tillage, they are more equally distributed over the soil profile and therefore generally sampled at a soil depth 0–20 cm (Pelosi & Römbke, 2016).

After collection, soil samples can be kept refrigerated at 4–6°C up to 2 weeks after collection (ISO, 2019). The ISO norm 23611-3 recommends to extract enchytraeids using a wet extraction approach which consists of submerging the soil samples in water on a sieve (mesh size approximatively 1 mm) for 2–7 days and collecting the sinking enchytraeids. Some variations of this approach exist, notably the wet funnel method that uses a light bulb to create a source of heat in order to increase the extraction efficiency and reduce the extraction time (Fig. 9) (Kobetičová & Schlaghamerský, 2003; O'connor, 1955). In contrast to earthworms, enchytraeids can only be identified alive as their transparent body becomes opaque after death, which makes identification impossible (Schmelz & Collado, 2010). Similar to earthworms, we strongly recommend sampling enchytraeids at multiple times throughout the year (e.g. in autumn and spring), in order to effectively evaluate the impacts of agricultural management on enchytraeids, alongside natural climatic fluctuations (Maraldo & Holmstrup, 2009).

6 How to assess earthworm and enchytraeid activity

It is often overlooked that earthworms and enchytraeids, in order to influence soil functions, not only need to be present but also to be active. In this section, we therefore want to discuss approaches specifically focused on assessing the activity of these soil biota (Fig. 10). While abundant earthworm populations are likely to reflect conditions where earthworms are active and reproducing, more direct measurements of earthworm activity can help us to better understand the effects of agricultural management on earthworms.

Quantifying bioturbation is particularly challenging due to the opaque nature of the soil matrix and the dynamic nature of bioturbation (Vidal et al., 2023). This might partly explain why directly assessing earthworm and enchytraeid activity have not yet been considered as indicators of soil health. Here, we argue that some methods could be quick and efficient enough to serve as biological indicators for soil health in the future.

6.1 How to assess earthworm activity

Ecotoxicological approaches – Earthworms have long been used as bioindicators in ecotoxicological studies due to their widespread presence in soil, sensitivity to contaminants and ease of handling (Sanchez-Hernandez, 2006; Velki & Ečimović, 2017). In standardized testing protocols, both acute (e.g. LC50 tests)

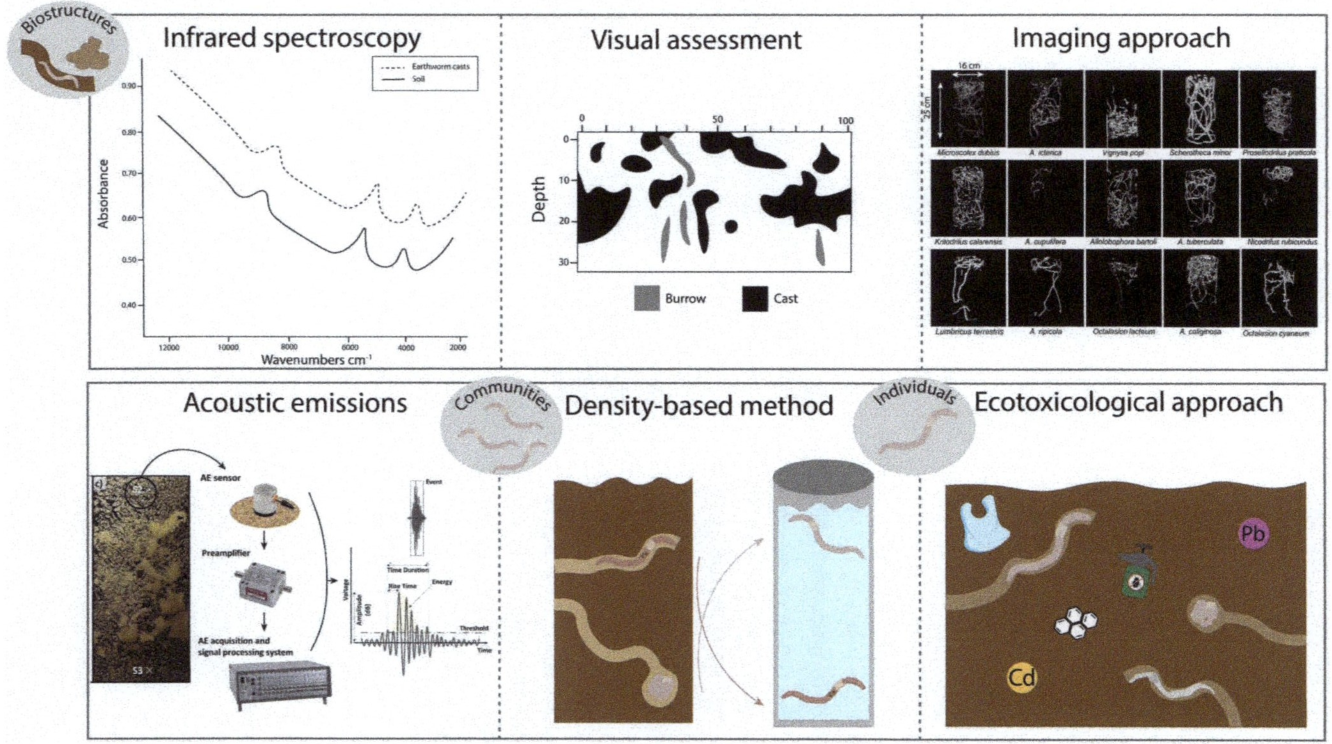

Figure 10 Different approaches used to quantify earthworm activity, classified based on their object of quantification: biostructures, earthworm communities and individual earthworms. The infrared spectroscopy and the visual assessment figures were created to depict hypothetical results; the imaging approach figure is from Vidal et al. (2023); the acoustic emissions figure is adapted from Lacoste et al. (2018) by combining parts of Fig. 1 and Fig. 4 from the original paper, which is licensed under a Creative Commons Attribution 4.0 International License (https://creativecommons.org/licenses/by/4.0/).

and chronic (e.g. reproduction tests) effects on earthworms can be assessed (ISO, 2023; OECD, 2016). For these tests, earthworms are exposed to either artificial or natural soils for 2 weeks in acute tests or for several weeks to months in chronic tests. Ecotoxicological assays typically focus on a few model species, with *Eisenia fetida*, *Eisenia andrei* and *Lumbricus terrestris* being the most frequently studied (e.g. Chelinho et al., 2011). These assays primarily assess the risks of soil pollution as an indicator of soil health, under controlled laboratory conditions. However, these tests can also be used to assess the habitat function of a soil using non-spiked soils. Native or ubiquitous earthworm species of interest can be kept in 500 g soil following the recommendations of ISO (2023). Growth and reproduction is assessed using juvenile or adult individuals, respectively, providing valuable information on their ability to survive in and colonize a particular habitat. Tests like this have been done using natural soils and several non-standard earthworm species like *A. caliginosa* in the temperate region (Bart et al., 2018a) and *P. corethrurus* in Brazil (Buch et al., 2017), though the duration of the test with the latter species had to be expanded to 90 days in order to ensure reproduction. Nonetheless, this approach does of course not fully capture the range of ecosystem functions that earthworms perform as a community in soil, such as their substantial influence on soil structure (Bartlett et al., 2010). Alternatively, the methods described below emphasize evaluating the impact of earthworm activity on soil through their bioturbation effects.

Visual assessment – Most existing methods for assessing the activity of earthworms focus on assessing earthworm-made structures including casts (Chevallier et al., 2006; Gerard, 1967; Perreault & Whalen, 2006) and burrows (Capowiez et al., 2021; Perreault & Whalen, 2006). Quantifying these structures can be done in the field using visual assessment of burrows and cast deposits on the soil surface (Mueller et al., 2009; Peigné et al., 2013; Pérès et al., 1998). A major drawback of this approach is that it fails to account for species that rarely produce surface casts, such as endogeic species (Capowiez et al., 2021) or for middens formed by anecics species in semi-permanent locations (Rossi & Nuutinen, 2004). Some studies have attempted to assess earthworm bioturbation at the profile scale, e.g. using the Soil-Structure-Pattern (SPP) method (Piron et al., 2012), but this method has proven too complex for larger monitoring streams or for providing practical management advice. Consequently, it has been proposed to add a 'earthworm bioturbation' item within the more widely used Visual Soil Structure Assessment (VSSA). This includes measuring the density and distribution of cast deposits, counting burrow openings and assessing the freshness of cast aggregates. This would result in a classification including four levels of bioturbation, from low to high (Piron et al., 2017).

Imaging approaches – To capture the activity of earthworms in soils, several imaging approaches with various degree of sophistication have been used to visually and quantitatively assess earthworm behavior, movement and their effect on soil structure. Image capturing technologies such as digital cameras or scanners have been used to monitor the temporal dynamics of soil biota, including earthworms. These methods include imaging devices placed directly in the field (Belaud et al., 2024), imaging within mesocosms (Djerdj et al., 2020; Jangorzo et al., 2015) or imaging after sample collection and preparation (Sauzet et al., 2023). X-ray tomography has often been used to create 3D images depicting how earthworms influence soil porosity and structure, allowing for monitoring burrow formation dynamics over time (Capowiez et al., 2011). A key drawback of this approach, however, is that it requires destructive sampling and access to advanced instrumentation.

Infrared spectroscopy – Approaches such as near infrared spectroscopy (NIRS) (Bottinelli et al., 2013; Zangerlé et al., 2014, 2016a) and mid-infrared spectroscopy (MIRS) (Bottinelli et al., 2021) show promise for rapidly and cost-effectively dating certain biostructures and linking their composition to specific earthworm species. In this approach, structures of interest can be collected from the field and exposed to an infrared light while intensity of adsorption is measured. The information extracted from the spectra enables to differentiate earthworm-processed soil from surrounding soil. While these approaches provide valuable insights into earthworm activity, they have not yet been adopted as direct indicators of soil heath and require substantial calibration measures. A more promising use of these infrared spectroscopic approaches as a soil health indicator may be to assess general soil quality and specific biological characteristics, such as soil microbial biomass or soil respiration (Cécillon et al., 2009).

Acoustic emissions – A few recent studies have explored the acoustic signatures of earthworms to monitor their activity (Keen et al., 2022; Lacoste et al., 2018; Robinson et al., 2024). Under controlled laboratory conditions, Lacoste et al. (2018) found a positive correlation between acoustic emissions and earthworm burrow length. More recently, Keen et al. (2022) were able to detect the acoustic emissions generated by invasive earthworm species in Tundra soils during a field mesocosm experiment. Although still in initial stages of development, this approach appears to be a promising, low-cost and non-invasive method to assess the bioturbation induced by earthworms, and possibly also other soil biota (Robinson et al., 2024).

Density-based method – Garamszegi et al. (2025) recently developed a density-based method to quantify earthworm activity. This promising, easy and quick method uses the fact that earthworms empty their gut when they enter in aestivation. As a result, active earthworms can be separated from inactive ones using a 1.08 g cm^{-3} glucose solution, with active earthworms sinking and inactive earthworm with empty guts floating. In contrast to the previously mentioned approaches, it has the advantage of capturing the current activity

Published by Burleigh Dodds Science Publishing Limited, 2025.

		Brief description	Object of quantification	Soil disturbance	Time	Realistic conditions	Equipment	Population representativity	Ethics
Ecotoxicological methods 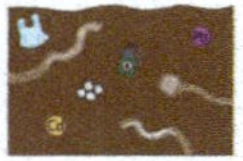	Acute tests	Earthworms are exposed to increasing concentrations of contaminants in an artificial or natural soil	Individual earthworms	Destructive			Laboratory	Few species	Earthworms exposed to toxic substance
	Chronic tests	Earthworm growth, reproduction and behaviour is monitored within a contaminated or not-contaminated soil	Individual earthworms	Destructive			Laboratory	Few species	Earthworms exposed to toxic substance
Visual assessment 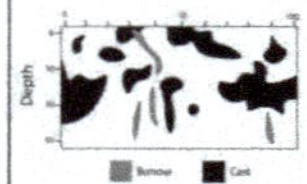	Soil surface	Quantification of surface casts and burrow openings	Biostructures	Non-destructive				Surface casting species	
	Profile scale	Quantification of casts, burrow openings and freshness of casts	Biostructures	Non-destructive				Entire population	
Imaging 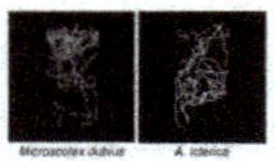	Cameras or scanners	Imaging earthworm behaviour, movement and their effect on soil structure	Biostructures	Non-destructive		Often, mesocosms experiments		Entire population	
	X-Ray (micro) tomography	Creating 3D images depicting how earthworms influence soil porosity and structure	Biostructures	Destructive				Entire population	Destructive sampling
Infrared spectroscopy 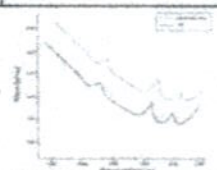		Detecting the chemical fingerprint of earthworm-processed soil	Biostructures	Mostly destructive				Entire population	
Acoustic emissions 		Detecting the acoustic signatures of earthworms	Earthworm communities	Non-destructive				Entire population	
Densit-based method 		Using a glucose solution to separate active earhtworms sinking from inactive earthworms with empty guts floating	Individual earthworms	Destructive				Entire population	Destructive sampling, earthworms exposed to a harmless glucose solution

Figure 11 A brief description of the main methods used to assess earthworm activity, along with their pros and cons. The color scale provides a relative comparison of methods based on key factors for assessing activity, ranging from yellow (positive) to red (negative). A detailed analysis of these approaches is beyond the scope of this review. Readers are directed to key references in the text for further information. The infrared spectroscopy and the visual assessment figures were created to depict hypothetical results; the imaging approach figure is from Vidal et al. (2023); the acoustic emissions figure is adapted from Lacoste et al. (2018) by combining parts of Figs 1 and 4 from the original paper, which is licensed under a Creative Commons Attribution 4.0 International License (https://creativecommons.org/licenses/by/4.0/).

of earthworms at the individual level, while other methods, when performed at the field scale, reflect the presence of semi-permanent structures mostly at the population level.

In conclusion, each method for assessing earthworm activity has its advantages and disadvantages (Fig. 11). The primary criterion for a soil health indicator is that it should be easy to implement in large-scale sampling campaigns, meaning that the method used should not be overly time-consuming or expensive while providing valuable information for practical management decisions.

6.2 How to assess enchytraeid activity?

Ecotoxicological approaches – Enchytraeids have been used as models in terrestrial ecotoxicology since the early 2000s as they were recognized as indicator organisms for chemical stress in terrestrial ecosystems (Didden & Römbke, 2001). Standardized protocols exist to assess survival and reproduction of enchytraeids when exposed to chemicals (ASTM, 2000; OECD, 2004). A few ecotoxicological studies deal with enchytraeid activity through the assessment of their feeding activity or avoidance behavior. For feeding behavior, the bait lamina method (ISO, 2016) can be used both under field (Amossé et al., 2018) and laboratory conditions (Bart et al., 2018b) as enchytraeids are considered to be among the main consumers of the bait in the topsoil (Helling et al., 1998). Moreover, avoidance test can be used to evaluate a specific locomotor response of enchytraeids when exposed to a chemical (Amorim et al., 2008; Kobetičová et al., 2010). These methods are informative and easy to implement but do not allow studying physical consequences of enchytraeid activity in soils.

Visual assessment and imaging approaches – Similar to endogeic earthworms, enchytraeids have low assimilation efficiencies which they compensate by high ingestion rates (Didden, 1993; Scheu, 1991). Van Vliet et al. (1995) calculated that every year, the enchytraeid community present at an agricultural site ingested and transported more than 2 kg of mineral soil per square meter, thereby considerably contributing to the evolution of soil structure through bioturbation and production of fecal pellets (Sandor & Schrader, 2012; Van Vliet et al., 1993). Some species such as *Enchytraeus lacteus* also contribute to the modification of soil surface roughness due to the production of casts (Schrader et al., 1997). Didden (1991) described the physical remnant of enchytraeids as a 'spongy' structure (Fig. 12a). The physical activity of enchytraeids in soils can be assessed through photographs (Van Vliet et al., 1993), determination of aggregate stability (Carrera et al., 2011) or using a laser relief meter for measuring surface roughness due to enchytraeids fecal pellets (Schrader et al., 1997). Otherwise, thin soil sections have been used *in*

Published by Burleigh Dodds Science Publishing Limited, 2025.

Figure 12 Effect of enchytraeids on soil structure. (a) Side and top views of a column experiment showing soil sieved to 2 mm, with a bulk density of 0.8 g cm^{-3}, incubated for 1 month without (left) and with (right) enchytraeids. The soil was inoculated with *Enchytraeus crypticus* (biomass equivalent to 240 000 individuals per m^2). Note the juveniles visible along the edges of the column in the side view. Source: Adapted from: Serbource et al. (2024). (b) X-ray microtomography images from another experiment showing soil cross-sections. Soil was incubated for 40 days with *Enchytraeus albidus* and *Enchytraeus crypticus*. Red squares represent the zones used to calculate indicators in Serbource et al. (2025).

situ to study the small aggregates made by enchytraeids (Carrera et al., 2011; Topoliantz et al., 2000). However, these structures can in turn be consumed by other soil invertebrates (e.g. earthworms, collembola), strongly limiting their visibility on thin sections. Another imaging approach that could, in principle, be used to assess enchytraeid activity is X-ray microtomography, a non-invasive and non-destructive imaging technique. So far, we are aware of only two studies that used this approach to analyze the enchytraeid pore network (Porre et al., 2016; Serbource et al., 2025) (Fig. 12b).

In conclusion, efforts to quantify the activity of enchytraeids in soils remain very limited. Most studies use model species, though wild enchytraeid species would provide more ecologically relevant insights. A major limitation remains that, unlike earthworms, many enchytraeid species remain challenging to breed due to the limited understanding of their specific needs.

7 Earthworms and enchytraeids as soil health indicators

7.1 Biological indicators, a recent history

Until recently, most soil health monitoring campaigns have primarily relied on chemical indicators, often overlooking physical and even more biological indicators (Bünemann et al., 2018; van Leeuwen et al., 2017). Soil organic carbon was the main biological indicator used until the late twentieth century. Early in the twenty-first century, however, some studies began to introduce more specific biological indicators (Bonilla-Bedoya et al., 2023), including earthworm-related ones and, to a lesser extent, enchytraeids (Pulleman et al., 2012). These soil animals, especially earthworms, were thus increasingly integrated within large programs to monitor soil quality (Fründ et al., 2010) (Section 7.3).

7.2 Earthworms and enchytraeids, good indicators of soil health?

A good indicator of soil health should (1) be sensitive to changes in land use, management and climate; (2) reflect general soil functioning by integrating its physical, chemical, and biological properties; and (3) be accessible and useful for stakeholders, such as farmers and policymakers (Doran & Parkin, 1997). The sensitivity of earthworms and enchytraeids to changes in land use, management and climate has been summarized in Sections 2 and 3, while the contributions of earthworms, and to a lesser extent enchytraeids, to soil functioning have been detailed in several reviews (Blouin et al., 2013; Briones, 2014; Coleman et al., 2024; Conti & Mulder, 2022; Vidal et al., 2023). However, while it is clear that earthworms and enchytraeids are sensitive to external factors and play essential roles in soil ecosystems, converting their functions into a practical and accessible indicator for stakeholders remains a challenge (Fig. 13). One reason for this is that effects on earthworms and enchytraeids are generally obvious at the genus or species level rather than at the level of total abundance, prohibiting easy access by non-specialists. Another challenge for stakeholders is to choose the most relevant among all the existing biological indicators. For instance, Zwetsloot et al. (2022) identified as many as 289 soil biology methods for assessing soil multifunctionality. Among these, it has been suggested that, for practical insights, measuring the key organisms (in this case, earthworms and enchytraeids) is often easier than measuring the processes they perform (Creamer et al., 2022). Based on this, Zwetsloot et al. (2022) recommend four methods each for assessing earthworms and enchytraeids, including morphological identification (Bouché, 1972; Schmelz & Collado,

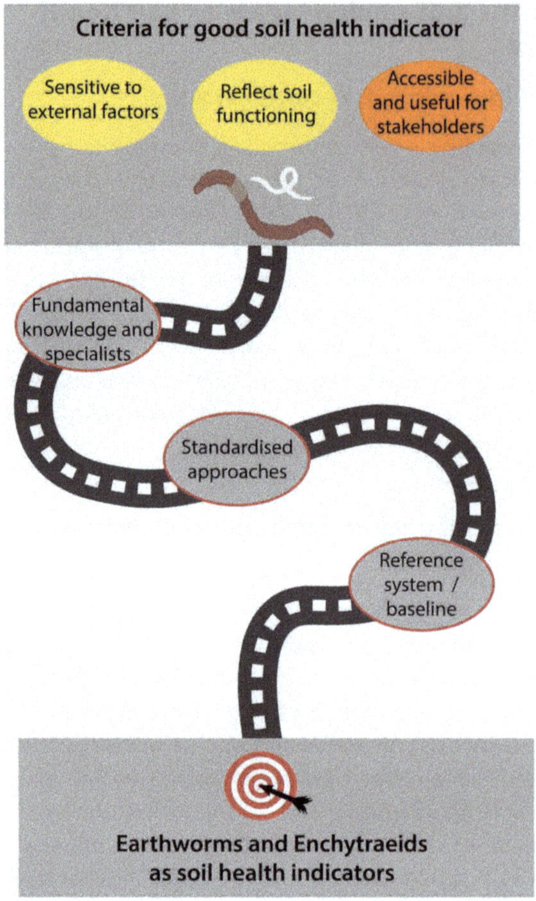

Figure 13 Opportunities and challenges of using earthworms and enchytraeids as indicators of soil health. Although earthworms and enchytraeids are sensitive to external factors and play essential roles in soil ecosystems (yellow circles), translating their functions into practical and accessible indicators for stakeholders remains a challenge (orange circle). To tackle this challenge, various aspects need to be considered (gray circles with red borders).

2010) counting/biomass or metabarcoding (Bienert et al., 2012; Creer et al., 2016; Pansu et al., 2015).

7.3 When to use earthworms and enchytraeids as indicators?

There is growing recognition that the selection of soil health indicators should be context-specific (Creamer et al., 2022; Schreefel et al., 2024), and that biological indicators alone are not sufficient to assess soil health and should be combined

Published by Burleigh Dodds Science Publishing Limited, 2025.

with physical and chemical indicators in order to capture soil multifunctionality (Coyne et al., 2022). Consequently, earthworms and enchytraeids should be incorporated as indicators only if they provide relevant insights for a specific purpose. So, when are earthworms and enchytraeids valuable as indicators? Creamer et al. (2022) note that earthworms and enchytraeids directly contribute to six and five key soil processes (out of 25), respectively, and play active roles in three of the four primary soil functions: 'water regulation and purification,' 'nutrient cycling,' and 'carbon and climate regulation'; without direct effect on 'disease and pest regulation' (Fig. 14). Earthworms, in particular, are considered essential for assessing soil functions, especially in relation to biodiversity (Griffiths et al., 2016). Although enchytraeids are less often considered, they contribute to soil biodiversity (van Leeuwen et al., 2017) and are recommended as soil quality indicators in cases where earthworms are absent (Pulleman et al., 2012). In addition, combining data on these two annelids can provide complementary information, as earthworms and enchytraeids do not necessarily respond the same way to stress in terms of abundance and diversity (Section 4) (Beylich &

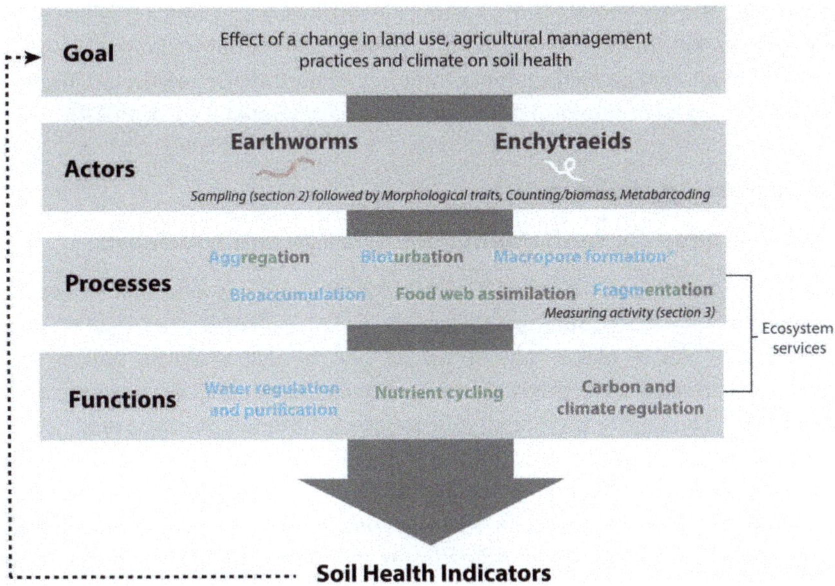

Figure 14 Why use earthworms and enchytraeids as soil health indicators? Main processes in which earthworm and enchytraeids are involved, determining their role on three key soil functions: water regulation and purification, nutrient cycling, and carbon and climate regulation. An asterisk (*) indicates a process carried out exclusively by earthworms, not enchytraeids. This classification is adapted from Creamer et al. (2022) and Zwetsloot et al. (2022); with a special focus on earthworms and enchytraeids. The function delivered thanks to one or more processes have the same color code. Processes that deliver more than one function are multi-colored according to the functions they deliver. The delivery of soil functions contributes to the provision of ecosystem services.

Graefe, 2009; Fründ et al., 2010). Although earthworms and enchytraeids are involved in fewer soil processes than microorganisms (Creamer et al., 2022), Delgado-Baquerizo et al. (2020) emphasize that larger invertebrates help keep to *multiple soil functions operating near peak capac*ity by regulating the flow of resources to smaller organisms that operate at lower functional levels.

As a result, earthworms, and to a lesser extent enchytraeids, are being incorporated in large frameworks to monitor soil health, as summarized in Fründ et al. (2010). For instance, the Biological Indicator system for Soil Quality (BISQ) of the Netherlands Soil Monitoring Network (NSMN) has been monitoring several biological indicators for 300 locations, including the abundance and diversity of earthworms and enchytraeids, from 1997 to 2014, every 6 years (Rutgers et al., 2009). In Germany, several federal states are involved in the long-term monitoring of earthworms and enchytraeids in some of the 800 permanent soil monitoring sites (Römbke et al., 2013). The French Soil Quality Monitoring Network has been monitoring earthworms in 109 locations in Britany through the RMQS Biodiv from 2006 to 2008 (Cluzeau et al., 2012). Yet, most of these initiatives take place in Europe, while there is still a strong data gap in, e.g., the Americas, North Australia, Asia, Russia and Africa (Phillips et al., 2019). To tackle this challenge, among others, the recent initiative 'Soil BON earthworm' aims at gathering the community of earthworm experts to improve the monitoring of earthworm distribution globally (Ganault et al., 2024).

7.4 Remaining challenges

7.4.1 The need for fundamental knowledge and specialists

To advance the development of biological indicators, it is essential to deepen scientific knowledge about the ecology and functional roles of earthworms and enchytraeids. For instance, only about 20% of the estimated 30 000 existing earthworm species have been identified (Decaëns et al., 2024) and the most useful ways to classify them are still debated (Bottinelli & Capowiez, 2021). Moreover, despite available identification keys (Schmelz and Collado, 2010), morphological identification of enchytraeids is challenging as they must be identified while alive, under a microscope. While the estimated diversity of enchytraeids is at least ten times lower than that of earthworms (Anthony et al., 2023), a significant share of these species remains unclassified (Decaëns et al., 2006). Traditional identification of earthworms and enchytraeids rely on dichotomous keys (Bouché, 1972; Schmelz & Collado, 2010; Sims & Gerard, 1999) which require specialist taxonomists – professionals who are increasingly rare (Andre et al., 2001). New DNA-based methods, such as barcoding, are increasingly used to assess biological diversity in soil samples (Dozsa-Farkas et al., 2012; Epp et al., 2012). Although these molecular identification methods offer promising alternatives, they are still in development and not yet sufficiently refined for widespread use (Decaëns et al., 2013; Dupont, 2009; Marchán et al., 2022). In the context of agroecology in which agriculture will have to rely on biodiversity to provide ecosystem services, a better quantification of the role of the different species or categories of earthworms and enchytraeids under different conditions is crucial. Moreover, understanding how land use, management practices and climate factors impact this soil biota (Sections 2 and 3), and thus soil functions and ecosystem services will enable the development of more sustainable agricultural systems that are adapted or resilient to future climate scenarios.

7.4.2 The need for standardized sampling and analytical approaches

Soil health cannot be effectively assessed using a single, standardized indicator; instead, choosing the right indicators requires a clear understanding of specific goals and context (Schreefel et al., 2024). When earthworms and enchytraeids are selected as indicators to address particular research questions, it is essential to define an appropriate measurement strategy for data collection. There is a strong consensus on the need to harmonize sampling and analysis methods across Europe and ideally worldwide (Ganault et al., 2024; Griffiths et al., 2016; van Leeuwen et al., 2017). Of particular importance is the spatial and temporal resolution of earthworm and enchytraeid sampling, which determines the reliability of the data. The distribution of earthworms and enchytraeids at field scale strongly varies between seasons (Section 5), but it is also strongly spatially clustered (Rossi, 2003; Van Vliet, 2000). Rossi (2003) showed that two earthworm species within a 45 × 45 m plot in a grass savanna in Côte d'Ivoire were clearly distributed in distinct patches within this plot. Valckx et al. (2009) showed that the distribution of several juvenile and adult earthworm species in a Belgium arable field was either overlapping or segregating. Rossi and Nuutinen (2004) demonstrated that sampling unit size was key in modelling the distribution of *L. terrestris* in a Finnish forest. While we fully support these harmonization efforts, it is important to recognize that sampling and analytical approaches for earthworms and enchytraeids may need adjustment based on specific context, the spatial and temporal resolution required, and research objectives (Section 4).

7.4.3 The need for reference systems/baseline/normal values

Once earthworm and/or enchytraeid indicators are chosen and implemented, it needs to be compared to a reference system, a baseline or normal values (Blair et al., 1997). For instance, once a management practice is applied, outcome

Published by Burleigh Dodds Science Publishing Limited, 2025.

indicators are used to illustrate the potential success of a change over time compared to a baseline (Schreefel et al., 2024). As trivial as this might sound, these types of values are often not available, with the high variability across seasons, land uses and soil types as a key constraint. The reference system should thus be well adapted to the context and goals of the study. Considering these needs, Krüger et al. (2018) laid the groundwork for establishing a reference system for biological indicators, including earthworms, in Wallonia, Belgium. To achieve this, the authors collected data on the number and biomass of earthworms at four different times throughout the year from 60 sites with varied land uses and soil types. Reference ranges were established by combining variances and calculating cumulative distribution functions. In France, the ongoing LANDWORM project aims to quantify the current and past (less than 50 years) effects of land use and management on earthworm communities. The project takes into account the divers soil and climatic conditions to better understand and predict earthworm community assembly and to identify land management practices that support their population. LANDWORM also aims to produce reference and threshold values to contribute to the bio-monitoring of the French territory, which could be used later at the European level. Ultimately, to efficiently use earthworms as soil health indicators, it is also crucial to link the reference values obtained in the types of projects mentioned above to their significance in terms of functions and ecosystem services.

8 Conclusion

To effectively use earthworms and enchytraeids as soil health indicators, several questions still need to be addressed in future research. Below, we conclude this chaper by highlighting key challenges that need to be addressed for advancing this field:

- *Understanding combined effects of management practices, land use, and climatic factors on earthworm and enchytraeid populations in order to mitigate climate change effects.* Specific management practices or land uses that promote greater abundance and diversity of these organisms are likely to enhance ecosystem resilience to adverse climatic conditions. However, this area remains largely understudied (Section 2).
- *Improving the accessibility and usefulness of indicators for stakeholders.* Stakeholders face a wide range of potential biological indicators to choose from. Identifying earthworms and enchytraeids to the genus or species level is challenging for non-specialists, despite being critical for evaluating the effects of land use, management practices and climate on

Published by Burleigh Dodds Science Publishing Limited, 2025.

soil health. Enhancing the user-friendliness of these indicators is therefore essential to their success (Section 7).

- *Developing methods to assess activity of earthworms and enchytraeids*. The presence of soil biota does not necessarily indicate high biological activity. In order to improve earthworm- and enchytraeid-based indicators, we therefore recommend the further development and adoption of methods to measure activity levels of earthworms and enchytraeids. These methods may be based on various approaches, such as density methods, visual assessments or acoustic approaches (Section 6).
- *Identifying unknown species of earthworms and enchytraeids*. The functionality of indicators is still severely hampered by lacks in the knowledge of earthworm and enchytraeid species. The identification of new species requires the expertise of specialist taxonomists, who are becoming increasingly rare. The development of new DNA-based methods, such as barcoding, offers promising support for species identification. However, this approach still needs further refinement (Section 7.4).
- *Studying interactions between earthworms and enchytraeids*. Both groups of soil biota coexist in natural environments but may respond differently to changes in land use, management and climate. There is still much knowledge on soil health to be gained by studying their interactions and complementary roles in the soil (Section 4).
- *Breeding more representative and relevant model species of earthworms and enchytraeids for laboratory testing*. Many studies on the effects of earthworms and enchytraeids on soil properties rely on a very limited number of model species. This is particularly true for enchytraeids, which are challenging to breed due to limited knowledge of their specific requirements. Developing better breeding techniques for various species of these faunal groups is essential for more representative laboratory testing (Section 6).
- *Testing effect of pesticides on earthworms and enchytraeids under realistic conditions, at ecologically relevant dosages and accounting for interactions between compounds*. Controlled tests assessing the impact of pesticides on these organisms often lack ecological relevance due to factors such as short test durations, the use of artificial soils, reliance on model species and single-species testing. More realistic test conditions are necessary for relevant indicators (Sections 2 and 3).

9 Where to look for further information

9.1 Key book chapters, articles, and guides

- Fründ, H.-C., et al. (2010). Earthworms as bioindicators of soil quality. In C. A. Edwards & J. R. Lofty (Eds.), *Biology of earthworms* (pp. 261–278). Springer.

- Pelosi, C. & Römbke, J. (2016). Are Enchytraeidae (Oligochaeta, Annelida) good indicators of agricultural management practices? *Soil Biology and Biochemistry*, 100, 255–263. https://doi.org/10.1016/j.soilbio.2016.06.030.
- Singh, J., et al. (2016). Extraction of earthworm from soil by different sampling methods: a review. *Environment, Development and Sustainability*, 18(6), 1521–1539. https://doi.org/10.1007/s10668-015-9703-5.
- Singh, J., et al. (2019). Climate change effects on earthworms-a review. *Soil Organisms*, 91(3), 114.
- Vidal, A., et al. (2023). Chapter One - The role of earthworms in agronomy: consensus, novel insights and remaining challenges. In D. L. Sparks (Ed.), *Advances in agronomy* (Vol. 181, pp. 1–78). Academic Press. https://doi.org/10.1016/bs.agron.2023.05.001.
- Guide for enchytraeid identification - Schmelz, R.M. & Collado, R. (2010). A guide to European terrestrial and freshwater species of Enchytraeidae (Oligochaeta). *Soil Organisms*, 82, 1–176. Document freely accessible: https://soil-organisms.org/index.php/SO/article/view/203.

9.2 International conferences

- The International Symposium on Earthworm Ecology – this conference is held every 4 years and provide updates on all aspects of earthworm research, as well as citizen science initiatives. The 13th edition of this symposium will be held in Wageningen University and Research, the Netherlands, in August 2026.
- The International Colloquium on Soil Zoology is held every 4 years and is one of the key scientific meetings on earthworms, enchytraeids as well as other soil faunal groups. The most recent colloquium was held in Cape town, South Africa, in August 2024.

9.3 Initiatives

- The Soil BON Earthworm – a global initiative aimed at gathering global data and information on earthworms. Initiated by the German Centre for Integrative Biodiversity Research (iDiv) Halle-Jena-Leipzig Soil BON Earthworm – A global initiative on earthworm distribution, traits, and spatiotemporal diversity patterns | Soil Organisms.

10 References

Abe, A. S. & Buck, N. (1985). Oxygen uptake of active and aestivating earthworm Glosso scolex pa ulistus (Oligochaeta, Glossoscolecidae). *Comparative Biochemistry and Physiology Part A: Physiology*, *81*, 63–66.

Abrahamsen, G. (1971). The influence of temperature and soil moisture on the population density of Cognettia sphagnetorum [Oligochaeta: Enchytraeidae] in cultures with homogenized raw humus. *Pedobiologia*, *11*(1), 417–424.

Akça, E., et al. (2024). *The state of soils in Europe*. Publications Office of the European Union.

Alvarez, M., et al. (2024). Effects of cropping sequences and rotational grazing on diversity, biomass, density and body mass of earthworms. *European Journal of Soil Biology*, *122*, 103647.

Amorim, M. J. B., et al. (2008). Enchytraeus albidus (Enchytraeidae): a test organism in a standardised avoidance test? Effects of different chemical substances. *Environment International*, *34*(3), 363–371.

Amossé, J., et al. (2018). Short-term effects of two fungicides on enchytraeid and earthworm communities under field conditions. *Ecotoxicology*, *27*, 300–312.

Andre, H. M., et al. (2001). Skilled eyes are needed to go on studying the richness of the soil. *Nature*, *409*(6822), 761. https://doi.org/10.1038/35057493

Andrén, O. & Lagerlöf, J. (1983). Soil fauna (microarthropods, enchytraeids, nematodes) in Swedish agricultural cropping systems. *Acta Agriculturae Scandinavica*, *33*(1), 33–52.

Anthony, M. A., et al. (2023). Enumerating soil biodiversity. *Proceedings of the National Academy of Sciences*, *120*(33), e2304663120. https://doi.org/10.1073/pnas.2304663120

Ashwood, F., et al. (2024). Earthworm records and habitat associations in the British Isles. *European Journal of Soil Biology*, *122*, 103642.

ASTM. (2000). *Standard guide for conducting laboratory soil toxicity or bioaccumulation tests with the lumbricid earthworm Eisenia fetida and the enchytraeid potworm Enchytraeus albidus*. ASTM Guideline No. E 1676– 97.

Awaknavar, J. S. & Karabhantanal, S. S. (2004). Toxicity of pesticides to earthworm, Polypheritima elongata (Michaelsen). *Journal of Entomological Research*, *28*(1), 63–72.

Awaknavar, J. S. & Karabhantanal, S. S. (2005). Effects of agro-chemicals on growth and development of earthworm, Pentoscolex corethrurus. *Journal of Ecotoxicology & Environmental Monitoring*, *15*(2), 179–187.

Ayres, I. & Guerra, R. A. T. (1981). Água como fator limitante na distribuição das minhocas (Annelida, Oligochaeta) da Amazônia Central. *Acta Amazonica*, *11*(1), 77–86.

Barbercheck, M. E., et al. (2009). Response of soil invertebrates to disturbance across three resource regions in North Carolina. *Environmental Monitoring and Assessment*, *152*, 283–298.

Bart, S., et al. (2017). Differences in sensitivity between earthworms and enchytraeids exposed to two commercial fungicides. *Ecotoxicology and Environmental Safety*, *140*, 177–184. https://doi.org/10.1016/j.ecoenv.2017.02.052

Bart, S., et al. (2018a). Aporrectodea caliginosa, a relevant earthworm species for a posteriori pesticide risk assessment: current knowledge and recommendations for culture and experimental design. *Environmental Science and Pollution Research*, *25*, 33867–33881.

Bart, S., et al. (2018b). How to assess the feeding activity in ecotoxicological laboratory tests using enchytraeids? *Environmental Science and Pollution Research*, *25*, 33844–33848.

Bartlett, M. D., et al. (2006). Inefficiency of mustard extraction technique for assessing size and structure of earthworm communities in UK pasture. *Soil Biology and Biochemistry*, *38*(9), 2990–2992.

Published by Burleigh Dodds Science Publishing Limited, 2025.

Bartlett, M. D., et al. (2010). A critical review of current methods in earthworm ecology: from individuals to populations. *European Journal of Soil Biology*, *46*(2), 67–73.

Bartz, M. L. C., et al. (2014). Earthworm richness in land-use systems in Santa Catarina, Brazil. *Applied Soil Ecology*, *83*, 59–70.

Bartz, M. L. C., et al. (2024). Earthworms as soil health indicators in no-tillage agroecosystems. *European Journal of Soil Biology*, *121*, 103605. https://doi.org/10.1016/j.ejsobi.2024.103605

Bayley, M., et al. (2010). Metabolic changes during estivation in the common earthworm Aporrectodea caliginosa. *Physiological and Biochemical Zoology*, *83*(3), 541–550.

Belaud, E., et al. (2024). In situ soil imaging, a tool for monitoring the hourly to monthly temporal dynamics of soil biota. *Biology and Fertility of Soils*. https://doi.org/10.1007/s00374-024-01851-8

Bell, N. L., et al. (2009). Best-practice pasture renewal for forage production and sustainability: description of a farmer-led study and initial findings. *Proceedings of the New Zealand Grassland Association*, 155–159.

Beylich, A. & Achazi, R. K. (1999). Influence of low soil moisture on enchytraeids. *Newsletter on Enchytraeidae*, *6*, 49–58.

Beylich, A. & Graefe, U. (2009). Investigations of annelids at soil monitoring sites in Northern Germany: reference ranges and time-series data. *Soil Organisms*, *81*(2), 175–196.

Beylich, A. & Graefe, U. (2010). Relationships between microannelid and earthworm activity. *Newsletter on Enchytraeidae No. 12*, *14*, 1.

Bienert, F., et al. (2012). Tracking earthworm communities from soil DNA. *Molecular Ecology*, *21*(8), 2017–2030.

Birkhofer, K., et al. (2008). Generalist predators in organically and conventionally managed grass-clover fields: implications for conservation biological control. *Annals of Applied Biology*, *153*(2), 271–280.

Blair, J. M., et al. (1997). Soil invertebrates as indicators of soil quality. In J. W. Doran & A. J. Jones (Eds.), *Methods for assessing soil quality* (pp. 273–291). Soil Science Society of America, Inc. https://doi.org/10.2136/sssaspecpub49.c16

Blanco-Canqui, H. (2022). Cover crops and soil ecosystem engineers. *Agronomy Journal*, *114*(6), 3096–3117.

Blankinship, J. C., et al. (2011). A meta-analysis of responses of soil biota to global change. *Oecologia*, *165*, 553–565.

Blouin, M., et al. (2013). A review of earthworm impact on soil function and ecosystem services. *European Journal of Soil Science*, *64*(2), 161–182. https://doi.org/10.1111/ejss.12025

Bohlen, P. J., et al. (2004). Ecosystem consequences of exotic earthworm invasion of north temperate forests. *Ecosystems*, *7*, 1–12.

Bonilla-Bedoya, S., et al. (2023). Mapping 50 years of contribution to the development of soil quality biological indicators. *Ecological Indicators*, *148*, 110091. https://doi.org/10.1016/j.ecolind.2023.110091

Bottinelli, N. & Capowiez, Y. (2021). Earthworm ecological categories are not functional groups. *Biology and Fertility of Soils*, *57*(2), 329–331. https://doi.org/10.1007/s00374-020-01517-1

Bottinelli, N., et al. (2013). Inability of Near Infrared Reflectance Spectroscopy (NIRS) to identify belowground earthworm casts in no-tillage soil. *Applied Soil Ecology*, *70*, 57–61.

Bottinelli, N., et al. (2021). Mid-infrared spectroscopy to trace biogeochemical changes of earthworm casts during ageing under field conditions. *Geoderma*, *383*, 114811.

Bouché, M. B. (1972). *Lombriciens de France. Ecologie et systématique* (Vol. 72, Issue HS). INRA Editions.

Bouche, P. M. B. (1969). Comparaison critique de méthodes dévaluation des populations de Lombricidés. *Pedobiologia*, *9*(1), 26–34.

Briones, M. J. I. (2014). Soil fauna and soil functions: a jigsaw puzzle. *Frontiers in Environmental Science*, *2*, 7.

Briones, M. J. I. & Schmidt, O. (2017). Conventional tillage decreases the abundance and biomass of earthworms and alters their community structure in a global meta-analysis. *Global Change Biology*, *23*(10), 4396–4419.

Briones, M. J. I., et al. (2007). Predicting potential impacts of climate change on the geographical distribution of enchytraeids: a meta-analysis approach. *Global Change Biology*, *13*(11), 2252–2269.

Brown, G. G., et al. (2006). Biodiversity and function of soil animals in Brazilian agroforestry systems. *Sistemas Agroflorestais: Bases Cientıficas Para o Desenvolvimento Sustentável. UENF, Campos Dos Goytacazes*, 217–242.

Brussaard, L., et al. (2004). Biological soil quality from biomass to biodiversity-importance and resilience to management stress and disturbance. In P. Schjønning, et al. (Eds.), *Managing soil quality: challenges in modern agriculture* (pp. 139–161). CABI Publishing Wallingford UK.

Buch, A. C., et al. (2013). Toxicity of three pesticides commonly used in Brazil to Pontoscolex corethrurus (Müller, 1857) and Eisenia andrei (Bouché, 1972). *Applied Soil Ecology*, *69*, 32–38.

Buch, A. C., et al. (2017). Ecotoxicology of mercury in tropical forest soils: impact on earthworms. *Science of the Total Environment*, *589*, 222–231.

Bünemann, E. K., et al. (2018). Soil quality – A critical review. *Soil Biology and Biochemistry*, *120*, 105–125. https://doi.org/10.1016/j.soilbio.2018.01.030

Burrows, L. A. & Edwards, C. A. (2004). The use of integrated soil microcosms to assess the impact of carbendazim on soil ecosystems. *Ecotoxicology*, *13*, 143–161.

Callaham Jr, M. A. & Hendrix, P. F. (1997). Relative abundance and seasonal activity of earthworms (Lumbricidae and Megascolecidae) as determined by hand-sorting and formalin extraction in forest soils on the southern Appalachian Piedmont. *Soil Biology and Biochemistry*, *29*(3–4), 317–321.

Cantelli, K. B. (2011). *Toxicidade aguda de carbofurano e carbendazim a minhocas em solo natural*. 43 f. Dissertação (Mestrado em Ciência do Solo) - Universidade Federal do Paraná, Curitiba.

Capowiez, Y., et al. (2021). Decreased burrowing activity of endogeic earthworms and effects on water infiltration in response to an increase in soil bulk density. *Pedobiologia*, *85*, 150728.

Capowiez, Y., et al. (2011). Using X-ray tomography to quantify earthworm bioturbation non-destructively in repacked soil cores. *Geoderma*, *162*(1–2), 124–131.

Carrera, N., et al. (2011). Interactive effects of temperature, soil moisture and enchytraeid activities on C losses from a peatland soil. *Pedobiologia*, *54*(5–6), 291–299.

Cécillon, L., et al. (2009). Assessment and monitoring of soil quality using near-infrared reflectance spectroscopy (NIRS). *European Journal of Soil Science*, *60*(5), 770–784. https://doi.org/10.1111/j.1365-2389.2009.01178.x

Chan, K. Y. & Heenan, D. P. (1995). Occurrence of enchytraeid worms and some properties of their casts in an Australian soil under cropping. *Soil Research*, *33*(4), 651–657.
Chan, K.-Y. & Munro, K. (2001). Evaluating mustard extracts for earthworm sampling. *Pedobiologia*, *45*(3), 272–278.
Chang, C.-H., et al. (2021). The second wave of earthworm invasions in North America: biology, environmental impacts, management and control of invasive jumping worms. *Biological Invasions*, *23*, 3291–3322.
Chang, Y. C. & Bruno, Z. (1970). The effect of piperazine on toad and earthworm muscle membrane potentials. *Pharmacology*, *4*(3), 143–151.
Chelinho, S., et al. (2011). Improving ecological risk assessment in the Mediterranean area: selection of reference soils and evaluating the influence of soil properties on avoidance and reproduction of two oligochaete species. *Environmental Toxicology and Chemistry*, *30*(5), 1050–1058.
Chevallier, A. J., et al. (2006). Mineral nitrogen cycling through earthworm casts in a grazed pasture under elevated atmospheric CO2. *Global Change Biology*, *12*(1), 56–60.
Christensen, B. (1956). Studies on Enchytraeidae 6. Technique for culturing Enchytraeidae, with notes on cocoon types. *Oikos*, *7*(Fasc. 2), 302–307.
Cluzeau, D., et al. (1992). Effects of intensive cattle trampling on soil-plant-earthworms system in two grassland types. *Soil Biology and Biochemistry*, *24*(12), 1661–1665.
Cluzeau, D., et al. (2012). Integration of biodiversity in soil quality monitoring: baselines for microbial and soil fauna parameters for different land-use types. *European Journal of Soil Biology*, *49*, 63–72.
Coleman, D. C., et al (2024). Soil fauna: occurrence, biodiversity, and roles in ecosystem function. In A. Paul & S. D. Frey (Eds.), *Soil microbiology, ecology and biochemistry* (pp. 131–159). Elsevier.
Conti, E. & Mulder, C. (2022). Chemistry-driven Enchytraeidae assemblages acting as soil and ecosystem engineers in edaphic communities. *Ecological Indicators*, *144*, 109529.
Cotton, D. C. F., & Curry, J. P. (1980). The response of earthworm populations (Oligochaeta, Lumbricidae) to high applications of pig slurry. *Pedobiologia*, *20*, 189–196.
Coyle, D. R., et al. (2017). Soil fauna responses to natural disturbances, invasive species, and global climate change: current state of the science and a call to action. *Soil Biology and Biochemistry*, *110*, 116–133.
Coyne, M. S., et al. (2022). Soil health – It's not all biology. *Soil Security*, *6*, 100051. https://doi.org/10.1016/j.soisec.2022.100051
Creamer, R. E., et al. (2022). The life of soils: integrating the who and how of multifunctionality. *Soil Biology and Biochemistry*, *166*, 108561. https://doi.org/10.1016/j.soilbio.2022.108561
Creer, S., et al. (2016). The ecologist's field guide to sequence-based identification of biodiversity. *Methods in Ecology and Evolution*, *7*(9), 1008–1018.
Curry, J. P. (1987). The invertebrate fauna of grassland and its influence on productivity. 1. The composition of the fauna. *Grass and Forage Science*, *42*(2), 103–120.
Curry, J. P. (1993). *Grassland invertebrates: ecology, influence on soil fertility and effects on plant growth*. Springer Science & Business Media.
Curry, J. P. (2004). Factors affecting the abundance of earthworms in soils. In C. A. Edwards (Eds.), *Earthworm ecology* (pp. 91–113). CRC Press.
Curry, J. P., et al. (2008). Relationships between earthworm populations and management intensity in cattle-grazed pastures in Ireland. *Applied Soil Ecology*, *39*(1), 58–64.

Published by Burleigh Dodds Science Publishing Limited, 2025.

Dalby, P. R., et al. (1998). Potential impact of an introduced lumbricid on a native woodland in South Australia. *Applied Soil Ecology*, *9*(1–3), 351–354.

Daugbjerg, P. (1988). Temperature and moisture preferences of three earthworm species (Oligochaeta, Lumbricidae). *Pedobiologia*, *32*(1–2), 57–64.

de Bruyn, L. A. L. & Kingston, T. J. (1997). Effects of summer irrigation and trampling in dairy pastures on soil physical properties and earthworm number and species composition. *Australian Journal of Agricultural Research*, *48*(7), 1059–1080.

de Lima E Silva, C. & Pelosi, C. (2024). Effects of glyphosate on earthworms: from fears to facts. *Integrated Environmental Assessment and Management*, *20*(5), 1330–1336.

Decaëns, T., et al. (2024). A can of worms: estimating the global number of earthworm species. *BioRxiv*, 2024.09.08.611896. https://doi.org/10.1101/2024.09.08.611896

Decaëns, T., et al. (2006). The values of soil animals for conservation biology. *European Journal of Soil Biology*, *42*, S23–S38.

Decaens, T., et al. (2016). DNA barcoding reveals diversity patterns of earthworm communities in remote tropical forests of French Guiana. *Soil Biology and Biochemistry*, *92*, 171–183.

Decaëns, T., et al. (2013). Potential of DNA barcoding for earthworm research in taxonomy and ecology. *Applied Soil Ecology*, *65*, 35–42. https://doi.org/10.1016/j.apsoil.2013.01.001

Delgado-Baquerizo, M., et al. (2020). Multiple elements of soil biodiversity drive ecosystem functions across biomes. *Nature Ecology & Evolution*, *4*(2), 210–220.

Demetrio, W. C., et al. (2020). Earthworms in Brazilian no-tillage agriculture: current status and future challenges. *European Journal of Soil Science*, *71*(6), 988–1005.

D'Hose, T., et al. (2018). Responses of soil biota to non-inversion tillage and organic amendments: an analysis on European multiyear field experiments. *Pedobiologia*, *66*, 18–28.

Dial, R. J., et al. (2016). The role of temperature in the distribution of the glacier ice worm, Mesenchytraeus solifugus (Annelida: Oligochaeta: Enchytraeidae). *Arctic, Antarctic, and Alpine Research*, *48*(1), 199–211.

Didden, W. A. M. (1991). *Population ecology and functioning of Enchytraeidae in some arable farming systems*. Wageningen University and Research.

Didden, W. A. M. (1993). Ecology of terrestrial Enchytraeidae. *Pedobiologia*, *37*, 2–29.

Didden, W. & Römbke, J. (2001). Enchytraeids as indicator organisms for chemical stress in terrestrial ecosystems. *Ecotoxicology and Environmental Safety*, *50*(1), 25–43.

Djerdj, T., et al. (2020). Observing earthworm behavior using deep learning. *Geoderma*, *358*, 113977. https://doi.org/10.1016/j.geoderma.2019.113977

Doran, J. W. & Parkin, T. B. (1997). Quantitative indicators of soil quality: a minimum data set. *Methods for Assessing Soil Quality*, *49*, 25–37.

Dózsa-Farkas, K. (1973). Saisondynamische Untersuchungen des Enchytraeiden-Beasatzes im Bodeneines ungarischen Quercetum petraeae cerris. *Pedobiologia*, *13*(5), 361–367.

Dózsa-Farkas, K. (1977). Beobachtungen über die Trockenheitstoleranz von Fridericia galba (Oligochaeta, Enchytraeidae). *Opuscula Zoologica Budapest*, *14*, 1–2.

Dozsa-Farkas, K., et al. (2012). Are Bryodrilus parvus Nurminen, 1970 and Bryodrilus librus (Nielsen and Christensen, 1959) (Annelida: Enchytraeidae) really different species? A revision based on DNA barcodes and morphological data. *Zootaxa*, *3276*(1), 38–50.

Drewry, J. J. (2006). Natural recovery of soil physical properties from treading damage of pastoral soils in New Zealand and Australia: a review. *Agriculture, Ecosystems & Environment*, *114*(2–4), 159–169.

Drumond, M. A., et al. (2015). The role of local knowledge and traditional extraction practices in the management of giant earthworms in Brazil. *PLoS One*, *10*(4), e0123913.

Dudas, R. T., et al. (2020). *Declínio das populações de minhocas no PD. A Granja, 859*, 47–49.

Dupont, L. (2009). Perspectives on the application of molecular genetics to earthworm ecology. *Pedobiologia*, *52*(3), 191–205.

Edwards, C. A. (1983). Earthworm ecology in cultivated soils. In C. A. Edwards (Eds.), *Earthworm ecology: from Darwin to vermiculture* (pp. 123–137). Springer.

Edwards, C. A. & Bohlen, P. J. (1996). *Biology and ecology of earthworms* (Vol. 3). Springer Science & Business Media.

Edwards, C. A. & Lofty, J. R. (1974). *The invertebrate fauna of the park grass plots. I.* Soil fauna.

Edwards, C. A. & Lofty, J. R. (1982). The effect of direct drilling and minimal cultivation on earthworm populations. *Journal of Applied Ecology*, 723–734.

Eggleton, P., et al. (2009). A six year study of earthworm (Lumbricidae) populations in pasture woodland in southern England shows their responses to soil temperature and soil moisture. *Soil Biology and Biochemistry*, *41*(9), 1857–1865.

Eisenhauer, N., et al. (2008). Efficiency of two widespread non-destructive extraction methods under dry soil conditions for different ecological earthworm groups. *European Journal of Soil Biology*, *44*(1), 141–145.

Epp, L. S., et al. (2012). New environmental metabarcodes for analysing soil DNA: potential for studying past and present ecosystems. *Molecular Ecology*, *21*(8), 1821–1833.

Fragoso, C., et al. (1997). Agricultural intensification, soil biodiversity and agroecosystem function in the tropics: the role of earthworms. *Applied Soil Ecology*, *6*(1), 17–35. https://doi.org/10.1016/S0929-1393(96)00154-0

Fragoso, C. & Lavelle, P. (1992). Earthworm communities of tropical rain forests. *Soil Biology and Biochemistry*, *24*(12), 1397–1408.

Fragoso, C., et al. (1999). Earthworm communities of tropical agroecosystems: origin, structure and influence of management practices. In P. Lavelle, et al. (Eds.), *Earthworm management in tropical agroecosystems* (pp. 27–55). CABI.

Fraser, P. M., et al. (2012). Influence of summer irrigation on soil invertebrate populations in a long-term sheep irrigation trial at Winchmore (Canterbury). *New Zealand Journal of Agricultural Research*, *55*(2), 165–180.

Fraser, P. M., et al. (1996). Earthworm species, population size and biomass under different cropping systems across the Canterbury Plains, New Zealand. *Applied Soil Ecology*, *3*(1), 49–57.

Fründ, H.-C., et al. (2010). Earthworms as bioindicators of soil quality. In C. A. Edwards & J. R. Lofty (Eds.), *Biology of earthworms* (pp. 261–278). Springer.

Gajda, Ł., et al. (2017). Food preferences of enchytraeids. *Pedobiologia*, *63*, 19–36.

Ganault, P., et al. (2024). *Soil BON earthworm: a global initiative on earthworm distribution, traits, and spatiotemporal diversity patterns. Soil Organisms, 96*(1), 2024.

Garamszegi, P., et al. (2025). A density-based method to objectively quantify earthworm activity. *Applied Soil Ecology, 206,* 105771. https://doi.org/10.1016/j.apsoil.2024.105771

Garcia, M. (2004). *Effects of pesticides on soil fauna: development of ecotoxicological test methods for tropical regions* (Vol. 19). Cuvillier Verlag.

García-Pérez, J. A., et al. (2014). Earthworm communities and soil properties in shaded coffee plantations with and without application of glyphosate. *Applied Soil Ecology, 83,* 230–237.

Garcia-Perez, J. A., et al. (2020). Interactive effect of glyphosate-based herbicides and organic soil layer thickness on growth and reproduction of the tropical earthworm Pontoscolex corethrurus (Müller, 1857). *Applied Soil Ecology, 155,* 103648.

Gerard, B. M. (1967). Factors affecting earthworms in pastures. *The Journal of Animal Ecology, 36,* 235–252.

Ghosh, S. (2021). Climate change and earthworms: a global perspective. *International Journal of Entomology Research, 6*(4), 167–171.

Goncharov, A. A., et al. (2023). A meta-analysis suggests climate change shifts structure of regional communities of soil invertebrates. *Soil Biology and Biochemistry, 181,* 109014. https://doi.org/10.1016/j.soilbio.2023.109014

Graefe, U. & Schmelz, R. M. (1999). Indicator values, strategy types and life forms of terrestrial Enchytraeidae and other microannelids. *Newsletter on Enchytraeidae, 6,* 59–67.

Griffiths, B. S., et al. (2016). Selecting cost effective and policy-relevant biological indicators for European monitoring of soil biodiversity and ecosystem function. *Ecological Indicators, 69,* 213–223. https://doi.org/10.1016/j.ecolind.2016.04.023

Grongroft, A. & Miehlich, G. (1983). Bedeutung der Bodenfeuchte fur die Populationsdynamik von Enchytraeidae (Oligochaeta) und Oribatei (Acari). *Abhandlungen Des Naturwissenschaftlichen Vereins in Hamburg.*

Gronstol, G. B., et al. (2000). A comparison of mustard, household detergent and formalin as vermifuges for earthworm sampling. *Fauna Norvegica, 20,* 27–30.

Guéi, A. M. & Tondoh, J. E. (2012). Ecological preferences of earthworms for land-use types in semi-deciduous forest areas, Ivory Coast. *Ecological Indicators, 18,* 644–651. https://doi.org/10.1016/j.ecolind.2012.01.018

Guéi, A. M., et al. (2019). Relationships between soil morpho-chemical parameters and earthworm community attributes in tropical agro-ecosystems in the Centre-West region of Côte d'Ivoire, Africa. *Tropical Ecology, 60,* 209–218.

Guerra, R. T. (1988). Ecologia dos oligochaeta da Amazônia. II. Estudo da estivação e da atividade de Chibui bari, através da produção de excrementos. *Acta Amazonica, 18*(1–2), 27–34.

Helling, B., et al. (1998). A comparison of feeding activity of collembolan and enchytraeid in laboratory studies using the bait-lamina test. *Applied Soil Ecology, 7*(3), 207–212.

Hendrix, P. F., et al. (1986). Detritus food webs in conventional and no-tillage agroecosystems. *Bioscience, 36*(6), 374–380.

Holland, J. E., et al. (2018). Liming impacts on soils, crops and biodiversity in the UK: a review. *Science of the Total Environment, 610,* 316–332.

Holmstrup, M. & Sjursen, H. (2001). Freeze induced glucose accumulation in the enchytraeid, Fredericia ratzeli from Greenland. *Cryo Letters, 22*(5), 273–276.

Holmstrup, M., et al. (2012). Increased frequency of drought reduces species richness of enchytraeid communities in both wet and dry heathland soils. *Soil Biology and Biochemistry*, *53*, 43–49.

Hughes, F. M., et al. (2018). Anticipating the response of the Brazilian giant earthworm (Rhinodrilus alatus) to climate change: implications for its traditional use. *Anais Da Academia Brasileira de Ciências*, *91*(1), e20180308.

ISO. (2016). *ISO 18311:2016 Soil quality – method for testing effects of soil contaminants on the feeding activity of soil dwelling organisms – Bait-lamina test.*

ISO. (2018). *DIN ISO 23611-1:2018(E), soil quality – sampling of soil invertebrates – hand-sorting and extraction of earthworms.*

ISO. (2019). *ISO 23611-3:2019 Soil quality – sampling of soil invertebrates, Part 3: sampling and extraction of enchytraeids.*

ISO. (2023). *European standard EN ISO 11268-2:2023 soil quality. Effects of pollutants on earthworms. Part 2: determination of effects on reproduction of Eisenia fetida/Eisenia andrei and other earthworm species.*

James, S. (1996). Earthworms. In Hall G. (Ed.), *Methods for the examination of organismal diversity in soils and sediments. IUBS methodology series*. CAB International.

Jangorzo, N. S., et al. (2015). Nondestructive monitoring of the effect of biological activity on the pedogenesis of a Technosol. *Journal of Soils and Sediments*, *15*(8), 1705–1715. https://doi.org/10.1007/s11368-014-1008-z

Kaneda, S., et al. (2006). Effects of ivermectin in dung pats on earthworm (Megascolecidae) populations and pat degradation in Japanese grassland. *Applied Soil Ecology*, *31*(3), 280–285. https://doi.org/10.1016/j.apsoil.2005.05.001

Karaban, K. & Uvarov, A. V. (2014). Non-trophic effects of earthworms on enchytraeids: an experimental investigation. *Soil Biology and Biochemistry*, *73*, 84–92.

Keen, S. C., et al. (2022). Non-native species change the tune of tundra soils: novel access to soundscapes of the Arctic earthworm invasion. *Science of The Total Environment*, *838*, 155976. https://doi.org/10.1016/j.scitotenv.2022.155976

Khalil, A. M. (2013). Ecotoxicological bioassays of the earthworms Allolobophora caliginosa Savigny and Pheretima hawayana Rosa treated with arsenate. *Journal of Biological Sciences*, *13*(3), 99–105. https://doi.org/10.3844/ojbsci.2013.99.105

Khalil, A. M. (2015). Neurotoxicity and biochemical responses in the earthworm Pheretima hawayana exposed to TiO2NPs. *Ecotoxicology and Environmental Safety*, *122*, 455–461.

King, K. L. & Hutchinson, K. J. (2007). Pasture and grazing land: assessment of sustainability using invertebrate bioindicators. *Australian Journal of Experimental Agriculture*, *47*(4), 392–403.

Kiss, T. B. W., et al. (2021). Dual stresses of flooding and agricultural land use reduce earthworm populations more than the individual stressors. *Science of the Total Environment*, *754*, 142102.

Klok, C., et al. (2006). Does reproductive plasticity in Lumbricus rubellus improve the recovery of populations in frequently inundated river floodplains? *Soil Biology and Biochemistry*, *38*(3), 611–618.

Kobetičová, K., et al. (2010). Ecotoxicity of wastes in avoidance tests with Enchytraeus albidus, Enchytraeus crypticus and Eisenia fetida (Oligochaeta). *Waste Management*, *30*(4), 558–564.

Kobetičová, K. & Schlaghamerský, J. (2003). On the efficiency of three schemes of enchytraeid wet. *Newsletter on Enchytraeidae*, *8*, 25–31.

Kopittke, P. M., et al. (2019). Soil and the intensification of agriculture for global food security. *Environment International*, *132*, 105078.

Krüger, I., et al. (2018). Defining a reference system for biological indicators of agricultural soil quality in Wallonia, Belgium. *Ecological Indicators*, *95*, 568–578. https://doi.org/10.1016/j.ecolind.2018.08.010

Lacoste, M., et al. (2018). Listening to earthworms burrowing and roots growing-acoustic signatures of soil biological activity. *Scientific Reports*, *8*(1), 10236.

Lagerlöf, J., et al. (1989). Dynamics and contribution to carbon flows of Enchytraeidae (Oligochaeta) under four cropping systems. *Journal of Applied Ecology*, *26*, 183–199.

Lagerlöf, J. & Strandh, M. (1997). Hatching of Enchytraeidae (Oligochaeta) from egg cocoons in agricultural soil exposed to different drought regimes-a laboratory study. *Pedobiologia*, *41*, 334–341.

Larsen, T., et al. (2016). Substantial nutritional contribution of bacterial amino acids to earthworms and enchytraeids: a case study from organic grasslands. *Soil Biology and Biochemistry*, *99*, 21–27.

Lavelle, P. (1983). The structure of earthworm communities. In C. A. Edwards (Eds.), *Earthworm ecology: from Darwin to vermiculture* (pp. 449–466). Springer.

Lavelle, P. & Lapied, E. (2003). Endangered earthworms of Amazonia: an homage to Gilberto Righi: the 7th international symposium on earthworm ecology· Cardiff· Wales· 2002. *Pedobiologia*, *47*(5–6), 419–427.

Lavelle, P., et al. (2006). Soil invertebrates and ecosystem services. *European Journal of Soil Biology*, *42*, S3–S15. https://doi.org/10.1016/j.ejsobi.2006.10.002

Lawrence, A. P. & Bowers, M. A. (2002). A test of the 'hot'mustard extraction method of sampling earthworms. *Soil Biology and Biochemistry*, *34*(4), 549–552.

Lee, K. E. (1961). Interactions between native and introduced earthworms. *Proceedings (New Zealand Ecological Society)*, *8*, 60–62.

Lee, K. E. (1985). *Earthworms: their ecology and relationships with soils and land use*. Academic Press Inc.

Leroy, B. L. M., et al. (2008). Earthworm population dynamics as influenced by the quality of exogenous organic matter. *Pedobiologia*, *52*(2), 139–150.

Liu, Q., et al. (2025). Grasslands support more diverse and resilient earthworm communities to climate change than croplands in Central Europe. *Agriculture, Ecosystems & Environment*, *377*, 109259.

Ma, Z., et al. (2023). A comprehensive study on the ecotoxicity of ivermectin to earthworms (Eisenia fetida). *Ecotoxicology and Environmental Safety*, *268*, 115709. https://doi.org/10.1016/j.ecoenv.2023.115709

Makulec, G. & Pilipiuk, I. (2000). Influence of plant diversity and earthworm casts on the abundance and species composition of the enchytraeids (Oligochaeta: Enchytraeidae) in a lysimetric experiment. *Polish Journal of Ecology*, *48*(3), 185–193.

Manetti, P. L., et al. (2010). Tillage system does not affect soil macrofauna in Southeastern Buenos Aires province, Argentina. *Spanish Journal of Agricultural Research*, *8*(2), 377–384.

Maraldo, K. (2009). *Enchytraeidae (Oligochaeta) in a changing climate: ecology and ecophysiology of enchytraeids exposed to climate changes*. National Environmental Reseach Institute, Aarhus University.

Maraldo, K. & Holmstrup, M. (2009). Recovery of enchytraeid populations after severe drought events. *Applied Soil Ecology*, *42*(3), 227–235.

Maraldo, K. & Holmstrup, M. (2010). Enchytraeids in a changing climate: a mini-review. *Pedobiologia*, *53*(3), 161–167.

Maraldo, K., et al. (2010). The counteracting effects of elevated atmospheric CO2 concentrations and drought episodes: studies of enchytraeid communities in a dry heathland. *Soil Biology and Biochemistry*, *42*(11), 1958–1966.

Maraldo, K., et al. (2015). Enchytraeids as indicator of soil quality in temporary organic grass-clover leys under contrasting management: a feasibility study. *Soil Biology and Biochemistry*, *91*, 32–39.

Marchán, D. F., et al. (2022). Perspectives in earthworm molecular phylogeny: recent advances in lumbricoidea and standing questions. *Diversity*, *14*(1), 30.

Mitchell, M. J., et al. (1978). *Effects of different sewage sludges on some chemical and biological characteristics of soil*. Wiley Online Library.

Morowati, M. (2000). Histochemical and histopathological study of the intestine of the earthworm (Pheretima elongata) exposed to a field dose of the herbicide glyphosate. *Environmentalist*, *20*, 105–111.

Mostert, M. A. (2001). *The ecotoxicity of five insecticides to the pheretima group (Oligochaeta) occurring on golf courses in the Pretoria Region (South Africa)*. University of Pretoria (South Africa).

Mostert, M. A., et al. (2002). The relative toxicities of insecticides to earthworms of the Pheretima group (Oligochaeta). *Pest Management Science*, *58*(5), 446–450.

Mueller, L., et al. (2009). Visual assessment of soil structure: evaluation of methodologies on sites in Canada, China and Germany: Part I: comparing visual methods and linking them with soil physical data and grain yield of cereals. *Soil and Tillage Research*, *103*(1), 178–187.

Nowak, E. (2004). Enchytraeids [Oligochaeta] in the agricultural landscape. *Polish Journal of Ecology*, *2*(52).

Nuutinen, V., et al. (2017). Soil faunal and structural responses to the settlement of a semi-sedentary earthworm Lumbricus terrestris in an arable clay field. *Soil Biology and Biochemistry*, *115*, 285–296.

O'connor, F. B. (1955). Extraction of enchytraeid worms from a coniferous forest soil. *Nature*, *175*(4462), 815–816.

OECD. (2004). *Test No. 220: enchytraeid reproduction test*. OECD.

OECD. (2016). *Test No. 222: earthworm reproduction test (Eisenia fetida/Eisenia andrei)*. OECD. https://doi.org/10.1787/9789264264496-en

Oliveira Filho, L. C. I., et al. (2018). Fauna edáfica em áreas com diferentes manejos e tempos de descarte de resíduos animais. *Scientia Agraria*, *19*(1), 113–123.

Oluko, P. S., et al. (2014). *Conservation agriculture effects on earthworm populations in Western Kenya and Eastern Uganda soils*. RUFORUM.

Panda, S. & Sahu, S. K. (2000). Assessment of recovery of population, biomass and reproduction of the earthworm Drawida willsi following the application of malathion under field conditions. *Biology and Fertility of Soils*, *32*, 82–88.

Pansu, J., et al. (2015). Landscape-scale distribution patterns of earthworms inferred from soil DNA. *Soil Biology and Biochemistry*, *83*, 100–105.

Parelho, C., et al. (2018). Biological endpoints in earthworms (Amynthas gracilis) as tools for the ecotoxicity assessment of soils from livestock production systems. *Ecological Indicators*, *95*, 984–990.

Patricio Silva, A. L., et al. (2014). Importance of freeze–thaw events in low temperature ecotoxicology of cold tolerant enchytraeids. *Environmental Science & Technology, 48*(16), 9790–9796.

Pedersen, P. G. & Holmstrup, M. (2003). Freeze or dehydrate: only two options for the survival of subzero temperatures in the arctic enchytraeid Fridericia ratzeli. *Journal of Comparative Physiology B, 173*, 601–609.

Peigné, J., et al. (2013). Assessment of soil structure in the transition layer between topsoil and subsoil using the profil cultural method. *Soil and Tillage Research, 127*, 13–25.

Pelosi, C., et al. (2014). Pesticides and earthworms: a review. *Agronomy for Sustainable Development, 34*, 199–228.

Pelosi, C., et al. (2021a). Comparison of the mustard oil and electrical methods for sampling earthworm communities in rural and urban soils. *Urban Ecosystems, 24*(2), 355–364.

Pelosi, C., et al. (2021b). Residues of currently used pesticides in soils and earthworms: a silent threat? *Agriculture, Ecosystems & Environment, 305*, 107167.

Pelosi, C., et al. (2009). Earthworm collection from agricultural fields: comparisons of selected expellants in presence/absence of hand-sorting. *European Journal of Soil Biology, 45*(2), 176–183. https://doi.org/10.1016/j.ejsobi.2008.09.013

Pelosi, C., et al. (2020). Soil Oligochaeta communities after 9 decades of continuous fertilization in a bare fallow experiment. *Soil Organisms, 92*(2), 129–141.

Pelosi, C. & Römbke, J. (2016). Are Enchytraeidae (Oligochaeta, Annelida) good indicators of agricultural management practices? *Soil Biology and Biochemistry, 100*, 255–263. https://doi.org/10.1016/j.soilbio.2016.06.030

Pelosi, C. & Römbke, J. (2018). Enchytraeids as bioindicators of land use and management. *Applied Soil Ecology, 123*, 775–779.

Pennock, D., et al. (2015). *Status of the world's soil resources*. Technical Summary FAO.

Pérès, G., et al. (1998). Earthworm activity and soil structure changes due to organic enrichments in vineyard systems. *Biology and Fertility of Soils, 27*(4), 417–424. https://doi.org/10.1007/s003740050452

Perreault, J. M. & Whalen, J. K. (2006). Earthworm burrowing in laboratory microcosms as influenced by soil temperature and moisture. *Pedobiologia, 50*(5), 397–403.

Phillips, H. R. P., et al. (2019). Global distribution of earthworm diversity. *Science, 366*(6464), 480–485.

Piron, D., et al. (2012). Morphological description of soil structure patterns produced by earthworm bioturbation at the profile scale. *European Journal of Soil Biology, 50*, 83–90.

Piron, D., et al. (2017). Indicators of earthworm bioturbation to improve visual assessment of soil structure. *Soil and Tillage Research, 173*, 53–63. https://doi.org/10.1016/j.still.2016.10.013

Plum, N. M. & Filser, J. (2005). Floods and drought: response of earthworms and potworms (Oligochaeta: Lumbricidae, Enchytraeidae) to hydrological extremes in wet grassland. *Pedobiologia, 49*(5), 443–453.

Porre, R. J., et al. (2016). Exploring the relationship between soil mesofauna, soil structure and N2O emissions. *Soil Biology and Biochemistry, 96*, 55–64.

Potapov, A. M., et al. (2022). Feeding habits and multifunctional classification of soil-associated consumers from protists to vertebrates. *Biological Reviews, 97*(3), 1057–1117.

Pulleman, M., et al. (2012). Soil biodiversity, biological indicators and soil ecosystem services–an overview of European approaches. *Current Opinion in Environmental Sustainability*, *4*(5), 529–538. https://doi.org/10.1016/j.cosust.2012.10.009

Raw, F. (1959). Estimating earthworm populations by using formalin. *Nature*, *184*(4699), 1661–1662. https://doi.org/10.1038/1841661a0

Ricci, F., et al. (2015). Positive effects of alternative cropping systems on terrestrial Oligochaeta (Clitellata, Annelida). *Soil Organisms*, *87*(2), 71–83.

Robinson, J. M., et al. (2024). Monitoring soil fauna with ecoacoustics. *Proceedings of the Royal Society B: Biological Sciences*, *291*(2030), 20241595. https://doi.org/10.1098/rspb.2024.1595

Römbke, J., et al. (2013). Die Beurteilung der Boden-Biodiversität: Ergebnisse eines UBA-Vorhabens. *Soil Organisms*, *28*(2), 123–146.

Römbke, J., et al. (2017). Effects of organic pesticides on enchytraeids (Oligochaeta) in agroecosystems: laboratory and higher-tier tests. *Frontiers in Environmental Science*, *5*, 20.

Rossi, J.-P. (2003). Clusters in earthworm spatial distribution: the 7th international symposium on earthworm ecology· Cardiff· Wales· 2002. *Pedobiologia*, *47*(5–6), 490–496.

Rossi, J.-P. & Nuutinen, V. (2004). The effect of sampling unit size on the perception of the spatial pattern of earthworm (Lumbricus terrestris L.) middens. *Applied Soil Ecology*, *27*(2), 189–196.

Rutgers, M., et al. (2016). Mapping earthworm communities in Europe. *Applied Soil Ecology*, *97*, 98–111.

Rutgers, M., et al. (2009). Biological measurements in a nationwide soil monitoring network. *European Journal of Soil Science*, *60*(5), 820–832. https://doi.org/10.1111/j.1365-2389.2009.01163.x

Sanchez-Hernandez, J. C. (2006). Earthworm biomarkers in ecological risk assessment. *Reviews of Environmental Contamination and Toxicology: Continuation of Residue Reviews*, 85–126.

Sandor, M. & Schrader, S. (2012). Interaction of earthworms and enchytraeids in organically amended soil. *North-Western Journal of Zoology*, *8*(1).

Satchell, J. E. (1969). Methods of sampling earthworm populations. *Pedobiologia*, *9*, 20–25.

Sauzet, O., et al. (2023). Long-term quantification of the intensity of clay-sized particles transfers due to earthworm bioturbation and eluviation/illuviation in a cultivated Luvisol. *Geoderma*, *429*, 116251. https://doi.org/10.1016/j.geoderma.2022.116251

Scheffczyk, A., et al. (2016). Nontarget effects of ivermectin residues on earthworms and springtails dwelling beneath dung of treated cattle in four countries. *Environmental Toxicology and Chemistry*, *35*(8), 1959–1969.

Scheu, S. (1991). Mucus excretion and carbon turnover of endogeic earthworms. *Biology and Fertility of Soils*, *12*(3), 217–220.

Schlaghamerský, J., et al. (2014). Earthworm invasion alters enchytraeid community composition and individual biomass in northern hardwood forests of North America. *Applied Soil Ecology*, *83*, 159–169.

Schmelz, R. M. & Collado, R. (2010). A guide to European terrestrial and freshwater species of Enchytraeidae (Oligochaeta). *Soil Organisms*, *82*(1), 1–176.

Schmidt, O. (2001a). Appraisal of the electrical octet method for estimating earthworm populations in Arable land. *Annals of Applied Biology*, *138*(2), 231–241.

Schmidt, O. (2001b). Time-limited soil sorting for long-term monitoring of earthworm populations. *Pedobiologia*, *45*(1), 69–83. https://doi.org/10.1078/0031-4056-00069

Scholz-Starke, B., et al. (2013). The response of soil organism communities to the application of the insecticide lindane in terrestrial model ecosystems. *Ecotoxicology*, *22*, 339–362.

Schon, N. L., et al. (2008). Soil fauna in grazed New Zealand hill country pastures at two management intensities. *Applied Soil Ecology*, *40*(2), 218–228.

Schon, N. L., et al. (2011). Effects of dairy cow treading pressures and food resources on invertebrates in two contrasting and co-occurring soils. *Soil Research*, *49*(8), 703–714.

Schon, N. L., et al. (2012). Vulnerability of soil invertebrate communities to the influences of livestock in three grasslands. *Applied Soil Ecology*, *53*, 98–107.

Schrader, S., et al. (1997). Impact of Collembola and Enchytraeidae on soil surface roughness and properties. *Biology and Fertility of Soils*, *25*, 396–400.

Schrader, S. & Seibel, C. (2001). Impact of cultivation management in an agroecosystem on hot spot effects of earthworm middens. *European Journal of Soil Biology*, *37*(4), 309–313.

Schreefel, L., et al. (2024). How to monitor the 'success' of agricultural sustainability: a perspective. *Global Food Security*, *43*, 100810. https://doi.org/10.1016/j.gfs.2024.100810

Senapati, B. K., et al. (1992). Ecotoxicological effects of malathion on earthworms. *Soil Biology and Biochemistry*, *24*(12), 1719–1722. https://doi.org/10.1016/0038-0717(92)90176-X

Senapati, B. K., et al. (1994). Impact of normal agricultural dose of malathion on a dominant paddy field earthworm Drawida willsi, Michaelsen. In G. K. Veeresh, et al. (Eds.), *Advances in management and conservation of soil fauna* (pp. 377–382). Oxford and IBH Publishing Co.

Senapati, B. K., et al. (1999). In-soil earthworm technologies for tropical agroecosystems. *Earthworm Management in Tropical Agroecosystems. CAB International, Wallingford, UK*, 199–237.

Serbource, C., et al. (2024). A meta-analysis to compare the sensitivities of earthworms and enchytraeids to different stressors. *European Journal of Soil Biology*, *122*, 103656. https://doi.org/10.1016/j.ejsobi.2024.103656

Serbource, C., et al. (2025). Enchytraeids: small but important ecosystem engineers. *Geoderma*, *453*, 117150. https://doi.org/10.1016/j.geoderma.2024.117150

Severon, T., et al. (2010). The impact of conventional and reduced tillage on the Enchytraeidae population in sandy soil and their correlation with plant residue and earthworms. *Newsletter on Enchytraeidae No. 12*, *14*, 45.

Silvertown, J., et al. (2006). The Park Grass Experiment 1856–2006: its contribution to ecology. *Journal of Ecology*, *94*(4), 801–814.

Sims, R. W. & Gerard, B. M. (1999). *Earthworms: notes for the identification of British species* (Vol. 31). Linnean Society of London and the Estuarine and Coastal Sciences Association.

Singh, J., et al. (2016). Extraction of earthworm from soil by different sampling methods: a review. *Environment, Development and Sustainability*, *18*(6), 1521–1539. https://doi.org/10.1007/s10668-015-9703-5

Singh, J., et al. (2019). Climate change effects on earthworms-a review. *Soil Organisms, 91*(3), 114.

Smith, P., et al. (2024). Status of the World's Soils. *Annual Review of Environment and Resources, 49*, 73–104.

Springett, J. A., et al. (1970). Vertical movement of Enchytraeidae (Oligochaeta) in moorland soils. *Oikos, 21*, 16–21.

Springett, J. A., et al. (1996). Influence of agriculture on Enchytraeidae fauna of soils in the south-west of the north island of New Zealand. *Pedobiologia, 5*, 461–466.

Springett, J. A. & Syers, J. K. (1984). Effect of pH and calcium content of soil on earthworm cast production in the laboratory. *Soil Biology and Biochemistry, 16*(2), 185–189.

Steffen. (2012). *Diversidade de minhocas e sua relação com ecossistemas naturais e alterados no estado do Rio Grande do Sul.* Universidade Federal de Santa Maria.

Steffen, G. P. K., et al. (2013). Earthworm extraction with onion solution. *Applied Soil Ecology, 69*, 28–31.

Stockdill, S. M. J. (1982). Effects of introduced earthworms on the productivity of New Zealand pastures. *Pedobiologia, 24*, 281–299.

Stroud, J. L. (2019). Soil health pilot study in England: outcomes from an on-farm earthworm survey. *PLoS One, 14*(2), e0203909.

Taheri, S., et al. (2018). Harmful or useful? A case study of the exotic peregrine earthworm morphospecies Pontoscolex corethrurus. *Soil Biology and Biochemistry, 116*, 277–289.

Thonon, I. & Klok, C. (2007). Impact of a changed inundation regime caused by climate change and floodplain rehabilitation on population viability of earthworms in a lower River Rhine floodplain. *Science of The Total Environment, 372*(2), 585–594. https://doi.org/10.1016/j.scitotenv.2006.10.028

Tondoh, J. E., et al. (2007). Can earthworms be used as bio-indicators of land-use perturbations in semi-deciduous forest? *Biology and Fertility of Soils, 43*(5), 585–592. https://doi.org/10.1007/s00374-006-0144-z

Topoliantz, S., et al. (2000). Earthworm and enchytraeid activity under different arable farming systems, as exemplified by biogenic structures. *Plant and Soil, 225*, 39–51.

Turbé, A., et al. (2010). *Soil biodiversity: functions, threats and tools for policy makers.* Bio Intelligence Service. http://ec.europa.eu/environment/archives/soil/pdf/biodiversity_report.pdf

Valckx, J., et al. (2009). Within-field spatial distribution of earthworm populations related to species interactions and soil apparent electrical conductivity. *Applied Soil Ecology, 41*(3), 315–328. https://doi.org/10.1016/j.apsoil.2008.12.005

van de Logt, R., et al. (2023a). Lumbricus terrestris abundance in grasslands on sandy soils in relation to soil texture, hydrology and earthworm community. *European Journal of Soil Biology, 119*, 103545.

van de Logt, R., et al. (2023b). The anecic earthworm Lumbricus terrestris can persist after introduction into permanent grassland on sandy soil. *European Journal of Soil Biology, 119*, 103536.

van Eekeren, N., et al. (2008). Soil biological quality after 36 years of ley-arable cropping, permanent grassland and permanent arable cropping. *Applied Soil Ecology, 40*(3), 432–446.

Van Eekeren, N., et al. (2009). A mixture of grass and clover combines the positive effects of both plant species on selected soil biota. *Applied Soil Ecology, 42*(3), 254–263.

van Leeuwen, J. P., et al. (2017). Gap assessment in current soil monitoring networks across Europe for measuring soil functions. *Environmental Research Letters*, *12*(12), 124007. https://doi.org/10.1088/1748-9326/aa9c5c

Van Vliet, P. C. J. (2000). *Enchytraeids*. In M. Sumner (Ed.), *Handbook of soil science* (pp. C70–C77). CRC Press.

Van Vliet, P. C. J., et al. (1993). The influence of Enchytraeidae (Oligochaeta) on the soil porosity of small microcosms. In L. Brussaard & M. J. Kooistra (Eds.), *Soil structure/soil biota interrelationships* (pp. 287–299). Elsevier.

Van Vliet, P. C. J., et al. (1995). Population dynamics and functional roles of Enchytraeidae (Oligochaeta) in hardwood forest and agricultural ecosystems. *Plant and Soil*, *170*, 199–207.

Velki, M. & Ečimović, S. (2017). Important issues in ecotoxicological investigations using earthworms. *Reviews of Environmental Contamination and Toxicology*, *239*, 157–184.

Vidal, A., et al. (2023). Chapter one - the role of earthworms in agronomy: consensus, novel insights and remaining challenges. In D. L. Sparks (Ed.), *Advances in agronomy* (Vol. 181, pp. 1–78). Academic Press. https://doi.org/10.1016/bs.agron.2023.05.001

Voisin, A. (1960). *Better grassland sward. Ecology, botany, management*. Crosby Lockwood.

Weyers, S. L., et al. (2008). Construction of an electrical device for sampling earthworm populations in the field. *Applied Engineering in Agriculture*, *24*(3), 391–397.

Whalen, J. K., et al. (1998). Population dynamics of earthworm communities in corn agroecosystems receiving organic or inorganic fertilizer amendments. *Biology and Fertility of Soils*, *27*, 400–407.

Willard, J. R. (1972). *Soil invertebrates: 1. Methods of sampling and extraction: Vol. Technical Report 7* (Matador Project).

Willard, J. R. (1974). *Soil Invertebrates: 7-Enchytraeidae (Annelida: Oligochaeta): populations and biomass: Vol. Technical Report 28* (Matador Project).

Yatoo, A. M., et al. (2022). Assessment of pesticide toxicity on earthworms using multiple biomarkers: a review. *Environmental Chemistry Letters*, *20*(4), 2573–2596.

Zaborski, E. R. (2003). Allyl isothiocyanate: an alternative chemical expellant for sampling earthworms. *Applied Soil Ecology*, *22*(1), 87–95.

Zangerlé, A., et al. (2014). Near Infrared Spectroscopy (NIRS) to estimate earthworm cast age. *Soil Biology and Biochemistry*, *70*, 47–53.

Zangerlé, A., et al. (2016a). Using Near Infrared Spectroscopy (NIRS) to identify the contribution of earthworms to soil macroaggregation in field conditions. *Applied Soil Ecology*, *104*, 138–147.

Zangerlé, A., et al. (2016b). The surales, self-organized earth-mound landscapes made by earthworms in a seasonal tropical wetland. *PLoS One*, *11*(5), e0154269.

Zwetsloot, M. J., et al. (2022). A flexible selection tool for the inclusion of soil biology methods in the assessment of soil multifunctionality. *Soil Biology and Biochemistry*, *166*, 108514. https://doi.org/10.1016/j.soilbio.2021.108514

Chapter 9

Microarthropods as soil health indicators

Junwei Hu and Stefaan De Neve, Ghent University, Belgium; and Zoë Lindo, University of Western Ontario, Canada

1 Classifying microarthropods

The classification of soil fauna has traditionally followed a very pragmatic approach, most commonly distinguishing groups based on body width (e.g. Swift et al., 1979):

- Microfauna: below 0.1 mm (e.g. nematodes (Nematoda), tardigrades (Tardigrada), rotifers (Rotifera)).
- Mesofauna: 0.1–2 mm (e.g. mites (Acari), springtails (Collembola), enchytraeids (Enchytraeidae)).
- Macrofauna: 2–20 mm (e.g. earthworms (Oligochaeta), beetles (Coleoptera)).
- Megafauna: over 20 mm (e.g. moles (Talpidae)).

This simple method is widely accepted and frequently used, and for good reason. Contrary to aboveground fauna, the habitat of belowground fauna is determined by the complex three-dimensional nature of the soil matrix, which contains pore spaces ranging in size from sub-micrometers to centimeters. The specific pore size that soil organisms can access is largely determined by their

http://dx.doi.org/10.19103/AS.2025.0159.10

body diameter (assuming they are not capable of modifying soil structure). Body size, therefore, influences both the movement of soil organisms and their interactions with other organisms and the soil environment. A key distinguishing feature of soil mesofauna is that they are 'terrestrial' animals, inhabiting the air-filled pore spaces, whereas microfauna tend to be 'aquatic', residing in water films and water-filled soil pore spaces. However, this classification based on body size is a topic of debate, with various limits of size classes for mesofauna proposed, for example 0.1–2 mm (Swift et al., 1979) and 0.2–2 mm (Lavelle et al., 1997).

The soil microarthropods are a major subgroup of mesofauna, distinguished by their arthropod characteristics (segmented bodies and exoskeletons) and their size range. They comprise a phylogenetically highly diverse set of organisms including:

- mites (Acari);
- springtails (Collembola);
- proturans (Protura);
- diplurans (Diplura);
- pseudoscorpions (Pseudoscorpionida);
- symphylans (Symphyla); and
- pauropods (Pauropoda).

Among these taxa, Collembola and Acari are the most widespread and often dominate the microarthropod population in most soils (e.g. Cortet et al., 2002; Dombos et al., 2017; Remelli et al., 2025). At the same time, several other taxonomic groups may be collected during specialized sampling for microarthropods and, depending on their body size, can be categorized as microarthropods. These include small-bodied arachnids (e.g. Opiliones and Araneae), myriapods such as millipedes (Diplopoda) and centipedes (Chilopoda), and insect larvae, particularly beetles (Coleoptera) and flies (Diptera). While these taxa are sometimes included in microarthropod surveys, they are often excluded or underrepresented. It is important to note that standard microarthropod sampling procedures are not optimized for these groups, and as a result, their true diversity and abundance may be underestimated. Similarly, other important mesofauna groups – most notably enchytraeids – are not addressed here, as these are ecologically significant but sampled very differently than microarthropods.

2 Occurrence of microarthropods in soils

Microarthropods are present in nearly all soils, with typically tens of species and tens of thousands of individuals per square meter (Bardgett and Van Der

Putten, 2014). Microarthropod abundance and diversity are extremely variable across spatial and temporal scales, depending on biome (Petersen and Luxton, 1982), which is characterized by distinct climate, soil type, and vegetation. Within biomes, their abundance and diversity are mainly determined by land use, edaphic properties and soil management in anthropogenic ecosystems, and microhabitat and vegetation in natural systems. This variability is attributed to the divergent sensitivity of microarthropod species to habitat conditions (Jiang et al., 2015; Mitchell et al., 2016), including

- the presence and depth of litter layers (Kaneda et al., 2012);
- soil organic matter content (Ducarme et al., 2004a);
- the availability of prey (Klironomos and Kendrick, 1995);
- soil bulk density and porosity (Vreeken-Buijs et al., 1998; Nielsen et al., 2008);
- temperature and moisture (Caruso et al., 2007);
- pH (Ducarme et al., 2004b);
- contamination (Khalil et al., 2009); and
- salinity (Klein and Ball, 2024).

Microarthropods are most abundant in undisturbed forest ecosystems, especially in temperate forests (hundreds of thousands of microarthropods per square meter) and less abundant in tropical forests where the litter layer is much thinner (Petersen and Luxton, 1982; Seastedt, 1984). Even in desert environments, soil microarthropod populations can be substantial, e.g., thousands of individuals per square meter in a desert soil in Southern California (Wallwork, 1972). Microarthropod abundance and diversity tends to decline in areas with high levels of soil disturbance, such as agricultural sites, compared to natural, undisturbed environments (Siepel, 1996; Rutgers et al., 2009). For instance, a nationwide monitoring programme in the Netherlands found that microarthropod abundance was highest in forests, whilst their abundance in agricultural ecosystems is generally much lower (Vreeken-Buijs et al., 1998), yet within agricultural systems abundance can vary significantly with management regimes such as tillage, fertilization, and irrigation (Menezes-Oliveira et al., 2021). Microarthropod abundance is generally higher in coarser-textured soils, primarily due to the greater volume of macropores in coarser soils (Vreeken-Buijs et al., 1998; Vanolli et al., 2023). Furthermore, microarthropod assemblages exhibit highly patchy patterns at fine spatial scales, as they tend to concentrate in microhabitats where conditions are particularly favorable (presence of litter, well-structured soil with large pores, moist, sufficient oxygen content, etc.). The population density and diversity of soil microarthropods can vary even at spatial scale as small as 10^{-1} m (Usher and Booth, 1986; Klironomos et al., 1999; Ducarme and Lebrun, 2004; Gao et al., 2014).

The vertical distribution of microarthropods within the soil profile is strongly influenced by the distribution of organic matter not only because organic matter serves as the basal energy source for primary decomposers such as fungi and bacteria but also because organic matter contributes to soil structure, enhancing porosity and habitat availability. As soil depth increases, there is generally an exponential decline in soil organic carbon (SOC) and macropore space, leading to reduced habitat suitability for many microarthropods (Ducarme et al., 2004b; Potapov et al., 2017). Microarthropods are typically most abundant in the organic (O horizon) soil layers (i.e. litter, fermentation, humus) and the upper few centimeters of mineral soil, leading to standard sampling practices focusing on these zones. However, deeper soil layers can contain significant numbers of microarthropods, e.g. Price and Benham (1977) found that 30–40% of the total extracted microarthropod community occurred below 30 cm in an agricultural soil in California. Microarthropods adapt to different soil depths through traits such as body size and pigmentation and are often grouped ecologically as epigeic (surface or litter dwellers), hemiedaphic (inhabiting both litter and soil), and euedaphic (true soil dwellers) (Faber, 1991).

Microarthropod abundance and diversity also fluctuate seasonally and annually, primarily due to changes in soil moisture, temperature, and litter input (Irmler, 2006; Eitminavičiūtė et al., 2008; Wu and Wang, 2019). These fluctuations are closely tied to climatic conditions and vegetation cycles, which vary across biomes and ecosystems (Petersen and Luxton, 1982). For example, microarthropod abundance was higher in October compared to April and August in deciduous and coniferous forests in Hengduan Mountains in southwestern China, likely due to favorable moderate temperatures (Wu et al., 2014). In tropical regions, where temperature remains relatively stable year-round and litter accumulation is typical low, seasonal dynamics are strongly influenced by rainfall patterns, which affect soil moisture. For example, Collembola abundance declined through the dry season (winter) and increased during the wet season in a dry evergreen forest in northeastern Thailand (Wiwatwitaya and Takeda, 2005). Agricultural and seasonal management practices can strongly modulate the natural temporal dynamics of soil microarthropods. For instance, ploughing can cause short-term declines in Acari and Collembola through habitat disturbance (Liu et al., 2016), while fertilization, crop rotations, and planting schedules alter the timing and availability of resources, amplifying or shifting normal seasonal fluctuations in abundance.

3 Functional roles of soil microarthropods

Soil microarthropods, by their feeding and movement, regulate the soil community, influence soil structure, and participate in carbon and nutrient cycling, therefore contributing to important soil-based ecosystem services, including primary production, water infiltration and storage, and global climate regulation. However, there is still a lack of quantification of their contribution to these soil processes and soil-based ecosystem services. While Jernigan et al. (2023) found that higher microarthropod abundance and diversity increased soil microbial activity and nutrient cycling suggesting a positive influence on soil health, studies suggest that microarthropods contribute only a small fraction (<1–10%) of total soil respiration (Petersen and Luxton, 1982). Although some studies have found that the effects of microarthropods on soil-based ecosystem services depend on their abundance and diversity, the nature of these relationships remains poorly understood (Seastedt, 1984), and further experimental research is needed to clarify how microarthropod abundance and diversity shape overall and specific aspects of soil health.

3.1 Soil community regulation

Most functional effects of microarthropods have been related to their feeding activity, with changes in these activities being primarily caused by changes in both available food sources and the microarthropod community (Wolters, 2001; Sechi et al., 2014). Microarthropods span a wide range of feeding types, and feeding behavior can vary considerably even within taxonomic groups.

Soil microarthropods exhibit feeding strategies ranging from primary and secondary decomposers that feed predominantly on litter or fungi to predators that consume nematodes and other microarthropods (Scheu, 2002; Potapov et al., 2022a). For example, Collembola are mainly fungivores but also include a number of detritivorous or herbivorous species, and a few predatory species (Chahartaghi et al., 2005; Malcicka et al., 2017). Within the Acari, Oribatida primarily feed on detritus and fungi but may opportunistically consume nematodes and (dead) arthropods (Heidemann et al., 2011); Mesostigmata are mainly predators of decomposers and microbivores (Koehler, 1997; Scheu, 2002); and Prostigmata span all major feeding guilds, including predators, fungivores-microbivores, plant feeders, and parasites, some of which also act as omnivores (Potapov et al., 2022a).

Through their feeding activity, microarthropods can influence various components of the soil community, such as nematodes (Kaneda and Kaneko, 2008; Taylor et al., 2010), although most studies have primarily focused on their effects on the microbial community. Microarthropods impact microbial abundance, diversity, and activity through their feeding activities in several ways (Janoušková et al., 2018; Brousseau et al., 2024):

1 by selective feeding;
2 by triggering compensatory growth;
3 by contributing to dispersal of bacteria and fungi;

4 by excreting materials rich in nutrients; and
5 by altering competition within the microbial community.

Although emerging evidence suggests that microarthropod abundance and diversity influence microbial communities, the relationship is complex, with underlying mechanisms yet to be fully elucidated. For example, an unvegetated mesocosm experiment with manipulated microarthropod abundance found that fungal species richness was positively correlated with microarthropod abundance, and that fungal community composition differed significantly between microarthropod abundance treatments (Salazar-Fillippo et al., 2025). In contrast, earlier studies did not reveal any effect of microarthropod abundance and diversity on microbial biomass using microcosms containing mineral soil (Kaneda and Kaneko, 2008) or a mixture of plant litter and soil in which seedlings of *Lolium perenne* were planted (Cole et al., 2003).

3.2 Soil structure modification

While the influence of soil microarthropods on soil structure is small compared to macrofauna (earthworms, termites, ants), it is certainly not negligible. Microarthropods can influence soil aggregation both by physically producing, modifying, and transporting organic matter that acts as binding agents, nuclei, or building blocks for aggregates, and by contributing organic materials such as fecal pellets, molts, and necromass, which serve as substrates for microbial activity and further enhance aggregate formation (Maaß et al., 2015). Experimental studies have demonstrated that Collembola can enhance soil macroaggregation (Siddiky et al., 2012a, 2012b). They have also been shown to contribute to the rehabilitation of sealed soil surfaces, producing a finely structured microrelief within just a few months (Langmaack et al., 2001). In addition, Collembola burrowing activities have been reported to increase surface roughness in compacted soils, further illustrating their role in soil structure amelioration (Schrader et al., 1997). Although microarthropods have been generally considered incapable of creating or modifying pore spaces, this assumption has rarely been experimentally verified due to methodological constraints. With the increasing capabilities and accessibility of micro X-ray CT and synchrotron facilities, such experimental proof can now be obtained.

3.3 Organic matter decomposition and stabilization

The role of microarthropods in litter decomposition has been extensively studied in field experiments primarily focusing on the comparison of difference in litter mass loss by exclusion methods (e.g. litter bags with different mesh sizes and/or insecticide treatments), which have yielded variable results. For example, Song et al. (2020) reported a 20% increase in litter decomposition in grassland with microarthropods present compared to no microarthropods using litterbags with different mesh sizes for microarthropods exclusion, whereas Pamminger et al. (2021) did not observe any effect of microarthropod presence on litter decomposition. A meta-analysis of 40 years of litter bag studies on this topic found a significantly negative microarthropod effect on litter decomposition after correction for an estimated mesh size effect (i.e. altered microclimatic conditions and the loss of fragmented but undecomposed litter), suggesting that their role has probably been overstated (Kampichler and Bruckner, 2009). Nonetheless, litter decomposition can be influenced by the abundance and diversity of microarthropods. For example, Collembola diversity has been shown to affect litter decomposition rates, although this relationship is strongly context-dependent (Eisenhauer et al., 2011). Importantly, the acceleration of litter decomposition by microarthropods appears limited to the early stages and does not alter the overall decomposable fraction of litter over longer periods. For instance, a mesocosm experiment found that monthly naphthalene application to suppress microarthropod abundance led to greater litter retention on the soil surface during the first 2 years, but both control and microarthropod suppressed soil lost 98% of the initial litter mass after 3 years (Soong et al., 2016).

The contribution of soil microarthropods to litter and soil organic matter degradation has also been assessed in controlled microcosm experiments (e.g. Lenoir et al., 2007; Wang et al., 2017). These studies have generally demonstrated a positive contribution of microarthropods to CO_2 emissions, with their contribution increasing with increasing population densities (Filser et al., 2002; Kaneda and Kaneko, 2008). Yet simulations from soil food web models have suggested that the direct contribution of soil microarthropods to CO_2 emissions is less than 2% of total carbon mineralization (Holtkamp et al., 2011; Barreto et al., 2024), which suggests that the contribution of microarthropods on SOC decomposition is likely largely indirect and attributed to litter fragmentation, carbon-enriched feces deposition, translocation of organic matter into deeper soil, and stimulation of microbial activity (Chamberlain et al., 2006).

Microarthropods may also contribute to the stabilization of organic matter by affecting both mineral-associated organic matter (MOAM) and particulate organic matter (POM) through promoting soil aggregation and altering soil organic matter chemistry (Soong and Nielsen, 2016). However, this potential role in organic matter stabilization has rarely been thoroughly evaluated in controlled experiments, likely because measuring such processes requires extended observation periods. Only one experiment traced the incorporation of litter-derived C from ^{13}C-labeled litter into the top 20 cm of a silty clay soil and showed that while the amount of litter-derived C in the soil increased over time, there was no difference between the control and microarthropod-suppressed soils after 3 years (Soong et al., 2016).

3.4 Nutrient cycling

The research on the role of soil microarthropods in nutrient cycling has focussed overwhelmingly on nitrogen (N) cycling. By feeding on bacteria, fungi, algae, nematodes, and soil organic matter, it is assumed that microarthropods increase soil N mineralization directly through their excreta and indirectly by interacting with microorganisms (Staddon et al., 2010; Lang and Russell, 2022). A recent meta-analysis (Lang et al., 2023) concluded that soil mesofauna (microarthropods and enchytraeids) increase N mineralization by 15% on average. Collembola have been shown to increase N mineralization at intermediate (−11.8 kPa) to high (−0.5 kPa) but not at low (−42.5 kPa) soil moisture content, consistent with the lower feeding activity at the lowest soil moisture (Kaneda and Kaneko, 2011). Increased N mineralization may also increase the risk of nitrate leaching (Bardgett and Chan, 1999), and a tendency for an increase in both processes has been found with higher microarthropod densities (Kaneda and Kaneko, 2008).

In contrast to the general positive effects of microarthropods on N mineralization, their influence on N_2O emissions is highly variable. Fungivorous Acari, e.g., have been reported to reduce soil N_2O emissions by consuming soil fungi that are unable to perform complete denitrification to N_2 (Shen et al., 2021). However, other studies have found no effect of Acari and Collembola on N_2O emissions (Kuiper et al., 2013; Porre et al., 2016), while some have reported an increase in emissions (Wu et al., 2015; Zhu et al., 2018). This variability likely stems from the exceedingly complex nature of N_2O production, reduction and emission processes, which are highly sensitive to even small changes in soil physical, chemical, and biological conditions and exhibit extreme spatial and temporal variability.

3.5 Plant growth

Microarthropods can influence plant growth both directly and indirectly, primarily through their feeding activities. Direct herbivory by microarthropods, especially Collembola, has long been recognized, but whether this has a generally detrimental effect remains ambiguous, likely depending on resource availability and feeding preferences (Endlweber et al., 2009; Jernigan et al., 2023). Indirect influence of soil microarthropods on plant growth has been demonstrated through their feeding on both pathogenic (Sabatini and Innocenti, 2001) and mycorrhizal fungi (Gange, 2000), as well as their role in nutrient supply mediated by litter fragmentation and microbivory on saprotrophic microbes (Jernigan et al., 2023). The effect of microarthropods on plant growth varies depending on both the soil conditions and plant species (Kuťáková et al., 2018), but overall, soil microarthropods exhibit a positive effect on plant growth (Jernigan et al., 2023). Importantly, both abundance and diversity of microarthropods influence plant growth, and a laboratory microcosm experiment reported that abundance has a greater effect than species richness (Cole et al., 2003). Not surprisingly, both microarthropod abundance and diversity exhibited a non-linear relationship with plant growth, highlighting their functional variability and redundancy, as well as the complexity of interpreting the role of microarthropods in soil functioning (Harris and Boerner, 1990; Liiri et al., 2002; Eisenhauer et al., 2011).

4 Management practices affecting soil microarthropods

The abundance and diversity of soil microarthropods are strongly affected by land use (change) and land management, and thus, not surprisingly, agricultural soils generally support smaller abundance and less diversity of microarthropods compared to natural sites like forests or grasslands on similar soils. While production forests are also managed to different extents (thinning, logging, replanting), this chapter focuses on management practices in agricultural land, because of the very large set of agricultural practices that strongly impacts microarthropod communities.

Agricultural management practices may strongly influence abiotic factors including:

- soil structure (e.g. tillage, organic matter application);
- soil moisture (e.g. tillage, mulching, irrigation);
- salinity (e.g. fertilization, irrigation); and
- toxicant levels (e.g. pesticide application).

and biotic factors including:

- the availability of food sources (e.g. tillage, organic matter application); and
- botanical diversity (e.g. crop rotation, intercropping, weed control).

In general, agricultural management practices that conserve a litter layer such as reduced and no tillage regimes and/or keep the soil covered (e.g. cover crops, Ren et al., 2024), and/or add exogenous organic matter to the soil (Betancur-Corredor et al., 2022), have been shown to increase microarthropod abundance. In contrast, pesticide application can negatively impact the soil microarthropod community (Menezes-Oliveira et al., 2021).

4.1 Reduced tillage

Reduced tillage minimizes the disturbance to soil structure, resulting in higher soil organic matter content in the topsoil, improved aggregate stability, and increased water-holding capacity, all of which are crucial for the activity of microarthropods. It has been reported that abundance and species richness of Collembola were 3–4 times higher in reduced tillage than in conventional tillage: differentiation between Collembola assemblages only occurred after 2 years but steadily increased until 4 years into the use of reduced tillage (Coulibaly et al., 2017). A meta-analysis on the effects of soil tillage on microarthropods revealed significantly higher total abundance of Collembola (+35%) and Acari (+23%) under reduced tillage compared to conventional tillage methods. This increase was largest for epedaphic (+81%) and hemiedaphic (+84%) Collembola and Oribatida (43%) and Mesostigmata (+57%) (Betancur-Corredor et al., 2022). The species richness and diversity of Collembola and Acari communities, however, were not significantly different between reduced and conventional tillage.

4.2 Plant diversification

Management practices that enhance plant functional diversity, including agroforestry, crop rotation, cover cropping, and intercropping, generally promote more heterogeneous microhabitats and support more abundant and diverse soil fauna (e.g. García-Tejero and Taboada, 2016). For example, the introduction of upland crops into flooded rice monocultures has been shown to support higher soil microarthropod abundance and diversity (John et al., 2021). Similarly, Ren et al. (2024) found that long-term inclusion of leguminous cover crops significantly increased the abundance, richness, and diversity of soil microarthropod communities as compared to the absence of cover crops. In addition, maize–peanut intercropping, compared to maize monoculture, doubled Collembola and Acari abundance, with most of the enriched species in the intercropping field identified as euedaphic collembolans and cleptoparasitic mites (Liu et al., 2025). Specific plant species can also affect the magnitude or even the direction of this effect (Marsden et al., 2020).

4.3 Organic amendments

The assessment of the impact of organic fertilization on soil microarthropods is challenging given the enormous diversity in available materials (e.g. liquid and solid manures, composts, municipal wastes, and digestates, amongst others). In a long-term experiment in Spain, a single application of farmyard manure over 3 years increased soil carbon and nitrogen but did not affect microarthropod abundance, whereas annual applications of green manures (straw and crop residues) increased microarthropod populations (Kautz et al., 2006). While the authors suggested that the effect was due to the provision of direct food sources, it is more likely that improvements in soil structure such as enhanced porosity, moisture retention, and microclimate stability associated with green manure addition mediated the positive response. Jiang et al. (2021) found that there was a positive relationship between the amount of surface corn stover mulch and soil microarthropod assemblage but that this varied with the season. In a global meta-analysis, Betancur-Corredor et al. (2023) reported overall positive effects of organic fertilization on the abundance of Collembola and Acari (approximately 2 times and 4 times higher, respectively) as compared to no organic fertilization.

4.4 Agrochemicals

Results from studies on the effects of mineral N fertilizer on soil microarthropods have been inconsistent. Ren et al. (2024) suggested the existence of a threshold effect, where moderate N application promotes soil faunal communities, but very high rates have negative impacts. In contrast, meta-analyses by Hu et al. (2022) and Betancur-Corredor et al. (2023) found that while mineral N fertilizer negatively affected soil nematodes, there were no significant effects on Collembola or Acari. Notably, Betancur-Corredor et al. (2023) also report that the effect of N fertilizer on microarthropods did not vary with application rate.

Pesticides, particularly insecticides, can have significant impacts on soil microarthropods, with effects depending on the type of chemical and its concentration. In a comprehensive review, Gunstone et al. (2021) summarized pesticide impacts on Collembola, Acari, and Protura across 394 research papers. They found that Acari were most affected by insecticides (65%), followed by fungicides (50%) and herbicides (39%), whereas Collembola were more strongly impacted by fungicides (82%) than by insecticides (75%) or herbicides (65%). This pattern may reflect the fungivorous diet of many Collembola, making them particularly sensitive to reductions in fungal food sources.

5 Methods for sampling, extraction, and identification

The monitoring of soil microarthropods relies on the processes of field sampling, extraction, and identification to establish microarthropod abundance and diversity. Variations in these methods can have a fundamental impact on the microarthropod data obtained, ultimately affecting subsequent interpretation.

5.1 Sampling

Given that soil microarthropods exhibit highly heterogeneous spatial distribution and seasonal fluctuations, collecting representative soil microarthropod communities at a specific temporal and spatial scale can be challenging. Therefore, careful consideration must be given to the selection of sampling locations and timing. Research scales can range from the field level (square meters to hectares) (e.g. Fusaro et al., 2018) to national level (thousands of square kilometers) (e.g. George et al., 2017). For large-scale sampling, grid-based designs or stratification based on land use, soil type, and agricultural practices are commonly employed to ensure representative site selection. Within each site or plot, local heterogeneity in microarthropod distribution is typically addressed using random or systematic sampling designs. Furthermore, due to seasonal fluctuations in microarthropod abundance and diversity, monthly or seasonal sampling is often conducted to monitoring temporal changes. Frequent sampling, such as once per season, is also beneficial when evaluating management practices, as it helps detect both short-term changes and long-term trends.

Ideally, the appropriate number of samples and the amount of soil per sample should be determined based on a priori knowledge of the spatial patchiness of microarthropods. However, in practice, sampling designs are often guided by pragmatic constraints and intuitive decision-making.

Sampling tools typically include augers and steel cores of variable sizes and spades, all with very different sampled area and depth. The choice of tool is often driven by practical considerations, such as equipment availability or the size of the extractor, rather than by the spatial patterns of microarthropods (Usher and Booth, 1986). However, for microarthropod sampling, augers are generally imprecise because they mix soil from different depths and disrupt natural distributions, whereas steel cores and spades provide a well-defined volume that better preserves vertical stratification and organism density.

Because widespread species are typically more abundant, while many rare species have limited distributions, comprehensive sampling of the community requires collecting multiple samples across the study area. For example, rarefaction curves revealed that at least 20 or more samples (2.5 cm diameter) were needed to accurately estimate Collembola species richness in pine forest soils of northern Latvia (Dirilgen et al., 2018). The smaller the area collected per sample unit, the larger the sample size required to accurately estimate population density and diversity, but a large series of small sample units provides a more representative coverage of the area and higher estimates of species richness than a smaller series of large samples with the same sampled total surface area (Koehler and Melecis, 2010). However, the number of samples that can be collected per site is often limited by the availability of sufficient extractors to process soil cores simultaneously, as well as constraints in time and labor required for the identification of microarthropods.

Potentially as important as accounting for horizontal spatial variability is considering vertical variability, notably for calculating and comparing soil health indicators. Key factors include whether the litter layer is included and the depth of the mineral soil sampled. This distinction is important as the organic (O) soil horizon that includes the LFH (litter, fermentation, humus) sublayers is technically part of the soil and contains very high microarthropod densities but is often treated separately from the mineral soil.

Generally, soil cores of 5–10 cm depth are taken, with some studies sampling to 30 cm deep (Vanolli et al., 2023). Because microarthropod densities often decrease exponentially with soil depth, even small differences in sampling depth can lead to substantial differences in observed abundances. Therefore, microarthropod counts should always be reported alongside the depth of soil sampled. After sampling, soil can be stored in plastic tubes to minimize disturbance to the soil structure and protected from heat during transportation. Soil samples should ideally be extracted as soon as possible after collection, as storage, even under refrigeration, can alter microarthropod abundances, prey availability, and soil moisture. While some studies have reported minimal changes in certain groups over several weeks (Bruckner, 2022), in practice, delays longer than a few days may compromise sample integrity.

5.2 Extraction

Microarthropods have been extracted from soil using a variety of methods, the majority of which can be divided into two broad categories, namely flotation and extraction.

Flotation methods involve dispersing the soil sample in a salt solution (Salt et al., 1948), saturated sucrose (Jenkins, 1983; Duncan et al., 2007), or a petroleum derivative (e.g. heptane floatation, Aucamp and Ryke (1964), where the apolar microarthropod cuticle has an affinity for heptane and not water). Flotation methods are based on physical and chemical properties (body density and hydrophobicity) of microarthropods versus soil particles. Flotation techniques are recommended for substrates with low organic matter content, where debris contamination is minimal, as separating animals from the debris can be time-consuming.

Extraction techniques are widely regarded as a standard technique in microarthropod assessments (e.g. Potapov et al., 2022b). Extraction methods for soil microarthropods generally rely on creating a physical or environmental gradient; most often heat, light, or desiccation that drives organisms to move downward through a mesh into a collecting vessel. Variants include active systems that use heat and/or light such as Berlese, Tullgren, Berlese–Tullgren, Kempson, and Macfadyen extractors, as well as passive methods like the Winkler sack. All these variants differ in scale and efficiency but are based on the same principle of exploiting behavioral responses to disturbance.

The Berlese-Tullgren funnel is the most widely used variant for single sample extraction of microarthropods. Typically, a soil core is placed upside down on a 2 mm mesh within the funnel. An incandescent lamp positioned above the sample creates a temperature and therefore moisture gradient. As the upper layers become increasingly hot and dry, microarthropods migrate downward to escape the unfavorable conditions. Eventually, they fall through the mesh into a collecting vial containing preservative liquid (e.g. 70% ethanol) beneath the funnel. Efficient extraction requires gradual drying of the sample from top to bottom. Most microarthropods will exit by the time the bottom of the sample has dried, which typically takes at least 5–7 days (Parisi et al., 2005; Kamczyc et al., 2020), though duration depends on soil moisture. Bulb intensity is critical: too much heat causes rapid drying and low efficiency, while moderate, steady heating maximizes recovery.

Extraction of microarthropods varies widely in efficiency depending on the method, soil type, and target organisms. For deeper mineral soils, modified Berlese–Tullgren methods with an additional flotation step have been tested, but they generally yield lower numbers and diversity indices than sucrose centrifugation and heptane floatation methods (Dritsoulas and Duncan, 2020).

Funnel-based methods such as the Berlese–Tullgren extract only active organisms, but they are far less labor-intensive than flotation techniques and co-extract less organic matter and minimize damage to organisms, resulting in better preservation of specimens for identification (Bruckner et al., 2023). They are therefore particularly suitable for large-scale surveys (Dritsoulas and Duncan, 2020). Numerous modifications to these extraction methods exist, including the Macfadyen extractor (Macfadyen, 1961), which creates steep temperature gradients by cooling the base of the sample. However, the classic Berlese–Tullgren method has been shown to be equally effective as high-gradient approaches (André et al., 2002; Bruckner et al., 2023).

Pitfall traps provide an alternative approach by capturing actively moving surface-dwelling microarthropods over time. These measure "activity abundance" (individuals per trap per trapping time), rather than true density or abundance (individuals per m^2), unless traps are deployed in standardized spatial arrays with appropriate numerical corrections (Zhao et al., 2013). While surface pitfall traps typically recover fewer individuals than Berlese–Tullgren funnels or flotation methods (Bruckner et al., 2023), belowground pitfall traps have proven effective for sampling euedaphic taxa, sometimes equaling or surpassing funnel extractions in species richness (Fioratti Junod et al., 2023). More recently, automated devices such as the EDAPHOLOG system, which integrates an opto-electronic sensor to count trapped microarthropods, allow for continuous, real-time monitoring of microarthropod activity in the field (Dombos et al., 2017).

Finally, extraction efficiency is influenced by subtle methodological choices, target organism groups, and soil type (Sandler et al., 2010; Bruckner et al., 2023). Even small modifications can alter outcomes, meaning that, ideally, methods should be calibrated for each soil and study design. While this is rarely feasible in large-scale surveys, awareness of these biases is essential, especially when comparing results between studies.

5.3 Identification

Extracted microarthropods are typically counted and identified morphologically based on microscopic observations at a variety of different levels of taxonomic resolution, ranging from order to species.

Taxonomic identification of microarthropods is often one of the most challenging steps in biodiversity studies. In many groups, even resolution to the level of suborder, superfamily, or family can require specialized training and detailed morphological examination. Achieving species-level identification is more demanding still but with access to reference collections, guidance from a taxonomic expert, and the use of reliable identification keys, these skills can be acquired over time. Independent self-learning is rarely sufficient; participation in introductory training courses is strongly recommended (Koehler and Melecis, 2010).

To support reliable morphological identification, slide mounting of specimens is typically necessary because many diagnostic characters can only be observed under compound microscopy. Mounts may be temporary (e.g. lactic acid), short-term (e.g. Hoyer's medium), or permanent (e.g. polyvinyl-lactophenol). Permanent slides along with the development of a digital photo database are useful for both documentation and training (Koehler and Melecis, 2010). Identification keys are available at multiple scales: global resources such as *A Manual of Acarology* (Krantz and Walter, 2009) and the *Checklist of the Collembola* (https://www.collembola.org/index.html) provide broad coverage, while regional or national keys should be used whenever possible. For specialized habitats or particular taxonomic groups, more detailed keys may be necessary to achieve consistent resolution (from family level to species level) in biodiversity assessments. However, limited access to keys, a shortage of trained specialists, and many unresolved groups collectively constitute the taxonomic impediment, posing significant challenges to soil microarthropod research.

New techniques are rapidly emerging that have the potential to enhance both the accuracy and efficiency of microarthropod analysis while significantly reducing labor and time. Machine learning approaches for automatic identification of soil microarthropods, mainly Collembola, from microscopic images have shown promise for estimating biomass and diversity (Schneider et al., 2022) and are attracting increasing attention (Sys et al., 2022; Oriol et al., 2024). To date, these techniques have achieved the highest accuracy at higher taxonomic levels such as order or class (e.g. Schneider et al., 2022; Filgueiras et al., 2023) or when applied to a limited number of species (e.g. Sys et al., 2022). Their performance depends critically on access to large, high-quality datasets, which are often limited, biased, or incomplete. Common challenges include imbalanced datasets with over-represented taxa, exclusion of rare groups, and noisy or poorly labeled images. By automating the identification of common or key species using state-of-the-art object detection, machine learning can support experts by allowing them to focus on identifying rarer taxa (Oriol et al., 2024). Combined with automated imaging systems and comprehensive data repositories, image-based recognition holds strong potential as a rapid, repeatable, robust, and standardized approach for future soil microarthropod identification.

Finally, DNA extraction from collected microarthropod samples, along with environmental DNA (eDNA), followed by high-throughput sequencing techniques, has been applied as an alternative to microscopic identification (Arribas et al., 2016; Ustinova et al., 2021; Lee et al., 2022). The cytochrome c oxidase subunit I (COI) mitochondrial DNA (mtDNA) marker has the largest number of reference sequences for taxonomic identification of animals and is frequently used for identifying microarthropods. Comparative studies between metabarcoding and morphological identification have shown that the number and identity of families recovered from mock communities are generally similar (Ustinova et al., 2021; Sotnikov et al., 2025). Species richness of Collembola was found to be comparable between methods in a tropical soil, although a large proportion remained unidentified, likely due to incomplete reference databases (Basset et al., 2022). While DNA metabarcoding can reliably detects species presence and richness, sequence read counts do not accurately reflect organism abundance due to biases in DNA extraction, amplification, and differences in biomass or genome copy number (Saitoh et al., 2016). Metabarcoding also faces challenges related to the reliability of reference databases, taxonomic bias, and systematic distortions introduced by different sequencing platforms and bioinformatic pipelines (Sotnikov et al., 2025), all of which need further improvement. Nevertheless, it has been suggested that metabarcoding can be particularly useful for rapid assessment of soil microarthropods for large scale monitoring (Oliverio et al., 2018).

6 Microarthropods as indicators of soil health

The use of microarthropods as indicators of soil quality or soil health is based on a number of basic assumptions (Pulleman et al., 2012):

- First, soil microarthropods are integral components of soil biodiversity and can be directly used as indicators of biodiversity conservation, a key soil-based ecosystem service.
- Second, their activities directly contribute to soil processes such as litter decomposition, organic matter turnover, nutrient cycling, and microbial community regulation. These processes play essential roles in the provision of soil-based ecosystem services, such as climate regulation and primary production, which may progressively be lost as microarthropod abundance and diversity decrease.
- Third, soil microarthropods are not randomly distributed but their abundance and diversity intrinsically reflect the suitability of soil physical, chemical, and biological conditions, such as appropriate soil pore space and available food resources.
- Fourth, as soil-dwelling microarthropods are small, and typically have fragile body structures, and limited dispersal abilities (Ojala and Huhta, 2001), they are sensitive to changes in environmental conditions and management practices that may not be noticed in other soil health indicators.

Effective soil health monitoring, particularly of soil biological properties, is often inaccessible to farmers due to the complexity or cost of conventional measurements; developing simple microarthropods-based indicators could provide a more accessible and cost-effective soil health assessment alternative.

Microarthropod-based soil health indicators have gained increasing popularity in the past two decades and have been used globally for assessing changes in soil health due to:

- differing types of land use (Ponge et al., 2003; Li et al., 2018; Crotty et al., 2023);
- management practices (Niwranski et al., 2002; Bedano et al., 2011; Arboláez et al., 2023);
- soil pollution (An et al., 2013);
- disturbance regimes (Meehan et al., 2019);
- fire (Mantoni et al., 2020); and
- soil restoration programmes (Andrés and Mateos, 2006).

They have also been integrated into several national and regional soil biodiversity monitoring networks across Europe (Pulleman et al., 2012), such as:

- the Dutch Soil Quality Network – Biological Indicator of Soil Quality (DSQN-BISQ) in the Netherlands;

- the Réseau de Mesures de la Qualité des Sols (RMQS-BioDiv) and the BioIndicator Programme in France; and
- the Countryside Survey-Scoping Biological Indicators of Soil Quality (SQID) in the United Kingdom.

Soil health indicators based on microarthropods can be broadly grouped into three main types: (1) based on microarthropod abundance (or those of indicator taxa), (2) based on microarthropod community characteristics (or those of subgroups), and (3) based on microarthropod traits, as reviewed below. It is important to be aware that these indicators vary significantly in the range of soil health aspects they address and demonstrate different levels of sensitivity (Van Straalen, 1998). It is also important to note that theoretical studies and practical applications of microarthropod-based indicators remain relatively limited compared to soil chemical or physical indicators. There is currently no consensus on the most suitable parameters that effectively meet the needs of farmers and policymakers. While the potential of microarthropod-based soil health indicators is evident, further research and work toward standardization is still needed to establish reliable, user-friendly metrics that can bridge the gap between scientific insight and real-world application.

6.1 Soil health indicators based on microarthropod abundance

Microarthropod-based indicators can serve two complementary purposes: as response indicators reflecting changes in the system caused by environmental conditions or management practices, or as effect indicators reflecting the contribution of microarthropods to ecosystem functions, such as litter decomposition, nutrient cycling, or primary production. When designing microarthropod-based soil health indicators, it is crucial to distinguish whether the goal is to detect system responses or to estimate functional effects.

Soil health indicators based on the abundance of bioindicator taxa can range from total microarthropods to specific groups like Collembola and Acari or even individual species. Broader indicators, such as total microarthropod abundance, may reflect general trends in abundance-dependent ecological functions, including decomposition and nutrient cycling or indicate system-level responses to changes in environmental conditions or management practices. In contrast, indicators based on specific taxa, like certain species of Acari or Collembola, may provide more in-depth insights into the relationships between these species and particular soil conditions.

Higher densities of microarthropods often correlate with enhanced soil health, as reflected by higher soil organic matter, improved soil structure, enhanced water infiltration, and increased primary productivity. For example, Cole et al. (2005) found that microarthropod abundance increased concurrently with improvements in soil fertility and plant productivity. Similarly, Wang et al. (2015) reported that the abundance of total microarthropod and Acari was strongly positively correlated with soil C and N levels in a poplar plantation in a coastal area of eastern China. Ma et al. (2024) also showed that microarthropod abundance consistently declined with decreasing soil quality, which was calculated as an index based on 11 properties including bulk density, organic matter, and various total and available nutrients, but this pattern was not observed for specific subgroups such as Oribatida and Entomobryidae. Higher abundance of Collembola has also been linked to increased fertilization intensity, possibly reflecting improved soil conditions (Cluzeau et al., 2012). Therefore, a decline in the abundance of microarthropods, Acari, or Collembola can indicate general soil degradation and may provide farmers with a simple tool to evaluate general soil health for management practices improvement.

Meanwhile, the abundance of microarthropods or subgroups such as Mesostigmata and Oribatida have been shown to be effective bioindicators of agricultural disturbance (Breure et al., 2005), but they are also sensitive to other types of disturbances, including wildfires, forest harvesting, heavy metal pollution, and habitat fragmentation (Gergócs and Hufnagel, 2009; Ivan and Vasiliu, 2009). The abundance of Acari and Collembola has been linked to agricultural management intensity and pollution (Cluzeau et al., 2012) but does not always exhibit a linear response (García-Segura et al., 2018). Changes in Collembola abundance also have been shown to be reliable bioindicators of soil compaction, as they are sensitive to alterations in soil structure (Niwranski et al., 2002; Greenslade, 2007). Additionally, soft-bodied microarthropods such as Collembola and Prostigmata, which possess limited cuticular sclerotization, have been found highly sensitive to changes in soil moisture and therefore proposed as potential rapid indicators of microhabitat water status (Convey et al., 2003). However, since soil bulk density, moisture content, and even certain pollutants can be measured directly and efficiently, using microarthropod abundance or that of subgroups at the order or suborder level as an indicator for these conditions may be considered indirect and of limited practical value (Gergócs and Hufnagel, 2009).

Certain microarthropod species are particularly sensitive or resistant to specific soil conditions, such as contamination, thus being suggested as suitable bioindicators for these conditions. For instance, the collembolan *Mesaphorura macrochaeta* and the oribatid mite *Tectocepheus velatus* have been reported to be particularly tolerant of heavy metal contamination. Studies have shown that *Mesaphorura macrochaeta* increases in abundance and often becomes dominant in contaminated sites (Niklasson et al., 2000; Pelaez-Sanchez et al., 2023). Several species of microarthropods (e.g. *Folsomia candida, Lobella sokamensis*) have been used in laboratory settings to evaluate soil contamination by assessing their survival when introduced to soil samples (An et al., 2013). Among oribatid mites, *Oppiella nova* is widely used in laboratory studies due to its abundance, ease of culture, and suitability for experiments on decomposition and soil processes. Other Collembola and Acari species have preference for acidic or alkaline soil conditions and have been proposed as bioindicators for soil pH. For example, *Tomocerus flavescens* is a strongly acidophilous Collembola, while *Pelops occultus* and *Platynothrus peltifer* are classified as strongly alkalophilous Acari (van Straalen and Verhoef, 1997). However, the use of individual species as soil health indicators in the field has remained very limited in practice. Although certain species have been theoretically proposed as indicators due to their correlation with specific soil factors, their reliability in field conditions is often compromised. In the field, species are influenced by a wide range of variables, making it difficult to attribute changes in abundance (or presence/absence) to a single target factor. Moreover, these indicator species may not be naturally present in many regions, and species level identification requires considerable effort and specialized expertise, which restricts their applicability.

6.2 Soil health indicators based on microarthropod community characteristics

Community-level data are generally considered more informative and reliable than abundance of individual taxa when evaluating soil conditions (Ruf and Beck, 2005). Changes in the microarthropod community provide a direct reflection of biodiversity and therefore of the soil functions they maintain. Key parameters for assessing soil health at the community level include diversity indices, abundance ratios, and community assemblage.

6.2.1 Diversity indices

Several studies have demonstrated that a range of ecosystem services depend on biodiversity, particularly species richness (Liiri et al., 2002; Allan et al., 2013). Among the most commonly applied diversity indices for assessing soil health using soil microarthropods communities are (Solascasas et al., 2022)

- Species richness;
- Shannon's diversity index; and
- Pielou's evenness index.

Species richness provides a straightforward count of the number of species present in a sample, representing a basic measure of biodiversity. Shannon's Diversity Index (often referred to as Shannon's diversity) integrates both species richness and individual species relative abundances, assigning greater weight to communities with greater species richness and a more even distribution of individuals among species. Pielou's Evenness Index is derived from Shannon's index and richness values, mathematically constraining values between 0 and 1, where higher values indicate species have more equal relative abundances, offering insights into the balance of the community.

Both species richness and Shannon's diversity measurements of soil microarthropods have been shown to positively correlate with a soil quality index based on soil chemical and microbiological properties (Heydari et al., 2020). They have also been found to decrease with increasing heavy metal contamination (Yu et al., 2021). However, in some cases, Shannon's diversity and species richness do not change along soil health or disturbance gradients, such as decreasing crop yield (Ma et al., 2024), management intensification (Ponge et al., 2003), and pollutant concentrations (Khalil et al., 2009; Winkler, 2014; García-Segura et al., 2018).

While species-level identification provides the highest ecological resolution, it demands significant taxonomic expertise and is labor-intensive. Calculating diversity indices at the genus level is often seen as a practical compromise between accuracy and feasibility. Research has also shown that Shannon's diversity and species richness were highly correlated between species and family levels of Mesostigmata and Oribatida (Meehan et al., 2019), suggesting that family-level identification for diversity indices may be a suitable option for rapid assessments and large-scale monitoring of microarthropod diversity.

6.2.2 Abundance ratios

The abundance ratio of Collembola to Acari (Collembola/Acari) has been used based on the idea that an increase in this ratio (via increased Collembola, decreased Acari, or both) is often associated with greater perturbations as Acari are more sensitive to disturbance (Joimel et al., 2017). While some studies have suggested that low total organic carbon, high bulk density, and elevated pH levels correlate with a higher Collembola/Acari ratio, this pattern is not consistently observed (Lupardus et al., 2021), and contrasting results have been reported (Tóth et al., 2025), suggesting that the relationship between soil properties and the Collembola-to-Acari ratio is context-dependent.

The (Oribatida+Mesostigmata)/(Prostigmata+Astigmata) index (OM/PA index) for soil degradation is based on the observation that Oribatida and Mesostigmata are more sensitive to agricultural operations than Prostigmata and Astigmata (Bedano et al., 2006), likely due to contrasting life-history traits (Siepel, 1995; Behan-Pelletier, 1999). Compared to Oribatida abundance alone, the OM/PA index has proven more effective at discriminating intensively managed sites (Bedano et al., 2011).

6.2.3 Community assemblage

Another approach is to consider the community holistically, rather than reducing it to a single parameter. In this approach, multivariate indicators (based on either the presence/absence or the abundance of each species in the community) can be subject to multivariate statistical analyses, such as permutational multivariate analysis of variance (PerMANOVA), cluster analysis, or ordination techniques (e.g. PCA).

At the assemblage level, community composition and structure of soil microarthropods and subgroup communities have been consistently shown to be sensitive to environmental changes and habitat disturbances (Yu et al., 2021; Pedro et al., 2023), thus serving as valuable bioindicators of soil health (de Filho et al., 2016). Meehan et al. (2019) found clear differences in Mesostigmata and Oribatida assemblages at the genus level between disturbed and undisturbed sites, with only minimal information loss compared to species-level identification. While multivariate analyses can effectively detect variations in community structure between a site of interest and reference sites, interpreting whether these deviations indicate degradation or improvement often remains a challenge.

6.3 Soil health indicators based on microarthropod traits

In addition to the variety of soil health indicators based on the abundance and taxonomic community characteristics of microarthropods, there are indices that focus on either microarthropod morphology or life history to assess soil health, offering trait-based perspectives that go beyond simple abundance metrics.

6.3.1 Qualità Biologica del Suolo – Artropode

The QBS-ar (Qualità Biologica del Suolo – Artropode) is a soil health index based on morphological and life-history traits rather than species-level identities for soil microarthropods (Menta et al., 2018). While QBS-ar is trait-based, it still requires identification to broad taxonomic groups (e.g. order) and, for some taxa like Collembola, to finer levels (often morphotype) to assign appropriate trait scores. The QBS-ar is based on the assumption that soils of higher quality – those with good physical condition, high organic matter content, and high biodiversity – tend to harbor a greater number of microarthropods that are morphologically well-adapted to soil environments (Parisi et al., 2005) as opposed to degraded or disturbed soils, which tend to support fewer, less specialized soil taxa.

The QBS-ar is a presence-based, trait-scored indicator in which each microarthropod group is assigned a score based on morphological traits reflecting their degree of adaptation to soil environments. Presence of a group contributes its score to the total regardless of its abundance, and summing these scores across all observed groups provides a single value that reflects the trait-based composition of the community as an indicator of soil quality.

The QBS-ar uses specific morphological adaptations – such as reduced pigmentation and visual structures, streamlined bodies, compact appendages (hairs, antennae, legs), and the loss of flying, jumping, or running abilities – that characterize euedaphic microarthropods and make them particularly sensitive to soil degradation. Building on these adaptations, the method classifies microarthropods at the order or class level, assigning each group an Ecological-Morphological Index (EMI) score. These scores range from 1 (epiedaphic, no adaptation to soil) to 20 (euedaphic, maximum adaptation).

Similar soil health indices based on the Collembola community (QBS-c and Collembola ecomorphological index) have been developed (Joimel et al., 2017; Yin et al., 2020). Like QBS-ar, these indices also rely solely on presence/absence data and do not incorporate species abundance. The QBS-ar has been shown to correlate with the number of eco-morphological groups (Blasi et al., 2013; Elmquist et al., 2023) and demonstrates high sensitivity to soil management practices, including tillage intensity, fertilization, and soil amendments, consistently detecting changes in soil conditions (Aspetti et al., 2010; Blasi et al., 2013; Menta et al., 2018), although some exceptions have been reported (Tabaglio et al., 2009). A weak but positive correlation between the QBS-ar and a soil health score representing nutrient cycling capacity (Haney et al., 2018) has been observed in a number of agroecosystem, but this pattern was not consistent across different sites (Elmquist et al., 2023). These findings suggest that QBS-ar has potential as a biological indicator of soil health, but its reliability may depend on environmental and management contexts. The QBS-ar is most effective in mineral soils, where the abundance of euedaphic taxa reflects favorable conditions. However, because QBS-ar assumes that a higher proportion of euedaphic traits indicates better soil quality, it may underestimate soil health in naturally litter-rich or organic soils, where epiedaphic and hemiedaphic taxa dominate without indicating degradation.

6.3.2 Maturity index

The Maturity Index (MI) was first developed to assess the response of soil nematode communities to soil environmental conditions and disturbance (Bongers, 1990) and has since become widely used (Du Preez et al., 2022). The MI approach classifies species according to *r*- vs *K*-select life history traits: *r*-strategists reproduce quickly, have high fecundity and short life spans, whereas K-strategists have slower life cycles, produce fewer offspring, and often have overlapping generations. The MI is calculated as the proportion of K-selected species relative to the total species in a sample, providing an indicator of soil stability or maturity. *K*-select species are generally more abundant in stable, less disturbed sols, while *r*-selected species dominate in disturbed or fluctuating environments.

A similar MI has been proposed for predatory Acari (Mesostigmata: Gamasina) as an indicator of soil health (Ruf, 1998), though it has not been widely applied. Overall, the MI reflects both anthropogenic and natural disturbance and is expected to be higher in less disturbed soils, a pattern observed in forests (Napierała and Błoszyk, 2021), urban green areas (Manu et al., 2021), and agroecosystems (Bedano and Ruf, 2007).

While the MI is well-established for nematodes, its application to other soil fauna, such as predatory Acari, should be approached with caution. Life-history classifications for non-nematode taxa are often less well-characterized, taxonomic resolution may be coarser, and responses to disturbance can differ from nematodes. Consequently, MI values may be less reliable or harder to interpret outside of nematode communities, particularly in complex or highly heterogeneous soils.

7 Case studies

Although microarthropod-based indicators have been widely used in numerous studies, evaluations of their effectiveness remain limited. For soil health indicators to be effective, their ability to provide meaningful insights into soil-based ecosystem services must be carefully evaluated. In this section, a number of microarthropod-based soil health indicators discussed in the previous sections will be illustrated by examples from the literature that exemplify best practice.

One example is the study by van Eekeren et al. (2022), which examined the restoration of soil microarthropods and their role in organic matter decomposition across 20 agricultural and 20 natural grasslands – former agricultural lands subject to ecological restoration under the Dutch National Ecological Network. The natural grasslands supported higher microarthropod diversity than agricultural grasslands. Microarthropod abundance increased with time after grassland establishment, but this trend was absent in grazed grasslands. Importantly, the litter stabilization factor – defined as the fraction of labile green material remaining after incubation in the Tea Bag Index approach – decreased with higher numbers of (herbo)fungivorous grazers. This study underscores the value of soil microarthropods as indicators of soil processes, linking their abundance to key soil-based ecosystem services such as litter decomposition.

Other studies have indirectly evaluated the effectiveness of microarthropods-based soil health indicators by examining their correlation with soil health indices derived from various physical, chemical, and microbiological properties. Heydari et al. (2020) investigated microarthropod community and soil chemical and microbial properties under three tree species and three shrub species in a semiarid oak forest in western Iran. Their findings showed that SOC, total nitrogen, available potassium and all of the studied soil microbiological characteristics (including microbial biomass carbon, microbial biomass nitrogen, basal respiration, and substrate induced respiration) were significantly higher under trees than under shrubs. The Shannon's diversity and species richness of the soil microarthropods were also significantly higher under trees than under shrubs. These indices for soil microarthropods were also positively and linearly related to the soil quality index (SQI) based on chemical and microbiological parameters. This study highlighted that Shannon's diversity and species richness of soil microarthropods can be used as effective soil bioindicators to inform restoration or conservation projects in semi-arid areas.

While indices based on soil microarthropods can be complex, it has been shown that even simple assessments of microarthropods may be sufficient to detect soil health changes with great sensitivity. Arboláez et al. (2023) assessed whether microarthropods (Acari and Collembola, further subdivided in 4 and 2 taxonomic groups, respectively) could be used as indicators of differences in soil health of 28 agricultural fields on (conventional) state farms, conventional private and organic private farms in Central Cuba, representing a gradient of decreasing external inputs and soil disturbance. Analyses of individual soil physical, chemical, and microbiological parameters (microbial biomass carbon, enzyme activities of dehydrogenase, β-glucosidase and β-glucosaminidase, and PLFA) could not discriminate the different farming systems. In contrast, total microarthropod abundance clearly discriminated conventional systems from organic systems. Similar patterns were observed when Acari and Collembola were analyzed separately, as well as when further considering Acari suborders (Mesostigmata, Oribatida, Prostigmata, Astigmata) and major Collembola families (Isotomidae and Entomobryidae). These results indicate that even a coarse, community-level assessment of microarthropod abundance can serve as a sensitive indicator of soil health differences resulting from agricultural management, without requiring detailed taxonomic resolution, and may detect differences that are not apparent in chemical, physical, or microbial parameters (Fig. 1).

These examples also illustrate good methodological practices essential for reliable results, including accounting for inherent soil properties. As for other soil health indicators, a crucial consideration is the impact of inherent soil properties on microarthropods abundance and community composition. In the study of van Eekeren et al. (2022), all 40 grasslands were located on sandy soil, and in the study of Heydari et al. (2020), all sites were on a sandy clay loam texture. Similarly, Arboláez et al. (2023) selected agricultural fields so as to represent only one soil type (Orthic Calcareous Cambisol, World Reference Base

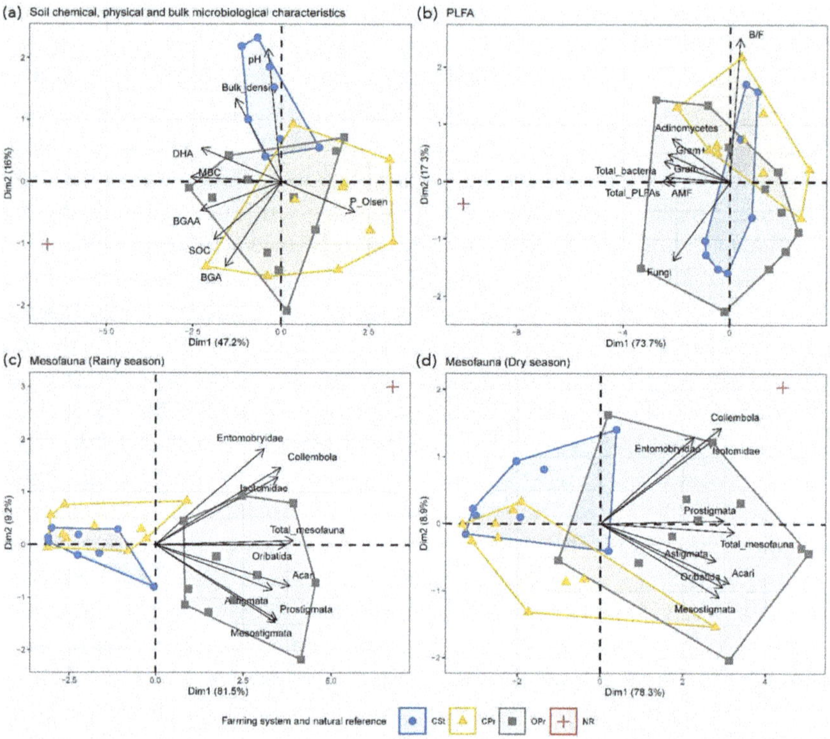

Figure 1 Principal component analysis (PCA) biplot of sampled agricultural fields in Central Cuba (CSt: conventional state farms; CPr: conventional private farms; OPr: organic private farms; NR: natural reference site) based on soil chemical, physical, and bulk microbiological characteristics (a) phospholipid fatty acid (PLFA) analysis, (b) and microarthropods community from rainy and (c) dry seasons. (d) DHA: dehydrogenase activity; MBC: microbial biomass carbon; BGAA: β-glucosaminidase activity; BGA: β-glucosidase activity; SOC: soil organic carbon; Gram+; Gram positive bacteria; Gram−: gram negative bacteria; AMF: arbuscular mycorrhizal fungi; B/F: bacteria-to-fungi ratio.

2015). Next to texture, historic land use and land management may have long lasting impacts on current soil health and ideally needs to be taken into account when evaluating the impact of current management. Arboláez et al. (2023) compiled detailed information with respect to the land use and management history going back to at least 15-20 years, next to detailed information on the crop and crop rotation and management (tillage, irrigation, fertilization, crop protection) at the time of sampling. Despite seasonal fluctuations in absolute abundance, patterns distinguishing management systems remained consistent across seasons, providing confidence in microarthropods as robust indicators (Heydari et al., 2020; Arboláez et al., 2023). Given their ecological significance and the availability of monitoring technologies, assessing the total number of soil microarthropods, particularly Collembola and Acari, is both practical and informative for large-scale soil monitoring and is feasible for farmers to implement, supporting informed soil management decisions. Recent guidance from the GLOSOB initiative by NETSOB (UN-FAO International Network on Soil Biodiversity) also advocates for including microarthropod assessments in soil monitoring and management. As illustrated by the case studies discussed above, GLOSOB recommends a tiered approach to soil biodiversity monitoring and assessment (Parnell et al., 2025): Tier 1 focuses on abundance of microarthropods; Tier 2 involves morphological identification at the class, order, and family levels; and Tier 3 incorporates functional diversity measures, such as the MI.

8 Conclusion and future trends

While substantial advancements have been made in the use of soil microarthropods for soil health indication, many challenges and opportunities for further research remain. Addressing these challenges is vital for enhancing our understanding and effective application of microarthropod-based soil health indicators. This section outlines some primary challenges in current soil microarthropod indicators research and offers perspectives for future studies and practical implementation.

1. The saying "Any analysis is only as good as the sample on which it is done" *a fortiori* applies to assessments of soil microarthropods. There is a clear lack of standardization in sampling, storage, pretreatment, and extraction methods for soil microarthropods. Pending such standardization, the sampling protocol (plot location, sampling equipment, sample dimensions, number of replicate samples, etc.) needs to be reported in detail, to better allow comparison between studies, especially at broader scales such as in regional or national monitoring. Research will be needed to fill the knowledge gap concerning (horizontal and vertical) spatial distribution of microarthropods and their dynamics in order to move away from purely pragmatic and "intuitive" approaches toward more representative sampling.
2. Although microarthropod-based soil health indicators have been increasingly applied and evaluated globally, very few studies, if any, have effectively correlated these indicators to actual quantified soil-based ecosystem services. Clearly, this should be the precondition. The contribution of soil microarthropod diversity to overall soil biodiversity, and the degree to which it correlates with other soil properties, remains unclear. Rather than attempting to link soil microarthropods to soil physical, chemical, and microbial indicators, many of which require their own validation, it is more effective to connect their abundance and diversity directly to key soil-based ecosystem services. Given the

multidimensional nature of soil health and the potential trade-offs among soil functions, microarthropods are likely indicators for some, but not all, soil functions. Determining which functions are best reflected by microarthropod parameters and which specific indicators are most appropriate for context-dependent management and policy objectives, requires further investigation.

3 Given the overriding influence of soil type and texture on soil properties, particularly biological properties, the use of microarthropods as a soil quality indicator is only meaningful when comparing soils that have similar inherent properties (e.g. mineralogy, texture). Failing to take account of this variable has led to erroneous interpretations and conclusions. Historical land management may also have a long-lasting influence on biological and other soil health parameters, and it is thus imperative for studies on soil health comparisons to provide detailed information on the land use history for a sufficient period of time, possibly several decades. As a result, large-scale monitoring frequently exhibits high variability in soil microarthropod data, complicating interpretation and conclusions. When assessing changes in soil health based on microarthropod indicators, baseline (benchmark) or reference points must be established accounting for differences in climate, soil type, and land-management history.

4 The research on microarthropods as indicators of soil health to date has been of a rather empirical nature. There is limited information on how microarthropods respond to varying levels of environmental stressors. Controlled experiments are needed to establish dose–response relationships and quantify the effects of specific stressors on microarthropod communities. However, caution should be exercised when extrapolating such results to field conditions, where multiple stressors usually act simultaneously and may impact the microarthropods very differently compared to a single stressor. Therefore, both controlled laboratory experiments examining the effects of individual and combined stressors on microarthropods, and field studies assessing microarthropods responses in natural settings where multiple interacting stressors coexist, should be encouraged to improve soil health diagnostics, especially for guiding management recommendations.

5 Microarthropod-based soil health indicators with fine taxonomic resolution have shown high sensitivity but remain relatively labor intensive as compared to other soil health indicators. A more widespread use of soil microarthropods in large scale, regional, or national soil health monitoring schemes would greatly benefit from automation of microarthropod counting and identification. This may be realized

in the future by further advances in image processing of extracted microarthropod samples, and progress in microarthropod DNA and eDNA analysis tools. Because indicators based on total abundance often perform as well as those requiring detailed taxonomic identification, soil microarthropod abundances at coarse taxonomic resolution could be applied as practical soil health indicators. This makes them accessible for use not only by researchers but also by farm advisors and farmers, with only limited training and effort.

9 Where to look for further information

The **Global Soil Biodiversity Atlas** (https://esdac.jrc.ec.europa.eu/content/global-soil-biodiversity-atlas) and the **European Atlas of Soil Biodiversity** (https://esdac.jrc.ec.europa.eu/content/european-atlas-soil-biodiversity) provide comprehensive overviews of soil biota, their geographical and temporal distribution, ecosystem functions, and the threats they face globally and in Europe, respectively, while also presenting strategies for sustainable management.

For a more focussed exploration of soil microarthropods, Stephen P. Hopkin's **Biology of the Springtails** (https://academic.oup.com/book/54269) offers an in-depth review of Collembola, covering their classification, physiology, behavior, ecology, and ecotoxicology. Similarly, **Mites: Ecology, Evolution & Behaviour** (https://link.springer.com/book/10.1007/978-94-007-7164-2) examines the life cycles, feeding habits, reproductive biology, and ecological significance of Acari, making it an essential resource for biologists and acarologists.

For taxonomic reference, the **Checklist of the Collembola** (https://www.collembola.org/index.html), compiled by Frans Janssens, provides an up-to-date database of valid species names and taxonomic revisions.

Lastly, **A Chaos of Delight** (https://www.chaosofdelight.org/) offers a visually rich exploration of soil microarthropods through a photographic guide that merges scientific insight with artistic appreciation, making it a valuable resource for both researchers and nature enthusiasts.

10 References

Allan, E., et al., 2013. A comparison of the strength of biodiversity effects across multiple functions. *Oecologia* 173, 223–237. https://doi.org/10.1007/s00442-012-2589-0

An, Y.J., et al., 2013. The collembola Lobella sokamensis juvenile as a new soil quality indicator of heavy metal pollution. *Ecological Indicators* 27, 56–60. https://doi.org/10.1016/j.ecolind.2012.11.017

André, H.M., et al., 2002. Soil biodiversity: myth, reality or conning? *Oikos* 96, 3–24. https://doi.org/10.1034/j.1600-0706.2002.11216.x

Andrés, P. and Mateos, E., 2006. Soil mesofaunal responses to post-mining restoration treatments. *Applied Soil Ecology* 33, 67–78. https://doi.org/10.1016/j.apsoil.2005.08.007

Arboláez, H.P.H., et al., 2023. Mesofauna as effective indicators of soil quality differences in the agricultural systems of central Cuba. *Applied Soil Ecology* 182, 104688. https://doi.org/10.1016/j.apsoil.2022.104688

Arribas, P., et al., 2016. Metabarcoding and mitochondrial metagenomics of endogean arthropods to unveil the mesofauna of the soil. *Methods in Ecology and Evolution* 7, 1071–1081. https://doi.org/10.1111/2041-210X.12557

Aspetti, G.P., et al., 2010. Assessment of soil-quality index based on microarthropods in corn cultivation in Northern Italy. *Ecological Indicators* 10, 129–135. https://doi.org/10.1016/j.ecolind.2009.03.012

Aucamp, J.L. and Ryke, P.A.J., 1964. Preliminary report on a grease film extraction method for soil micro-arthropods. *Pedobiologia* 4, 77–79. https://doi.org/10.1016/s0031-4056(22)00142-1

Bardgett, R.D. and Chan, K.F., 1999. Experimental evidence that soil fauna enhance nutrient mineralization and plant nutrient uptake in montane grassland ecosystems. *Soil Biology and Biochemistry* 31, 1007–1014. https://doi.org/10.1016/S0038-0717(99)00014-0

Bardgett, R.D. and Van Der Putten, W.H., 2014. Belowground biodiversity and ecosystem functioning. *Nature* 515, 505–511. https://doi.org/10.1038/nature13855

Barreto, C., et al., 2024. Restructuring of soil food webs reduces carbon storage potential in boreal peatlands. *Soil Biology and Biochemistry* 193. https://doi.org/10.1016/j.soilbio.2024.109413

Basset, Y., et al., 2022. Comparison of traditional and DNA metabarcoding samples for monitoring tropical soil arthropods (Formicidae, Collembola and Isoptera). *Scientific Reports* 12, 1–16. https://doi.org/10.1038/s41598-022-14915-2

Bedano, J.C., et al., 2006. Influence of three different land management practices on soil mite (Arachnida: Acari) densities in relation to a natural soil. *Applied Soil Ecology* 32, 293–304. https://doi.org/10.1016/j.apsoil.2005.07.009

Bedano, J.C., et al., 2011. Assessment of soil biological degradation using mesofauna. *Soil and Tillage Research* 117, 55–60. https://doi.org/10.1016/j.still.2011.08.007

Bedano, J.C. and Ruf, A., 2007. Soil predatory mite communities (Acari: Gamasina) in agroecosystems of Central Argentina. *Applied Soil Ecology* 36, 22–31. https://doi.org/10.1016/j.apsoil.2006.11.008

Behan-Pelletier, V.M., 1999. Oribatid mite biodiversity in agroecosystems. *Agriculture, Ecosystems and Environment* 74, 411–423.

Betancur-Corredor, B., et al., 2022. Reducing tillage intensity benefits the soil micro- and mesofauna in a global meta-analysis. *European Journal of Soil Science* 73, 1–15. https://doi.org/10.1111/ejss.13321

Betancur-Corredor, B., et al., 2023. Organic nitrogen fertilization benefits selected soil fauna in global agroecosystems. *Biology and Fertility of Soils* 59, 1–16. https://doi.org/10.1007/s00374-022-01677-2

Blasi, S., et al., 2013. Soil microarthropod communities from Mediterranean forest ecosystems in Central Italy under different disturbances. *Environmental Monitoring and Assessment* 185, 1637–1655. https://doi.org/10.1007/s10661-012-2657-2

Bongers, T., 1990. The maturity index: an ecological measure of environmental disturbance based on nematode species composition. *Oecologia* 83, 14–19. https://doi.org/10.1007/BF00324627

Breure, A.M., et al., 2005. Ecological classification and assessment concepts in soil protection. *Ecotoxicology and Environmental Safety* 62, 211–229. https://doi.org/10.1016/j.ecoenv.2005.03.025

Brousseau, P.M., et al., 2024. Functional traits of plant roots and Collembola determine their tri-trophic interactions with soil microbes. *Functional Ecology* 462–477. https://doi.org/10.1111/1365-2435.14714

Bruckner, A., et al., 2022. Effects of soil core handling, transport and storage on numbers and body sizes of edaphic predatory mites (Gamasina). *Soil Organisms* 94, 163–169. https://doi.org/10.25674/so94iss3id304

Bruckner, A., et al., 2023. No indication of methodological biases in tullgren and macfadyen extraction of edaphic microarthropods. *European Journal of Soil Biology* 115, 103464. https://doi.org/10.1016/j.ejsobi.2022.103464

Caruso, T., et al., 2007. Modelling local-scale determinants and the probability of microarthropod species occurrence in Antarctic soils. *Soil Biology and Biochemistry* 39, 2949–2956. https://doi.org/10.1016/j.soilbio.2007.06.010

Chahartaghi, M., et al., 2005. Feeding guilds in Collembola based on nitrogen stable isotope ratios. *Soil Biology and Biochemistry* 37, 1718–1725. https://doi.org/10.1016/j.soilbio.2005.02.006

Chamberlain, P.M., et al., 2006. Translocation of surface litter carbon into soil by Collembola. *Soil Biology and Biochemistry* 38, 2655–2664. https://doi.org/10.1016/j.soilbio.2006.03.021

Cluzeau, D., et al., 2012. Integration of biodiversity in soil quality monitoring: Baselines for microbial and soil fauna parameters for different land-use types. *European Journal of Soil Biology* 49, 63–72. https://doi.org/10.1016/j.ejsobi.2011.11.003

Cole, L., et al., 2003. Effect of density and species richness of soil mesofauna on nutrient mineralisation and plant growth. *Biology and Fertility of Soils* 1, 1–1. https://doi.org/10.1007/s00374-003-0702-6

Cole, L., et al., 2005. Relating microarthropod community structure and diversity to soil fertility manipulations in temperate grassland. *Soil Biology and Biochemistry* 37, 1707–1717. https://doi.org/10.1016/j.soilbio.2005.02.005

Convey, P., et al., 2003. Soil arthropods as indicators of water stress in Antarctic terrestrial habitats? *Global Change Biology* 9, 1718–1730. https://doi.org/10.1046/j.1365-2486.2003.00691.x

Cortet, J., et al., 2002. Impacts of different agricultural practices on the biodiversity of microarthropod communities in arable crop systems. *Agriculture, Ecosystems & Environment* 38, 6–11.

Coulibaly, S.F.M., et al., 2017. Effect of different crop management practices on soil Collembola assemblages: a 4-year follow-up. *Applied Soil Ecology* 119, 354–366. https://doi.org/10.1016/j.apsoil.2017.06.013

Crotty, F.V., et al., 2023. Evaluating the impact of long-term land use change and age since disturbance on soil faunal diversity. *Forests* 14, 1–11. https://doi.org/10.3390/f14091882

de Filho, L.C.I.O., et al., 2016. Collembola community structure as a tool to assess land use effects on soil quality. *Revista Brasileira de Ciencia Do Solo* 40, 1–18. https://doi.org/10.1590/18069657rbcs20150432

Dirilgen, T., et al., 2018. Analysis of spatial patterns informs community assembly and sampling requirements for Collembola in forest soils. *Acta Oecologica* 86, 23–30. https://doi.org/10.1016/j.actao.2017.11.010

Dombos, M., et al., 2017. EDAPHOLOG monitoring system: automatic, real-time detection of soil microarthropods. *Methods in Ecology and Evolution* 8, 313–321. https://doi.org/10.1111/2041-210X.12662

Dritsoulas, A. and Duncan, L.W., 2020. Optimizing for taxonomic coverage: a comparison of methods to recover mesofauna from soil. *Journal of Nematology* 52, 1–9. https://doi.org/10.21307/JOFNEM-2020-104

Ducarme, X., et al., 2004a. Comparison of endogeic and cave communities: Microarthropod density and mite species richness. *European Journal of Soil Biology* 40, 129–138. https://doi.org/10.1016/j.ejsobi.2004.10.003

Ducarme, X., et al., 2004b. Are there real endogeic species in temperate forest mites? *Pedobiologia* 48, 139–147. https://doi.org/10.1016/j.pedobi.2003.10.002

Ducarme, X. and Lebrun, P., 2004. Spatial microdistribution of mites and organic matter in soils and caves. *Biology and Fertility of Soils* 39, 457–466. https://doi.org/10.1007/s00374-004-0730-x

Duncan, L.W., et al., 2007. Food web responses to augmenting the entomopathogenic nematodes in bare and animal manure-mulched soil. *Journal of Nematology* 39, 176–189.

Du Preez, G., et al., 2022. Nematode-based indices in soil ecology: Application, utility, and future directions. *Soil Biology and Biochemistry* 169, 108640. https://doi.org/10.1016/j.soilbio.2022.108640

Eisenhauer, N., et al., 2011. Collembola species composition and diversity effects on ecosystem functioning vary with plant functional group identity. *Soil Biology and Biochemistry* 43, 1697–1704. https://doi.org/10.1016/j.soilbio.2011.04.015

Eitminavičiūtė, I., et al., 2008. Dynamic and seasonal fluctuations of microarthropod complex in coniferous forest soil. *Ekologija* 54, 201–215. https://doi.org/10.2478/v10055-008-0031-z

Elmquist, D.C., et al., 2023. Linking agricultural diversification practices, soil arthropod communities and soil health. *Journal of Applied Ecology* 60, 1952–1963. https://doi.org/10.1111/1365-2664.14453

Endlweber, K., et al., 2009. Collembola switch diet in presence of plant roots thereby functioning as herbivores. *Soil Biology and Biochemistry* 41, 1151–1154. https://doi.org/10.1016/j.soilbio.2009.02.022

Faber, J.H., 1991. Functional cclassification of soil fauna: a new approach. *Oikos* 62, 110–117.

Filgueiras, C.C., et al., 2023. The smart soil organism detector: an instrument and machine learning pipeline for soil species identification. *Biosensors and Bioelectronics* 221, 114417. https://doi.org/10.1016/j.bios.2022.114417

Filser, J., et al., 2002. The role of Collembola in carbon and nitrogen cycling in soil. *Pedobiologia* 46, 234–245. https://doi.org/10.1078/0031-4056-00130

Fioratti Junod, M., et al., 2023. Below-ground pitfall traps for standardised monitoring of soil mesofauna: Design and comparison to Berlese/Tullgren funnels. *Pedobiologia* 101, 150911. https://doi.org/10.1016/j.pedobi.2023.150911

Fusaro, S., et al., 2018. Functional biodiversity, environmental sustainability and crop nutritional properties: a case study of horticultural crops in north-eastern Italy. *Applied Soil Ecology* 123, 699–708. https://doi.org/10.1016/j.apsoil.2017.06.023

Gange, A., 2000. Arbuscular mycorrhizal fungi, Collembola and plant growth. *Trends in Ecology and Evolution* 15, 369–372. https://doi.org/10.1016/S0169-5347(00)01940-6

Gao, M., et al., 2014. Relative roles of spatial factors, environmental filtering and biotic interactions in fine-scale structuring of a soil mite community. *Soil Biology and Biochemistry* 79, 68–77. https://doi.org/10.1016/j.soilbio.2014.09.003

García-Segura, D., et al., 2018. Macrofauna and mesofauna from soil contaminated by oil extraction. *Geoderma* 332, 180–189. https://doi.org/10.1016/j.geoderma.2017.06.013

García-Tejero, S. and Taboada, Á., 2016. Microhabitat heterogeneity promotes soil fertility and ground-dwelling arthropod diversity in Mediterranean wood-pastures. *Agriculture, Ecosystems and Environment* 233, 192–201. https://doi.org/10.1016/j.agee.2016.09.004

George, P.B.L., et al., 2017. Evaluation of mesofauna communities as soil quality indicators in a national-level monitoring programme. *Soil Biology and Biochemistry* 115, 537–546. https://doi.org/10.1016/j.soilbio.2017.09.022

Gergócs, V. and Hufnagel, L., 2009. Application of oribatid mites as indicators. *Applied Ecology and Environmental Research* 7, 79–98.

Greenslade, P., 2007. The potential of Collembola to act as indicators of landscape stress in Australia. *Australian Journal of Experimental Agriculture* 47, 424–434. https://doi.org/10.1071/EA05264

Gunstone, T., et al., 2021. Pesticides and soil invertebrates: a hazard assessment. Frontiers in Environmental Science 9, 1–21. https://doi.org/10.3389/fenvs.2021.643847

Haney, R.L., et al., 2018. The soil health tool–Theory and initial broad-scale application. *Applied Soil Ecology* 125, 162–168. https://doi.org/10.1016/j.apsoil.2017.07.035

Harris, K.K. and Boerner, R.E.J., 1990. Effects of belowground grazing by collembola on growth, mycorrhizal infection, and P uptake of Geranium robertianum. *Plant and Soil* 129, 203–210. https://doi.org/10.1007/BF00032414

Heidemann, K., et al., 2011. Molecular detection of nematode predation and scavenging in oribatid mites: laboratory and field experiments. *Soil Biology and Biochemistry* 43, 2229–2236. https://doi.org/10.1016/j.soilbio.2011.07.015

Heydari, M., et al., 2020. Soil quality and mesofauna diversity relationship are modulated by woody species and seasonality in semiarid oak forest. *Forest Ecology and Management* 473, 118332. https://doi.org/10.1016/j.foreco.2020.118332

Holtkamp, R., et al., 2011. Modelling C and N mineralisation in soil food webs during secondary succession on ex-arable land. *Soil Biology and Biochemistry* 43, 251–260. https://doi.org/10.1016/j.soilbio.2010.10.004

Hu, J., et al., 2022. Effects of nitrogen addition on soil faunal abundance: a global meta-analysis. Global Ecology and Biogeography 31, 1655–1666. https://doi.org/10.1111/geb.13528

Irmler, U., 2006. Climatic and litter fall effects on collembolan and oribatid mite species and communities in a beech wood based on a 7 years investigation. *European Journal of Soil Biology* 42, 51–62. https://doi.org/10.1016/j.ejsobi.2005.09.016

Ivan, O. and Vasiliu, A., 2009. Oribatid mites (Acari, Oribatida) - Bioindicators of forest soils pollution with heavy metals and fluorine. *Annals of Forest Research* 52, 11–18.

Janoušková, M., et al., 2018. Microarthropods influence the composition of rhizospheric fungal communities by stimulating specific taxa. *Soil Biology and Biochemistry* 122, 120–130. https://doi.org/10.1016/j.soilbio.2018.04.016

Jenkins, W.R., 1983. A rapid centrifugal-flotation technique for separating nematodes from soil. *Plant Disease Reporter* 48, 692.

Jernigan, A., et al., 2023. Soil microarthropod effects on plant growth and development. *Plant and Soil* 483, 27–45. https://doi.org/10.1007/s11104-022-05766-x

Jiang, Y., et al., 2015. Composition and spatial distribution of soil mesofauna along an elevation gradient on the north slope of the Changbai Mountains, China. *Pedosphere* 25, 811–824. https://doi.org/10.1016/S1002-0160(15)30062-X

Jiang, Y., et al., 2021. Relationship between the amounts of surface corn stover mulch and soil mesofauna assemblage varies with the season in cultivated areas of northeastern China. *Soil and Tillage Research* 213, 105091. https://doi.org/10.1016/j.still.2021.105091

John, K., et al., 2021. Soil fauna groups respond differentially to changes in crop rotation cycles in rice production systems. *Pedobiologia* 84, 150703. https://doi.org/10.1016/j.pedobi.2020.150703

Joimel, S., et al., 2017. Urban and industrial land uses have a higher soil biological quality than expected from physicochemical quality. *Science of the Total Environment* 584–585, 614–621. https://doi.org/10.1016/j.scitotenv.2017.01.086

Kamczyc, J., et al., 2020. An alternative, portable method for extracting microarthropods from forest soil. *Acta Oecologica* 109, 103655. https://doi.org/10.1016/j.actao.2020.103655

Kampichler, C. and Bruckner, A., 2009. The role of microarthropods in terrestrial decomposition: a meta-analysis of 40 years of litterbag studies. *Biological Reviews* 84, 375–389. https://doi.org/10.1111/j.1469-185X.2009.00078.x

Kaneda, S. and Kaneko, N., 2008. Collembolans feeding on soil affect carbon and nitrogen mineralization by their influence on microbial and nematode activities. *Biology and Fertility of Soils* 44, 435–442. https://doi.org/10.1007/s00374-007-0222-x

Kaneda, S. and Kaneko, N., 2011. Influence of collembola on nitrogen mineralization varies with soil moisture content. *Soil Science and Plant Nutrition* 57, 40–49. https://doi.org/10.1080/00380768.2010.551107

Kaneda, S., et al., 2012. Significance of litter layer in enhancing mesofaunal abundance and microbial biomass nitrogen in sweet corn-white clover living mulch systems. *Soil Science and Plant Nutrition* 58, 424–434. https://doi.org/10.1080/00380768.2012.699881

Kautz, T., et al., 2006. Abundance and biodiversity of soil microarthropods as influenced by different types of organic manure in a long-term field experiment in Central Spain. *Applied Soil Ecology* 33, 278–285. https://doi.org/10.1016/j.apsoil.2005.10.003

Khalil, M.A., et al., 2009. Identification of metal-responsive oribatid mites in a comparative survey of polluted soils. *Pedobiologia* 52, 207–221. https://doi.org/10.1016/j.pedobi.2008.10.002

Klein, R.R. and Ball, B.A., 2024. Soil resources vs. physicochemical soil properties as drivers of abundance and diversity of low Arctic soil mesofauna communities. Polar Biology 47, 617–627. https://doi.org/10.1007/s00300-024-03254-9

Klironomos, J.N. and Kendrick, B., 1995. Relationships among microarthropods, fungi, and their environment. *Plant and Soil* 170, 183–197. https://doi.org/10.1007/BF02183066

Klironomos, J.N., et al., 1999. Designing belowground field experiments with the help of semi-variance and power analyses. *Applied Soil Ecology* 12, 227–238. https://doi.org/10.1016/S0929-1393(99)00014-1

Koehler, H.H., 1997. Mesostigmata (Gamasina, Uropodina), efficient predators in agroecosystems. *Agriculture, Ecosystems and Environment* 62, 105–117.

Koehler, H.H. and Melecis, V., 2010. Long-term observations of soil mesofauna, in: F. Müller (ed.) *Long-Term Ecological Research: Between Theory and Application*. Springer, Dordrecht, pp. 203–220.

Krantz, G.W. and Walter, D.E., 2009. *A Manual of Acarology*, 3rd ed. Texas Tech University Press, Lubbock, Texas, USA. https://doi.org/10.1111/j.1439-0469.2009.00546.x

Kuiper, I., et al., 2013. Soil invertebrate fauna affect N2O emissions from soil. *Global Change Biology* 19, 2814–2825. https://doi.org/10.1111/gcb.12232

Kuťáková, E., et al., 2018. Soil microarthropods alter the outcome of plant-soil feedback experiments. *Scientific Reports* 8, 1–11. https://doi.org/10.1038/s41598-018-30340-w

Lang, B., et al., 2023. Soil mineral nitrogen content is increased by soil mesofauna and nematodes – a meta-analysis. *Soil Organisms* 95, 117–128–117–128. https://doi.org/10.25674/so95iss2id310

Lang, B. and Russell, D.J., 2022. Excretion of nitrogenous waste by soil fauna and assessment of the contribution to soil nitrogen pools. *Soil Organisms* 94, 59–83–59–83. https://doi.org/10.25674/so94iss2id182

Langmaack, M., et al., 2001. Effect of mesofaunal activity on the rehabilitation of sealed soil surfaces. *Applied Soil Ecology* 16, 121–130. https://doi.org/10.1016/S0929-1393(00)00108-6

Lavelle, P., et al., 1997. Soil function in a changing world: The role of invertebrate ecosystem engineers. *European Journal of Soil Biology* 33, 159–193.

Lee, Y.S., et al., 2022. Species-specific environmental DNA analysis of the index species in soil ecosystem, Allonychiurus kimi (Collembola: Onychiuridae). *Ecology and Evolution* 12, 1–12. https://doi.org/10.1002/ece3.9598

Lenoir, L., et al., 2007. Bottom-up or top-down control in forest soil microcosms? Effects of soil fauna on fungal biomass and C/N mineralisation. *Biology and Fertility of Soils* 43, 281–294. https://doi.org/10.1007/s00374-006-0103-8

Li, F.R., et al., 2018. Land-use change alters patterns of soil biodiversity in arid lands of northwestern China. *Plant and Soil* 428, 371–388. https://doi.org/10.1007/s11104-018-3673-y

Liiri, M., et al., 2002. Relationship between soil microarthropod species diversity and plant growth does not change when the system is disturbed. *Oikos* 96, 137–149. https://doi.org/10.1034/j.1600-0706.2002.960115.x

Liu, W., et al., 2016. Impact of grassland reseeding, herbicide spraying and ploughing on diversity and abundance of soil arthropods. *Frontiers in Plant Science* 7, 1–9. https://doi.org/10.3389/fpls.2016.01200

Liu, Y., et al., 2025. Intercropping outweighs straw incorporation driving community and functional diversity of microarthropods after 5 years of tillage practices. *Agriculture, Ecosystems and Environment* 377, 109251. https://doi.org/10.1016/j.agee.2024.109251

Lupardus, R.C., et al., 2021. Can soil invertebrates indicate soil biological quality on well pads reclaimed back to cultivated lands? *Soil and Tillage Research* 213, 105082. https://doi.org/10.1016/j.still.2021.105082

Ma, C., et al., 2024. Effects of soil quality decline on soil-dwelling mesofaunal communities in agricultural lands of the Mollisols Region, China. *Agriculture (Switzerland)* 14, 766. https://doi.org/10.3390/agriculture14050766

Maaß, S., et al., 2015. Functional role of microarthropods in soil aggregation. *Pedobiologia* 58, 59–63. https://doi.org/10.1016/j.pedobi.2015.03.001

Macfadyen, A., 1961. Improved funnel-type extractors for soil arthropods. Journal of Animal Ecology 30, 171. https://doi.org/10.2307/2120

Malcicka, M., et al., 2017. Ecomorphological adaptations in Collembola in relation to feeding strategies and microhabitat. *European Journal of Soil Biology* 78, 82–91. https://doi.org/10.1016/j.ejsobi.2016.12.004

Mantoni, C., et al., 2020. Use of microarthropods to evaluate the impact of fire on soil biological quality. *Journal of Environmental Management* 266, 110624. https://doi.org/10.1016/j.jenvman.2020.110624

Manu, M., et al., 2021. Soil mite communities (Acari: Mesostigmata) as indicators of urban ecosystems in Bucharest, Romania. *Scientific Reports* 11, 1–14. https://doi.org/10.1038/s41598-021-83417-4

Marsden, C., et al., 2020. How agroforestry systems influence soil fauna and their functions - a review. *Plant and Soil* 453, 29–44. https://doi.org/10.1007/s11104-019-04322-4

Meehan, M.L., et al., 2019. Soil mites as bioindicators of disturbance in the boreal forest in northern Alberta, Canada: testing taxonomic sufficiency at multiple taxonomic levels. *Ecological Indicators* 102, 349–365. https://doi.org/10.1016/j.ecolind.2019.02.043

Menezes-Oliveira, V.B., et al., 2021. Changes in soil mesofauna structure due to different land use systems in south Minas Gerais, Brazil. *Environmental Monitoring and Assessment* 193. https://doi.org/10.1007/s10661-021-09214-8

Menta, C., et al., 2018. Soil Biological Quality index (QBS-ar): 15 years of application at global scale. *Ecological Indicators* 85, 773–780. https://doi.org/10.1016/j.ecolind.2017.11.030

Mitchell, R.J., et al., 2016. Relative importance of local- and large-scale drivers of alpine soil microarthropod communities. *Oecologia* 182, 913–924. https://doi.org/10.1007/s00442-016-3688-0

Napierała, A. and Błoszyk, J., 2021. The maturity index for Uropodina (Acari: Mesostigmata) communities as an indicator of human-caused disturbance in selected forest complexes of Poland. *Experimental and Applied Acarology*, 84(2), 263–277. https://doi.org/10.1007/s10493-021-00607-5

Nielsen, U.N., et al., 2008. Soil pore volume and the abundance of soil mites in two contrasting habitats. *Soil Biology and Biochemistry* 40, 1538–1541. https://doi.org/10.1016/j.soilbio.2007.12.029

Niklasson, M., et al., 2000. Environmental stress and reproductive mode in Mesaphorura macrochaeta (Tullbergiinae, Collembola). *Pedobiologia* 44, 476–488. https://doi.org/10.1078/S0031-4056(04)70065-7

Niwranski, K., et al., 2002. Effects of vehicle disturbance and soil compaction on Arctic collembolan abundance and diversity on Igloolik Island, Nunavut, Canada. *European Journal of Soil Biology* 38, 193–196. https://doi.org/10.1016/S1164-5563(02)01142-1

Ojala, R. and Huhta, V., 2001. Dispersal of microarthropods in forest soil. *Pedobiologia* 45, 443–450. https://doi.org/10.1078/0031-4056-00098

Oliverio, A.M., et al., 2018. A DNA metabarcoding approach to characterize soil arthropod communities. *Soil Biology and Biochemistry* 125, 37–43. https://doi.org/10.1016/j.soilbio.2018.06.026

Oriol, T., et al., 2024. Automatic identification of Collembola with deep learning techniques. *Ecological Informatics* 81. https://doi.org/10.1016/j.ecoinf.2024.102606

Pamminger, T., et al., 2021. Investigating the role of soil mesofauna abundance and biodiversity for organic matter breakdown in arable fields. *Integrated Environmental Assessment and Management* 18, 1423–1433. https://doi.org/10.1002/ieam.4563

Parisi, V., et al., 2005. Microarthropod communities as a tool to assess soil quality and biodiversity: a new approach in Italy. *Agriculture, Ecosystems and Environment* 105, 323–333. https://doi.org/10.1016/j.agee.2004.02.002

Parnell, J.J., et al., 2025. Combining science and policy for a unified global soil biodiversity observatory. *Nature Ecology and Evolution* 9, 1–5. https://doi.org/10.1038/s41559-025-02754-z

Pedro, P.M., et al., 2023. Ecological monitoring using Collembola metabarcoding with extremely low bycatch amplification. *Ecological Indicators* 157, 111202. https://doi.org/10.1016/j.ecolind.2023.111202

Pelaez-Sanchez, S., et al., 2023. Invertebrate communities (Collembola and Acari) in soil cover treatments for mine tailings in a long-term field experiment. *Land Degradation and Development* 34, 4728–4740. https://doi.org/10.1002/ldr.4805

Petersen, H. and Luxton, M., 1982. A comparative analysis of soil fauna populations and their role in decomposition processes. *Oikos* 39, 287–388. https://doi.org/10.2307/3544689

Ponge, J.F., et al., 2003. Collembolan communities as bioindicators of land use intensification. *Soil Biology and Biochemistry* 35, 813–826. https://doi.org/10.1016/S0038-0717(03)00108-1

Porre, R.J., et al., 2016. Exploring the relationship between soil mesofauna, soil structure and N2O emissions. *Soil Biology and Biochemistry* 96, 55–64. https://doi.org/10.1016/j.soilbio.2016.01.018

Potapov, A.M., et al., 2017. Arthropods in the subsoil: Aabundance and vertical distribution as related to soil organic matter, microbial biomass and plant roots. *European Journal of Soil Biology* 82, 88–97. https://doi.org/10.1016/j.ejsobi.2017.09.001

Potapov, A.M., et al., 2022a. Feeding habits and multifunctional classification of soil-associated consumers from protists to vertebrates. *Biological Reviews* 97, 1057–1117. https://doi.org/10.1111/brv.12832

Potapov, A.M., et al., 2022b. Global monitoring of soil animal communities using a common methodology. *Soil Organisms* 94, 55–68. https://doi.org/10.25674/so94iss1id178

Price, D.W. and Benham, G.S., 1977. Vertical distribution of soil-inhabiting microarthropods in an agricultural habitat in California. *Environmental Entomology* 6, 575–580. https://doi.org/10.1093/ee/6.4.575

Pulleman, M., et al., 2012. Soil biodiversity, biological indicators and soil ecosystem services-an overview of European approaches. *Current Opinion in Environmental Sustainability* 4, 529–538. https://doi.org/10.1016/j.cosust.2012.10.009

Remelli, S., et al., 2025. Below-ground biodiversity in agricultural systems: the role of crop-specific management in shaping soil arthropod communities. *Agriculture, Ecosystems and Environment* 393, 109850. https://doi.org/10.1016/j.agee.2025.109850

Ren, H., et al., 2024. Green manure provides growth benefits for soil mesofauna by promoting soil fertility in agroecosystems. *Soil and Tillage Research* 238, 106006. https://doi.org/10.1016/j.still.2024.106006

Ruf, A., 1998. A maturity index for predatory soil mites (Mesostigmata: Gamasina) as an indicator of environmental impacts of pollution on forest soils. *Applied Soil Ecology* 9, 447–452. https://doi.org/10.1016/S0929-1393(98)00103-6

Ruf, A. and Beck, L., 2005. The use of predatory soil mites in ecological soil classification and assessment concepts, with perspectives for oribatid mites. *Ecotoxicology and Environmental Safety* 62, 290–299. https://doi.org/10.1016/j.ecoenv.2005.03.029

Rutgers, M., et al., 2009. Biological measurements in a nationwide soil monitoring network. *European Journal of Soil Science* 60, 820–832. https://doi.org/10.1111/j.1365-2389.2009.01163.x

Sabatini, M.A. and Innocenti, G., 2001. Effects of collembola on plant-pathogenic fungus interactions in simple experimental systems. *Biology and Fertility of Soils* 33, 62–66. https://doi.org/10.1007/s003740000290

Saitoh, S., et al., 2016. A quantitative protocol for DNA metabarcoding of springtails (Collembola). *Genome* 59, 705–723. https://doi.org/10.1139/gen-2015-0228

Salazar-Fillippo, A.A., et al., 2025. Effects of microarthropod density on soil fungal community composition in nutrient-poor ecosystems. *Functional Ecology* 39, 432–445. https://doi.org/10.1111/1365-2435.14733

Salt, G., et al., 1948. The arthropod population of pasture soil. *Journal of Animal Ecology* 17, 139–150.

Sandler, R.V., et al., 2010. Berlese-Tullgren funnel efficiency for soil arthropod extraction from typic argiudoll soils in Buenos Aires Province (Argentina). *Suelo* 28, 1–7.

Scheu, S., et al., 2002. The soil food web: structure and perspectives. *European Journal of Soil Biology* 38, 11–20. https://doi.org/10.1016/S1164-5563(01)01117-7

Schneider, S., et al., 2022. Bulk arthropod abundance, biomass and diversity estimation using deep learning for computer vision. *Methods in Ecology and Evolution* 13, 346–357. https://doi.org/10.1111/2041-210X.13769

Schrader, S., et al., 1997. Impact of Collembola and Enchytraeidae on soil surface roughness and properties. *Biology and Fertility of Soils* 25, 396–400. https://doi.org/10.1007/s003740050331

Seastedt, T.R., 1984. The role of microarthropods in decomposition and mineralization processes. *Annual Review of Entomology* 29, 25–46.

Sechi, V., et al., 2014. Collembola feeding habits and niche specialization in agricultural grasslands of different composition. *Soil Biology and Biochemistry* 74, 31–38. https://doi.org/10.1016/j.soilbio.2014.02.019

Shen, H., et al., 2021. Mitigating N2O emissions from agricultural soils with fungivorous mites. *The ISME Journal* 15, 2427–2439. https://doi.org/10.1038/s41396-021-00948-4

Siddiky, M.R.K., et al., 2012a. Soil biota effects on soil structure: interactions between arbuscular mycorrhizal fungal mycelium and collembola. *Soil Biology and Biochemistry* 50, 33–39. https://doi.org/10.1016/j.soilbio.2012.03.001

Siddiky, M.R.K., et al., 2012b. Arbuscular mycorrhizal fungi and collembola non-additively increase soilaggregation. *Soil Biology and Biochemistry* 47, 93–99. https://doi.org/10.1016/j.soilbio.2011.12.022

Siepel, H., 1995. Applications of microarthropod life-history tactics in nature management and ecotoxicology. *Biology and Fertility of Soils* 19, 75–83. https://doi.org/10.1007/BF00336351

Siepel, H., 1996. Biodiversity of soil micro arthropods: the filtering of species. *Biodiversity and Conservation* 5, 251–260. https://doi.org/10.1007/BF00055834

Solascasas, P., et al., 2022. Edaphic arthropods as indicators of the ecological condition of temperate grassland ecosystems: a systematic review. *Ecological Indicators* 142, 109277. https://doi.org/10.1016/j.ecolind.2022.109277

Song, X., et al., 2020. The contributions of soil mesofauna to leaf and root litter decomposition of dominant plant species in grassland. *Applied Soil Ecology* 155, 103651. https://doi.org/10.1016/j.apsoil.2020.103651

Soong, J.L. and Nielsen, U.N., 2016. The role of microarthropods in emerging models of soil organic matter. *Soil Biology and Biochemistry* 102, 37–39. https://doi.org/10.1016/j.soilbio.2016.06.020

Soong, J.L., et al., 2016. Soil microarthropods support ecosystem productivity and soil C accrual: evidence from a litter decomposition study in the tallgrass prairie. *Soil Biology and Biochemistry* 92, 230–238. https://doi.org/10.1016/j.soilbio.2015.10.014

Sotnikov, I.V., et al., 2025. Metabarcoding of a mock community of soil invertebrates: DNA extraction, false positives, and data filtration. *Pedobiologia* 109, 151022. https://doi.org/10.1016/j.pedobi.2025.151022

Staddon, P., et al., 2010. Connectivity, non-random extinction and ecosystem function in experimental metacommunities. *Ecology Letters* 13, 543–552. https://doi.org/10.1111/j.1461-0248.2010.01450.x

Swift, M.J., et al. (Eds.), 1979. *Decomposition in Terrestrial Ecosystems*. University of California Press, Berkely, California. https://doi.org/10.1525/9780520407114

Sys, S., et al., 2022. CollembolAI, a macrophotography and computer vision workflow to digitize and characterize samples of soil invertebrate communities preserved in fluid. *Methods in Ecology and Evolution* 13, 2729–2742. https://doi.org/10.1111/2041-210X.14001

Tabaglio, V., et al., 2009. Physico-chemical indicators and microarthropod communities as influenced by no-till, conventional tillage and nitrogen fertilisation after four years of continuous maize. *Soil and Tillage Research* 105, 135–142. https://doi.org/10.1016/j.still.2009.06.006

Taylor, A.R., et al., 2010. Impact of microarthropod biomass on the composition of the soil fauna community and ecosystem processes. *European Journal of Soil Biology* 46, 80–86. https://doi.org/10.1016/j.ejsobi.2009.11.003

Tóth, Z., et al., 2025. An arthropod-based assessment of biological soil quality in winter wheat fields across Hungary. *Agriculture, Ecosystems and Environment* 378. https://doi.org/10.1016/j.agee.2024.109325

Usher, M.B. and Booth, R.G., 1986. Arthropod communities in a maritime Antarctic moss-turf habitat: multiple scales of pattern in the mites and Collembola. *Journal of Animal Ecology* 55, 155. https://doi.org/10.2307/4699

Ustinova, E.N., et al., 2021. Soil arthropod communities are not affected by invasive Solidago gigantea Aiton (Asteraceae), based on morphology and metabarcoding analyses. *Soil Biology and Biochemistry* 159, 108288. https://doi.org/10.1016/j.soilbio.2021.108288

van Eekeren, N., et al., 2022. Microarthropod communities and their ecosystem services restore when permanent grassland with mowing or low-intensity grazing is installed. *Agriculture, Ecosystems and Environment* 323, 107682. https://doi.org/10.1016/j.agee.2021.107682

Van Straalen, N.M., 1998. Evaluation of bioindicator systems derived from soil arthropod communities. *Applied Soil Ecology* 9, 429–437. https://doi.org/10.1016/S0929-1393(98)00101-2

van Straalen, N.M. and Verhoef, H.A., 1997. The development of a bioindicator system for soil acidity based on arthropod pH preferences. *Journal of Applied Ecology* 34, 217. https://doi.org/10.2307/2404860

Vanolli, B.S., et al., 2023. Edaphic mesofauna responses to land use change for sugarcane cultivation: insights from contrasting soil textures. *Frontiers in Ecology and Evolution* 11, 1–14. https://doi.org/10.3389/fevo.2023.1305115

Vreeken-Buijs, M.J., et al., 1998. Relationships of soil microarthropod biomass with organic matter and pore size distribution in soils under different land use. *Soil Biology and Biochemistry* 30, 97–106. https://doi.org/10.1016/S0038-0717(97)00064-3

Wallwork, J.A., 1972. Distribution patterns and population dynamics of the micro-arthropods of a desert soil in southern California. *Journal of Animal Ecology* 41, 291. https://doi.org/10.2307/3470

Wang, M., et al., 2017. Effect of Collembola on mineralization of litter and soil organic matter. *Biology and Fertility of Soils* 53, 563–571. https://doi.org/10.1007/s00374-017-1200-6

Wang, S., et al., 2015. Responses of soil microarthropods to inorganic and organic fertilizers in a poplar plantation in a coastal area of eastern China. *Applied Soil Ecology* 89, 69–75. https://doi.org/10.1016/j.apsoil.2015.01.004

Winkler, D., 2014. Collembolan response to red mud pollution in Western Hungary. *Applied Soil Ecology* 83, 219–229. https://doi.org/10.1016/j.apsoil.2013.07.006

Wiwatwitaya, D. and Takeda, H., 2005. Seasonal changes in soil arthropod abundance in the dry evergreen forest of north-east Thailand, with special reference to collembolan communities. *Ecological Research* 20, 59–70. https://doi.org/10.1007/s11284-004-0013-x

Wolters, V., 2001. Biodiversity of soil animals and its function. *European Journal of Soil Biology* 37, 221–227. https://doi.org/10.1016/S1164-5563(01)01088-3

Wu, H., et al., 2015. Interactions between earthworms and mesofauna has no significant effect on emissions of CO2 and N2O from soil. *Soil Biology and Biochemistry* 88, 294–297. https://doi.org/10.1016/j.soilbio.2015.06.005

Wu, P. and Wang, C., 2019. Differences in spatiotemporal dynamics between soil macrofauna and mesofauna communities in forest ecosystems: the significance for soil fauna diversity monitoring. *Geoderma* 337, 266–272. https://doi.org/10.1016/j.geoderma.2018.09.031

Wu, P., et al., 2014. Composition and spatio-temporal variation of soil microarthropods in the biodiversity hotspot of northern Hengduan Mountains, China. *European Journal of Soil Biology* 62, 30–38. https://doi.org/10.1016/j.ejsobi.2014.02.013

Yin, R., et al., 2020. Soil functional biodiversity and biological quality under threat: Intensive land use outweighs climate change. *Soil Biology and Biochemistry* 147. https://doi.org/10.1016/j.soilbio.2020.107847

Yu, S., et al., 2021. Soil mesofauna community changes in response to the environmental gradients of urbanization in Guangzhou city. *Frontiers in Ecology and Evolution* 8, 1–11. https://doi.org/10.3389/fevo.2020.546433

Zhao, Z.H., et al., 2013. Solving the pitfalls of pitfall trapping: a two-circle method for density estimation of ground-dwelling arthropods. *Methods in Ecology and Evolution* 4, 865–871. https://doi.org/10.1111/2041-210X.12083

Zhu, X., et al., 2018. Interactions between earthworms and mesofauna affect CO2 and N2O emissions from soils under long-term conservation tillage. *Geoderma* 332, 153–160. https://doi.org/10.1016/j.geoderma.2017.09.007

Chapter 10

Nematodes as indicators of soil health

Deborah A. Neher, University of Vermont, USA; and Krisztina N. Mosdossy, Cornell University, USA

1 Introduction

The sheer abundance of nematodes, their biodiversity, and occupation of every trophic niche and ecosystem type make them one of the most useful ecological indicators amongst all soil microinvertebrates (Freckman, 1988; Neher, 2001a). Globally, there are an estimated $4.4 \pm 0.64 \times 10^{20}$ nematodes (total biomass of approximately 0.3 Gt) inhabiting surface soils, with greater abundance in sub-Arctic regions (38%) than in temperate (24%) or tropical (21%) (van den Hoogen et al., 2019). Nematodes are not only ubiquitous but also serve as key indicators of important soil ecosystem functions, in part because they occupy at least five functional or trophic groups, ranging from grazers of primary decomposers to predators of other nematodes and plant-parasites (Yeates et al., 1993). Their central position in the detrital food web allows them to infer soil carbon (C) and nutrient cycling (Moore and de Ruiter, 1991). All nematodes release excess ammonium (NH_4^+) into soil, a form readily available to plants. Bacterivores and fungivores graze upon and disperse bacteria and fungi, respectively, which tends to maintain microflora populations in the rhizosphere at a 'youthful' state

http://dx.doi.org/10.19103/AS.2025.0159.11

with fast rates of growth, selectively shaping the microbial community in ways that promote plant growth (Ingham et al., 1985). The breadth of ecological function that nematodes influence makes them excellent soil health indicators.

The concept of nematodes as indicators of soil health and their predictive use in soil management represents a major shift in emphasis in the science of nematology (Beaumelle et al., 2021). Formerly a science focused on the control of parasitic and harmful species, now nematodes are recognized for their contribution to biological processes in the soil. By the 1980s, nematologists were beginning to recognize and promote the use of nematode community indicators, including free-living (beneficial) nematodes, as metrics for assessing terrestrial soil health (Freckman, 1988). Nematodes make effective indicators because they occupy a trophic level above microbes in the food chain and integrate physical, chemical, and biological properties related with their food resources (Neher, 2001a). Thus, soil chemical properties (sand, clay, pH) and fungal: bacterial biomass ratio account for only 23% of the variation explained by nematode communities (Neher and Campbell, 1994). Initial attempts explored various ratios and individual trophic groups (Freckman and Ettema, 1993) but it was really Bongers' (1990) paper that introduced the 'maturity index' that was a major advance; the index, which based on family level assessment, was more practical than indices based on species (Ritz and Trudgill, 1999). Through time, improvements to calculating nematode indices of soil health evolved into a sophisticated and rigorous system that reflects resource availability, disturbance, and ecological succession.

Ecological succession is a helpful concept in understanding the maturity stage of ecosystems, including agroecosystems. In nematode ecology, the use of family-based 'maturity' indices provides insights into the ecological successional status of a soil community, which can reflect its history of disturbance (Neher, 2010). A major disturbance, such as fumigation or tillage, pushes the community back to early stages of succession, where recovery can be tracked with increased values of maturity indices (Ettema and Bongers, 1993). In these early successional stages, nematode communities are dominated by organisms with short generation times, small body sizes, rapid dispersal, and generalist feeding habits, such as bacteria and bacterivorous nematodes (Beare et al., 1992). If left undisturbed, the soil community transitions to include fungi and fungivorous nematodes and eventually omnivores and higher-level predators.

Originally, the maturity index (MI) was based solely on free-living nematodes, while plant-parasites were separated into a separate plant-parasitic maturity index (PPI). Both MI and PPI are univariate indices, distilling the community composition into a single numeric value, but the concept can be partitioned into dimensions that represent structural complexity (SI) and nutrient enrichment (EI) (Ferris et al., 2001). Since 2014, the popularity of EI and SI has increased, likely because of their associated graphic and suggested interpretations (Du Preez et al., 2022). This faunal analysis diagram (Fig. 1) is particularly useful for explaining differences between locations, cropping systems, or ecosystems but not changes in management in a single location (Du Preez et al., 2022). Finally, approaches focusing on biomass as a metric of a metabolic footprint (MF) have been introduced (Ferris, 2010). The footprint approach assigns biomass units to published widths and lengths of adult nematodes (e.g. Ferris, 2010; Neher and Weicht, 2013). Once converted to carbon units, estimates of C fluxes can be estimated and linked to decomposition and climate change models or incorporated into energetic food web models (Buchkowski et al., 2023).

In this chapter, we will cover several key topics: methods for sampling and extraction to aid standardization, identification techniques (comparing both morphological and molecular approaches), index calculations and interpretation. For each main topic, we provide recommendations based on current practices and highlight approaches for improving the detection of changes that are crucial for quantifying soil health, such as the role of sentinel

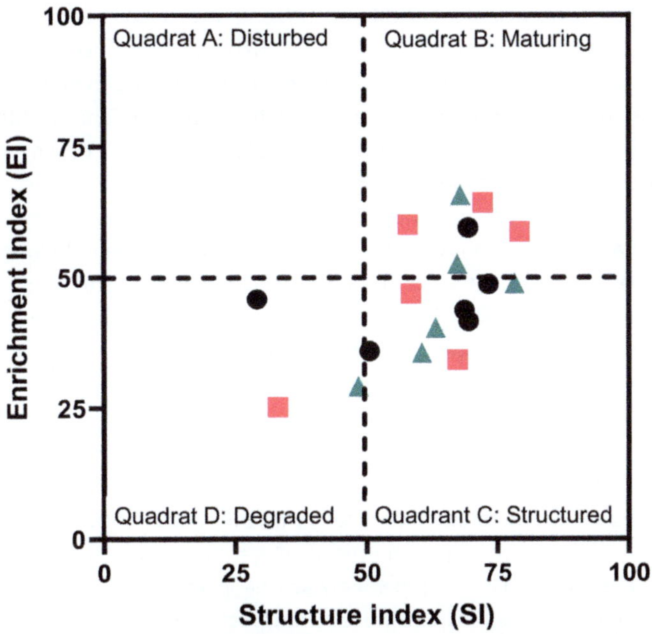

Figure 1 Plot of enrichment vs. structure indices of different management practices of field corn (*Zea mays* L.) at harvest. Experimental plots varied in EI and SI but these differences were not explained by corn treatment. Black circle: a Coleopteran Bt hybrid expressing the Cry3Bb1 protein, Red square: a non-Bt isoline treated with insecticide tefluthrin as positive control; Teal triangle: a non-Bt Isoline without insecticide as negative control. Data from Neher et al. (2014) *Soil Biol. Biochem.*, 76, 127–135.

species, ecological interactions, and links to ecosystem function. Finally, example case studies will be described to illustrate the application of nematode indicators as metrics of soil health.

2 Developments in nematode detection

Most commercial and public laboratories only process plant-parasitic nematodes for agronomic and quarantine purposes. However, indices of nematode communities rely on free-living nematodes that constitute the majority of nematodes found in most soil samples. Commercial laboratories that handle large numbers of samples for identification of free-living nematodes are rare (Neher, 2010). Successional indices necessitate the identification of entire communities, at least to the family level, and an agricultural sample may contain over 25 families and 50 genera of free-living nematodes (e.g. Neher et al., 1995; Neher et al., 2014; Neher et al., 2019). Unfortunately, morphological identification to an adequate taxonomic level is time-consuming and poses a high barrier to entry for novices. The challenge is to simplify and expedite sample processing while enhancing the accuracy and precision of detecting changes in management practices and environmental conditions. Several approaches can address these challenges and yield synergistic results. For instance, focusing on sentinel taxa and developing molecular tools can significantly streamline the process.

Successful use of nematodes as indicators requires careful attention to sampling (Barker, 1985a; Neher and Campbell, 1996) and extraction methods (Barker, 1985b; Neher et al., 1995). The primary goal of these processes is to collect and extract a sample that accurately represents the nematode community at any given time without bias related to specific locations where cores are taken. Importantly, no method of collecting or extracting nematodes can be used in all situations (Freckman and Baldwin, 1990). Regardless of the method employed, consistency is key: samples should be taken at the same time of year, from the same depth and volume, and using the same technique.

2.1 Sampling

Nematode communities shift dynamically with the seasons and plant phenology (cropping cycle), reflecting climatological variations more accurately than calendar dates. A single sample taken at one point in time will not capture the temporal dynamics and should be interpreted within its specific context rather than generalized (Freckman and Baldwin, 1990). For annual crops, the ideal sampling times are spring before planting or pre-plant field operations; mid-season, around anthesis when root growth stops; and fall, near harvest. If only one sample can be taken in a year, and the focus is on plant-parasitic nematodes, the best time to sample is near harvest when plants are mature and potentially more susceptible to pathogens.

Distribution of nematodes varies both horizontally and vertically, influenced by plant roots, litter, and soil type. Therefore, it is essential to collect multiple cores across the area of interest to ensure representativeness and obtain sufficient volume for extraction. These cores are then combined to form a composite sample, which is subsampled before extraction and identification. The minimum number of cores to be collected for one soil sample is as follows: 10 cores for small plots (1–5 m^2), 20+ cores for medium plots (5–100 m^2), and 30+ cores for large plots (>100 m^2). Mixing the cores within a composite sample is best done by hand because mechanical mixing can damage nematodes (Neher et al., 1995; van Bezooijen, 2006).

Given the spatial aggregation or clustering of biological and abiotic soil characteristics, systematic sampling yields more reliable results than random sampling (Barker et al., 1985a). The importance of a systematic pattern, combined with a random starting point, lies in providing a representative sample and reducing bias. A systematic approach means collecting cores at multiple locations equal distance apart. The pattern can be a line, zigzag, serpentine, or points on a grid (Neher et al., 1995). The placement of cores should depend on the specific research question. For instance, studies focused on plant productivity may require sampling in the plant rhizosphere, whereas regional monitoring programs necessitate sampling across different soil types, vegetation, and both rhizosphere and bulk soil (Neher et al., 1998). Another useful approach is to use a stratified random pattern where strata represent different soil textures, vegetation type, or management practices.

Soil samples for nematode analysis are typically collected using an Oakfield core (2.0 cm diameter) to a depth of 15–20 cm (van Bezooijen, 2006). This depth was historically chosen to represent the plow pan, indirectly reflecting rooting depth. Nematodes reside in soil wherever there are roots or decaying detritus. Therefore, in ecosystems with deep-rooting vegetation, such as grasslands or forests, sampling may extend deeper, e.g. to 40 cm.

Samples should be stored in plastic bags to avoid dehydration and kept in an insulated container in the shade to maintain the temperature they were at in the field prior to sampling. Extreme temperatures will damage the samples. The standard practice is to store samples at 4°C (van Bezooijen, 2006); however, there are valid reasons to store samples at 10°C (Coyne et al., 2014) or 15°C (Neher et al., 1995). At 15°C, nematodes may reproduce, but the community structure remains relatively constant (Barker et al., 1969).

For regional and national scale environmental monitoring programs, special considerations regarding sampling design and number of samples are crucial. To enable quantitative inferences about the proportion of land area, the distribution of selected fields across the region should correspond to the hectarage of each cropping system from the previous year (Neher et al., 1998). Based on this design, Neher and Campbell (1996) established recommendations for how many composite samples (transects) to collect per field and the number of subsamples to assay for each transect to achieve a high degree of sample reliability.

2.2 Extraction and their biases

Extraction should occur as soon as possible, but it is important to note that nematodes tend to store better in soil than after extraction. Samples can be stored in soil for up to 6 months without significant effects on nematode populations (van Bezooijen, 2006). Once extracted, sample integrity has a limited shelf life unless nematodes are preserved with a fixative.

Both morphological and molecular identification techniques necessitate the extraction of nematodes from soil to obtain a clean sample. Attempts to extract DNA directly from soil samples often lead to the amplification of non-target taxa such as fungi, plants, and other metazoans, which can reduce the sequence reads for nematodes (Sapkota and Nicolaisen, 2015).

No extraction method guarantees 100% efficiency. It is essential to choose the most appropriate method for each situation while maintaining consistency and standardization. Extraction methods vary in their efficacy for retrieving nematodes of differing mobility and trophic groups (Table 1). Additionally, variations in extraction methods can be compounded by differences between laboratories. Therefore, results should be interpreted in the context of laboratory and extraction method interactions rather than considering the extraction method alone.

Extraction methods were developed first for plant-parasitic nematodes focusing on high throughput, possible with elutriation methods that rely on size, shape, and sedimentation rate differences between nematodes and soil particles. Unfortunately, these elutriation methods, while efficient for extracting plant-parasitic nematodes, tend to overlook important ecological groups. Specifically, they extract omnivores and predators an average of 5.5 and 6.6 times less frequently, respectively, than the Cobb's sifting and gravity method when applied to subsamples of the same composite soil sample (Neher et al., 1995).

Ecological studies have largely neglected the extraction of endoparasites from roots, because this often necessitates a separate method and longer incubation period (Freckman and Baldwin, 1990). Nematodes can be extracted from litter/roots using a mist extraction (Ayoub, 1980) or funnel spray (van Bezooijen, 2006) method. While these methods yield clean samples, they require the construction of a misting apparatus (Ayoub, 1980).

Most extraction methods are indirect, relying on various properties to separate nematodes from their surrounding medium, such as weight and settling rate, size and shape, and mobility. The Baermann method has been adopted by international programs and global surveys (Guerra et al., 2021) due to its simplicity and low cost. However, the method relies on the active mobility of nematodes, which can result in under-sampling slow-moving or sedentary species. In this technique, samples are either wrapped and placed in a funnel (Baermann) or placed on top of a filter in a collection tray or pan. Nematodes must then swim through the material into free water, where they settle.

Flotation methods involve using an extraction fluid with a higher specific gravity than nematodes, allowing the nematodes to float during centrifugation. The supernatant is then passed through a fine sieve (25–35 μm) to concentrate the sample. Common extraction fluids include sugar, $MgSO_4$, or $ZnSO_4$, all of which have their unique challenges and can distort the shape of nematodes making identification difficult (van Bezooijen, 2006). For example, sugar is inexpensive but sticky to work with and can create an osmotic gradient that may cause nematodes to shrivel if not rinsed properly.

Generally, Cobb's method and centrifugal flotation with pre-extraction is more efficient than Baermann or centrifugal flotation without pre-extraction (van Bezooijen, 2006). Pre-extraction is typically performed using a sieving method. Sieves and flotation techniques capture both mobile and sedentary nematodes, whether dead or alive.

Sieving approaches are effective for extracting of all types of nematodes but require slightly more specialized equipment and skill. The approach is also known as the Modified Cobb's decanting and sieving method or Cobb's sifting and gravity method (Ayoub, 1980). Although more time consuming, the extraction efficiency increases significantly (up to 90–95% efficiency) when more sieves are used and each sieve is passed through three times (Neher et al., 1995). However, sieves with finer mesh sizes can be expensive and are prone to damage. After the final pass, the sample can be concentrated and cleaned using centrifugal flotation (e.g. Bekker et al., 2022) or a cotton-wool filter extraction tray, where the sample is left for 48 h (e.g. Darby et al., 2007).

The final step in all extraction methods is to create a suspension of nematodes in water. Because nematodes settle at a rate of approximately 1″ (2.54 cm) per hour, it is crucial to remix the suspension before counting and identification to ensure a representative sample. One effective way to suspend and mix the aqueous samples is by using an aquarium bubbler.

2.3 Morphology

Fresh specimens are valuable for identification because movement patterns are a key feature. However, tracking a moving target can be challenging, and

Table 1 Extraction method comparison

	Baermann 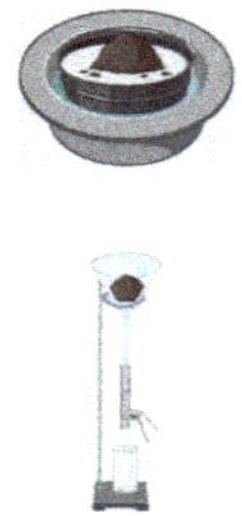	Centrifugal flotation 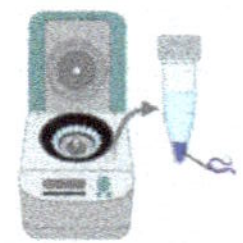	Sieving 	Misting 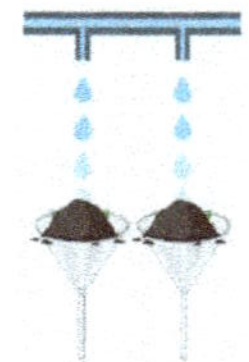	Elutriator 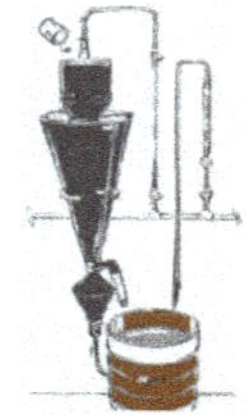
Principal	Nematodes crawl out of material into the water; use facial tissue, milk filters, napkins, or cheesecloth to contain sample	Based on differential specific gravity so nematodes float; supernatant through fine sieve	Wet slurry containing nematodes passed through series of sieves to progressively smaller mesh opening	Nematodes crawl out of material into water; Every 10 min sprinkler switched on for 1.5 min.	Makes use of difference in size, shape, and sedimentation rate between nematodes and soil particles
Variations	Funnel, pie pan, tray, filter, Whitehead tray, cotton-wool extraction tray	Sugar, magnesium sulfate, zinc sulfate	Cobb's sifting and gravity; Modified Cobb's decanting and sieving 500–1000 μm, 350–375 μm, 175 μm, 100 μm, 45 μm	Synonym: funnel spray; means to handle roots & litter	Oostenbrink
Bias	Mobile, standardized time (72 h; harvest every 24 h to avoid decay and deterioration)	Both mobile and inmobile	Both active and inactive	Active only	Omnivores and predators under-represented

Sample size	25–50 g		100–500 g		≤ 250 mL
Advantages	Simple, inexpensive, fairly clean sample	Immobile nematodes; ~ 1 h and relatively clean suspension	High extraction efficiency using only small set of sieves and little water; tapwater okay	Avoids oxygen depletion, clean samples	Efficient and easy to standardize
Disadvantages	Can go anaerobic; 72 hours (harvest every 24 hr); avoid tapwater; kimwipes have a mesh that reduce efficiency; Poor extraction of large & non-mobile, temperature and oxygen dependent	Requires swinging bucket centrifuge	Nematodes can be easily damaged; sieves can be expensive and require an experienced worker; relatively time consuming	May take a few weeks for all nematodes to crawl out of roots or litter	Expensive equipment requires large amounts of water
Van Bezooijen, J. (2006)	Figure 9	Figure 19	Figure 11	Figure 11	Figures 16 and 17

Van Bezooijen, J. (2006) *Methods and Techniques for Nematology*. Wageningen University, The Netherlands. https://www.scribd.com/document/234069625/Methods-and-TechniquesforNematology_

the shelf-life of samples is limited. To prevent decay in aqueous solutions, samples are typically fixed in a preservative, such as formalin or a DMSO/EDTA/saturated NaCl solution (DESS; Yoder et al., 2006). While temporary mounts are typically used for ecological studies, permanent mounts in anhydrous glycerol are essential for archiving samples (Freckman and Baldwin, 1990). Temporary mounts can be created using an 8% formalin solution (Darby et al., 2007). Formalin is a known carcinogen so good ventilation or use in a fume hood is recommended. Individual nematodes can be mounted on microscope slides, but mass mounts of mixed taxa onto slides is a time-saving step. The paraffin wax ring method (van Bezooijen, 2006) is recommended, utilizing a round cover slip with edges sealed with fingernail polish.

Ecologists typically identify at least 10% or up to 150 representative individuals per sample to the genus level (Darby et al., 2007). If fewer than 100 nematodes are present in a sample, all individuals should be identified (Neher et al., 2017). When counting a subsample, it is necessary to determine the total number of nematodes within a sample to convert subsample identifications into proportions of the full sample. Any count values should be adjusted to reflect numbers per unit dry weight or volume of soil.

2.3.1 Resolution

Identifying nematodes to at least the family or genus level requires use of a compound microscope. Optical filters, such as differential interference contrast (DIC), can increase visibility by increasing contrast, making it easier to distinguish fine features in these transparent organisms. Additionally, specific stains like lactophenol cotton blue, acid fuchsin, or Lugol's solution can further enhance certain features. Freckman and Baldwin (1990) provide a comprehensive key for classifying classes and orders of terrestrial nematodes; however, this resolution is often insufficient for computing community indices. Beginners may benefit from a simpler pictorial key that classifies nematodes into eight groups based on the presence a stylet and the nature of the stylet or stoma (Fig. 2 and Fig. 3 in Neher, 2023).

2.3.2 Useful keys

Unfortunately, there is no single key available to identify all groups of free-living nematodes within a sample. Within the Nematoda phylum there are 5 classes, 39 orders, and 241 families (Hodda, 2007). About 14 000 species of free-living, invertebrate- and plant-associated nematode species are known, described and accepted. There are an additional 12 000 species of nematode parasites on vertebrates (Hodda et al., 2009). Different keys specialize in different taxa and geography. Some popular keys include Jairajpuri and Ahmad (1992) or Andrássy (2009) for Dorylaimida; Hunt (1993) for Aphelenchida, Longidoridae, and Trichodoridae; de Ley et al. (2003) for Secernentea; and Siddiqi (2000) for Tylenchida. There are keys that focus on specific countries such as The Netherlands (Bongers, 1994), Hungary (Andrássy, 2005; Andrássy, 2007; Andrássy, 2009), or northern Great Plains of the US (Thorne, 1974). General keys are also available, such as Nickle (1991) for agriculture and Abebe et al. (2006) for freshwater and soil nematodes. However, there is a pressing need for a comprehensive key that covers free-living nematodes across all orders, which would eliminate the necessity of consulting multiple keys.

2.4 Molecular

2.4.1 Universal barcode

While morphological identification provides real-time, genus-specific, and accurate community-level results, nematodes exhibit phenotypic plasticity which complicates the identification process. In contrast, molecular methods hold promise for characterizing entire communities using standard polymerase chain reaction (PCR) techniques. However, these methods face limitations due to the scarcity of DNA probes and primers (Table 2). For example, primers designed for *Caenorhabditis elegans* do not work effectively on other bacterivorous nematodes, including members of the families Cephalobidae and Panagrolaimidae, as well as the genus *Plectus* (Neher and Stürzenbaum, 2006). The mitochondrial protein coding gene, cytochrome *c* subunit I (COX-1), emerged during the quest for a universal barcode marker for most animals (Ratnasingham and Hebert, 2007); however, as with most animals, it fails to recover many nematode taxa (Ahmed et al., 2019). The most commonly used DNA barcodes for family, genus, and sometimes species-level identification are 18S rRNA and 28S rRNA genes, targeting the hypervariable V4 – V8 and D2 – D3 regions, respectfully (Porazinska et al., 2009; Sapkota and Nicolaisen, 2015). However, amplifying 18S and 28S rRNA genes offer limited taxonomic resolution, particularly when samples contain closely related species (Powers, 2004). In contrast, mitochondrial protein coding genes, such as COX-1, offer higher resolution. The lack of comprehensive DNA libraries hinders the recovery of many taxa, thus calling for a need to expand DNA libraries regardless of whether using mitochondrial protein coding genes or rRNA genes (Porazinska et al., 2009; Powers et al., 2021). Nevertheless, both molecular techniques are capable of detecting all life stages (Derycke et al., 2007; Powers et al., 2021), an improvement from morphology, which relies mainly on vermiform and adult life stages.

Consequently, most molecular methods currently only enable accurate characterization of nematode communities to the family level. Individual nematodes can be identified to the species level using single primers (Powers, 2004), but this approach has low throughput (Gendron et al., 2024). Extensive

Table 2 Table of commonly used and emerging primer pairs

Target genes	Primer	Nucleotide sequence (5′-3′)	Fragment length (bp)	Reference
COX-1	JB5GED	ACCTAAACTTARWACRTARTGAAAATG	420	Bowles et al., 1992
	JB3	TTTTTTGGGCATCCTGAGGTTTAT		
COX-1	JB4.5	TAAAGAAAGAACATAATGAAAATG	442	Bowles et al., 1992
	JB3	TTTTTTGGGCATCCTGAGGTTTAT		
COX-1	COIFGED	CCTTTGGGCATCCNGARGTNTAT	450	Ren et al., 2024
	JB5GED	ACCTAAACTTARWACRTARTGAAAATG		
18S	18Sr2b	GCCTTGCCAGCCCGCTCAGTACAAAGGGCAGGGACGTAAT	380	Porazinska et al., 2009
	NF1	GCCTCCCTCGCGCCATCAGGGTGGTGCATGGCCGTTCTTAGTT		
28S		CTGGTGCCAGCAGCCGCGGYAA		Machida and Knowlton, 2012
		TCCGTCAATTYCTTTAAGTT		
28S		GGGAAAGAAGACCCTGTTGAG		Machida and Knowlton, 2012
		GCTTGGCBGCCACAAGCCAGTTA		

efforts are underway to develop metabarcoding techniques that allow for the detection of entire nematode communities, thereby generating sufficient information to use these organisms as indicators of soil health.

Results from morphological and molecular techniques can sometimes yield inconsistent findings. For example, one study reported only weak correlations between morphology-based abundance and DNA biomass metrics derived from COX-1 methods (Gendron et al., 2024). Conversely, another study using the nematode MI found that the two methods produced similar results (Herren et al., 2020). The comparability of results between morphological and molecular methods often hinges on the analytical approach used. When examining trophic groups, bacterivorous and plant-parasitic nematodes showed relatively consistent results between morphology and COX-1 detection. In contrast, omnivores and predators were only detected through morphological methods and not with COX-1 (Gendron et al., 2024). Further, when analyzing across replicates and sites, the nematode trophic groups identified through COX-1 exhibited inconsistencies. Using both molecular and morphological methods on the same sample to characterize nematode communities likely produces the most accurate results compared to one or the other method.

2.4.2 New, mitochondrial-based metagenomics

Mitometagenomics (mtMG) offers an alternative to shotgun sequencing, allowing for high throughput analysis while bypassing the need for taxon-specific primers and overcoming the poor phylogenetic resolution of early methods (Gendron et al., 2023). The mtMG approach can recover up to 95% of all expected species across trophic groups with a mean 99% sequence identity in mock communities *in vitro*. This method was further validated in an agricultural field study under strawberry production, where mtMG was particularly effective in detecting crop-specific plant-parasitic nematodes and evaluating species-level effects of fumigants (Gendron et al., 2023).

While mtMG is proving to increase the accuracy and depth of information on nematode taxa, the scarcity of primer libraries remains a challenge. The most common reference database for nematodes, SILVA, contains only about 5600 nematode sequences from the 18S rRNA gene (Hodda, 2022). Major efforts are underway to synthesize existing primers and curate nematode-specific databases, such as the open-source 18S-NemaBase, drawing from broader databases like SILVA that encompass all living taxa (Gattoni et al., 2023). Although mtMG provides good coverage for some taxa, like Rhabditidae (Gendron et al., 2023), even the most recently designed primers fail to amplify other important taxa, such as *Merlinius*, a globally distributed and significant plant-parasite. Specialized primers remain necessary for these taxa (Ren et al., 2024).

In addition to the scarcity of primer libraries, unexplained discrepancies between mtMG and traditional morphological techniques hinder the widespread adoption of mtMG. For example, the diversity of trophic groups at various depths varies greatly depending on whether mtMG or morphological methods are used (Gendron et al., 2024). Moreover, mtMG tends to yield a simpler and more homogenous nematode community composition across samples compared to morphology-based approaches. Notably, mtMG often fails to detect omnivores, predators, and some key plant-parasites, suggesting that further refinement is needed before this technique can be fully relied upon for accurate bioindication of soil health.

2.5 Limitations and future research

Many nematologists are shifting from traditional morphological methods to molecular methods, finding advantages such as speed and cost-effectiveness, though concerns over the accuracy of these newer techniques persist. Despite these advancements, expertise in nematode taxonomy remains essential, as molecular methods should complement, not replace, morphological approaches to achieve a comprehensive understanding of nematode communities (Geisen et al., 2018). While morphological identification is labor intensive, it delivers immediate data, unlike molecular methods which involve multiple steps, including DNA extraction and sequencing, that can delay results (e.g. Powers et al., 2009). Empirical studies that evaluate differences in time and resources required for molecular and morphological methods are urgently needed (Du Preez et al., 2022).

Molecular techniques offer semi-quantitative estimates of diversity but fall short of measuring absolute abundances or biomass, key metrics needed to link nematodes to ecosystem functions and global biogeochemical cycles (van den Hoogen et al., 2019; Schenk et al., 2019). Factors like body size and amplicon size vary among nematode taxa, which biases read counts and skews taxon-specific abundances (Darby et al., 2013). Additionally, the number of rRNA gene copies correlates with genome size in nematodes, like other eukaryotes (Prokopowich et al., 2003; Bik et al., 2013). However, biomass correlates with read abundances when using 18S metabarcoding, while variation in primer binding sites for COX-1 results in low recovery, even for large nematodes like *Eudorylaimus* and *Mylonchulus* genera (Ren et al., 2024).

Molecular techniques also face challenges in distinguishing between living (active or dormant) and dead organisms, limiting their ability to quantify the active components of nematode communities. Furthermore, rRNA polymorphism is common in nematodes and can be shaped by genomic factors like mutation, genetic drift, and selection or environmental conditions. For example, rRNA copy numbers may vary with ecological factors like food availability (e.g. Bik et al., 2013). Despite these limitations, the ongoing refinement of molecular methods is improving the accuracy of nematode community identification. These advances are increasing the utility of nematodes as indicators for assessing ecosystem properties and soil health.

3 Developments in community analysis

3.1 Ecology

Nematode community indices are limited by the incomplete understanding of many nematode taxa's ecology (du Preez et al., 2022). Ecological traits of free-living nematodes are often inferred from morphological structures, particularly mouthparts which help predict their diet and lifestyles (Yeates et al., 1993; Hodda, 2022) (Table 3). These traits are central to the colonizer-persister (*cp*) scale, a framework used to categorize nematodes based on their responses to environmental disturbances (Bongers, 1999). The *cp* scale, ranging from 1 to 5, aligns with the *r-K* type strategist scale in ecology, which describes organisms' life history strategies along a gradient from rapid colonizers (*r*-strategists) to long-term persisters (*K*-strategists). Nematodes with lower *cp* values (1–2) are early colonizers, such as Rhabditidae, whose generation time span only 2 weeks to a year. Populations of *cp1* nematodes are resilient to disturbances due to their dauer larva stage, a non-feeding phase that allows them to survive unfavorable conditions. As 'enrichment opportunists,' their

Table 3 Colonizer-persister (*cp*) value assignments for select nematode families (Bongers and Bongers (1998) unless a different year in parentheses where 1995 or 1991 refer to Bongers et al., (1995), or Bongers et al., (1991), respectively

cp value	Family
1 Enrichment opportunists[b]	Rhabditidae[a], Diplogasteridae[a] (*s.l.*), Panagrolaimidae[a], Bunonematidae, Myloaimidae (Bongers, 1995), Monhysteridae (FB)
2 Basal fauna[b] General opportunists	Cephalobidae, Plectidae, Monhysteridae (Bongers, 1995), Aphelenchoididae, Leptolaimidae (Bongers, (1991), Aphelenchidae (FB), Seinuridae (FB), Microlaimidae (FB)
3 Rudimentary food web structure[b]	Teratocephalidae, Chromadoridae, Diphtherophoridae, Prismatolaimidae, Tripylidae (FB), Achromadoridae (FB)
4 Greater food web structure[b]	Alaimidae, Mononchidae, Leptonchidae, Qudsianematidae, Dorylaimidae, Tylencholaimidae (FB)
5 Highest food web structure[b]	Aporcelaimidae, Actinolaimidae, Thornenematidae, Belondiridae, Discolaimidae (FB)

[a]Omit dauer larvae because they are non-feeding or inactive stages, indicating something happened in the 'past' (survival stage)
Those with FB or [b] are from Ferris and Bongers (2006).

populations can spike rapidly after a disturbance like a nutrient influx but often decline just as quickly (Freckman, 1988; Bongers et al., 1995). Nematodes with *cp*2 values are 'generalist opportunists' exist in low-resource environments and can adapt to moderate environmental fluctuations without dauer larvae. They reflect ecosystems in intermediate states, making them useful indicators of moderate disturbances. Nematodes at the higher end of the *cp* scale (3–5) include *K*-strategist omnivores and predators, longer generation times (sometimes spanning several years), larger body sizes, fewer offspring, and sensitivity to disturbance. These nematodes indicate more stable, undisturbed environments.

3.2 Calculations

To compute nematode community indices, the first step is assigning a *cp* value (Table 3) and a trophic group (Yeates et al., 1993). The following is a summary of how the most common indices are calculated, providing a numerical basis and general interpretation. For more detailed explanations, including options and associated risks, refer to Neher and Darby (2006; 2009; 2025) and Schmidt et al. (2025).

There are a family of maturity indices that generally serve as an indicator of disturbance, with lower values reflecting more disturbed environments and higher values indicating less disturbance or stable environments. All versions use identical *cp* assignments and math. They vary in whether plant-parasitic

Table 4 Maturity index variations

Index	Trophic groups	*CP*[a]	Interpretation
MI	Free-living (i.e. bacterivores, fungivores, omnivores, predators)	1–5	Smaller values represent environments disturbed by nutrient enrichment or pollution-induced stress; larger values represent recovered or stable environments
MI25	Free-living	2–5	Same as MI but avoids spikes of ephemeral *cp*1 nematodes that spike after applications of fertilizers or heavy metals; less temporal fluctuation
PPI	Plant-parasites	2–5	Associated with plant growth vigor; Sensitive to whether *cp*2 Tylenchidae nematodes are included (as plant-parasites) or excluded (as fungivores)
ΣMI	Free-living and plant-parasites	1–5	Hybrid of MI and PPI and avoids confusion of trophic assignment of Tylenchidae. Acknowledges that complete nematode communities provide integral information
ΣMI25	Free-living and plant-parasites	2–5	Same as ΣMI except less temporal fluctuation

[a]CP = colonizer-persister values (Bongers, 1990).
These indices all reflect the state of succession of a system whereby disturbance and its consequent enrichment effects result in a setback of succession to an earlier state.

nematodes or *cp*1 groups are included (Table 4). The original MI represents free-living taxa only and excludes plant-parasites. MI tends to decrease with increasing microbial activity and pollution-induced stress. One can complement the MI by computing a separate plant parasite index (PPI) or a version that combines both free-living and plant-parasites (ΣMI25). All variations of maturity indices are computed as a weighted mean frequency,

$$MI\,or\,PPI = \sum \frac{v_i \times f_i}{n}$$

where v_i = *cp* value assigned to family, f_i = frequency of family *i* in sample, *n* = total number of individuals in a sample. Given that populations of *cp*1 nematodes increase rapidly but briefly after intense disturbance, index values are considered more stable if one excludes *cp*1 whether for free-living nematodes alone (MI25) or combined with plant-parasites (ΣMI25). Combining MI and PPI is logical if they both correlate to disturbance in the same direction (Neher and Campbell, 1994), which is not always the case (Bongers et al., 1997). The benefit of doing so avoids any confusion of trophic assignment.

The fungivore to bacterivore ratio (F:B) is a simple yet effective index that also reflects ecological succession (Neher and Campbell, 1994). Early stages of colonization are dominated by bacteria and their predators, followed by fungi and fungivores. The recommended method for calculating this ratio is F/(F+B), where the numerals represent actual abundance rather than proportions (Neher and Campbell, 1994). Reporting this equation in publications is essential for consistency, as other ratios like F/B or B/F are not true proportions and can lead to misinterpretation. This index can be used to infer decomposition rates and pathways but that aspect can be better portrayed with the Channel Index (CI).

The CI is analogous to the F:B ratio but is interpreted as reflecting the predominant pathway or rate of decomposition, being either bacterial or fungal (Du Preez et al., 2022). Small values of CI indicate bacterivores are prevalent, indicating fast decomposition and nutrient turnover because of readily available C sources, such as root exudates and simple sugars. Whereas large CI values indicate fungivores are relatively prevalent and indicate slower decomposition and less readily available C such as woody plant litter with high lignin content. Recent studies challenge the simplistic 'slow vs. fast' decomposition model (de Vries et al., 2016), urging caution in interpreting index values. For example, plant photosynthesis and the release of labile C as root exudates can stimulate populations of arbuscular and ectomycorrhizal fungi, which in turn may boost fungivore nematode populations (de Vries et al., 2016). The CI is calculated as follows:

$$Channel\,Index(CI) = 100 \times \frac{0.8Fu_2}{3.2Ba_1 + 0.8Fu_2}$$

where Fu_2, Ba_1, and Ba_2 represent the percentage of *cp*2 fungivores, *cp*1 bacterivores, and *cp*2 bacterivores, respectively.

The Enrichment Index (EI) represents the flow of resources into the food web by enrichment opportunistic bacterivore and fungivore nematodes. The EI weights are the inverse of *cp* values, focusing on the colonizers or *cp* values of 1 and 2 for bacterivores and fungivores representing the portion of food web that responds ephemerally to bursts of food availability and EI (Du Preez et al., 2022). The EI is calculated as follows:

$$Enrichment(e) = 0.8 * Fu_2 + 3.2 * Ba_1$$

where Fu_2 and Ba_1 represent the percentage of *cp*2 fungivores and *cp*1 bacterivores, respectively.

The Structural Index (SI) infers food web structure or complexity (Du Preez et al., 2022), thus analogous to the persister portion of the *cp* scale. The SI focuses on the larger *cp* values or more like the MI35 axis of *cp* triangles, a precursor to SI (de Goede et al., 1993). The SI is calculated as follows:

$$Structure(s) = 0.8 * Ca_2 + 1.8cp_3 + 3.2cp_4 + 5.0cp_5$$

where Ca_2 represents the percentages of carnivores (predators) with a *cp*2, and the other variables represent the percentage of free-living nematodes weighted as *cp*3, *cp*4, and *cp*5.

Graphing the enrichment trajectory as a function of the structural direction creates a graphical representation of food web condition (Ferris et al., 2001). This graphic is an evolution of the original *cp* triangle precursor, which featured three axes: *cp*1, *cp*2, and *cp*3–5 (De Goede et al., 1993). Each quadrant of the graphic represents a different state of condition (Fig. 1). Each axis is computed relative to *cp*2 bacterivores (basal opportunists):

$$Enrichment\,trajectory = 100 \times \frac{e}{e+b}$$

$$Structure\,trajectory = 100 \times \frac{s}{s+b}$$

$$Basal(b) = 0.8 * Fu_2 + 0.8 * Ba_2$$

There are a variety of other graphic illustrations that are less popular, but useful for various applications related to basic ecology or assessment of soil health across regions. For example, multivariate approaches illustrate community assembly in static (canonical correspondence analysis or redundancy analysis) or time series (principal response curves), illustrated in Fig. 2 and Fig. 3,

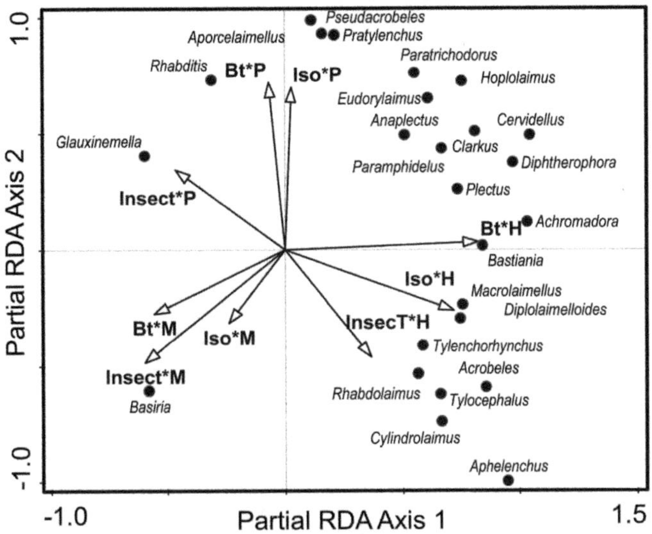

Figure 2 The overall community structure and nematode genera varied more by seasonal phenology than corn treatment. Constrained-Partial redundancy analysis (RDA) biplot of nematode genera in soil with the two-way interaction of treatment and phenology as explanatory variables and year as a covariable treated as a block (Bt: Bt hybrid, Iso: non-Bt isoline without insecticide, Insect: non-Bt isoline with insecticide tefluthrin; P: pre-plant, M: midseason at anthesis, and H: harvest phenology). Circles represent each of the 25 of 73 nematode genera that explained the most variation. Eigenvalues (lambda) are 0.0785 (pseudo-F = 4.7, $P = 0.006$), 0.0504, 0.0223, and 0.0142 for the first (horizontal), second (vertical), third and fourth axes respectively. The first two axes represent 64.48% of the fitted variation. Sample size was 18 per treatment after averaging subsamples (3 phenology times x 2 years x 3 replicates). The analysis was performed using Canoco software, version 5 (Microcomputer Power, Ithaca, New York, United States). Reprinted from Soil Biology and Biochemistry, vol 76, Neher, D.A., Muthumbi, A.W.N. and Dively, G.P., Impact of coleopteran-active Bt corn on non-target nematode communities in soil and decomposing corn roots, pp. 127–135., Copyright (2014), with permission from Elsevier.

respectively. Cumulative Distribution Functions are tools used to illustrate the distribution of index values on a land area basis (Neher et al., 1998, Fig. 4). Co-correspondence (Neher et al., 2017) or canonical correlation (Neher and Campbell, 1994) are tools useful to examine multivariate associations between groups of plants or microbes in relationship to nematode communities. Tools such as Principal Components Analysis (Neher et al., 2012) or Factor Analysis (Martin et al., 2022) can be useful to create new and fewer variables that represent communities with plethora taxa. Computational techniques such as Artificial Neural Networks (Lek and Guegan, 1999) and null models (Gotelli, 2001) in conjunction with current multivariate statistical techniques may help to link individual taxa with habitat conditions.

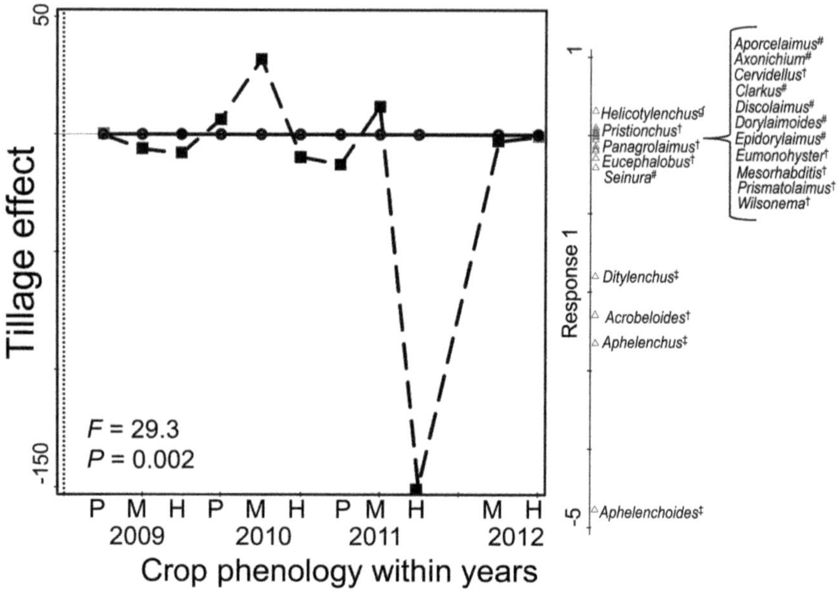

Figure 3 Compared to no-till, conventional tillage increased abundances of *Aphelenchoides*, *Aphelenchus*, *Acrobeloides*, and *Ditylenchus*, especially at harvest in the first three years of the experiment. Principal response curve biplot of PRC (*y*-axis, till) and crop phenology (P: planting, M: midseason, H: harvest in 2009 to 2011, and M and H in 2012) are shown. Dashed line represents conventional tillage and solid line for no tillage as reference line. Genus names are coded by trophic group (†: bacterivores, ‡: fungivores, ɠ: plantparasites, #: omnivores-predators). For simplicity, the biplot is restricted to illustrate only the 20 genera that explained the most variation. Statistical significance was computed by Monte Carlo permutation of both first ordination axis and all axes together. Reprinted from Applied Soil Ecology, vol 140, Neher, D.A., Nishanthan, T., Grabau, Z.J. and Chen, S.Y., Crop rotation and tillage affect nematode communities more than biocides in monoculture soybean', pp. 89–97, Copyright (2019), with permission from Elsevier.

Standard diversity index values (e.g. Shannon, Simpson) have limited utility because they do not retain taxon identity, thus incapable of distinguishing native and introduced species. In nematode communities, diversity indices are typically calculated at the genus or trophic level. Genus is usually the finest resolution used for free-living nematodes. Trophic diversity reflects the complexity of food webs, with a maximum value equal to the number of trophic groups present. Higher values represent food webs containing more omnivores and predators that reside relatively high in the food chain (Neher and Campbell, 1994)

Neher conducted a series of experiments to evaluate the statistical reliability of a range of indices, based on their ability to detect a 10% change between two time points and 80% statistical power (1-β) where β is Type 2

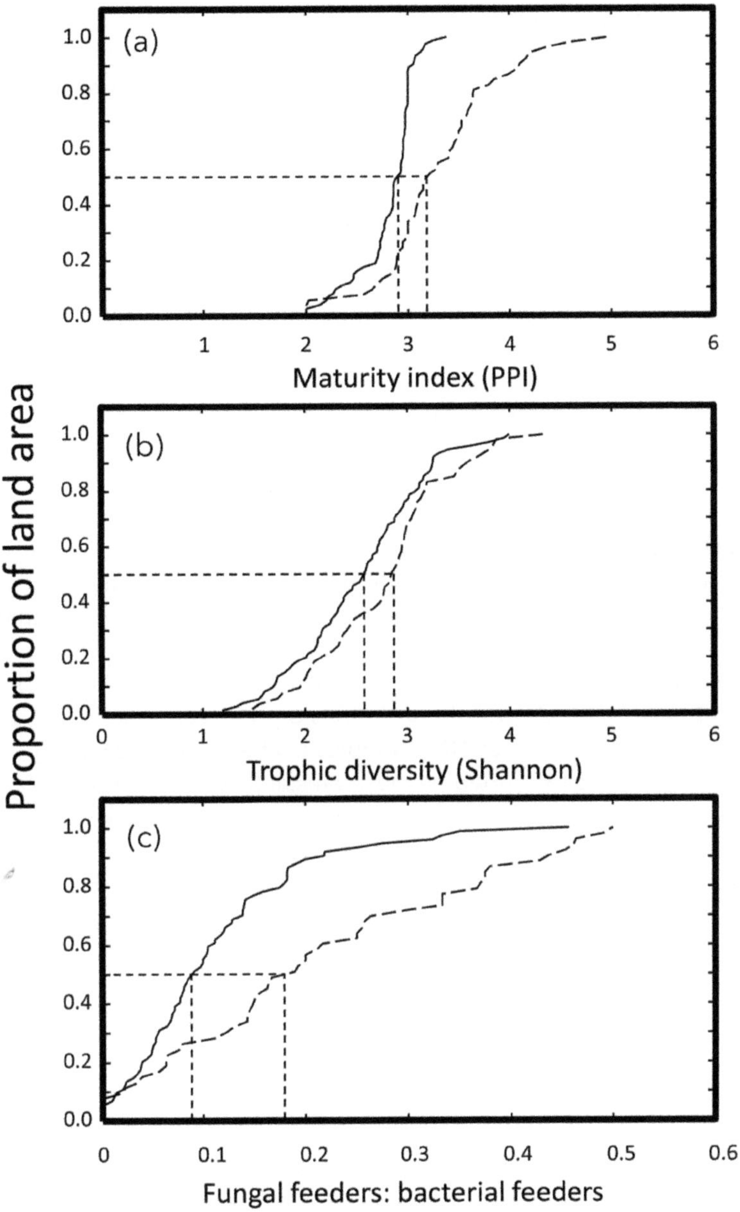

Figure 4 Values of PPI and F:B were greater in perennial than annual crops, but there was no difference in trophic diversity values. Cumulative distribution function of the (a) maturity index of plant-feeding nematodes, (b) Shannon trophic diversity, and (c) fungal-to bacterial-feeding (= fungivores to bacterivores) nematodes for soybean (solid line) and perennial (dashed line) crops sampled in North Carolina. Dotted lines represent median values. Reprinted from Applied Soil Ecology, vol 1, Neher, D.A. and Campbell, C. L., Nematode communities and microbial biomass in soils with annual and perennial crops, pp. 17–28, Copyright (1994), with permission from Elsevier.

error (Neher et al., 1995; Neher and Campbell, 1996). Based on those studies, maturity (ecological succession) indices differentiated among sampling sites better and more efficiently than diversity indices and are more appropriate for a regional and/or national monitoring program (Neher et al., 1995). Better performing indices require fewer samples which save time and cost.

3.3 Sentinel taxa

Maturity indices depend on accurate assignment of *cp* values, as these classifications form the basis for understanding nematode community responses to environmental changes. Specific nematode taxa may behave or respond differently than expected based on their assigned trophic or *cp* group (Table 5). Inaccurate *cp* assignment clouds the interpretation of indices. Empirical confirmation of hypothesized ecology has potential to greatly improve the sensitivity and interpretation of community indices.

The solution is to validate these assignments and identify sentinel taxa. Potential sentinel taxa can be screened first by using meta-analysis of published data (e.g. see Table 6) to determine which taxa respond consistently to specific type(s) and magnitude of disturbance (e.g. Fiscus and Neher, 2002, Zhao and Neher, 2013). Candidate taxa can be tested empirically by natural history experiments (e.g. Li et al., 2005) or independent assessments of sensitivity to disturbance that do not rely on life history characteristics, e.g., general biomarker approach such as DNA adducts (Neher and Stürzenbaum, 2006). Adducts are a complex formed by the chemical binding of DNA to pollutants such as polycyclic aromatic hydrocarbons. If mis-repaired, they can lead to deleterious mutations. This approach was tested on *C. elegans* as a model system but needs to be extended to a full range of nematodes within and among *cp* categories.

Table 5 Discrepancies between *CP* values and response to disturbance

Genus	*CP*	Trophic group	Sensitive	Tolerant	Reference[a]
Eumonhystera	1	Bacterivore	√		2
Acrobeles	2	Bacterivore	√		1
Plectus	2	Bacterivore	√		1
Clarkus	4	Predator		√	2
Enchodelus	4	Omnivore		√	2
Tylencholaimus	4	Fungivore		√	2
Aporcelaimus	5	Predator		√	2

[a]1: Korthals et al. 1998, 2: Fiscus and Neher 2002.

Table 6 Sentinel taxa that consistently or directly respond to specific types of disturbance

Type of disturbance	Decrease	Increase
Cultivation	*Diphtherophora*[4], *Eucephalobus*[2], *Eudorylaimus*[2], *Heterocephalobus*, *Prismatolaimus*[4], *Tylenchorhynchus*[4], *Wilsonema*[2]	*Anatonchus*[2], *Chiloplacus*[2], *Clarkus*[2], *Ditylenchus*[4], *Dorylaimoides*[4], *Psilenchus*[4], *Tylencholaimellus*[2]
Synthetic chemical fertilizers	*Helicotylenchus*[3], *Plectus*[4]	
Organic fertilization	*Cephalobus*[3]	*Cruznema*[4], *Mesodorylaimus*[4], *Mesorhabditus*[4], *Nygolaimus*[4], *Odontolaimus*[2], *Prismatolaimus*[2], *Pratylenchus*[3]
Herbicides/ insecticides	*Alaimus*[2]	*Diphtherophora*[2], *Eucephalobus*[2], *Monhystera*[2]
Cadmium	*Heterocephalobus*[4]	*Diploscapter*[4], *Eucephalobus*[4], *Meloidogyne*[4], *Pratylenchus*[4], *Prismatolaimus*[4]
Chromium	*Aporcelaimellus*[4]	*Criconemoides*[1], *Paratylenchus*[1]
Copper	*Heterocephalobus*[4]	*Aphelenchoides*[4], *Chiloplacus*[1], *Pratylenchus*[1], *Seinura*[4]
Nickle	*Cephalobus*[4]	*Plectus*[1], *Wilsonema*[4]
Zinc		*Aphelenchoides*[4], *Cephalobus*[1], *Filenchus*[4], *Malenchus*[4], *Prismatolaimus*[4], *Tylenchus*[1], *Tylenchorhynchus*[4]

[1]Ekschmitt and Korthals (2006).
[2]Fiscus and Neher (2002).
[3]Neher and Olson (1999).
[4]Zhao and Neher (2013).

3.4 Limitations and future research

Trophic assignments are relatively coarse and can be ambiguous because genera within a trophic group or family can respond differently to the same disturbance (Fiscus and Neher, 2002). Additional tools such as stable isotope analysis have proved useful to study nematode trophic interactions, especially in complex ecosystems where direct observations are difficult. For example, Darby and Neher (2012) used stable isotopes to demonstrate that nematodes in biological soil crusts were consuming biologically fixed N from cyanobacteria.

Feeding habits and reproductive potential are known to vary within families (Yeates 2003). In Tylenchidae, e.g., there are genera (e.g. *Tylenchus*) that are often considered plant associates (Yeates 2010) and other genera (*Filenchus*) that are fungivores (Okada et al., 2005). Whether Tylenchidae is grouped with plant-parasitic or free-living nematodes in maturity indices can heavily influence the results and interpretations of soil health and ecosystem disturbance.

Subsequent studies are needed to verify the sentinel status of specific taxa and to determine the geographic or ecological range of their utility. These experiments should be conducted to quantify impacts of multiple and interacting management practices on biodiversity, nutrient cycling, pest populations, and plant productivity (Neher, 2010). Once sentinel taxa are validated, one can disregard the inconsistent and more cosmopolitan taxa from index calculations. Including them just makes it more difficult to detect true changes ('signal') separate from natural variation ('noise'). The result is fewer taxa that need to be identified and indices that have greater precision and accuracy. Furthermore, this information is necessary to improve molecular kits, so they are approachable across a broader spectrum of scientists, complete with interpretation tools. There could even be designer kits based on ecosystem type and by major land resource region (Neher et al., 1998).

4 Relating analyses to soil health

Healthy soils typically exhibit a high abundance and biodiversity of free-living nematodes in complex food webs with long chains and feedback loops, while containing a relatively low proportion of herbivores or pathogens (Neher, 2001a). The utility of using nematodes as indicators of soil health is contingent on their ability to reflect ecological processes such as N mineralization (Neher, 2010) or decomposition. There have been modeling attempts to link flows of C and N to predict function (Hunt et al., 1987, Ferris et al. 2001), as well as experiments with empirical measures of nematode communities and function (Neher et al., 2012).

Calibration and interpretation of nematode community indices require some sort of relatively 'undisturbed' reference or healthy soil when used for environmental monitoring, conservation, and restoration. The choice of baselines for intensively managed agricultural practices on soil health should reflect the native vegetation of a region. For example, consensus is building for use of long-term pastures or managed grasslands as baselines for agricultural fields in the USA (Neher et al., 1998), South Africa (Girgan et al., 2020), and New Zealand (Yeates and Bongers, 1999). In addition, soil properties such as texture, salinity, pH, and organic matter influence nematode community assemblages. If ignored, they can inflate inherent variability and thus interfere with detection of treatment differences. However, quantifying them and treating them as covariables in statistical analysis can increase the likelihood of detecting treatment effects (Barbercheck et al., 2009; van den Hoogen et al., 2019).

4.1 Linking to N mineralization

Nematodes make N available to plants by excreting excess NH_4^+ and indirectly by grazing decomposers including bacteria and fungi (Ingham et al., 1985; Mosdossy et al., 2004b). Under field conditions, bacterivorous and predatory nematodes are estimated to contribute (directly and indirectly) about 8% and 19% of N mineralization in conventional and integrated farming systems, respectively (Beare, 1997). These estimates were confirmed in a regional study (Neher et al., 2012). Net mineral N is regulated by a multi-trophic relationship among C:N of plant litter, microbial biomass, and the successional status of nematode communities. More immobilization occurred when *cp*3 and *cp*4 nematodes were more abundant than *cp*1 or *cp*2 nematodes regardless of abundant microbial biomass (Mosdossy et al., 2024a). Intensive disturbance decouples the contribution of available N to plants by nematodes (de Ruiter et al., 1993).

Notably, there are time lags between relative abundance of nematodes and subsequent availability of N. For example, negative correlations were found with 1, 2, or more months of lag time between nitrate or NH_4^+concentrations and nematode indices (e.g. Neher, 1999; Neher et al., 2012). Nematode community structure may thus depend more on the past than current environmental conditions. Understanding these time lags is crucial for developing effective sampling schedules to assess grazer-food resource relationships. Ignoring them may lead to misinterpretations, suggesting an opposite or negligible relationship.

Analyses of N mineralization in food webs typically focus on bottom-up effects. For example, abundances of bacterivores are correlated positively with N fertility (e.g. Zhang et al., 2017), as are plant-parasites (Todd, 1996). However, top-down interactions influence N mineralization and energy flux (Thakur and Geisen, 2019). For example, predators may eat bacterivores through a 'trophic cascade' (Wardle and Yeates, 1993), which explains why they are responsible for a notable amount of N mineralization. Indeed, inclusion of network topology in food web analysis can drastically affect interpretation of functional indicators, including nematode trophic groups (Potapov et al., 2022).

4.2 Linking to decomposition and climate change models

Given their abundance and ubiquity, nematodes are major contributors to C cycling in soil and general ecosystem function (van den Hoogen et al., 2019). Studying nematodes in the context of the soil food web under different agroecological management systems can indicate soil C levels, plant productivity, and effects on processes that drive function (Zhang et al., 2017; Jochum and Eisenhauer, 2022). Indeed, decomposition of litter in cultivated soil can result in litter-specific fungal and nematode communities that affect bioavailable N for crops depending on whether litter was surface-mulch or incorporated into the soil (Mosdossy et al., 2024a, 2024b).

Historically, climate change models relied mostly on soil bacteria and fungi and litter chemistry to estimate carbon pools and flows and have largely ignored the contribution of soil fauna. However, soil fauna are increasingly identified as important drivers of soil organic matter formation, C mineralization, and soil health in general. Therefore, new ways of incorporating them into C models, particularly through the use of food web models are becoming apparent (Filser et al., 2016; Grandy et al., 2016; Buchkowski et al., 2022; Barreto et al., 2024). Recently, there has been an effort to map the biogeography of nematodes at a global scale as the first steps toward global biogeochemical models and predicting elemental cycling under current future climate scenarios (van den Hoogen et al., 2019).

Climate change can come in the form of increased temperature or altered precipitation leading to water stress. In desert experiments, nematodes are sensitive to increased precipitation frequency and elevated temperatures (more than protozoa) so climate change could change their relative contribution to N mineralization (Darby et al., 2011). Nematodes can survive temperatures up to 60°C as long as they are allowed to go into an anhydrobiotic state (Darby, 2008). If forced to be active because of recent precipitation, their sensitivity to elevated temperatures increases (Darby, 2008). Thus, nematodes are valuable indicators of changes in soil ecosystem processes as climate extremes become more common.

The EI, CI, SI approach can be translated from numbers to units of C creating a MF which can be used to estimate C flux within food webs (Ferris, 2010). MF defines metabolism as the sum of production and respiration derived from biomass estimates expressed in C units. Body mass is established based on weight, length, and diameter of the vermiform body shape using a formula proposed by Andrássy (1956). Body mass estimates are available through publications (e.g. Ferris et al., 2001; Neher and Weicht, 2013) that populate the Nemaplex database, available online (see Section 8). Flux can then be estimated with tools such as the R package, *fluxweb* allowing relatively easy analysis of energetic food webs (Gauzens et al., 2019). Estimates of nematode C fluxes can be linked to decomposition and climate change models and can be a powerful approach to quantify ecosystem services by accounting for C that is being transferred through trophic interactions (Barnes et al., 2018). For example, herbivory may be detected by loss of root biomass but C flux from the plant is not accounted for comprehensively compared to quantifying C assimilation and respiration by plant-parasitic nematodes (Barnes et al., 2018). Carbon and nitrogen fluxes of nematodes and the broader soil food web can be modeled using the R package, *soilfoodwebs* (Buchkowski et al., 2023). This is a true energetic food web model (sensu Moore and de Ruiter 2012), thus improving upon *fluxweb* to more accurately quantify C flux through nematodes, which can allow for more accurate predictions of ecosystem function in the face of climate change.

5 Impacts of particular agronomic practices (crops/soil and water management)

Nematode community indices are sensitive to management practices (Neher and Olson, 1999). For example, nematode MF fluctuate based on management techniques (e.g. litter mulching, tillage) and seasonal changes (Ferris, 2010), making them strong indicators of soil organic C, microbial biomass, and respiration, especially in agricultural systems (Schmidt et al., 2020; Luo et al., 2021). Bacterivorous nematode MF tends to increase with litter amendments, particularly in fall, reflecting accelerated C flux (Ewald et al., 2020). In contrast, the MF of fungivorous nematodes remains stable regardless of litter amendments, indicating a constant C flux (Ewald et al., 2020). However, organic amendments and mineral fertilizers can push the soil food web toward earlier successional stages (Ferris and Bongers, 2006). Greater diversity and abundance often correspond to greater C utilization and more efficient energy transfer within the nematode food web (Gauzens et al., 2019).

There are two major types of disturbances in agricultural ecosystems, physical and chemical, which have contrasting effects on the soil food web.

5.1 Physical

Cultivation is more damaging to the soil food web than chemical applications (Neher, 2010; Bongiorno et al., 2019). Mechanical tillage shreds organic matter, increasing its surface area for colonization by decomposer bacteria and fungi, which accelerates decomposition. Decomposition rates are 1–1.4 times faster in tilled than in no-till systems, releasing nutrients more quickly and decreasing food web complexity (Beare, 1997). Tillage also disrupts soil structure and faunal habitats, decreasing populations of fungi and fungivores (Neher, 2010). In contrast, no-till practices support later ecological succession and greater trophic diversity within nematode communities (Neher, 2010).

In a case study, no-tillage and a monoculture of susceptible soybeans developed natural suppression to soybean cyst nematode (*Heterodera glycines*) (Chen, 2007). However, once the soil was cultivated or rotated to corn, this suppression was lost (Neher et al., 2019, Fig. 3). Mechanisms of suppression have been attributed to nematode-trapping fungi (*Hirsutella rhossiliensis*) (Chen, 2007) and disruption of a free-living nematode community containing a common core of bacterivores (*Wilsonema*) and omnivore-predators (*Aporcelaimus* or *Aporcelaimum*, *Clarkus*, *Dorylamoides*, *Eudorylaimus*, and *Paraxonchium*) (Neher et al., 2019). Traditionally, monocultures of a susceptible host typically promote disease. In response, rotation to a non-host is recommended. Rotation may decrease populations of the plant-parasitic nematode, but it also reduced food web complexity (trophic diversity, SI), and increased values of EI compared to a naturally suppressive soil (Neher et al., 2019). Relatively later stages of succession in the nematode community are congruent with hypotheses about soils with high organic matter content (e.g. 6.64 ± 0.97% in Neher et al., 2019) supporting high densities or diversities of soil microbes and most notably, high abundance and diversity of antagonistic populations (Weller et al., 2002; Bonanomi et al., 2010). Numerous studies have found that organic matter correlates directly with increased numbers of nematode antagonists (Oka, 2010).

5.2 Chemical

Chemicals can be in the form of nutrients (e.g. synthetic fertilizers, organic matter) or pollutants (e.g. pesticides, metals, petroleum products). Application of either can be disruptive to soil food webs and result in a rapid increase in bacteria and *cp*1 bacterivorous nematodes, thus increasing EI values (De Goede et al., 1993; Ferris et al., 2001). This enrichment affect is short-lived (e.g. Ettema and Bongers, 1993; Yeates and Bird, 1994), having minimal impact on nematode communities over the long term (Dmowska and Ilieva, 1995).

The type of fertilizer and pest management strategies are factors that differentiate conventional from organic agriculture, but these differences might not be reflected in nematode indices if the frequencies of disturbance are similar in both management systems. For example, similar values of MI25, ΣMI, and F:B were observed between soils managed organically or conventionally with cultivation and annual crops (Neher, 1999). Another study also observed similar proportions of *cp*1, *cp*4, and *cp*5 in organic or conventional cropping systems (Yeates et al., 1997).

Understanding the individual and combined effects of cultivation and chemical applications on nematode communities is essential, as these factors often produce contradictory results. In an 18-year farming trial in southeast Nebraska, chemical treatments were organized into a disturbance hierarchy: O (manure only), F (mineral fertilizer only), HF (mineral fertilizer plus herbicides), and HFI (mineral fertilizer, herbicides, and insecticides) (Neher and Olson, 1999). All fields were cultivated but varied in organic inputs and pesticide use. The relative abundance of bacterivores, plant-parasites, and PPI were greater and MI values smaller after organic than synthetic fertilizer applications. It appears that herbicide and insecticide applications shift nematode communities to earlier successional stages more than cultivation or organic amendments do. This study highlights the distinct effects of pesticides and nutrient sources on nematode community composition.

5.3 Limitations and future research

Caution is needed when using MF as indicators of ecosystem function, as online databases rely on the size and biomass of adult females, without accounting for age-structured populations of variability in size or ecology. This is true for the user-friendly online tool, The Nematode Indicator Joint Analysis (NINJA), for calculating various nematode community indices (Sieriebriennikov et al., 2014). However, NINJA-computed MF reflects the potential lifetime C utilization, not the C utilization or estimate of ecosystem function at one point in time. Furthermore, Nemaplex is the most comprehensive database available to retrieve biometrics data as well as respiration and production rates and other life history traits, all of which are used to model C flux. Yet, this database contains a limited number of species.

Nematodes are excellent candidates for monitoring energy flux dynamics because they occupy multiple trophic levels (Moore and De Ruiter, 2012). Yet community and ecosystem ecology have not fully embraced the use of nematodes as indicators of energy flux, which links food web theory with biodiversity-ecosystem functioning theory (Jochum et al., 2021). Embracing ubiquitousness, diversity, and positioning in the soil food web by nematodes could lead to a more comprehensive understanding of energy flux in agroecosystems, thus informing best practices for soil health.

6 Future trends

Metabarcoding is an exciting new technique showing promise of solving the conundrum of not currently having barcodes or universal markers for free-living nematodes. Tailoring the technique to nematodes could overcome barriers that prevent nematodes from being utilized widely in soil health programs and monitoring programs.

Empirical confirmation of hypothesized ecology has potential to greatly improve the sensitivity and interpretation of community indices. For example, reports of taxa that do not respond to disturbance as predicted necessitate empirical testing of *cp* assignments. Metadata analyses of existing data identify potential sentinel taxa that can be verified by natural history experiments and general biomarkers.

The multitrophic nature of nematode communities make them good candidates for testing ecological theory initially limited to above-ground systems. For example, they also make good candidates for energy flux models which can link food web theory with biodiversity-ecosystem functioning theory (Jochum et al., 2021).

7 Summary

The concept of nematodes as indicators of soil health represents a major shift in emphasis in the science of nematology. Nematodes make good indicators because they reside a trophic level above microbes in the food chain and integrate physical, chemical, and biological properties. The best performing indices are based on successional maturity and food web complexity. There is a paucity of laboratories that can morphologically identify free-living nematodes at least to family. Therefore, many nematologists are moving away from traditional morphological methods and toward molecular methods citing benefits like speed and cost-effectiveness, but some important challenges need to be overcome before those methods are readily available. For example, we need metabarcoding that detects all nematode genera and aligns with morphological identification and ecological traits. The precision of the indices would be improved with empirical verification to generate a proven set of sentinel taxa. Community indices are useful as indicators of soil health because they are responsive to management practices, including cultivation, fertilization, and pest management practices. Beyond applications in soil health, nematode communities can make good test subjects for applications including ecological theory and global climate change models.

8 Where to look for further information

The following articles provide a good overview of the subject:

- Du Preez, G., et al. (2022), 'Nematode-based indices in soil ecology: application, utility, and future directions', *Soil Biol. Biochem.*, 169, 108640. https://doi.org/10.1016/j.soilbio.2022.108640.
- Ferris, H. (1993), 'New frontiers in nematode ecology', *J. Nematol.*, 25(2), 374–382.
- Hodda, M. (2022), 'Phylum nematoda: feeding habits for all valid genera using a new, universal scheme encompassing the entire phylum, with descriptions of morphological characteristics of the stoma, a key, and discussion of the evidence for trophic relationships', *Zootaxa*, 5114(1), 318–451. https://doi.org/10.11646/zootaxa.5114.1.3.
- Kakouli-Duarte, T. Korthals G.W., Sánchez Moreno, S., du Preez, G. and de Goede, R. (2025) *Nematodes as Environmental Indicators: From Theory to Practice*, 2nd edition. CABI, Wallingford, UK.
- Neher, D.A. (1999), 'Soil community composition and ecosystem processes: comparing agricultural ecosystems with natural ecosystems', *Agroforest. Syst.,* 45, 159–185. https://doi.org/10.1023/A:1006299100678.
- Neher, D.A. (2001a), 'Role of nematodes in soil health and their use as indicators', *J. Nematol.,* 33(4), 161–168.
- Neher, D.A. (2001b), 'Nematode communities as ecological indicators of agroecosystem health'. In S.R. Gliessman (Ed), *Agroecosystem*

Sustainability: Developing Practical Strategies. CRC Press, Boca Raton, FL, pp. 105–120.

- Neher, D.A. (2010), 'Ecology of plant and free-living nematodes in natural and agricultural soil', *Annu. Rev. Phytopathol.,* 48(1), 371–394. https://doi.org/10.1146/annurev-phyto-073009-114439.
- Neher, D.A. and Powers, T.O. (2023), 'Nematodes', *Encyclopedia of Soils in the Environment*, 11, 105–111. https://doi.org/10.1016/B978-0-12-822974-3.00038-0.
- Neher, D.A. (2023), 'Moving up within the food web: protists, nematodes and other microfauna'. In N. Uphoff and J. Thies (Eds), *Biological Approaches to Regenerative and Resilient Soil Systems*, 2nd edition. CRC, Boca Raton, FL, pp. 157–168.
- Yeates, G.W. (2010), 'Nematode in ecological webs'. In Wiley (Ed), *Encyclopedia of Life Sciences*. John Wiley & Sons, Chichester, UK. https://doi.org/10.1002/9780470015902.a0021913.

Following are seminal articles or books which have shaped the subject:

- Bongers T. (1990), 'The maturity index: an ecological measure of environmental disturbance based on nematode species composition', *Oecologia,* 83(1), 14–19. https://doi.org/10.1007/BF00324627.
- Ettema C.H. and Bongers T. (1993), 'Characterization of nematode colonization and succession in disturbed soil using the maturity index', *Biol. Fert. Soils* 16, 79–85. https://doi.org/10.1007/Bf00369407.
- Ferris, H., et al. (2001), 'A framework for soil food web diagnostics: extension of the nematode faunal analysis concept', *Appl. Soil Ecol.* 18(1), 13–29. https://doi.org/10.1016/S0929-1393(01)00152-4.
- Ferris, H. (2010), 'Form and function: Metabolic footprints of nematodes in the soil food web', *Eur. J. Soil Biol.*, 46(2), 97–104. https://doi.org/10.1016/j.ejsobi.2010.01.003.
- Freckman D.W. and Caswell E.P. (1985), 'The ecology of nematodes in agroecosystems', *Annu. Rev. Phytopathol.* 23(1), 275–296.
- Freckman D.W. (1988), 'Bacterivorous nematodes and organic-matter decomposition', *Agricult. Ecosys. Environ*. 24(1–3), 195–217. https://doi.org/10.1016/0167-8809(88)90066-7.
- Ingham, R.E., et al. (1985), 'Interactions of bacteria, fungi, and their nematode grazers: effects on nutrient cycling and plant growth', *Ecol. Monogr.,* 55(1), 119–140. https://doi.org/10.2307/1942528.
- Wardle, D.A., et al. (1995), 'The detritus food web and the diversity of soil fauna as indicators of disturbance regimes in agroecosystems', *Plant Soil,* 170, 35–43.

- Wasilewska L. (1979), 'The structure and function of soil nematode communities in natural ecosystems and agrocenoses', *Pol. Ecol. Stud.*, 5, 97–145.
- Yeates G.W. (1994), 'Modification and qualification of the nematode maturity index', *Pedobiologia,* 38, 97–101. https://doi.org/10.1016/s0031-4056(24)00113-6.
- Yeates G.W., et al. (1993), 'Feeding habits in soil nematode families and genera—an outline for soil ecologists', *J. Nematol.,* 25(3), 315–331.

References that provide useful methods are as follows:

- Buchkowski, R.W., et al. (2023), 'soilfoodwebs: an R package for analyzing and simulating nutrient fluxes through food webs', *Eur. J. Soil Biol.*, 119, 103556. https://doi.org/10.1016/j.ejsobi.2023.103556.
- Coyne, D.L., et al. (2014), *Practical Plant Nematology: A Field and Laboratory Guide*, 2nd edition. SP-IPM Secretariat, International Institute of Tropical Agriculture (IITA), Cotonou, Benin. https://cgspace.cgiar.org/server/api/core/bitstreams/48d92e60-d14b-4cb1-8ba8-2b008da74189/content; accessed September 18, 2024.
- Freckman, D.W. and Baldwin, J.G. (1990), 'Nematodes'. In D.L. Dindal (Ed), *The Biology Guide*, John Wiley, New York, NY, pp. 155–200.
- Gauzens, B., et al. (2019), 'fluxweb: An R package to easily estimate energy fluxes in food webs', *Methods Ecol. Evol.*, 10(2), 270–279. https://doi.org/10.1111/2041-210x.13109.
- Neher, D.A. and Darby, B.J. (2006), 'Computation and application of nematode community indices: general guidelines'. In A. Eyualem, et al. (Eds), *Freshwater Nematodes: Ecology and Taxonomy.* CABI, Wallingford, UK, pp. 211–222. https://doi.org/10.1079/9780851990095.0211.
- Neher, D. and B. Darby. (2009), 'General community indices that can be used for analysis of nematode assemblages'.l In M. Wilson and T. Kakouli-Duarte (Eds), *Nematodes as Environmental Indicators.* CABI, Wallingford, UK, pp. 107–123.
- Neher, D.A. and Darby, B.J. (2025), 'General community indices that can be used for analysis of nematode assemblages' In T. Kakouli-Duarte Korthals G.W., Sánchez Moreno, S., du Preez, G. and de Goede, R (Eds), *Nematodes as Environmental Indicators: From Theory to Practice*, 2nd edition. CABI, Wallingford, UK. https://doi.org/10.1079/9781845933852.01.
- Nemaplex.UCDavis.edu: Revision Date: 10/07/2024; accessed October 10, 2024.
- Schmidt, J.H., et al. (2025), 'Indices developed specifically for analysing nematode assemblages'. In T. Kakouli-Duarte Korthals G.W., Sánchez Moreno, S., du Preez, G. and de Goede, R (Eds), *Nematodes as Environmental Indicators: From Theory to Practice*, 2nd edition. CABI, Wallingford, UK.

- Sieriebriennikov, B., et al. (2014), 'NINJA: an automated calculation system for nematode-based biological monitoring', *Eur. J. Soil Biol.* 61, 90–93. https://doi.org/10.1016/j.ejsobi.2014.02.004. Within the tool (https://shiny.wur.nl/ninja/), one can compute all the indices described in this chapter, and there is a simple introductory video, https://www.youtube.com/watch?v=l7njoxvMW6Q.
- Van Bezooijen, J. (2006), *Methods and Techniques for Nematology*. Wageningen University, The Netherlands. https://www.scribd.com/document/234069625/Methods-and-TechniquesforNematology; accessed October223, 2025.

Key journals are as follows:

- Applied Soil Ecology
- Soil Biology and Biochemistry
- European Journal of Soil Biology
- Pedobiologia
- Journal of Nematology, https://sciendo.com/journal/JOFNEM
- Nematology (Brill), https://brill.com/view/journals/nemy/nemy-overview.xml?language=en
- Nematropica, https://journals.flvc.org/nematropica

Key conferences are as follows:

- Soil Ecology Society, https://www.soilecologysociety.com/
- European Society of Nematologists, https://www.esn-online.org/
- ONTA: Organization of Nematologists of Tropical Americas, https://ontaweb.org/
- SON: Society of Nematologists, https://www.nematologists.org/
- Soil Ecology section of Ecological Society of America, https://www.esa.org/soilecology/home/

Major international research projects are as follows:

- Global Soil Biodiversity Initiative (https://www.globalsoilbiodiversity.org/)
- NETSOB, International Network on Soil Biodiversity (van der Putten et al., 2023)
- SoilBON (Guerra et al., 2021)
    - https://soilbonfoodweb.org/ – protocols & Manuals https://soilbonfoodweb.org/protocols-and-manuals/
    - https://soilbonfoodweb.org/wp-content/uploads/2024/02/SBF-Team-Nematode-extraction_v1.6.pdf
    - https://youtu.be/8Ennbh0KkUQ?si=-I4TV7sapPuRzMuC

Top research center to keep up with research trends are as follows:

- Ghent University – Belgium
- Wageningen University – The Netherlands
- University of Florida – Gainesville, USA
- University of California – Davis, USA
- University of California – Riverside, USA

9 References

Abebe, E., et al. (2006), *Freshwater Nematodes: Ecology and Taxonomy*. CABI, Wallingford, UK.

Ahmed, M., et al. (2019), 'Metabarcoding of soil nematodes: the importance of taxonomic coverage and availability of reference sequences in choosing suitable marker(s)', *Metabarcoding and Metagenomics*, 3, e36408. https://doi.org/10.3897/mbmg.3.36408.

Andrássy, I., (1956), 'The determination of volume and weight of nematodes', *Acta Zoologica Academiae Scientiarium Hungaricae*, 2, 1–15.

Andrássy, I. (2005), *Free-living nematodes of Hungary, (Nematoda Errantia) Vol. 1*. Hungarian Natural History Museum and Systematic Zoology Research Group of the Hungarian Academy of Sciences, Budapest.

Andrássy, I. (2007), *Free-living nematodes of Hungary, (Nematoda Errantia) Vol. 2*. Hungarian Natural History Museum and Systematic Zoology Research Group of the Hungarian Academy of Sciences, Budapest.

Andrássy, I. (2009), *Free-living nematodes of Hungary, (Nematoda Errantia) Vol. 3*. Hungarian Natural History Museum and Systematic Zoology Research Group of the Hungarian Academy of Sciences, Budapest.

Ayoub, S.M. (1980), *Plant Nematology, An Agricultural Training Aid*. California Department of Food and Agriculture, NemaAid Publications, Sacramento, California.

Barbercheck, M.E., et al. (2009), 'Response of soil invertebrates to disturbance across three resource regions in North Carolina', *Environ. Monit. Assess.*, 152, 283–298. https://doi.org/10.1007/s10661-008-0315-5.

Barker, K.R., et al. (1969), 'Effects of storage temperature and extraction procedure on recovery of plant-parasitic nematodes from field soils', *J. Nematol.*, 1(3), 240–247.

Barker, K.R. (1985a), 'Sampling nematode communities'. In K.R. Barker, et al. (Eds), *An Advanced Treatise on Meloidogyne Vol. II: Methodology*. North Carolina State University Graphics, Raleigh, NC, pp. 2–17.

Barker, K.R. (1985b), 'Nematode extraction and bioassays'. In K.R. Barker, et al. (Eds), *An Advanced Treatise on Meloidogyne Vol. II: Methodology*. North Carolina State University Graphics, Raleigh, NC, pp. 19–35.

Barnes, A.D., et al. (2018), 'Energy flux: the link between multitrophic biodiversity and ecosystem functioning', *Trends Ecol. Evol.*, 33(3), 186–197.

Barreto, C., et al. (2024), 'Restructuring of soil food webs reduces carbon storage potential in boreal peatlands', *Soil Biol. Biochem.*, 193, 109413. https://doi.org/10.1016/j.soilbio.2024.109413.

Beare, M.H. (1997), 'Fungal and bacterial pathways of organic matter decomposition and nitrogen mineralization in arable soil'. In L. Brussaard and R. Ferrera-Cerrato (Eds), *Soil Ecology in Sustainable Agricultural Systems*. Lewis, Boca Raton, FL, pp. 37–70.

Beare, M.H., et al. (1992), 'Microbial and faunal interactions and effects on litter nitrogen and decomposition in agroecosystems', *Ecol. Monogr.*, 62(4), 569–591. https://doi.org/10.2307/2937317.

Beaumelle, L., et al. (2021), 'Soil fauna diversity and chemical stressors: a review of knowledge gaps and roadmap for future research', *Ecography* 44(6), 845–859. https://doi.org/10.1111/ecog.05627.

Bekker, S., et al. (2022), Nematode assemblages in conservation and conventional maize fields: a South African baseline study. *J. Plant Dis. Protect.*, 129(5), 1235–1248. https://doi.org/10.1007/s41348-022-00619-2.

Bik, H.M., et al. (2013), 'Intra-genomic variation in the ribosomal repeats of nematodes', *PLoS One*, 8(10), e78230. https://doi.org/10.1371/journal.pone.0078230.

Bonanomi, G., et al. (2010), 'Identifying the characteristics of organic soil amendments that suppress soilborne plant diseases', *Soil Biol. Biochem.*, 42(2), 136–144. https://doi.org/10.1016/j.soilbio.2009.10.012.

Bongers, T. (1990), 'The maturity index: an ecological measure of environmental disturbance based on nematode species composition', *Oecologia,* 83(1), 14–19. https://doi.org/10.1007/BF00324627.

Bongers, T. (1994), *De Nematoden van Nederland. Koninkleijke Nederlandse Natuuhistorische Vereniging*. Utrecht, The Netherlands (an English translation of the key itself is available).

Bongers, T. (1999), 'The maturity index, the evolution of nematode life history traits, adaptive radiation and cp-scaling', *Plant Soil*, 212(1), 13–22. https://doi.org/10.1023/A:100457190042.

Bongers, T. and Bongers, M. (1998), 'Functional diversity of nematodes', *Appl. Soil Ecol.*, 10(3), 239–251.

Bongers, T., et al. (1991), 'Interpretation of disturbance-induced maturity decrease in marine nematode assemblages by means of the Maturity Index', *Mar. Ecol. Prog. Ser.*, 76(2), 135–142. https://doi.org/10.3354/Meps076135.

Bongers, T., et al. (1995), 'Proposed changes of c-p classification for nematodes', *Russ. J. Nematol.*, 3(1), 61–62.

Bongers, T., et al. (1997), 'Inverse relationship between the nematode maturity index and plant parasite index under enriched nutrient conditions', *Appl. Soil Ecol.* 6(2), 195–199. https://doi.org/10.1016/S0929-1393(96)00136-9.

Bongiorno, G., et al. (2019), 'Reduced tillage, but not organic matter input, increased nematode diversity and food web stability in European long-term field experiments', *Mol Ecol.*, 28(22), 4987–5005. https://doi.org/10.1111/mec.15270.

Bowles, J., et al. (1992), 'Genetic-variants within the genus *Echinococcus* identified by mitochondrial-DNA sequencing', *Mol. Biochem. Parasit.* 54(2), 65–174. https://doi.org/10.1016/0166-6851(92)90109-W.

Buchkowski, R.W., et al. (2022), 'Cannibalism has its limits in soil food webs', *Soil Biol. Biochem.*, 172, 108773. https://doi.org/10.1016/j.soilbio.2022.108773.

Buchkowski, R.W., et al. (2023), 'soilfoodwebs: an R package for analyzing and simulating nutrient fluxes through food webs', *Eur. J. Soil Biol.*, 119, 103556. https://doi.org/10.1016/j.ejsobi.2023.103556.

Chen, S. (2007), 'Suppression of *Heterodera glycines* in soils from fields with long-term soybean monoculture', *Biocontrol Sci. Techn.*, 17(2), 125–134. https://doi.org/10.1080/09583150600937121.

Coyne, D.L., et al. (2014), *Practical Plant Nematology: A Field and Laboratory Guide*, 2nd edition. SP-IPM Secretariat, International Institute of Tropical Agriculture (IITA), Cotonou, Benin. https://cgspace.cgiar.org/server/api/core/bitstreams/48d92e60-d14b-4cb1-8ba8-2b008da74189/content; accessed September 18, 2024.

Darby, B.J. (2008), *Influence of Altered Temperature and Precipitation on Desert Microfauna and Their Role in Mediating Soil Nutrient Availability* (Doctoral dissertation, University of Nevada, Las Vegas). Graduate College Dissertations and Theses, 64. https://scholarworks.uvm.edu/graddis/64.

Darby, B.J. and Neher, D.A. (2012), 'Stable isotope composition of microfauna supports the occurrence of biologically fixed nitrogen from cyanobacteria in desert soil food webs', *J. Arid Environ.* 85, 76–78. https://doi.org/10.1016/j.jaridenv.2012.06.006.

Darby, B.J., et al. (2007), 'Ecological maturity of biological soil crust affects belowground nematode community composition', *Appl. Soil Ecol.* 35, 203–212. https://doi.org/10.1016/j.apsoil.2006.04.006.

Darby, B.J., et al. (2011), 'Few short-term effects of elevated soil temperature and increased frequency of summer precipitation on the abundance and taxonomic diversity of desert soil micro- and meso-fauna', *Soil Biol. Biochem.*, 43(7), 1474–1481. https://doi.org/10.1016/j.soilbio.2011.03.020.

Darby, B.J., et al. (2013), 'High-throughput amplicon sequencing of rRNA genes requires a copy number correction to accurately reflect the effects of management practices on soil nematode community structure', *Mol. Ecol.* 22(21), 5456–5471. https://doi.org/10.1111/mec.12480.

de Goede, R.G.M., et al. (1993), 'Graphical presentation and interpretation of nematode community structure, CP triangles', *Meded. Fac. Landbouww. Univ. Gent*, 58, 743–750.

de Ley, P., et al. (2003), *Identification of Freeliving Nematodes (Secernentea).* University of California Extension, Riverside, California, USA.

de Ruiter, P.C., et al. (1993), 'Calculation of nitrogen mineralization in soil food webs', *Plant Soil*, 157(2), 263–273. www.jstor.org/stable/42939067.

Derycke, S., et al. (2007), 'Spatiotemporal analysis of population genetic structure in *Geomonhystera disjuncta* (Nematoda, Monhysteridae) reveals high levels of molecular diversity' *Mar. Biol.*, 151, 1799–1812.

de Vries, F.T. and Caruso, T. (2016), Eating from the same plate? Revisiting the role of labile carbon inputs in the soil food web. *Soil Biol. Biochem.*, 102, 4–9. https://doi.org/10.1016/j.soilbio.2016.06.023.

Dmowska, E. and Ilieva, K. (1995), 'The effect of prolonged diverse mineral fertilization on nematodes inhabiting the rhizosphere of spring barley', *Eur. J. Soil Biol.*, 31(4), 189–198.

Du Preez, G., et al. (2022), 'Nematode-based indices in soil ecology: application, utility, and future directions', *Soil Biol. Biochem.*, 169, 108640. https://doi.org/10.1016/j.soilbio.2022.108640

Ekschmitt, E. and Korthals, G.W. (2006), 'Nematodes as sentinels of heavy metals and organic toxicants in the soil', *J. Nematol.* 38(1), 13–19.

Ettema, C.H. and Bongers, T. (1993), 'Characterization of nematode colonization and succession in disturbed soil using the Maturity Index', *Biol. Fert. Soils* 16, 79–85. https://doi.org/10.1007/Bf00369407.

Ewald, M., et al. (2020), 'Effects of resource manipulation on nematode community structure and metabolic footprints in an arable soil across time and depth', *Nematology*, 22(9), 1025–1043.

Ferris, H. (1993), 'New frontiers in nematode ecology', *J. Nematol.*, 25(2), 374–382.

Ferris, H., et al. (2001), 'A framework for soil food web diagnostics: extension of the nematode faunal analysis concept', *Appl. Soil Ecol.* 18(1), 13–29. https://doi.org/10.1016/S0929-1393(01)00152-4.

Ferris, H. and Bongers, T. (2006), 'Nematode indicators of organic enrichment', *J. Nematol.* 38(1), 3–12.

Ferris, H. (2010), 'Form and function: metabolic footprints of nematodes in the soil food web', *Eur. J. Soil Biol.*, 46(2), 97–104. https://doi.org/10.1016/j.ejsobi.2010.01.003.

Filser, J., et al. (2016), 'Soil fauna: Key to new carbon models', *Soil*, 2(4), 565–582.

Fiscus, D.A. and Neher, D.A. (2002), 'Distinguishing sensitivity of free-living soil nematode genera to physical and chemical disturbances', *Ecol. Appl.*, 12(2), 565–575. https://doi.org/10.1890/1051-0761(2002)012[0565:Dsofls]2.0.Co;2.

Freckman, D.W. (1988), 'Bacterivorous nematodes and organic-matter decomposition', *Agricult. Ecosys. Environ.* 24(1–3), 195–217. https://doi.org/10.1016/0167-8809(88)90066-7

Freckman, D.W. and Baldwin, J.G. (1990), 'Nematodes'. In D.L. Dindal (Ed), *The Biology Guide*, John Wiley, New York, NY, pp. 155–200.

Freckman D.W. and Caswell E.P. (1985), 'The ecology of nematodes in agroecosystems', *Annu. Rev. Phytopathol.* 23(1), 275–296.

Freckman, D.W. and Ettema, C.H. (1993), 'Assessing nematode communities in agroecosystems of varying human intervention', *Agricult. Ecosys. Environ.*, 45(3–4), 239–261. https://doi.org/10.1016/0167-8809(93)90074-Y.

Gattoni, K., et al. (2023), '18S-NemaBase: curated 18S rRNA database of nematode sequences', *J. Nematol.*, 55(1), e2023-1. https://doi.org/10.2478/jofnem-2023-0006

Gauzens, B., et al. (2019), 'fluxweb: an R package to easily estimate energy fluxes in food webs', *Methods Ecol. Evol.*, 10(2), 270–279. https://doi.org/10.1111/2041-210x.13109.

Geisen, S., et al. (2018), 'Integrating quantitative morphological and qualitative molecular methods to analyse soil nematode community responses to plant range expansion', *Methods Ecol. Evol.*, 9(6), 1366–1378. https://doi.org/10.1111/2041-210x.12999.

Gendron, E.M., et al. (2023), 'Nematode mitochondrial metagenomics: a new tool for biodiversity analysis', *Mol. Ecol. Resour.*, 23(5), 975–989. https://doi.org/10.1111/1755-0998.13761.

Gendron, E.M., et al. (2024), 'Improving understanding of nematode communities in agricultural settings: A comparison of mitometagenomics and morphology', *Metabarcoding Metagenom.*, 8, e123387. https://doi.org/10.3897/mbmg.8.123387.

Girgan, G., et al. (2020), 'Nematodes and the effect of seasonality in grassland habitats of South Africa', *J Nematol.* 52(1), 1–22. https://doi.org/10.21307/jofnem-2020-118.

Grandy, A.S., et al. (2016), 'Beyond microbes: Are fauna the next frontier in soil biogeochemical models?', *Soil Biol. Biochem.*, 102, 40–44. https://doi.org/10.1016/j.soilbio.2016.08.008.

Gotelli, N.J. (2001), 'Research frontiers in null model analysis', *Global Ecol. Biogeogr.*, 10(4), 337–343. https://doi.org/10.1046/j.1466-822X.2001.00249.x.

Guerra, C.A., et al. (2021), 'Tracking, targeting, and conserving soil biodiversity: a monitoring and indicator system can inform policy', *Science* 371(6526), 239–241. https://doi.org/10.1126/science.abd7926. https://youtu.be/8EnnbhOKkUQ?si=-I4TV7sapPuRzMuC

Herren, G.L., et al. (2020), 'Effects of synthetic fertilizer and farm compost on soil nematode community in long-term crop rotation plots: a morphological and metabarcoding approach', *PLoS One*, 15(3), e0230153. https://doi.org/10.1371/journal.pone.0230153.

Hodda M. (2007), 'Phylum nematoda', *Zootaxa*, 1668, 265–293. https://doi.org/10.11646/zootaxa.1668.1.14

Hodda, M., et al. (2009), 'Nematode diversity in terrestrial, freshwater aquatic and marine systems'. In M.J. Wlson and T. Kakouli-Duarte (Eds), Nematodes as Environmental Indicators, CAB International, Cambridge, UK, pp. 45–93.

Hodda, M. (2022), 'Phylum Nematoda: Feeding habits for all valid genera using a new, universal scheme encompassing the entire phylum, with descriptions of morphological characteristics of the stoma, a key, and discussion of the evidence for trophic relationships', *Zootaxa*, 5114(1), 318–451. https://doi.org/10.11646/zootaxa.5114.1.3.

Hunt, D.J. (1993), *Aphelenchida, Longidoridae and Trichodoridae: Their Systematics and Bionomics*. CAB International, Cambridge, UK.

Hunt, H.W., et al. (1987), 'The detrital food web in a shortgrass prairie', *Biol. Fertil. Soils* 3(1–2), 57–68. https://doi.org/10.1007/BF00260580

Ingham, R.E., et al. (1985), 'Interactions of bacteria, fungi, and their nematode grazers: effects on nutrient cycling and plant growth', *Ecol. Monogr.*, 55(1), 119–140. https://doi.org/10.2307/1942528.

Jairajpuri, M.S. and Ahmad, W. (1992), *Dorylaimida: Free-living, Predaceous and Plant-Parasitic Nematodes*. E. J. Brill, New York, NY.

Jochum, M., et al. (2021), 'For flux's sake: General considerations for energy-flux calculations in ecological communities', *Ecol. Evol.*, 11(19), 12948–12969. https://doi.org/10.1002/ece3.8060.

Jochum, M. and Eisenhauer, N. (2022), 'Out of the dark: using energy flux to connect above-and belowground communities and ecosystem functioning', *Eur. J. Soil Sci.*, 73(1), e13154. https://doi.org/10.1111/ejss.13154.

Kakouli-Duarte, T., Korthals, G.W., Sánchez Moreno, S., du Preez, G. and de Goede, R. (2025) *Nematodes as Environmental Indicators: From Theory to Practice*, 2nd edition. CABI, Wallingford (Oxford), UK.

Korthals, G.W., et al. (1998), 'Influence of perennial ryegrass on a copper and zinc affected terrestrial nematode community', *Appl. Soil Ecol.*, 10(1–2), 73–85. https://doi.org/10.1016/S0929-1393(98)00039-0.

Lek, S. and Guegan, J F. (1999), 'Artificial neural networks as a tool in ecological modelling, an introduction', *Ecol. Model.*, 120 (2–3), 65–73. https://doi.org/10.1016/S0304-3800(99)00092-7.

Li, F., et al. (2005), 'Contrasting effects of copper and benzo(a)pyrene concentration on life history characteristics of *Aphelenchus* and *Acrobeloides*', *Ecotoxicology,* 14(4), 419–429. https://doi.org/10.1007/s10646-004-1347-4.

Luo, J., et al. (2021), 'Effects of residue mulching amounts on metabolic footprints based on production and respiration of soil nematodes in a long-term no-tillage system', *Land Degrad. Dev.*, 32(7), 2383–2392. https://doi.org/10.1002/ldr.3918.

Machida, R.J. and Knowlton, N. (2012), 'PCR primers for metazoan nuclear 18S and 28S ribosomal DNA sequences', *PloS One* 7, e46180. https://doi.org/10.1371/journal.pone.0046180.

Martin, T., et al. (2022), 'The integration of nematode communities into the soil biological health framework by factor analysis', *Ecol. Indic.*, 136, 108676. https://doi.org/10.1016/j.ecolind.2022.108676.

Moore, J.C. and de Ruiter, P.C. (1991), 'Temporal and spatial heterogeneity of trophic interactions within below-ground food webs', *Agricult. Ecosys. Environ.*, 34(1–4), 371–397.

Moore, J.C. and de Ruiter, P.C. (2012), 'Soil food webs in agricultural ecosystems'. In T.E. Cheeke, et al. (Eds), *Microbial Ecology of Sustainable Agroecosystems*. CRC Press, Boca Raton, FL, pp. 63–88.

Mosdossy, K., et al. (2024a), 'Adding ligneous litter to cultivated organic soil changes the soil micro-food web and alters soil nitrogen availability', *Appl. Soil Ecol.*, 196, 105271. https://doi.org/10.1016/j.apsoil.2024.105271.

Mosdossy, K., et al. (2024b), *Miscanthus* litter additions induce a successional change in the soil micro-food web with apparent decreases in soil nitrogen. *Soil Biol. Biochem.*, 199, 109589. https://doi.org/10.1016/j.soilbio.2024.109589.

Neher, D.A. (1999), 'Soil community composition and ecosystem processes: comparing agricultural ecosystems with natural ecosystems', *Agroforest. Syst.*, 45, 159–185. https://doi.org/10.1023/A:1006299100678.

Neher, D.A. (2010), 'Ecology of plant and free-living nematodes in natural and agricultural soil', *Annu. Rev. Phytopathol.*, 48(1), 371–394. https://doi.org/10.1146/annurev-phyto-073009-114439.

Neher, D.A. (2001a), 'Role of nematodes in soil health and their use as indicators', *J. Nematol.*, 33(4), 161–168.

Neher, D.A. (2001b), 'Nematode communities as ecological indicators of agroecosystem health'. In S.R. Gliessman (Ed), *Agroecosystem Sustainability: Developing Practical Strategies*. CRC Press, Boca Raton, FL, pp. 105–120.

Neher, D.A. (2023), 'Moving up within the food web: Protists, nematodes and other microfauna'. In N. Uphoff and J. Thies (Eds), *Biological Approaches to Regenerative and Resilient Soil Systems*, 2nd edition. CRC, Boca Raton, FL, pp. 157–168.

Neher, D.A. and Campbell, C.L. (1994), 'Nematode communities and microbial biomass in soils with annual and perennial crops', *Appl. Soil Ecol.*, 1(1), 17–28. https://doi.org/10.1016/0929-1393(94)90020-5.

Neher, D.A. and Campbell, C.L. (1996), 'Sampling for regional monitoring of nematode communities in agricultural soils', *J. Nematol.*, 28(2), 196.

Neher, D.A. and Darby, B.J. (2006), 'Computation and application of nematode community indices: general guidelines'. In A. Eyualem, et al. (Eds), *Freshwater Nematodes: Ecology and Taxonomy*, CABI, Wallingford, UK, pp. 211–222. https://doi.org/10.1079/9780851990095.0211.

Neher, D. and Darby, B.J. (2009), 'General community indices that can be used for analysis of nematode assemblages'. In M. Wilson and T. Kakouli-Duarte (Eds), *Nematodes as Environmental Indicators*, CABI, Wallingford, UK, pp. 107–123. https://doi.org/10.1079/9781845933852.01.

Neher, D.A. and Darby, B.J. (2025), 'General community indices that can be used for analysis of nematode assemblages'. In Kakouli-Duarte, T., Korthals G. W., Sánchez

Moreno, S., di Preez, G. and de Goede, R. (Eds), *Nematodes as Environmental Indicators: From Theory to Practice*, 2nd Edition, CABI, Wallingford (Oxford), UK.

Neher, D.A. and Olson, R.K. (1999), 'Nematode communities in soils of four farm cropping management systems', *Pedobiologia*, 43(5), 430–438.

Neher, D.A. and Powers, T.O. (2023), 'Nematodes', *Encyclopedia of Soils in the Environment*, 11, 105–111. https://doi.org/10.1016/B978-0-12-822974-3.00038-0.

Neher, D.A. and Stürzenbaum, S.R. (2006), 'Extra-long PCR, an identifier of DNA adducts in single nematodes (*Caenorhabditis elegans*)', *Comp. Biochem. Phys. C,* 144(3), 279–285. https://doi.org/10.1016/j.cbpc.2006.10.005.

Neher, D.A. and Weicht, T.R. (2013), 'Nematode genera in forest soil respond differentially to elevated CO_2', *J. Nematol.*, 45(3), 214–222.

Neher, D.A., et al. (1995), 'Measures of nematode community structure and sources of variability among and within agricultural fields', *Plant Soil,* 170, 167–181. https://doi.org/10.1007/Bf02183065.

Neher, D.A., et al. (1998), 'Comparison of nematode communities in agricultural soils of North Carolina and Nebraska', *Ecol. Appl.*, 8(1), 213–223. https://doi.org/10.1890/1051-0761(1998)008[0213:Concia]2.0.Co;2.

Neher, D.A., et al. (2012), 'Linking invertebrate communities to decomposition rate and nitrogen availability in pine forest soils', *Appl. Soil Ecol.,* 54, 14–23. https://doi.org/10.1016/j.apsoil.2011.12.001.

Neher, D.A., et al. (2014), 'Impact of coleopteran-active Bt corn on non-target nematode communities in soil and decomposing corn roots', *Soil Biol. Biochem.*, 76, 127–135. https://doi.org/10.1016/j.soilbio.2014.05.019.

Neher, D.A., et al. (2017), 'Environmental indicators reflective of road design in a forested landscape', *Ecosphere,* 8(3), e01734. https://doi.org/10.1002/ecs2.1734.

Neher, D.A., et al. (2019), 'Crop rotation and tillage affect nematode communities more than biocides in monoculture soybean', *Appl. Soil Ecol.,* 140, 89–97. https://doi.org/10.1016/j.apsoil.2019.03.016.

Nemaplex.UCDavis.edu: Revision Date: 10/07/2024; Accessed October 10, 2024.

Nickle, W.R. (1991), *Manual of Agricultural Nematology*. Marcel Dekker, New York, NY.

Oka, Y. (2010), 'Mechanisms of nematode suppression by organic soil amendments: a review', *Appl. Soil Ecol.*, 44(2), 101–115. https://doi.org/10.1016/j.apsoil.2009.11.003.

Okada, H., et al. (2005), 'Fungal-feeding habits of six nematode isolates in the genus *Filenchus*', *Soil Biol. Biochem.,* 37(6), 1113–1120. https://doi.org/10.1016/j.soilbio.2004.11.010.

Porazinska, D.L., et al. (2009), 'Evaluating high-throughput sequencing as a method for metagenomic analysis of nematode diversity', *Mol. Ecol. Resour.*, 9(6), 1439–1450. https://doi.org/10.1111/j.1755-0998.2009.02611.x.

Potapov, A.M. (2022), Multifunctionality of belowground food webs: resource, size and spatial energy channels. *Biol. Rev.*, 97(4), 1691–1711.

Powers, T.O. (2004), 'Nematode molecular diagnostic: from bands to barcodes', *Annu. Rev. Phytopathol.,* 42(1), 367–383. https://doi.org/10.1146/annurev.phyto.42.040803.140348.

Powers, T.O., et al. (2009), 'Tropical nematode diversity: vertical stratification of nematode communities in a Costa Rican humid lowland rainforest', *Mol. Ecol.*, 18(5), 985–996. https://doi.org/10.1111/j.1365-294X.2008.04075.x.

Powers, T.O., et al. (2021), 'Nematode biodiversity assessments need vouchered databases: a BOLD reference library for plant-parasitic nematodes in the superfamily Criconematoidea', *Genome*, 64(3), 232–241. https://doi.org/10.1139/gen-2019-0196.

Prokopowich, C.D., et al. (2003), 'The correlation between rDNA copy number and genome size in eukaryotes', *Genome*, 46(1), 48–50. https://doi.org/10.1139/G02-103.

Ratnasingham, S. and Hebert, P.D., (2007), 'BOLD: The barcode of life data system' *Mol. Ecol. Notes*, 7(3), 355–364.

Ren, Y., et al. (2024), 'A single degenerated primer significantly improves COX1 barcoding performance in soil nematode community profiling', *Soil Ecol. Lett.*, 6(2), 230204. https://doi.org/10.1007/s42832-023-0204-4.

Ritz, K. and Trudgill, D.L. (1999), 'Utility of nematode community analysis as an integrated measure of the functional state of soils: Perspectives and challenges - discussion paper', *Plant Soil*, 212(1), 1–11. https://doi.org/10.1023/A:1004673027625.

Sapkota, R. and Nicolaisen, M. (2015), 'High-throughput sequencing of nematode communities from total soil DNA extractions', *BMC Ecology*, 15, 1–8. https://doi.org/10.1186/s12898-014-0034-4.

Schenk, J., et al. (2019), 'Metabarcoding data allow for reliable biomass estimates in the most abundant animals on earth', *Metabarcoding Metagenom.*, 3, e46704. https://doi.org/10.3897/mbmg.3.46704.

Schmidt, J.H., et al. (2020), 'Bacterivorous nematodes correlate with soil fertility and improved crop production in an organic minimum tillage system', *Sustainability*, 12(17), 6730. https://doi.org/10.3390/su12176730.

Schmidt, J.H., et al. (2025), 'Indices developed specifically for analysing nematode assemblages'. In Kakouli-Duarte T., Korthals G. W., Sánchez Moreno, S., di Preez, G. and de Goede, R. (Eds), *Nematodes as Environmental Indicators: From Theory to Practice*, 2nd Edition, CABI, Wallingford (Oxford), UK.

Siddiqi, M.R. (2000), *Tylenchida: Parasites of Plants and Insects*, 2nd edition. CABI, Cambridge UK.

Sieriebriennikov, B., et al. (2014), 'NINJA: An automated calculation system for nematode-based biological monitoring', *Eur. J. Soil Biol.* 61, 90–93. https://doi.org/10.1016/j.ejsobi.2014.02.004.

Thakur, M.P. and Geisen, S. (2019), Trophic regulations of the soil microbiome. *Trends Microbiol.*, 27(9), 771–780. https://doi.org/10.1016/j.tim.2019.04.008.

Thorne, G. (1974), 'Nematode of the Northern Great Plains. Part II. Dorylaimoidea in part (Nemata: Adenophorea)', Tech. Bull. 41. Agricultural Experiment Station, South Dakota State University, Brookings, SD.

Todd, T.C. (1996), 'Effects of management practices on nematode community structure in tallgrass prairie', *Appl. Soil Ecol.*, 3, 235–246. https://doi.org/10.1016/0929-1393(95)00088-7.

van Bezooijen, J. (2006), *Methods and Techniques for Nematology*, Wageningen University, The Netherlands. https://www.scribd.com/document/234069625/Methods-and-TechniquesforNematology; accessed October 23, 2025.

van den Hoogen, J., et al. (2019), 'Soil nematode abundance and functional group composition at a global scale', *Nature*, 572(7768), 194–198. https://doi.org/10.1038/s41586-019-1418-6.

van der Putten, W.H., et al. (2023), 'Soil biodiversity needs policy without borders', *Science* 379, 32–34. https://doi.org/10.1126/science.abn7248.

Wardle, D.A. and Yeates, G.W. (1993), 'The dual importance of competition and predation as regulatory forces in terrestrial ecosystems – evidence from decomposer food-webs', *Oecologia*, 93(2), 303–306. https://doi.org/10.1007/BF00317685.

Wardle, D.A., et al. (1995), 'The detritus food web and the diversity of soil fauna as indicators of disturbance regimes in agroecosystems', *Plant Soil,* 170, 35–43.

Wasilewska, L. (1979), 'The structure and function of soil nematode communities in natural ecosystems and agrocenoses', *Pol. Ecol. Stud.,* 5, 97–145.

Weller, D.M., et al. (2002), 'Microbial populations responsible for specific soil suppressiveness to plant pathogens', *Annu. Rev. Phytopathol.*, 40(1), 309–348. https://doi.org/10.1146/annurev.phyto.40.030402.110010.

Yeates, G.W. (1994), 'Modification and qualification of the nematode maturity index', *Pedobiologia,* 38, 97–101. https://doi.org/10.1016/s0031-4056(24)00113-6.

Yeates, G.W. (2003), 'Nematodes as soil indicators: functional and biodiversity aspects', *Biol. Fertil. Soils,* 37(4), 199–210. https://doi.org/10.1007/s00374-003-0586-5.

Yeates, G.W. (2010), 'Nematodes in ecological webs', eLS. https://doi.org/10.1002/9780470015902.a0021913.

Yeates, G.W. and Bird, A.F. (1994) 'Some observations on the influence of agricultural practices on the nematode faunae of some South Australian soils', *Fund. Appl. Nematol.*, 17(2), 133–145.

Yeates, G.W. and Bongers, T. (1999), 'Nematode diversity in agroecosystems', *Agricult. Ecosys. Environ.*, 74(1–3), 113–135. https://doi.org/10.1016/S0167-8809(99)00033-X.

Yeates, G.W., et al. (1993), 'Feeding habits in soil nematode families and genera–an outline for soil ecologists', *J. Nematol.,* 25(3), 315–331.

Yeates, G.W., et al. (1997), 'Faunal and microbial diversity in three Welsh grassland soils under conventional and organic management regimes', *J. Appl. Ecol.*, 34, 453–470. https://doi.org/10.2307/2404890.

Yoder, M., et al. (2006), DESS: a versatile solution for preserving morphology and extractable DNA of nematodes', *Nematology*, 8(3), 367–376.

Zhang, X., et al. (2017), 'Ecosystem services of the soil food web after long-term application of agricultural management practices', *Soil Biol. Biochem.*, 111, 36–43. https://doi.org/10.1016/j.soilbio.2017.03.017.

Zhao, J. and Neher, D.A. (2013), 'Soil nematode genera that predict specific types of disturbance', *Appl. Soil Ecol.,* 64, 135–141. https://doi.org/10.1016/j.apsoil.2012.11.008.

Part 3

Applications

Chapter 11

Biological indicators in laboratory-based soil health testing programmes

Joseph P. Amsili and Harold van Es, Cornell University, USA

1 Introduction

The growing interest in soil health reflects a heightened appreciation of the role that soils play in food provisioning and other essential ecosystem services, as well as concerns about physical, chemical and biological soil degradation. Healthy soil supports growing high-yielding crops, utilizing nutrients and water efficiently, fostering soil biology, increasing resilience to extreme weather events, and providing a range of other ecosystem services such as protection of air and water quality (Reicosky, 2018a, 2018b). Ways of improving soil health are increasingly seen as central to sustainable crop production (Horwath, 2022). However, sustainable soil management clearly requires accurate, standardized and accessible ways to measure and monitor soil health (van Es & Thies, 2023).

There have been major advances in technologies measuring a wide range of soil physical, chemical and biological properties, including lab-based, field-based, remote and proximal sensing techniques (Lobsey & Biswas, 2023; Otten, 2021). Key advances in measuring soil biology are the subject of this book, and lab-based biological indicators are the focus of this chapter (Norton et al., 2025). But most methods for assessing soil biology have a mix of strengths and weaknesses in terms of what they can or cannot measure, the degree of precision and accuracy, and how easy they are to use in practice. Field methods can be compromised by problems such as non-controllable environmental conditions, potential user bias, or standardization issues, which

http://dx.doi.org/10.19103/AS.2025.0159.12

undermine robustness and interpretability (Moebius-Clune et al., 2017). While all approaches have a role to play, it is generally accepted that laboratory-based tests with well-defined protocols and controlled conditions remain the "gold standard" in soil health testing.

Building on well-established soil fertility testing methods (e.g. nutrients and pH), there is now a range of laboratory-based biological testing methods designed specifically to support holistic soil management. These can be categorized into two main groups (Thies et al., 2023): (1) indirect methods that quantify various soil organic matter fractions (from which soil organic matter dynamics and soil biological processes can be inferred) and (2) direct methods that measure the abundance, activity or diversity of soil biota.

Discussion of these methods is the focus of this book and many other studies (Chenu et al., 2015; Kandeler, 2015; Thies, 2015; Thies et al., 2023). However, many were initially designed for research and may not be fully validated nor necessarily amenable to high-throughput commercial testing that land managers need. For example, while there is a lot of potential for DNA-based microbial metrics, concerns remain about uncertainties in data interpretation (Fierer et al., 2021). Advances in DNA-based methods for assessing microbial abundance and diversity are explored in Yarwood (2025).

2 Selecting soil health indicators

In recent years, significant progress was made in creating interpretation frameworks using soil health indicators to translate measurement outcomes into actionable information. Measurement of pH and extractable soil nutrient concentrations have been the cornerstone of measuring and managing soil fertility and fertilizer application since the 1920s (Stockdale et al., 2013). However, the concept of soil health is more holistic and refers to the overall well-being of the soil environment. Interest is also rooted in a recognition that soil biology, soil physics, and soil chemistry need to be considered in an integral manner (Bünemann et al., 2018). This has inevitably complicated development and use of measures of soil health that encompass different soil properties, functions, and ecosystems services. It requires a sound foundation of scientific knowledge in an appropriate biophysical context (e.g. region or soil type) and set of management factors.

Several commonly agreed criteria have been used to evaluate useful soil health indicators that could be integrated into an interpretative framework. It has been suggested that a soil health indicator needs to meet the following general criteria (Moebius-Clune et al., 2017; Stott, 2019):

- be sensitive to changes in soil management;
- represent agronomically and environmentally important processes;

- allow for ease of sampling (i.e. samples can be collected by a farmer or agricultural service provider);
- have low cost of analysis per sample;
- be accurate, repeatable, and have low analytical variability; and
- be interpretable in relevant contexts.

Accordingly, the US Department of Agriculture released a recommended set of soil health indicators and methods to encourage standardization (Stott, 2019) in the context of a range of soil processes:

- soil organic matter dynamics;
- soil structural stability;
- general microbial activity: short-term C mineralization and enzyme activity;
- carbon food source;
- bioavailable nitrogen; and
- microbial diversity.

In each case, the report reviews relevant indicators, indicator measurement methods and how well they meet key criteria for a good soil health indicator and method. Table 1 provides an updated and expanded summary of key biological soil health indicators (Stott, 2019; van Es & Thies, 2023).

Subsequently, there have been significant advances in biological soil health testing through the application of machine learning and artificial intelligence and diffuse reflectance spectroscopy. For example, recently we demonstrated that autoclaved-citrate extractable (ACE) protein could be effectively predicted using a random forest model (a machine learning technique) developed from other lower-cost soil health measurements included in the Comprehensive Assessment of Soil Health (CASH; Amsili et al., 2025). Diffuse reflectance spectroscopy in the visible and near-infrared (VNIR) or mid-infrared (MIR) can predict soil organic carbon, carbonates, clay, sand, and novel biological soil health indicators, like soil respiration (Comstock et al., 2019; Ghimire et al., 2025; Sanderman et al., 2020). These pedotransfer approaches can greatly reduce analysis costs once the initial database investments have been made.

In addition, the revolution in high-throughput microbiome sequencing, driven by the exponential decline in cost, is another new approach for assessing biological soil health and providing microbial metrics of diversity and function. Two recent studies showed that bacterial microbiome 16S rRNA data could be used to predict biological soil health indicators and provide novel community weighted microbiome genomic traits that indicate microbial life-history strategies (Wilhelm et al., 2023; Wilhelm et al., 2022). Since then, several companies have begun offering soil microbiome testing to the public using amplicon sequencing of 16S rRNA or metagenomic techniques to provide

Table 1 Soil processes and recommended laboratory soil health indicators and methodologies

Biological soil process	Soil health indicator	Method
Organic matter cycling and carbon storage	Soil organic matter, soil organic carbon, total nitrogen	Dry combustion* Loss on ignition*
General microbial activity	Short-term C mineralization	CO_2 respired—4 day*CO_2 respired—3 day* CO_2 respired—24 hours*
General microbial activity	Enzyme activity	β-glucosidase* N-acetyl-β-D-glucosaminidase* Phosphomonoesterases* and Phosphatases* Arylsulfatase*
Carbon food source	Available organic C fractions	Permanganate oxidizable C* Particulate OM 28-day mineralization Water-extractable organic C* Soluble carbohydrates Substrate-induced respiration Microbial biomass C*
Bioavailable nitrogen	Available organic N fractions	Autoclaved-citrate-extractable (ACE) protein* Water-extractable organic N* Short-term mineralization Potentially mineralizable N • 7-day anaerobic* • 28-day aerobic Illinois soil N test N-acetyl-glucosaminidase Protease
Microbial diversity	Community structure	Phospholipid fatty acid (PLFA)* Ester-linked fatty acid methyl ester profile (EL-FAME) Bacterial metagenomics* Amplicon sequencing of 16S rRNA gene

Methods with an asterisk following them are being offered by major commercial soil health testing laboratories in North America.
Source: Expanded from Stott (2019) and adapted from van Es and Thies (2023).

detailed functional microbial interpretations. But while there is high potential for DNA-based microbial metrics, there is also skepticism among scientists because detailed functional microbial metrics remain poorly validated (Fierer et al., 2021). In all, the assessment of soil biological indicators is an active research area, but unfortunately thus far little consensus exists on the most relevant processes and the most useful indicators for practical soil health testing.

Measurement of soil health indicators is only one aspect of soil health assessment. Several commercial soil laboratories stop there and provide only the measured outcomes. To realize the full purpose of soil health testing, which is to assess the effects of human management on soil health and provide guidance for management, proper interpretation of the measured values is required.

3 From measurement to action: interpreting and benchmarking biological indicator results

Although quality standards exist to protect water and air (EPA, 2023), few analogous metrics exist for soil. Soil health indicators need to be integrated into standardized interpretative frameworks that translate measured outcomes into actionable information for land managers. This involves addressing questions related to (Amsili et al., 2023; Bünemann et al., 2018; Moebius-Clune et al., 2017; Nunes et al., 2024):

- the minimum number of indicators required to provide a practical and actionable assessment of soil functions that characterize soil health;
- the value of an indicator within an acceptable range in the context of inherent pedoclimatic factors like climate, soil taxonomy, topography, land use, etc.;
- reasonable targets in the context of inherent factors (i.e. benchmarking);
- the degree to which an out-of-range value is likely to result in impairment of soil functions and processes; and
- short-term and long-term management solutions to address a soil constraint.

Soil health scoring functions can be used to calculate benchmarks, which in turn are useful metrics for developing realistic soil health targets. Early soil health assessment frameworks, including the Soil Management Assessment Framework (SMAF) and the Comprehensive Assessment of Soil Health (CASH), integrated multiple biological, physical, and chemical indicators, and provided contextual interpretation of measured values (Nunes et al., 2021; Stockdale et al., 2021). These frameworks interpret soil health indicators in the context of key site-specific pedoclimatic factors that exert control on carbon and nitrogen indicators, microbial activity, etc. (Nunes et al., 2024; Nunes et al., 2021; von Lützow et al., 2006). SMAF interprets soil health indicators in the context of region, climate, soil texture, soil taxonomy, and inherent organic matter content but was calibrated using a relatively small dataset (Andrews et al., 2004; Nunes et al., 2021). Whereas CASH interprets soil health indicators in the context of three soil texture groups (coarse, medium, and fine) and is based on a larger database, but the scoring functions are most appropriately applied to specific regions (Fine et al., 2017).

More recently, a Bayesian interpretation approach called the Soil Health Assessment Protocol and Evaluation (SHAPE) was developed from a large national US database. It provides a robust update to SMAF and CASH scoring functions for soil organic carbon, permanganate oxidizable C, autoclaved citrate extractable carbon, and 4-day soil respiration (short-term C mineralization). SHAPE scores soil health indicators for a soil's peer group based on combinations of five soil texture and five soil taxonomy classes that are adjusted for mean annual temperature and precipitation (Nunes et al., 2024; Nunes et al., 2021). A spatially explicit version of SHAPE is also available for soil organic carbon, which incorporates proximity to other samples in the database (Veum et al., 2025; https://shapesoilhealth.shinyapps.io/shape_app_new/).

Furthermore, soil health outcomes have been shown to be contextual to land use systems, mostly related to biomass inputs, perenniality, and levels of soil disturbance (Amsili et al., 2023; Rubio et al., 2025). A more granular system was therefore developed to interpret soil health data in the context of cropping systems, which facilitates the establishment of soil health benchmarks (Amsili et al., 2023). This offers a farm-level interpretation and management framework that accounts for both the inherent pedoclimatic environment and the cropping system, jointly referred to as "production environment". Hence, the Production Environment Soil Health (PESH) benchmark approach.

4 Conclusion

It is widely recognized that understanding and optimizing biological processes in soil is fundamental to improving soil health and the range of ecosystem services that it can deliver. Laboratory-based soil health testing allows for consistent indicator analysis and has emerged as a tool to assess these processes to support land management decisions. In recent years, significant progress was made in integrating appropriate soil health indicators into commercial testing laboratories and advancing interpretation frameworks for those indicators. However, further convergence of soil health indicator methodologies and interpretation tools is needed to increase the utility and adoption of soil health testing. Advances in the use of machine learning and artificial intelligence, diffuse reflectance spectroscopy, and high-throughput sequencing of the soil microbiome will likely further advance soil health testing programmes in the future.

5 Where to look for further information

The following articles and books provide a good overview of the subject:

- van Es, H. M. & Thies, J. E. (2023) Measuring and Monitoring Soil Health, in Uphoff, N. & Thies, J. E. (eds), *Biological Approaches to Regenerative Soil Systems*. Boca Raton: CRC Press, 537–550.
- Thies, J. E., et al. (2023) Laboratory Methods for Assessing Soil Health, in Uphoff, N. & Thies, J. E. (eds), *Biological Approaches to Regenerative Soil Systems*. Boca Raton: CRC Press, 551–566.
- Magdoff, F. R. & van Es, H. (2021) *Building Soils For Better Crops: Ecological Management for Healthy Soils* (4th ed.). Waldorf: The Sustainable Agriculture Research and Education (SARE) Program.
- Moebius-Clune, B. N., et al. (2017). *Comprehensive Assessment of Soil Health - The Cornell Framework. Edition 3.2.* Available from: www.soilhealth.cals.cornell.edu/training-manual/.

The following societies and professional organizations provide a good overview of the subject:

- Soil Science Society of America: https://www.soils.org/.
- Soil and Water Conservation Society: https://www.swcs.org/.
- British Society of Soil Science: https://soils.org.uk/.
- International Union of Soil Sciences: https://www.iuss.org/.

Major international research projects to keep an eye on:

- Soil Health Benchmarks: https://soilhealthbenchmarks.eu/.

Top research centers to keep up with:

- Soil Health Institute: https://soilhealthinstitute.org/.
- Cornell Soil Health Program: https://soilhealth.cals.cornell.edu/
- New York Soil Health Initiative: https://www.newyorksoilhealth.org/.

6 References

Amsili, J. P., et al. (2023) Empirical approach for developing production environment soil health benchmarks. *Geoderma Regional*, 34, e00672.

Amsili, J. P., et al. (2025) Pedotransfer functions for soil protein based on random forest modeling for routine soil health analysis. *Communications in Soil Science and Plant Analysis*, 56(9), 1381–1393.

Andrews, S. S., et al. (2004) The soil management assessment framework. *Soil Science Society of America Journal*, 68(6), 1945–1962.

Bünemann, E. K., et al. (2018) Soil quality: a critical review. *Soil Biology and Biochemistry*, 120, 105–125.

Chenu, C., et al. (2015) Chapter 13 - Methods for Studying Soil Organic Matter: Nature, Dynamics, Spatial Accessibility, and Interactions with Minerals, in Paul, E. A. (ed), *Soil Microbiology, Ecology and Biochemistry* (4th ed.). Boston: Academic Press, 383–419.

Comstock, J. P., et al. (2019) Carbonate determination in soils by mid-IR spectroscopy with regional and continental scale models. *PLoS One*, 14(2), e0210235.

EPA (2023) *National Primary Drinking Water Regulations*. Available online: https://www.epa.gov/ground-water-and-drinking-water/national-primary-drinking-water-regulations [Accessed 8-14-2025]

Fierer, N., et al. (2021) How microbes can, and cannot, be used to assess soil health. *Soil Biology and Biochemistry*, 153, 108111.

Fine, A. K., et al. (2017) Statistics, scoring functions, and regional analysis of a comprehensive soil health database. *Soil Science Society of America Journal*, 81(3), 589–601.

Ghimire, S., et al. (2025) Using mid-infrared spectroscopy to estimate soil microbial properties at the continental scale. *Applied Soil Ecology*, 211, 106110.

Horwath, W. (2022) *Improving Soil Health* (1st ed.). Cambridge: Burleigh Dodds Science Publishing.

Kandeler, E. (2015) Chapter 7 - Physiological and Biochemical Methods for Studying Soil Biota and Their Functions, in Paul, E. A. (ed), *Soil Microbiology, Ecology and Biochemistry* (4th ed.). Boston: Academic Press, 187–222.

Lobsey, C. & Biswas, A. (2023) *Advances in Sensor Technology for Sustainable Crop Production* (1st ed.). Cambridge: Burleigh Dodds Science Publishing.

Moebius-Clune, B. N., et al. (2017) *Comprehensive Assessment of Soil Health - The Cornell Framework. Edition 3.2*. Geneva: Cornell University. Available from: http://soilhealth.cals.cornell.edu/training-manual/

Norton, J., et al. (2025) *Measuring and Assessing the Biological Health of Soils* (1st ed.). Cambridge: Burleigh Dodds Science Publishing.

Nunes, M. R., et al. (2021) The soil health assessment protocol and evaluation applied to soil organic carbon. *Soil Science Society of America Journal*, 85(4), 1196–1213.

Nunes, M. R., et al. (2024) SHAPEv1.0 Scoring curves and peer group benchmarks for dynamic soil health indicators. *Soil Science Society of America Journal*, 88(3), 858–875.

Otten, W. (2021) *Advances in Measuring Soil Health* (1st ed.). Cambridge: Burleigh Dodds Science Publishing.

Reicosky, D. (2018a) *Managing Soil Health for Sustainable Agriculture Volume 1: Fundamentals* (1st ed.). Cambridge: Burleigh Dodds Science Publishing.

Reicosky, D. (2018b) *Managing Soil Health for Sustainable Agriculture Volume 2: Monitoring and Management* (1st ed.). Cambridge: Burleigh Dodds Science Publishing.

Rubio, V., et al. (2025) Biomass inputs drive agronomic management impacts on soil health. *Agriculture, Ecosystems & Environment*, 378, 109316.

Sanderman, J., et al. (2020) Mid-infrared spectroscopy for prediction of soil health indicators in the United States. *Soil Science Society of America Journal*, 84(1), 251–261.

Stockdale, E. A., et al. (2013) Soil Fertility, in Gregory, P. J. & Nortcliff, S. (eds), *Soil Conditions and Plant Growth*. Hoboken: Wiley-Blackwell.

Stockdale, E. A., et al. (2021) Developing Soil Health Indicators for Improved Soil Management on Farm, in Otten, W. (ed), *Advances in Measuring Soil Health* (1st ed.). Cambridge: Burleigh Dodds Science Publishing.

Stott, D. E. (2019) *Recommended Soil Health Indicators and Associated Laboratory Procedures*. Soil Health Technical Note No. 450-03. U.S. Department of Agriculture, Natural Resources Conservation District. Available from: https://directives.sc.egov.usda.gov/OpenNonWebContent.aspx?content=44475.wba

Thies, J. E. (2015) Chapter 6 - Molecular Approaches to Studying the Soil Biota, in Paul, E. A. (ed), *Soil Microbiology, Ecology and Biochemistry* (4th ed.). Boston: Academic Press, 151–185.

Thies, J. E., et al. (2023) Laboratory Methods for Assessing Soil Health, in Uphoff, N. & Thies, J. E. (eds), *Biological Approaches to Regenerative Soil Systems*. Boca Raton: CRC Press, 551–566.

van Es, H. M. & Thies, J. E. (2023) Measuring and Monitoring Soil Health, in Uphoff, N. & Thies, J. E. (eds), *Biological Approaches to Regenerative Soil Systems*. Boca Raton: CRC Press, 537–550.

Veum, K. S., et al. (2025) Spatially explicit heteroskedastic modeling for the soil health assessment protocol and evaluation version 1.0S. *Soil Science Society of America Journal*, 89(3), e70065.

von Lützow, M., et al. (2006) Stabilization of organic matter in temperate soils: mechanisms and their relevance under different soil conditions – a review. *European Journal of Soil Science*, 57(4), 426–445.

Wilhelm, R. C., et al. (2022) Predicting measures of soil health using the microbiome and supervised machine learning. *Soil Biology and Biochemistry*, 164, 108472.

Wilhelm, R. C., et al. (2023) Ecological insights into soil health according to the genomic traits and environment-wide associations of bacteria in agricultural soils. *ISME Communications*, 3(1), 1.

Yarwood, S. (2025) Advances in DNA based methods for assessing microbial abundance and diversity, in Norton, J., et al. (eds), *Measuring and Assessing the Biological Health of Soils* (1st ed.). Cambridge: Burleigh Dodds Science Publishing.

Chapter 12

Using biological indicators of soil health to assess the impact of particular cropping practices

Shamim Gul, McGill University, Canada and University of Balochistan, Pakistan; and Joann K. Whalen, McGill University, Canada and Mohammed VI Polytechnic University, Morocco

1 Introduction

Soil covers the Earth in a thin layer, accounting for just one ten-millionth of the Earth's radius (Kopittke et al., 2024). Soil provides 99% of the food that sustains human nutrition and also supplies medicines courtesy of medicinal wild herbs and plants, as well as antibiotics derived from soil microorganisms (Kopittke et al., 2024; Zhang et al., 2024). Soil also provides diverse ecosystem services that sustain healthy people on a healthy planet. The principal soil ecosystem services are biomass production, carbon pool regulation and provision of habitat to 25% of the Earth's biodiversity, in addition to nutrient and water cycling upon which terrestrial systems depend. There is also a growing sense that healthy soils self-regulate their biological activity and either constrain the growth of pathogens or reduce their disease-causing effects, such as in

http://dx.doi.org/10.19103/AS.2025.0159.14

suppressive soils, which provide an infection-free environment for seeds to germinate, establish and grow (Sahu et al., 2019). The general consensus is that a healthy soil has a continued ability to provide these essential ecosystem functions for plants, animals and humans. Furthermore, the notion of soil health connects soil science and agriculture to land use policy, considering stakeholder requirements and sustainable supply chain management (Lehmann et al., 2020). Since healthy soil has a strong positive relationship with crop yields, soil health is a vital measure of our ability to feed the world's population.

Soil is a complex porous substrate containing mineral particles (silt, clay) coated with well-decomposed organic matter that exists in aggregates. It also contains sand, larger rock fragments, organic residues, roots, microorganisms, meso and macrofauna. The pore space is occupied by water that contains dissolved inorganic and organic substances, along with atmospheric gases and volatiles. Soil functions depend upon energy flow, fundamentally through the biogeochemical cycling of soil organic carbon (SOC) and other nutrients. Energy and nutrients fuel the activities of the living soil biota – microorganisms and soil fauna – which are a vital component of soil health and the ecosystem's multifunctionality supported by soil.

The concept of soil health relies on an understanding of soil physical, chemical and biological processes. For instance, SOC concentration regulates soil health by providing carbon and nutrients to microorganisms. Furthermore, being highly charged, SOC is involved in the formation of organo-mineral complexes and aggregates. Increasing SOC concentration is typically linked to a lower soil bulk density, increased soil porosity and greater nutrient- and water-holding capacity (Garratt et al., 2018). Microbial abundance, microbial biomass carbon (MBC) and microbial community structure are involved in key soil functions such as enzyme activities, nutrient cycling, soil aggregation and soil suppression against plant pathogens (Hartmann and Six, 2023; Philippot et al., 2024; Tonon-Debiasi et al., 2024). In particular, biological indicators of soil health are related to the microbial biomass, microbial community structure and the abundance and diversity of soil fauna (Ologunde et al., 2024).

Microbial activities are constrained by predation from soil microfauna. Furthermore, microorganisms are redistributed by the soil mesofauna and macrofauna that moderate the soil structure, creating porosity that permits gas exchange and supports root expansion, as well as the aerobic respiration of roots and other soil organisms. Soil fauna also transport organic residues and soil organic matter in horizontal and vertical directions, thus bringing nutrient-rich materials into the vicinity of growing roots and helping to increase soil fertility at the plant scale (Al-Maliki et al., 2021; Demetrio et al., 2022).

Farming practices such as conventional tillage versus minimum tillage, monocropping versus crop rotation, agroforestry and residue mulching, and organic fertilizer inputs are often selected to achieve crop yield and economic targets. These farming practices also have a significant influence on soil biological functions under diverse environmental conditions (Nunes et al., 2020; Agyei et al., 2024). This review aims to demonstrate the influence of farming systems on biological indicators that are relevant to soil health. Empirical data from published studies informs the hypothetical model that soil health outcomes are related to more diverse cropping systems, coupled with high inputs of organic amendments and less frequent and shallow tillage, due to the ability of such practices to promote soil organic carbon accumulation. This view is supported by the fact that conservation agriculture promotes soil health under diverse climatic conditions.

2 Major cropping practices across different climatic zones

Farming practices worldwide can be classified as follows:

1. conventional agriculture
2. conservation agriculture

A simple definition of conventional agriculture is that it relies on mechanical tillage that greatly disturbs the soil structure (mouldboard ploughing) and application of inorganic nitrogen, phosphorus and potassium (NPK) fertilizers to produce grain or fiber in a monocropping system, possibly with a simple rotation of two or three annual crops. In contrast, conservation agriculture has minimal soil mechanical disturbance (no tillage, shallow tillage, reduced tillage) and uses both NPK fertilizers and organic residuals (manures, compost, biochar, etc.) to support a diversified cropping system that can include polyculture (e.g. with trees, intercropping) and continuous vegetation cover (e.g. cover crop) (IFOAM, 2005; FAO, 2014; Knapp and van der Heijden, 2018).

Tillage has been synonymous with agriculture for thousands of years, although the addition of inorganic fertilizers and more intense soil disturbance with advanced machinery/vehicles began to expand significantly from the middle of the twentieth century (Farooq and Siddique, 2015). Today, conventional agriculture is the dominant agricultural practice worldwide (Kassam et al., 2019; Vidaller and Dutoit, 2022). Conservation agriculture emerged as an alternative practice in the 1990s (CIMMYT, 2020), although the idea of minimizing tillage to protect soil was first described by Edward H. Faulkner (1945), after the catastrophe of the "Dust Bowls" that hit a wide area of the midwestern United States of America in 1930s. Thereafter, soil conservation was promoted by encouraging farmers to adopt minimum tillage, plant cover crops and leave the crop residues in the field after crop harvest. These practices and other methods to protect soil against water and wind erosion are known collectively as "conservation agriculture" (Farooq and Siddique, 2015). The conservation agriculture movement began in North America and has gradually spread worldwide. By 2015/2016, conservation agriculture was reported to be underway in 78 countries, covering about 180 M ha of cropland, which is approximately 12.5% of the global cropland (Kassam et al., 2019). At present, South and North America have the greatest adoption of conservation agriculture, followed by Australia, New Zealand, Asia, Russia, Ukraine, Europe and Africa (Kassam et al., 2019).

Conservation agriculture continues to expand worldwide, without restriction to a particular climatic zone. In their metadata analysis, Steward et al. (2018) found 715 independent studies that demonstrated the effectiveness of conservation agriculture in tropical and subtropical environments with rainy to dry (arid and semi-arid) climates. Conservation agriculture practices are reported for the arid lands of the Middle East and North African countries such as Morocco, Iran, Iraq, Egypt, Syria, southern Turkey, Lebanon and Tunisia (Devkota et al., 2022). The suitability of conservation agriculture is thus confirmed for arid lands with Mediterranean climate (warm to hot dry summers and rainy cool winters) that receive 100–600 mm rainfall per year (Devkota et al., 2022). Conservation agriculture is practiced in arid to semi-arid temperate regions of Inner Mongolia, China (Zhang et al., 2022), the Indo-Gangetic Plains, India (Patra et al., 2019) and the continental semi-arid monsoon climate of the eastern part of the Loess Plateau (Jia et al., 2022). Conservation agriculture is also widespread in rainy temperate regions, such as Bangladesh (Alam et al., 2020), south-eastern Kenya (Kiboi et al., 2021), southern New South Wales, Australia (Conyers et al., 2019). The positive influence of conservation agriculture has been reported for the cold temperate regions in Switzerland (Oliveira et al., 2024), Denmark (Rocco et al., 2024), multiple locations in the United States (Rui et al., 2022; McDaniel et al., 2023) and Canada (Takam Fongang et al., 2024).

3 Soil biology and soil health

Virtually all terrestrial plants grow in soil, which makes the biological interactions in the soil-plant system important for soil health assessment. Soil biological functions are often differentiated between:

1. the rhizosphere: the root-associated soil that begins at the root surface, extending into the rhizoplane and rhizosheath where root secretions and soil microorganisms intermingle; and
2. bulk soil, where roots are absent.

The rhizosphere is a zone of intense biological activity with a diverse community of soil bacteria, fungi, archaea and other microorganisms, that far exceeds the

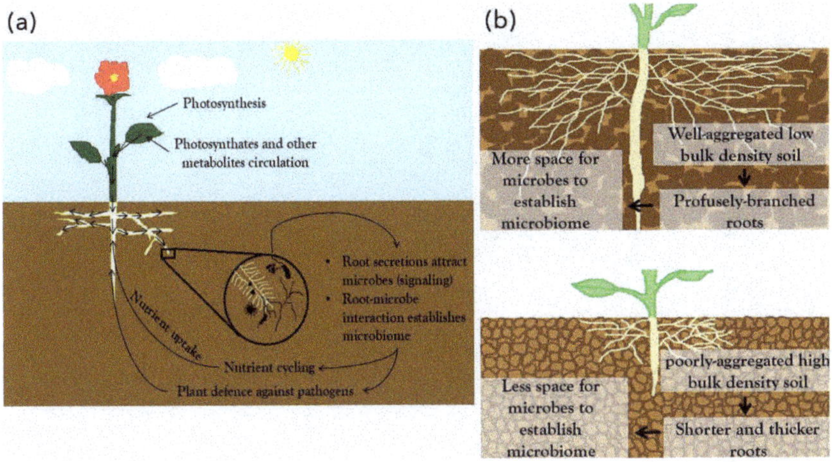

Figure 1 Plant root–microbial interaction (A) in the rhizosphere and (B) controlled by soil structure, particularly porosity and aggregation.

bulk soil (Whalen and Gul, 2023). Soil microorganisms connect plant roots to soil elemental cycles via the decomposition and mineralization of organic matter, solubilisation of inorganic phosphorus, mobilization of potassium and atmospheric nitrogen fixation. Through soil microorganisms, plants can acquire nutrients for their growth and development (Sahu et al., 2019; Fig. 1).

Besides facilitating plant nutrient acquisition, microorganisms also secrete plant growth-promoting hormones (indole acetic acid, gibberellins and cytokinins) that encourage root system elongation, which enhances structural support for the shoots and flower and grain development, along with greater surface area to capture water and nutrients. Although the rhizosphere contains beneficial and pathogenic microorganisms, biotic feedback in the soil microbial community generally strengthens the plant defence system by inhibiting the growth of pathogens, such as by secreting antibiotics and promoting systemic induced resistance in plants (Sahu et al., 2019; Whalen and Gul, 2023). The abundance of beneficial microbial genera in the rhizosphere is considered a biological indicator of soil health because it reflects a favorable environment for crop growth (Sahu et al., 2019; Shi et al., 2021).

Farming practices can modify the soil environment, which affects plant–microbial interactions. Conventional farming reduces, whereas conservation agriculture promotes, the formation of soil macroaggregates and accrual of soil organic matter (Erktan et al., 2020; Dutta et al., 2024; Han et al., 2024; Table 1). Well-aggregated soil facilitates gas exchange and aerobic respiration of plant roots and microorganisms, and a high concentration of organic matter retains more energy, nutrients and water for plants and microorganisms (Fig. 2). Such a soil promotes root growth and activity of diverse microorganisms in rhizosphere

Table 1 Experimental location, climatic zone, cropping system, age of agricultural field, tillage system, soil organic carbon (SOC) (g kg^{-1}) and soil microbial biomass carbon (MBC) (g kg^{-1}) of agricultural fields from diverse climatic zones across the world under conventional or conservation agricultural practices. Gains in soil organic carbon (SOC, in g kg^{-1}) and soil microbial carbon (MBC, in g kg^{-1}) are associated with conservation agriculture practices, including reduced tillage and complex cropping patterns. Data are from a global systematic review of diverse experimental locations, climatic zones, cropping systems, age of agricultural fields and tillage systems. Tillage systems are abbreviated as CT; conventional tillage, NT; no tillage, RT; reduced tillage. Within a study case, values with different letters indicate significant differences at $P< 0.05$.

Experimental location	Climatic zone	Cropping system	Age (year)	Tillage system	SOC	MBC	References
Mills River, North Carolina, United States of America	Humid subtropical	Sweet corn, cucumber, tomato, cabbage, pepper, summer squash, broccoli	16	CT_{NPK}	7.3^c	100^c	Wang et al. (2011)
			16	CTorg.	8.5^c	190^c	
			16	RT_{NPK}	13^b	245^b	
			16	RTorg.	17^a	398^a	
Cuttack, India	Tropical	Rice	3	CT	8.7^b	53.1^e	Neogi et al. (2014)
			3	RT	10.5^a (CT and RT data are based on pooled samples)	69.4^d	
		Maize	3	CT		126.9^b	
			3	RT		139.7^a	
		Cowpea	3	CT		107.7^c	
			3	RT		129.7^a	
Wuxue, Hubei, China	Humid mid-subtropical monsoon	Rice-wheat rotation	2	CT_0	19.60^d	1846^d	Guo et al. (2016)
			2	CTs	21.29^b	2366^b	
			2	NT_0	20.33^c	2024^c	
			2	NTs	21.75^a	2657^a	
Bourne, Melton and Oakham, United Kingdom	Humid temperate	Mostly wheat	Older than no till fields	CT	14.1^b	425^b	Mangalassery et al. (2015)
			7	NT	15.3^a	650^a	

Buenos Aires province, Argentina	Humid temperate	Soybean-maize rotation	10	CT_0	22.0[a]	0.26[b]	Chavarria et al. (2018)
		Soybean-maize rotation with oat, vetch and radish as cover crops	10	$CT_{cover\ crops}$	23.2[a]	0.29[b]	
		Soybean-maize rotation with triticale and vetch as cover crops	10	$RT_{cover\ crops}$	19.9[b]	0.47[a]	
Haryana, India	Semi-arid sub-tropical	Wheat-rice rotation	3	CT_0	6.8[d]	646[e]	Choudhary et al. (2018a)
			3	CTs	9.4[b]	1113[c]	
			3	NT_0	7.3[c]	890[d]	
			3	NTs	10.5[a]	1182[c]	
		Maize-wheat rotation	3	CT_0	7.5[c]	895[d]	
			3	CTs	10.6[a]	1500[b]	
			3	NT_0	9.0[b]	1278[c]	
			3	NTs	10.2[a]	1990[a]	
Karnal, India	Semi-arid sub-tropical	Rice-wheat rotation	5	CT	4.7[b]	170.0[b]	Choudhary et al. (2018b)
		Rice-wheat-mungbean rotation	5	NT	6.9[a]	232.0[a]	
Karnal, Haryana, India	Semi-arid, sub-tropical	Rice-wheat rotation	4	CT_0	16.2[b]	441[b]	Jat et al. (2019)
			4	NTs	20.3[a]	626[a]	
Mymensingh, Bangladesh	Tropical monsoon	Rice-wheat-mung bean rotation wheat-mung bean-–rice rotation	7	CT_{LS}	11.6[b]	28.02[a]	Jahangir et al. (2021)
				CT_{HS}	14.9[a]	29.89[a]	
				Strip tillage $_{LS}$	14.4[a]	35.43[a]	
				Strip tillage $_{HS}$	16[a]	36.96[a]	

(Continued)

Table 1 (*Continued*)

Experimental location	Climatic zone	Cropping system	Age (year)	Tillage system	SOC	MBC	References
Fars, Iran	Semi arid Mediterranean	Wheat-corn rotation	2 years of CA and 7 years of CT practices	CT_0	10.3[a]	306.37[c]	Mirzavand et al. (2022)
				CTs		339.74[c]	
				RT_0	10.6[a]	479.27[b]	
				RTs		660.19[a]	
Diversified cropping system as crop rotation under conservation agriculture							
La Pampa, Argentina	Humid temperate	Sorgham	3	NT	692.5[b]	90[b]	Frasier et al. (2016)
		Sorgham-rye	3	NT	697.0[b]	95[b]	
		Sorgham-vetch	3	NT	843.5[a]	105[ab]	
		Sorgham-rye + vetch	3	NT	811.0[a]	120[a]	
Iowa, United States of America	Humid temperate	Corn-soybean rotation	2	RT	29.49[a]	350[b]	King and Hofmockel (2017)
		Corn-soybean-small grain + red clover	3	RT	28.02[a]	408[ab]	
		Corn-soybean-small grain + alfalfa	4	RT	29.20[a]	508[a]	
Patna, Bihar, India	Humid subtropical	Rice-wheat-fallow	7	RT	21.0[b]	69.87[c]	Samal et al. (2017)
		Rice-wheat-greengram	7	NT	22.5[ab]	83.76[b]	
		Rice-wheat-cowpea	7	NT	24.5[a]	89.32[a]	
		Rice-Potato+maize-cowpea	7	RT	19.5[c]	88.17[a]	
Karnal, India	Semi-arid, subtropical	Rice-wheat rotation	5	CT	4.7[b]	170.0[c]	Choudhary et al. (2018b)
		Rice-wheat-mung bean	5	RT	6.1[a]	207.0[b]	
		Rice-wheat-mung bean	5	NT	6.9[a]	232.0[b]	
		Maize-wheat-mungbean	5	NT	6.7[a]	265.9[a]	

Kirinyaga, Kenya	Tropical	Potato	3	–	–	98^{c}	Nyawade et al. (2019)
		Potato intercropped with lima bean	3	–	–	106^{b}	
		Potato intercropped with dolichos	3	–	–	198^{a}	
Zengcheng, China	Tropical	Rice	3	–	10.5^{b}	330^{b}	Wang et al. (2021)
		Rice intercropped with *Thalia dealbata*	3	–	12.8^{a}	490^{a}	
		Rice intercropped with *Acorus calamus*	3	–	12.5^{a}	405^{a}	
Hoshiarpur, Punjab, India	Tropical semi-arid	Fodder-wheat rotation	3	–	4.39^{c}	0.074^{c}	Singh et al. (2021)
		Eucalyptus, fodder-wheat rotation	3	–	7.49^{b}	0.113^{b}	
		Poplar, fodder-wheat rotation	3	–	8.24^{a}	0.147^{a}	
Arlington, WI USA	Cold humid temperate	Continuous maize	29	CT	21.3^{b}	321^{c}	Rui et al. (2022)
		Maize-soybean rotation	29	RT	21.3^{bc}	309^{c}	
		Maize-soybean-wheat rotation	29	RT	20.1^{c}	591^{b}	
		Maize-alfalfa-alfalfa-alfalfa rotation	29	RT	22.5^{bc}	543^{b}	
		Maize-oats/alfalfa-alfalfa rotation	29	RT	23.7^{ab}	842^{a}	
Laddakh, Himalaya, India	Cold desert	Barley, mustard, oats as monocrops	10	CT	4.09^{c}	244.4^{c}	Sharma et al. (2022)
		Barley–buckwheat or barley–turnip rotation	10	CT	4.97^{b}	337.5^{b}	

(*Continued*)

Table 1 (*Continued*)

Experimental location	Climatic zone	Cropping system	Age (year)	Tillage system	SOC	MBC	References
		Agroforestry (poplar/willow)	10	RT	6.20[a]	446.4[a]	
		Orchards of apple, pear, apricot, grapes or walnut	10	RT	6.01[a]	358.4[b]	
New Delhi, India	Semi-arid subtropical	Maize-rajmash-buckwheat	3	CT	11.85[b]	325[c]	Babu et al. (2023)
			3	RT	12.08[ab]	352[b]	
			3	NT	12.62[a]	383[a]	
Hickory Corners, United States of America	Cold humid temperate	Maize–soy rotation	12	RT	7.70	257[b]	McDaniel et al. (2023)
		Maize–soy–wheat rotation + Cereal rye and clover cover crops	12	RT	9.63	427[ab]	
		Spring fallow – after tillage, the regrowth of natural grassland species with up to 10 species of plants	12	RT	8.74	682[a]	
New Delhi, India	Subtropical semi-arid	Rice-wheat-mung bean	10	CT	5.43[c]	548.0[c]	Das et al. (2024)
		Rice-wheat rotation	10	NT	6.57[b]	716.0[b]	
		Rice-wheat rotation + residue return	10	NT	6.66[b]	755.0[b]	
		Rice-wheat rotation + residue return + Sesbania bispinosa co-cultivation	10	NT	6.64[b]	763.0[b]	

		Rice-wheat-mung bean rotation	10	NT	7.44[a]	844.0[a]	
		Rice-wheat-mung bean rotation + residue return	10	NT	7.46[a]	846.0[a]	
Pendleton, Oregon, USA	Semi-arid Mediterranean	Winter wheat-fallow rotation	28	NT	17.6[b]	120[b]	Domnariu et al. (2024)
		Winter wheat-pea rotation	11	NT	18.9[a]	180[a]	

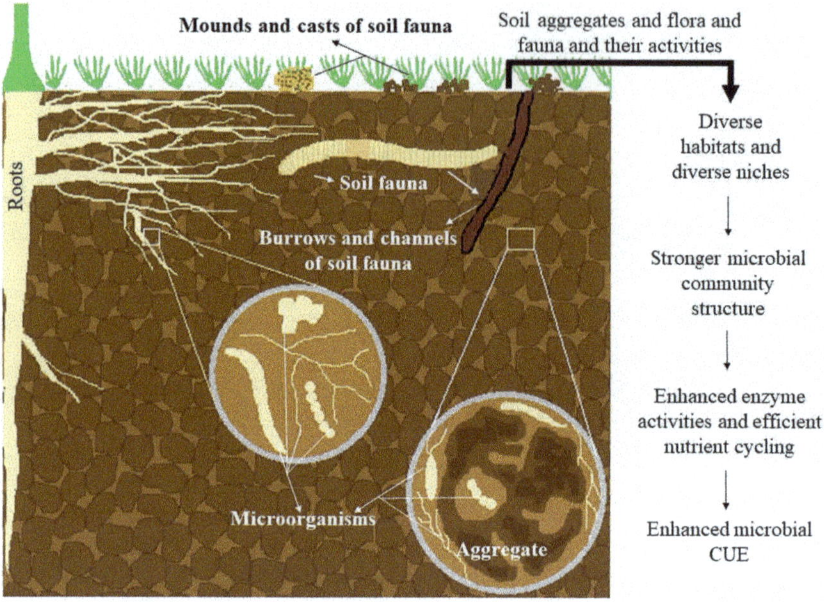

Figure 2 Soil microhabitats created by growing roots and burrowing soil fauna are vital niches for soil microorganisms that control microbial abundance, activities and carbon use efficiency (CUE)

and bulk soil (Bhattacharyya et al., 2021, 2022; Fig. 1). For example, Shi et al. (2021) reported more beneficial microorganisms of the genus *Sphingomonas* and the phylum *Proteobacteria* in the rhizosphere of Lanzhou lily (*Lilium davidii*) with an increase in soil organic matter. The bulk soil with lower soil organic matter concentration had more pathogenic bacteria of *Bacillus simplex* and the genus *Streptomyces* and fewer beneficial bacteria of the genera *Hyphomicrobium* and *Sphingomonas*. Likewise, Hu et al. (2021) reported a greater abundance of *Clonostachys* and *Fusarium* in the rhizosphere of maize in soil with low organic matter concentration following 7 years of mouldboard tillage practice, compared to the soil with high organic matter concentration under reduced tillage practice in a temperate continental monsoon climate in jilin Province, China. They also found that more pathogens in the rhizosphere soil were negatively related with crop yield.

Farming practices also influence soil microbial community structure (Table 2), which in turn influence the biotic control of crop pathogens (Mackay et al., 2023). As evident from Table 2, conservation agriculture enhanced the abundances of bacteria from 3% to 329% and the abundances of fungi from 527% (P0.05) compared to neighboring conventional fields, with variation due to the length of time since the conservation agriculture practices were adopted and other factors (e.g. net primary productivity, climatic conditions and other

Table 2 Experimental location, its climate, cropping systems of the research field, age of agricultural field, tillage system, bacterial and fungal abundances of soil. Gains in soil microbial (bacteria, fungi, actinobacteria) abundances are associated with conservation agriculture practices including reduced tillage and complex cropping patterns. Data are from a global systematic review of diverse experimental locations, climatic zones, cropping systems, age of agricultural fields and tillage systems. Tillage systems are abbreviated as CT, conventional tillage; NT, no tillage; RT; reduced tillage. 0, no residue return; r, residue return, units for bacterial and fungal abundances are given in the first cell of a given study' G^{+}, gram positive bacteria; G^{-}, gram negative bacteria; A;,actinobacteria; AM;,arbuscular mycorrhiza. Within a study case, values with different letters indicate significant differences at $P < 0.05$.

Experimental location	Climatic zone	Cropping system	Age (year)	Tillage system	Bacterial abundance	Fungal abundance	References
Douglas, Georgia, United States of America	Humid temperate	Cotton	30	CT	$G^{+} = 8.19^{b}$ (PLFA, units not specified)	9.08^{b} (PLFA, units not specified)	Simmons and Coleman (2008)
					$G^{-} = 6.52^{b}$		
				NT	$G^{+} = 8.81^{a}$	17.74^{a}	
					$G^{-} = 8.92^{a}$		
Belle Mina, Alabama, USA	Subtropical humid	9–13 years of cotton monocrop, 4 years of corn monocrop	14	CT	Bacteria = 20.9^{b} (nmol g^{-1} soil PLFA)	Saprophytic fungi = 1.40^{b} (nmol g^{-1} soil PLFA)	Mathew (2012)
					Actinobacteria = 0.99^{b}	AM fungi = 1.17^{b}	
		9–13 years of cotton monocrop, 4 years of corn monocrop	18	NT	Bacteria = 50.9^{a}	Saprophytic fungi = 4.47^{a}	
					Actinobacteria = 2.09^{b}	AM fungi = 4.32^{a}	
La Cage near Versailles, France	Oceanic temperate	Wheat-pea-wheat-oilseed rape	14	CT_r	10.02^{c} (Log10 gene copy number, g^{-1} soil)	8.51^{b} (Log10 gene copy number, g^{-1} soil)	Henneron et al. (2015)

(*Continued*)

Table 1 (*Continued*)

Experimental location	Climatic zone	Cropping system	Age (year)	Tillage system	Bacterial abundance	Fungal abundance	References
		Wheat-pea-wheat-oilseed rape-(alfalfa-alfalfa	14	NT_r	10.20[b]	8.52[b]	
		Wheat-pea-wheat–maize + permanent plant cover of red fescue	14	NT_r	10.35[a]	8.75[a]	
Haryana, India	Semi-arid sub-tropical	Maize-wheat rotation	3	CT_0	81.6 Total bacteria (CFU ×104 g^{-1} soil)[ef]	54.3[de]	Choudhary et al. (2018a)
			3	CT_r	94.5[ab]	73.2[a]	
			3	NT_0	88.8[bc]	66.3[b]	
			3	NT_r	96.2[a]	77.3[a]	
		Rice-wheat rotation	3	CT_0	74.7[f]	45.3[f]	
			3	CT_r	84.0[cde]	58.8[cd]	
			3	NT_0	79.3[ef]	52.0[e]	
			3	NT_r	86.7[cd]	64.3[bc]	
Karnal, Haryana, India	Semi-arid, sub-tropical	Rice-wheat rotation	4	CT_0	Total bacteria = 11.4 (c.f.u × 10^5 g^{-1} soil) Actinobacteria = 4.4	6.1[c]	Jat et al. (2019)

		Rice-wheat-mungbean rotation	4	$NT_{\text{residue of only rice returned}}$	Total bacteria = 14.9* Actinobacteria = 7.3*	7.4[b]	
		Rice-wheat-mungbean	4	NT_r	Total bacteria = 12.3 Actinobacteria = 7.2*	6.2[c]	
		Maize-wheat-mungbean	4	NT_r	Total bacteria = 16.9* Actinobacteria = 8.4*	11.7[a]	
Umbauba, Sergipe State, Brazil	Tropical with dry summers	Maize monocrop	3	–	G^+ = 40.8 (% mol PLFA) G^- = 29.2[a] A = 2.2[b]	Saprophytic fungi = 10.6[b] (% mol PLFA) AM fungi = 4.1[a]	Lopes et al. (2020)
		Maize with pigeonpea (*Cajanus cajan*) as intercrop	10	–	G^+ = 42.1	Saprophytic fungi = 11.7[a]	
Zengcheng city, China	Tropical monsoon	Rice monocrop	3	NT	G^+ = 20 n mol g^{-1} soil G^- = 38 A = 11	15[b]	Wang et al. (2021)

(*Continued*)

Table 1 (*Continued*)

Experimental location	Climatic zone	Cropping system	Age (year)	Tillage system	Bacterial abundance	Fungal abundance	References
		Rice intercropped with *Thalia dealbata*	3	NT	$G^+ = 55$*	29^a	
					$G^- = 87$*		
					A = 24*		
Zarghan Field Station, Fars, Iran	Semi arid Mediterranean	Wheat monocrop	6	CT0	74.5×10^5 (no. $cm^2)^d$ (total bacterial abundance)	21.7×10^2 (no. $cm^2)^d$	Mirzavand et al. (2022)
			6	CTr	119×10^{5d}	83.83×10^{2c}	
			6	RT0	130×10^{5d}	136.2×10^{2b}	
			6	RTr	375.3×10^{5b}	188.2×10^{2a}	
			6	NT0	268.8×10^{5c}	169.1×10^{2a}	
			6	NTr	463.5×10^{5a}	205.3×10^{2a}	
		Corn monocrop	6	CT0	57.00×10^{5c}	57.26×10^{2c}	
			6	CTr	97.50×10^{5c}	76.15×10^{2c}	
			6	RT0	127.2×10^{5c}	104.6×10^{2b}	
			6	RTr	319.7×10^{5b}	150.8×10^{2a}	
			6	NT_0	402.2×10^{5b}	115.5×10^{2b}	
			6	NTr	419.2×10^{5a}	151.6×10^{2a}	

Chuka, Kenya	Humid temperate	Maize-corn-vegetables-legumes-potato	12	Residue removed + synthetic fertilizer + manure	3.76 x 10^9 16S rRNA gene copies [g^{-1} DW soil][b]	–	Krause et al. (2023)
		Maize-corn-vegetables-legumes-potato	12	Residue returned + organic fertilizers (composted manure, *Tithonia* tea and mulch)	1.31 x 10^{10}[a]	–	
Mymensingh and Ishurdi, Pabna, Bangladesh	Subtropical humid, monsoon	Wheat-Aman rice-lentil-Aman rice-mustard and Wheat-lentil-mustard	3	CT	5.14E+05[b] (CFU g^{-1} soil)	4.38E+03[c] (CFU –[1] soil)	Sarker et al. (2023)
				RT	5.69E+05[b]	7.44E+03[b]	
				NT	7.56E+05[a]	1.02E+04[a]	
New Delhi, India	Subtropical semi-arid	Rice-wheat-mung bean	10	CT	1.40[d] (CFU ×105 g^{-1})	1.07[b] (CFU ×105 g^{-1})	Das et al. (2024)
		Rice-wheat rotation	10	NT	1.67[cd]	1.33[b]	
		Rice-wheat rotation + residue return	10	NT	2.40[b]	1.45[b]	
		Rice-wheat-mung bean rotation	10	NT	3.12[a]	2.60[a]	
		Rice-wheat-mung bean rotation + residue return	10	NT	3.17[a]	2.45[a]	

factors) (Table 2). Conservation tillage encourages bacterial-fungal interactions that reduce pathogen infections in crops (Hu et al., 2021). This suggests that the well-aggregated soil with high organic matter resulting from conservation agricultural practices can increase the abundance of beneficial fungi and bacteria (Tautges et al., 2016). Thus, we consider the abundance of pathogens as a key biological indicator of soil health, and for conservation agriculture to potentially suppress pathogens to improve crop growth and yield.

4 Soil fauna as indicators of soil health: microfauna, mesofauna and macrofauna

4.1 Soil microfauna

Soil is inhabited by a multitude of soil animals, which are classified according to their size. This includes microfauna (<100 μm in size) such as nematodes and protists; mesofauna (100–2000 μm in size) like the acari (mites), enchytraeids, diplura, symphyla, collembola, isoptera and diptera; and macrofauna (>2000 μm in size), such as oligochaete (earthworms), Enchytraeidae, isopoda, termites, ants, mollusca, coleoptera (beetles), arachnida (spiders, scorpions) and myriapoda (millipedes and centipedes) (Coleman et al., 2024).

Among the soil animals, the protists (amoeba, flagellates and ciliates) are the most abundant unicellular eukaryotes. More than 10 000 protists are commonly found in a single gram of soil (Stefan et al., 2014), with hundreds of species of soil protists in a typical sample (Geisen et al., 2018). Among them are phototrophs, bacterivores, fungivores, plant pathogens, aboveground animal parasites, algal herbivores and predators of soil nematodes, although they are commonly known as dominant predators that regulate the size and function of soil bacterial populations (Potapov et al., 2022). Interestingly, all functional groups are present in every soil worldwide (Potapov et al., 2022). Nematodes are another type of soil microfauna. With more than 25 000 species of soil nematodes, they are the most abundant multicellular organisms found in soil (van den Hoogen et al., 2019). They occupy all the trophic levels of the food web in the rhizosphere and bulk soil as plant parasites, fungivores, bacterivores, omnivores and predators. Although their direct benefit to plants is not yet documented, they are well-known as root herbivores that induce systemic resistance in plants against pathogens (Topalovic and Geisen, 2023). The diversity and abundance of soil nematodes are indicators of healthy soil (Zhang et al., 2017). By regulating soil bacteria and fungal populations, protists and nematodes are involved in nutrient cycling and indirectly contribute to plant nutrition (Chandarana and Amaresna, 2021; Yang et al., 2021). As the population densities of bacterial or fungal species increase, these microfauna prey upon those bacterial or fungal populations, providing habitat and substrates that allow other microbial species to grow (Henneron et al., 2015). Therefore, soil microfauna help maintain a resilient and dynamic microbial community structure, which is important for nutrient cycling and the development of suppressive soils that benefit plants.

The diversity and abundance of individual groups of protists and nematodes is linked to agricultural practices. Conventional agricultural practices tend to reduce, whereas conservation agricultural practices tend to increase microfauna diversity (Henneron et al., 2015; Zhang et al., 2017). A study in Zurich, Switzerland detected more than 100 different nematode genera in soil samples from vegetable fields (sample size = 20 under conventional agriculture, *n* = 20 with no tillage, no pesticide and no synthetic fertilizer input) and grasslands (*n* = 20; Yang et al., 2021). Nematode diversity was negatively affected by the land use intensification associated with conventional tillage and monocropping (Yang et al., 2021). This suggests that conservation agriculture should promote microfaunal diversity and, by consequence, increase this biological indicator of soil health.

4.2 Soil mesofauna

Soil mesofauna are found in all terrestrial ecosystems including Antarctica. Nearly 10 000 species of Oribatida, 6000 species of Astigmata, 26 000 species of Trombidiformes (including Prostigmata), 12 000 species of Mesostigmata, 3500 species of Pseudoscorpions, 201 species of Symphyla, 900 species of Pauropoda, more than 1000 species of Diplura and 9000 species of Collembola are reported (Potapov et al., 2022). The soil mesofauna occupy all trophic levels as primary, secondary and tertiary consumers. They feed on freshly fallen and decomposing litter, bacteria, fungi (including mycorrhiza), mosses, lichens, algae and some are predators of other mesofauna, nematodes, rotifers and protozoans (Potapov et al., 2022).

The abundance and diversity of soil mesofauna is an indicator of soil health due to their involvement in soil physico-chemical and biological processes. Soil mesofauna facilitate the decomposition of organic matter, regulate the microbial community structure, distribute organic matter in the soil and promote soil aggregation (Erktan et al., 2020; Fig. 2). Farming practices affect the abundance and diversity of soil mesofauna, and they tend to be sensitive to soil mechanical disruption and land use. Compared to conventional tillage, long-term conservation tillage increased the abundance of two collembolan species; *Ceratophysella spp.* and *Entomobrya sibirica* and the diversity of bacteria associated with these mesofauna, specifically increasing the abundance of *Pelomonas and Sphingomonas* bacterial genera (Liu et al., 2024a). As many species of *Sphingomonas* are plant growth promoting microorganisms (Lee et al., 2023; Kaur and Karnwal, 2023; Sultana et al., 2024), they could be associated with more collembola in conservation agriculture for the benefit of crops.

4.3 Soil macrofauna

Among soil macrofauna, 6700 species of spiders (Araneae), 8000 species of millipedes (Diplopoda), 3300 species of centipedes (Chilopoda), 4000 species of woodlice (Oniscidea), 1900 species of earwigs (Dermaptera), 26 000 species of Orthoptera (grasshopper), 460 species of Embioptera (webspinners or footspinners), 4640 species of cockroaches, 3100 species of termites (Termitoidae), 6300 species of thrips (Thysanoptera), more than 100 000 species of true bugs (Hemiptera), 6000 species of booklice, psocids (Psocoptera), 16 000 species of ants (Formicidae), 400 000 species of beetles (Coleoptera), 160 000 species of butterflies (Lepidoptera), 7000 species of earthworms (Annelida), 760 species of pot worms (Enchytraeidae) and 25 000 species of gastropods (Mollusca) are reported (Potapov et al., 2022). They are widespread in global soils, with localized patchy distribution according to their ecological niche.

Earthworms, termites and ants are considered to be soil ecosystem engineers. They distribute freshly fallen plant tissues, partially decomposed or soil-humus-complex organic substances throughout the soil profile and across the soil surfaces. Furthermore, they facilitate nutrient cycling and carbon transformations. The burrows and channels formed by earthworms are macropores, which not only help in soil aeration, it also provide habitat for other soil organisms. The most important role of soil macrofauna is creating the soil structure, and their abundance in soil is positively related to the formation and stability of soil macroaggregates (Erktan et al., 2020). Macroaggregates provide refuge to soil microorganisms and mesofauna from predators, since they can live in pores corresponding to their body size within aggregates and between aggregates (Fig. 2). Since they facilitate soil biological functions, therefore, the diversity and abundance of individuals classified as soil macrofauna are important indicators of soil health (Velasquez and Lavelle 2019; Erktan et al., 2020). Farming practices greatly influence the diversity and abundance of the macrofauna. Conventional tillage practices often damage earthworms and other macrofauna because the tillage abrades their bodies and exposes them to predators, while also destroying their habitat by filling burrows with soil, chopping organic residues at the soil surface and mixing them into the profile. Consequentially, tillage reduce the habitat and niches of soil macrofauna, and is negatively related to several soil functions such as macroaggregate structure, bulk density, soil aeration and water infiltration (Erktan et al., 2020; Fig. 3). How conventional and conservation agricultural practices influence soil fauna is explained further in the following sections.

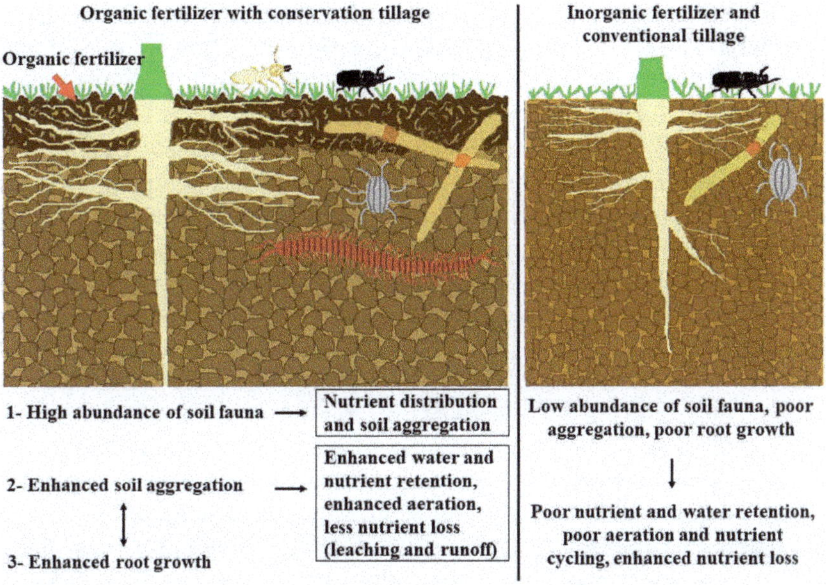

Figure 3 Fewer nutrient losses are anticipated from organically fertilized agroecosystems under conservation tillage practices, relative to conventional agroecosystems, in part due to soil fauna that support greater nutrient retention in the soil-plant system, including aggregates.

5 Influence of cropping practices on soil microbial abundance and community structure

5.1 Soil microbial abundance

Soil microbial abundance is either the number of microorganism (i.e. individuals of distinct species identified by direct counting or estimated from gene copies in DNA extracts, in other words alpha diversity) or their biomass, typically determined by the low-cost chloroform fumigation-direct extraction procedure as microbial biomass carbon (MBC) (Vance et al., 1987; Joergensen and Mueller, 1996). Microbial abundance is directly linked to soil enzyme production and the associated nutrient cycling (Das et al., 2024; Domnariu et al., 2024). Furthermore, the MBC is positively related with soil macroaggregate formation and stability (Jat et al., 2019; Jahangir et al., 2021; McDaniel et al., 2023). As the soil MBC concentration is associated with soil structure, linked to gas exchange, water infiltration and root proliferation, as well as nutrient cycling and soil organic carbon storage in connection to crop production, it is a suitable biological indicator of soil health.

Agricultural management significantly influences the abundance of soil microorganisms. In general, conservation tillage (reduced or no till) increases the soil MBC concentration relative to conventional tillage in adjacent plots or fields (Table 1). Interestingly, the 8–162% greater ($P < 0.05$) MBC concentration in conservation tillage than conventional tillage is consistent in agroecosystems across climatic zones of the world (Table 1). When reduced or no tillage is accompanied by soil amendment with organic materials, such as residues retained after crop harvest (termed 'residue return'), this further enhances the abundance of soil microorganisms by 31–56% (Table 1). Interestingly, the positive influence of conservation tillage, coupled with residue return or crop diversification also exerted a significant positive influence on the yield of crop as compared to conventional tillage with residue removal and monocropping system (Samal et al., 2017; Choudhary et al., 2018b; Das et al., 2024). For example, 2 years of conservation tillage of a wheat-corn rotation in Fars, Iran, increased the abundance of microbial biomass by 56% in the upper 0–30 cm soil layer, compared to 7 years of conventional tillage in the same cropping system. Interestingly, the experimental plots under conservation tillage, in which residues were returned, the MBC was significantly higher than conventional tillage by 94% and by 38% than the plots under conservation tillage with residue removal (Mirzavand et al., 2022). They also found a strong positive correlation between the yield of wheat and corn and the concentration of MBC ($R^2 = 0.91$ for wheat and $R^2 = 0.95$ for corn). As conceptualized in Fig. 4, the soil MBC concentration is likely responding to the delivery of more plant-based substrates to multiple niches at the soil surface and throughout the soil profile, which is associated with improved ecological functions such as greater crop production, more soil organic carbon storage and enhanced soil health.

5.2 Soil microbial community structure

Soil microbial community structure refers to assemblage of soil microbial species, also termed beta diversity. Microbial communities that contain an abundance of all microbial groups (bacteria, actinobacteria, archaea, saprophytic and arbuscular mycorrhizal fungi) is indicative of healthy soil (Hartmann and Six, 2023). The inherent redundancy within the soil microbial community means that more beta diversity assures full coverage of soil functions regardless of the environmental conditions, to sustain enzyme activities for nutrient cycling, plant–microorganisms interaction for better plant growth performance and disease resistance, and soil aggregation (Hartmann and Six, 2023).

Agricultural practices are important for structuring soil microbial communities. In general, conservation tillage can significantly increase the abundance of various soil microbial groups, relative to conventional tillage, with additional diversity achieved through crop diversification (Table 2). For example, 8 years of maize-soybean rotation increased the relative abundances of bacterial groups, i.e. Actinobacteria, Proteobacteria, Nacardioides, and

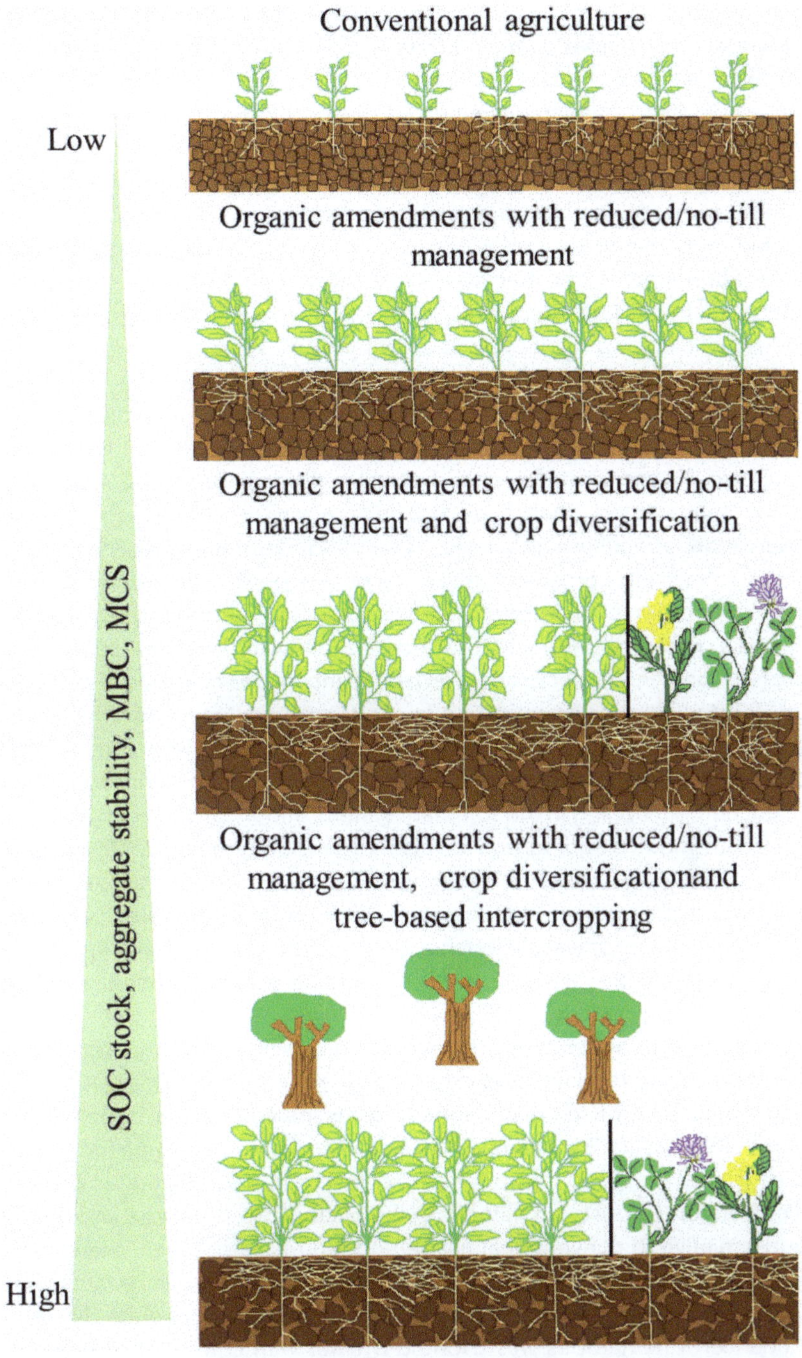

Figure 4 Cropping system diversification delivers more plant-based substrates to multiple niches in the soil profile, leading to a greater SOC stock, more aggregate stability, larger microbial biomass (MBC) and more complex microbial community structure (MCS)

the relative abundances of bacterial genera *Sphingomonas, Arthrobacter, Cryobacterium, Haliangium, Marmoricola, Methylibium, Ramlibacter, Rhizobacter,*and *Variovorax* compared to soybean monocropping in a cold temperate continental climate (Xu et al., 2024). Interestingly, a positive relationship was also observed between the yield of crop and the abundances of these microbial groups. Not only did crop rotation increase the abundance of these microbial groups, but it also increased the abundances of genes involved in phosphorus and potassium metabolism. The implication is that greater abundance of soil bacteria is associated with more efficient uptake of essential plant nutrients like phosphorus and potassium, although this may be a site-specific finding that depends upon the soil fertility and soybean nutrient requirements.

The connection between soil microbial communities and crop production was also documented in Europe. The field sites (Lelystad, the Netherlands, Gembloux, Belgium and Hamerstorf, Germany) compared maize monoculture and an annual crop (maize, potato) with a winter cover crop to several conservation agriculture practices: diversified crop rotations, cover crops, residue return and organic fertilizer (pig slurry). The diversified cropping system with organic amendments (cover crops, residue return, pig slurry) not only had a greater abundance of soil fungi, bacteria and archaea, but there was more N cycling and larger proportion of N-fixing microorganisms in the soil (Liu et al., 2024b). Further investigation is needed to evaluate over time, under a given climatic condition, the amendment rate and the timing of organic inputs, required to improve the microbial community structure and consequently the efficient nutrient cycling in the soil-plant system, under conservation agriculture under both monocropping and diversified cropping systems. Moreover, the influence of such practices on crop growth performance such as biomass and yield production, nutrient uptake and nutrient use efficiency also needs to be monitored on temporal scale. Such a study will help scientists adapt the best management practice under a given climatic condition for better soil health and high crop production.

Conservation agriculture that includes diversified cropping systems with cover crops, residue return and organic fertilizer also has a positive influence on the abundance of soil microfauna (protists, nematodes) in diverse climatic regions of the world. For instance, 10 years of conservation tillage (ridge tillage and no tillage) increased the abundance of soil bacteria and fungi, as well as the soil microfauna (protists, nematodes in the bacterivore, omnivore and predator groups) relative to conventional tillage under a maize-soybean rotation in the temperate monsoon climate of Dehui County, Jilin Province, China (Zhang et al., 2015). In a farm at La Cage, France, which has an oceanic temperate climate, the abundance of nematodes was two-fold greater with conservation tillage and red fescue cover cropping (14 year history) than conventional tillage (Henneron et al., 2015). Sarker et al. (2023) reported the soil fungal abundance increased significantly and the nematode abundance was similar or greater after 3 years of conservation tillage, relative to conventional tillage, in wheat-based cropping systems (Wheat-Aman rice-lentil-Aman rice-mustard and Wheat-lentil-mustard rotation) in the subtropical humid monsoon climate of an agricultural farm at Mymensingh and Ishurdi, Pabna, Bangladesh. Four years of reduced tillage with or without residue return increased the abundances of soil microfauna in the upper soil layer (0–5 cm depth) as compared to conventional tillage under similar cropping system (spring pea-winter wheat-rapeseed-spring barley-maize-winter wheat rotation) in an agricultural field at Estrees-Mons, Northern France with a temperate semi-arid climate (Hedde et al., 2024). These findings are expected, since soil microfauna are predators of soil microorganisms and should be more numerous when more microbial prey are present in the soil ecosystem.

6 Influence of cropping practices on soil mesofauna

Agricultural practices are expected to affect the soil mesofauna, which is another biological indicator of soil health. Well-structured soil, formed when soil minerals aggregate with fungal hyphae and organic biopolymers and plant roots through the activities of soil fauna (earthworms, ants, termites, etc.), provides well-aerated favorable habitat to soil fauna (Tisdall and Oades, 1982; Or et al., 2021; Coleman et al., 2024). Deep and frequent mechanical disturbance of soil through tillage practices disrupts soil structure and reduces bulk density, which in return reduces the soil fauna abundance (Betancur-Corredor et al., 2022; Tschanz et al., 2024).

An extensive meta-analysis by Betancur-Corredor et al. (2022) suggests that reduced tillage increases the abundance of soil mesofauna in agroecosystems across the world. For instance, 10 years of conservation tillage and crop diversification at the agricultural research farms at Bethlehem and Reitz, South Africa (regions with arid to semi-arid climatic conditions) significantly increased the diversity of arthropods (including mesofauna and macrofauna) by approximately 45% relative to fields under conventional tillage with monocroping of soybean, maize, oats, sunflower and wheat (Mamabolo et al., 2024). In this study, crop diversification involved growing those crops in mixed cultivation (Mamabolo et al., 2024). Likewise, Rieff et al. (2020) found no-till practice as more protective for collembolans, mites and enchytraeids than conventional tillage for the soil under a maize monocropping system at the research farm at Penacova Municipality, Portugal, with temperate arid climate. Likewise, three croplands with 7, 10 and 17 years of conservation tillage practices at Paris basin, France with cold humid temperate climate, on average, increased the abundances of Euedaphic and Hemiedaphic collembola by

33% and 492% respectively than the soils of 6 croplands under 20 years of conventional agriculture ($P < 0.05$) (Chassain et al., 2024). Another study that was conducted in Oakpark Research Centre, Carlow, Ireland, with cool temperate oceanic climate, revealed a 35–37.5% significant increase in the abundance of collembola in the 0–9 cm soil layer under reduced than conventional tillage in a winter wheat monocropping system (Brennan et al., 2006). However, using litter bag technique to collect collembola, Hanisch et al. (2022) found five-fold higher abundance of collembola in soil under conventional than minimum tillage practice with similar cropping system (cereals and oilseed rape) in the agricultural field at Lower Saxony, Germany, with semi-arid temperate climate. Contrary to the findings of Hanisch et al. (2022), the soil texture of the study of Rieff et al. (2020) was sandy loam (51% sand, 35% silt and 13% clay); whereas, the soil related to the study of Hanisch et al. (2022) was fine-textured (silty clay loam with 12% sand, 73% silt and 15% clay). It is possible that deep tillage in fine-textured high organic matter soils may promote the abundance of collembola by providing them with more well-aerated microhabitats to deeper soil depths (Hanisch et al., 2022). The study of Hanisch et al. (2022) was related to the placement of litterbags at different depths in conventional and reduced tillage systems. They placed litter bags at 20 cm depth in conventional tillage system and at 10 cm depth in conservation tillage system. This might be another reason why the abundances of collembola were found higher under conventional than conservational tillage systems. Another study that was conducted in the research farm under wheat monocropping system, at Rugballegaard, Denmark, with warm humid temperate climate, conventional tillage in the upper soil layer (0–4 cm) strongly reduced the abundance of collembola but the collective effect of tillage on the abundance of collembola from the depth of 0–4 to the depth of 4–32 cm soil was not different between no-till and conventional tillage (Petersen, 2002).

In general, conservation agriculture increases the abundance of soil mesofauna. This is possibly due to the extra habitat and resources associated with more plant cover and more diverse plant species, which promote the biodiversity of soil fauna (Elmquist et al., 2023; Mamabolo et al., 2024). For instance, Ren et al. (2024) reported that the abundances of Acarina, Collembola, Diplura, Hymenoptera and Thysanoptera at 0–20 cm soil depth were associated with soil nutrients, with greater soil organic carbon and total nitrogen found in the winter wheat cropping system that had a summer legume crop (Huai bean, soybean and mung bean) for 14 years, rather than a summer fallow. Also, an agricultural farm at Changwu County, Shaanxi Province, China with continental semi-arid climate, there was 27–57% more MBC and unique soil mesofauna such as Elateridae and Eosentomidae in the diversified cropping system (Ren et al., 2024). In the humid oceanic climate of Halland county, Sweden, the conservation system had perennial leys grown for three years in a 7-year old field (that was previously conventionally tilled) with a crop rotation of barley, oat, winter wheat, fava bean, kale, rye and pea, with organic fertilizer amendment; whereas, the conventional practice used conventional tillage, inorganic fertilizer amendment and with no perennial ley (Heinen et al., 2023). In this study, a significant increase in the abundances of collembola, oribatida, mesostigmata, Acari juvenile and other arthropods were found in the upper 0–10 cm soil layer of the conservation system compared to conventional agriculture (Heinen et al., 2023). In general, these empirical evidences support our assertion that soil mesofauna abundance increases with conservation agricultural practices such as reduced or no tillage, crop diversification and cover cropping.

7 Influence of cropping practices on soil macrofauna

Agricultural management greatly influences the abundance of soil macrofauna, another biological indicator of soil health (Table 3). In general, there are more soil macrofauna when fields are minimally tilled (reduced or no tillage) or experience other conservation agriculture practices such as cover cropping, residue return, crop rotation and organic fertilizers. For instance, the abundance of total soil macrofauna increased by 163–218% in a conservation agriculture systems compared to conventional tillage with no residue return (i.e. there were 29, 35 and 11 individuals m^{-2} soil in rotational cropping, no till with residue return and conventional tillage system with no residue return; Sithole et al., 2018); this finding was observed after 13 years of these cropping practices for wheat-soybean production under the semi-arid climate at a research farm in KwaZulu-Natal, South Africa. Switching to reduced tillage, and leaving more residue on the soil surface were important to increase the macrofauna in two agricultural fields in Zimbabwe at Southeast Lowveld, a sub-tropical arid climate (rainfall <450 mm $year^{-1}$) where sorghum was the main crop, and the south of Kadoma town, a humid sub-tropical climate (650–800 mm rainfall $year^{-1}$), where maize was the main crop (Mutema et al., 2013). This study revealed a significant positive relationship between the residue return (0, 2, 4 and 6 t ha^{-1}) that covered the soil surface under conservation agriculture and the abundances of termites, ants, centipedes and beetle larva (Mutema et al., 2013). The fewest macrofauna were recovered from soil under conventional tillage, where residues were fragmented and incorporated by tillage implements into the soil profile after crop harvest (Mutema et al., 2013).

Soil macrofauna are similarly responsive to conservation agriculture in temperate regions of the world. After 14 years of conservation agriculture (crop rotation of wheat-pea-wheat-maize, no tillage and grass cover crop), the soil macrofauna abundance increased significantly by at least two-fold and by orders of magnitude for all trophic levels (Geophagous, Detritivores, Predators,

Table 3 Experimental location, its climate, cropping systems of the research field, age of agricultural field, tillage system, type of macrofauna and abundances macrofauna in soil. Gains in soil macrofauna abundances are associated with conservation agriculture practices including reduced tillage and complex cropping patterns. Data are from a global systematic review of diverse experimental locations, climatic zones, cropping systems, age of agricultural fields and tillage systems. Tillage systems are abbreviated as CT, conventional tillage; NT, no tillage; RT, reduced tillage; ST, strip tillage. Within a study case, values with different letters indicate significant differences at $P < 0.05$; * indicates the significant difference ($P < 0.05$) of CT treatment than other treatments.

Experimental location	Climatic zone	Cropping system	Age (years)	Tillage system	Macrofauna type	Macrofauna abundance (individuals m^{-2})	Reference
Cabrera-Deheza, Argentine Pampas	Semi-arid		Wheat-soybean-peanut-soybean-maize-soybean	CT	Lumbricina	9.6	Domínguez et al. (2018)
					Formicidae	44.8	
					Coleoptera	3.2	
					Araneae	9.6	
					Enchytraeidae	28.8	
			Wheat-soybean-maize-soybean-soybean-soybean	NT	Lumbricina	3.2	
					Formicidae	489.6	
					Coleoptera	0	
					Araneae	3.2	
					Enchytraeidae	6.4	
			Maize-soybean-soybean-maize-peanut-rye	NT	Lumbricina	60.8	
					Formicidae	3.2	
					Coleoptera	0	
					Araneae	6.4	
					Enchytraeidae	0	
Pergamino, Argentine Pampas	Humid temperate	10	Soybean monocrop	CT	Lumbricina	60.8	
					Formicidae	140.8	

					Coleoptera	16	
					Araneae	12.8	
					Enchytraeidae	556.8	
			Soybean monocrop	NT	Lumbricina	166.4	
					Formicidae	76.8	
					Coleoptera	22.4	
					Araneae	6.4	
					Enchytraeidae	483.2	
			Maize-soybean-wheat-soybean-maize-soybean.	NT	Lumbricina	240	
					Formicidae	54.4	
					Coleoptera	9.6	
					Araneae	12.8	
					Enchytraeidae	182.4	
Monze, Zambia	Subtropical humid	3	Maize monocrop	CT	Termite	`90[b]	Muoni et al. (2019)
		3	Maize-cotton rotation	CT	Termite	`73[b]	
		3	Maize monocrop	RT (direct seeding with animal-drawn seeder)	Termite	`466[a]	
		3	Maize-cotton-sunn hemp rotation	RT (direct seeding with animal-drawn seeder)	Termite	`681[a]	

(Continued)

Table 1 (*Continued*)

Experimental location	Climatic zone	Cropping system	Age (years)	Tillage system	Macrofauna type	Macrofauna abundance (individuals m^{-2})	Reference
Colorado State University, United States of America	Semi-arid	Corn monocrop	4	CT with 0.5 t ha^{-1} residue return	Lumbricina	~150[b]	Deleon et al. (2020)
				RT with 2.1 t ha^{-1} residue return		~175[b]	
				ST with 4 t ha^{-1} residue return		~425[a]	
Stratton, Colorado	Semi-arid	32	wheat-fallow	NT	Hymenoptera	0	Kelly et al. (2020)
					Coleoptera	28	
					Araneae	0	
					Hemiptera	0	
					Lepidoptera	0	
					Earthworms	144	
			wheat-corn-fallow	NT	Hymenoptera	40	
					Coleoptera	64	
					Araneae	8	
					Hemiptera	0	
					Lepidoptera	4	
					Earthworms	124	

			Wheat, corn, proso millet, sorghum	NT	Hymenoptera	100	
					Coleoptera	28	
					Araneae	4	
					Hemiptera	16	
					Lepidoptera	0	
					Earthworms	222	
California, United States of America	Semi-arid Mediterranean	16	14 years Tomato-cotton and 2 years of sorghum-garbanzo bean in all tillage systems	CT with no cover crop	Earthworms	~49*	Kelly et al. (2021)
					Isopods	~10*	
					Total macrofauna	~218*	
				CT with winter cover crop	Earthworms	~98	
					Isopods	~75	
					Total macrofauna	~378	
				NT with no cover crop	Earthworms	~24	
					Isopods	~1	
					Total macrofauna	~183	
				NT with no cover crop	Earthworms	~90	
					Isopods	~18	
					Total macrofauna	~245	
Uppsala, eastern Sweden	Humid continental	10	Winter wheat-peas-oilseed rape	CT	Anecic earthworm density	4[b]	Torppa and Taylor (2022)
		10		RT		24[a]	
		10		NT		50[a]	

(*Continued*)

Table 1 (*Continued*)

Experimental location	Climatic zone	Cropping system	Age (years)	Tillage system	Macrofauna type	Macrofauna abundance (individuals m^{-2})	Reference
Paraná state, Brazil	Subtrophic, humid	40	Black oat monocrop	CT	Total macrofauna	52[b]	Ramos et al. (2023)
		30		NT		231[a]	
Bethlehem and Reitz, South Africa	Arid to semi-arid	50	Oats or wheat as monocrop	CT	Earthworms ind. m^{-2}	~25[b]	Mamabolo et al. (2024)
		22	Maize-soybean, maize-sunflower, soybean-sunflower, cowpea-maize	RT		~80[a]	

Phytophagous, Herbivores, Rhizophagous, Granivores and Frugivores), compared to adjacent plots with conventional practices (deep tillage and the use of inorganic fertilizer; Henneron et al., 2015). This study occurred in an oceanic temperate climate at a research farm in Versailles, France. Conservation agriculture increased earthworm abundance significantly, with 272% more Lumbricidae, 950% more anecic and 8900% more epigeic earthworms than with conventional practices. The abundance of Isopoda was seven-fold greater, Diplopoda was 10 times more numerous, Araneae was nine-fold higher and Gastropoda were 19 times more abundant with conservation agriculture than conventional practices (Henneron et al., 2015). These and other studies listed in Table 3 demonstrate the more soil macrofauna associated with conservation agriculture practices (reduced and no tillage, soil cover with residues, cover cropping and crop rotation), consistently across various climatic regions of the world.

8 Farming practices to optimize soil biological health: recommendations and conclusions

As biological indicators of soil health, the soil microorganisms and fauna are directly implicated in crop growth, nutrient cycling, soil organic carbon storage, soil structure, water infiltration and gas exchange (Fig. 4). All of these processes are part of sustaining the soil health, to ensure its continued function and resilience in the face of degradative forces. Farming practices have a profound influence on the abundance and diversity of these biological indicators. Conventional tillage disrupts the soil structure, leading to the abrasion and burial of soil fauna, or their exposure to predators (i.e. birds) at the soil surface. This can be fatal for mesofauna and macrofauna, reducing their abundance in soil (Arai et al., 2018; Frazao et al., 2018). The removal of surface litter eliminates the preferred habitat and food supply of soil mesofauna and macrofauna, another disruption to their populations. This is why conservation tillage (reduced or no tillage) coupled with organic amendments are effective in promoting an abundant and diverse soil fauna (Table 2, 3). The positive influence of such practices on the abundance of earthworms is reported frequently (Arai et al., 2018; Bai et al., 2018; Frazao et al., 2018).

While general statements can be made about how soil microorganisms and soil fauna are impacted by farming practices (conventional versus conservation and tillage), cover cropping, residue return and crop rotation, we lack the ability to benchmark the biological indicators of soil health. Currently, biological indicators are assessed in side-by-side plots or fields, to determine their response to contrasting farming practices. This is due to the fact that biological indicators of soil health are context-specific, depending on the soil texture and soil organic matter within a given climatic condition. Establishing a baseline or reference database, which is normalized for the soil texture or initial soil organic matter level (either of a long-term conventional agriculture field or an undisturbed reference plot) will help to predict how the biological indicator responds for a given soil textural class within a climatic condition.

Another way to approach biological indicators of soil health is to calibrate them relative to crop yield. This may be challenging, since conservation agriculture may initially reduce crop yields significantly, even if it supports a greater abundance of soil microorganisms, when fields are first converted from conventional to conservation tillage (Neogi et al., 2014). With time, we might expect a larger number of microorganisms and higher crop yield under conservation agriculture than conventional agriculture practices (Samal et al., 2017; Choudhary et al., 2018b; Das et al., 2024). Furthermore, as is indicated from Tables 1 and 2, the response of biological indicators to conservation agriculture is less pronounced in cold humid temperate regions than warm monsoon, tropical or dry regions. Even if conservation agriculture generally improves the abundance and diversity of soil microorganisms and soil fauna, it may not always increase crop growth and yield because there are additional factors, including climatic events, agronomic factors and pests that affect crop performance. The relative importance of soil health for crop resilience exposed to fluctuating climatic conditions and non-soil stressors merits further investigation.

In conclusion, agricultural practices exert a significant influence on biological indicators of soil health around the world. Compared to conventional practices (deep tillage and residue removal after crop harvest), conservation agriculture that emphasizes reduced tillage along with residue return, use of organic fertilizers, cover cropping and crop rotation increases the abundance of soil microorganisms (fungi, bacteria, actinobacteria, archaea) and fauna (microfauna, mesofauna and macrofauna) in the upper 0–15 cm of the soil profile, where >80% of biological activities occur. Interestingly, this positive influence of conservation agriculture on the biological indicators of soil health is observed across regions with diverse climatic conditions. Biological indicators of soil health are responsive to reduced tillage intensity, and further increases in their abundance and diversity are achieved with practices such as cover cropping, residue return, organic fertilizers and crop diversification. Overall, the abundance and diversity of soil biota at the species-level and within communities are involved in multiple soil ecosystem functions, making them important biological indicators of soil health.

9 Where to look for further information

Further reading:

- Reports on soil biodiversity, including the soil organisms, soil conservation and agriculture, and research in soil biodiversity. Food and Agriculture

Organization of the United Nations. Available from: https://www.fao.org/soils-portal/soil-biodiversity/facts-and-figures/en/
- Soil health indicators related to biological properties. Soil Health Institute. Available from: https://soilhealthinstitute.org/our-work/initiatives/measurements/
- Soil biodiversity assessment. Global Soil Partnership. Available from: https://www.fao.org/global-soil-partnership/areas-of-work/soil-biodiversity/en/

Key journals/ conferences:

- Soil Biology and Biochemistry
- European Journal of Soil Biology
- Global Soil Biodiversity Conference and other events/ resources available from the Global Soil Biodiversity Initiative. Available from: https://www.globalsoilbiodiversity.org/

10 References

Agyei, B., et al. (2024), Farm-level variability in soil biological health indicators in Michigan is dependent on management and soil properties, *Soil Sci. Soc. Am. J.* 88(2): 326–338. https://doi.org/10.1002/saj2.20630

Alam, M. K., et al. (2020), Soil nitrogen storage and availability to crops are increased by conservation agriculture practices in rice–based cropping systems in the Eastern Gangetic Plains, *Field Crop. Res.* 250: 107764. https://doi.org/10.1016/j.fcr.2020.107764

Al-Maliki, S., et al. (2021), Earthworms and eco-consequences: Considerations to soil biological indicators and plant function: a review, *Acta Ecologica Sinica*, 41(6): 512–523. https://doi.org/10.1016/j.chnaes.2021.02.003

Arai, M., et al. (2018), Two-year responses of earthworm abundance, soil aggregates, and soil carbon to no-tillage and fertilization, *Geoderma* 332: 135–141. https://doi.org/10.1016/j.geoderma.2017.10.021

Babu, S., et al. (2023), Soil carbon dynamics under organic farming: Impact of tillage and cropping diversity, *Ecol. Indic.* 147: 109940. https://doi.org/10.1016/j.ecolind.2023.109940

Bai, Z., et al. (2018), Effects of agricultural management practices on soil quality: a review of long-term experiments for Europe and China, *Agr. Ecosyst. Environ.* 265: 1–7. https://doi.org/10.1016/j.agee.2018.05.028

Betancur-Corredor, B., et al. (2022), Reducing tillage intensity benefits the soil micro-and mesofauna in a global meta-analysis, *Eur. J. Soil Sci.* 73(6): e13321. https://doi.org/10.1111/ejss.13321

Bhattacharyya, R., et al. (2022), Soil organic carbon is significantly associated with the pore geometry, microbial diversity and enzyme activity of the macro-aggregates under different land uses, *Sci. Total Environ.* 778: 146286. https://doi.org/10.1016/j.scitotenv.2021.146286

Bhattacharyya, S. S., et al. (2021), Soil carbon sequestration – An interplay between soil microbial community and soil organic matter dynamics, *Sci. Total Environ.* 815: 152928. https://doi.org/10.1016/j.scitotenv.2022.152928

Brennan, A., et al. (2006), Collembola abundances and assemblage structures in conventionally tilled and conservation tillage arable systems, *Pedobiologia* 50(2): 135–145. https://doi.org/10.1016/j.pedobi.2005.09.004

Chandarana, K. A. and Amaresna, N. (2021), Soil protists: An untapped microbial resource of agriculture and environmental importance, *Pedosphere* 32(1): 184–197, https://doi.org/10.1016/S1002-0160(21)60066-8

Chassain, J., et al. (2024), Indicators of practice intensity unearth the effects of cropping systems on soil mesofauna. *Agri. Ecosyst. Environ.* 362: 108854, https://doi.org/10.1016/j.agee.2023.108854

Choudhary, M., et al. (2018a), Changes in soil biology under conservation agriculture based sustainable intensification of cereal systems in Indo-Gangetic Plains, *Geoderma* 313: 193–204. https://doi.org/10.1016/j.geoderma.2017.10.041

Choudhary, M., et al. (2018b), Soil biological properties and fungal diversity under conservation agriculture in Indo-Gangetic Plains of India, *J. Soil Sci. Plant Nut.* 18(4): 1142–1156. http://dx.doi.org/10.4067/S0718-95162018005003201

Chavarria, D. N., et al. (2018), Response of soil microbial communities to agroecological versus conventional systems of extensive agriculture, *Agr. Ecosyst. Environ.* 264: 1–8. https://doi.org/10.1016/j.agee.2018.05.008

Coleman, D. C., et al. (2024), Chapter 5 - Soil fauna: occurrence, biodiversity, and roles in ecosystem function. In Paul, E. A., and Frey, S. D. (eds) *Soil Microbiology, Ecology and Biochemistry* (5th ed.). Amsterdam: Elsevier, pp. 131–159. https://doi.org/10.1016/B978-0-12-822941-5.00005-3

CYMMYT. (2020), https://www.cimmyt.org/news/what-is-conservation-agriculture/#:~:text=The%20term%20%E2%80%9Cconservation%20agriculture%E2%80%9D%20was,wheat%20systems%20in%20Latin%20America

Conyers, M., et al. (2019), The strategic use of minimum tillage within conservation agriculture in southern New South Wales, Australia, *Soil Till. Res.* 193: 17–26. https://doi.org/10.1016/j.still.2019.05.021

Das, P., et al. (2024), conservation agriculture based rice-wheat cropping system in Indo-Gangetic Plain, *Appl. Soil Ecol.* 193: 105128. https://doi.org/10.1016/j.fcr.2017.10.021

Deleon, E., et al. (2020), Conservation tillage supports soil macrofauna communities, infiltration, and farm profits in an irrigated maize-based cropping system of Colorado, *Soil Sci. Soc. Am. J.* 84(6): 1943–1956. https://doi.org/10.1002/saj2.20167

Demetrio, W., et al. (2022), Macrofauna communities and their relationship with soil structural quality in different land use systems, *Soil Res.* 60: 648–660. https://doi.org/10.1071/SR21157.

Devkota, M., et al. (2022), Conservation agriculture improves agronomic, economic, and soil fertility indicators for a clay soil in a rainfed Mediterranean climate in Morocco, *Agr. Syst.* 201: 103470. https://doi.org/10.1016/j.agsy.2022.103470

Domínguez, A., et al. (2018), Soil macrofauna diversity as a key element for building sustainable agriculture in Argentine Pampas, *Acta Oecologica* 92: 102–116. https://doi.org/10.1016/j.actao.2018.08.012

Domnariu, H., et al. (2024), Legume cover cropping and nitrogen fertilization influence soil prokaryotes and increase carbon content in dryland wheat systems, *Agr. Ecosyst. Environ.* 367: 108959. https://doi.org/10.1016/j.agee.2024.108959

Dutta, A., et al. (2024). Impact of conservation agriculture on soil organic carbon sequestration and enzyme activity under rice–wheat cropping system in a Vertisol, *Agric. Res.* 13(4): 789–801. https://doi.org/10.1007/s40003-024-00734-6

Elmquist, D. C., et al. (2023), Linking agricultural diversification practices, soil arthropod communities and soil health, *J. Appl. Ecol.* 60(9): 952–1963. https://doi.org/10.1111/1365-2664.14453

Erktan, A., et al. (2020), The physical structure of soil: determinant and consequence of trophic interactions, *Soil Biol. Biochem.* 148: 107876. https://doi.org/10.1016/j.soilbio.2020.107876

FAO. (2014), What is conservation agriculture? FAO CA website. http://www.fao.org/ag/ca/1a.html (Accessed 15 December 2020).

Faulkner, E. H. (1945), *Ploughman's Folly*. London: Michael Joseph.

Farooq, M. and Siddique, K. (2015). Conservation Agriculture: Concepts, Brief History, and Impacts on Agricultural Systems. In: Farooq, M. and Siddique, K. (eds) *Conservation Agriculture*. Cham: Springer. https://doi.org/10.1007/978-3-319-11620-4_1

Frasier, I., et al. (2018), High quality residues from cover crops favor changes in microbial community and enhance C and N sequestration, *Global Ecol. Conserv.* 6: 242–256. https://doi.org/10.1016/j.gecco.2016.03.009

Frazao, J., et al. (2018), Soil structure formation and organic matter distribution as affected by earthworm species interactions and crop residue placement, *Geoderma* 338: 453–463 https://doi.org/10.1016/j.geoderma.2018.07.033

Garratt, M. P., et al. (2018), Enhancing soil organic matter as a route to the ecological intensification of European arable systems. *Ecosystems* 21(7): 1404–1415.

Geisen, S., et al. (2018), Soil protists: a fertilefrontier in soil biology research. *FEMS Microbiol. Rev.* 42: 293–323. https://doi.org/10.1093/femsre/fuy006

Guo, L.-J., et al. (2016), Effects of conservation tillage on topsoil microbial metabolic characteristics and organic carbon within aggregates under a Rice (*Oryza sativa* L.) –Wheat (*Triticum aestivum* L.) cropping system in Central China, *PLoS One* 11(1): e0146145. https://doi.org/10.1371/journal.pone.0146145

Han, Z., et al. (2024), Conservation tillage enhances the sequestration and iron-mediated stabilization of aggregate-associated organic carbon in Mollisols, *CATENA* 243: 108197. https://doi.org/10.1016/j.catena.2024.108197

Hanisch, J., et al. (2022) The role of Collembola for litter decomposition under minimum and conventional tillage, *J. Plant Nut. Soil Sci. 185*(4): 529–538. https://doi.org/10.1002/jpln.202200077

Hartmann, M. and Six, J. (2023), Soil structure and microbiome functions in agroecosystems, *Nat. Rev. Earth Environ.* 4: 4–18. https://doi.org/10.1038/s43017-022-00366-w

Hedde, M., et al. (2024), Crop management strategies shape the shared temporal dynamics of soil food web structure and functioning, *Agr. Ecosyst. Environ.* 370: 109058. https://doi.org/10.1016/j.agee.2024.109058

Heinen, J., et al. (2023), Combining organic fertilisation and perennial crops in the rotation enhances arthropod communities, *Agr. Ecosyst. Environ.* 349: 108461. https://doi.org/10.1016/j.agee.2023.108461

Henneron, L., et al. (2015), Fourteen years of evidence for positive effects of conservation agriculture and organic farming on soil life, *Agron. Sustain. Dev.* 35: 169–181. https://doi.org/10.1016/s0167-8809(99)00028-6

Hu, X., et al. (2021), Conventional and conservation tillage practices affect soil microbial co-occurrence patterns and are associated with crop yields, *Agr. Ecosyst. Environ.* 319: 107534. https://doi.org/10.1016/j.agee.2021.107534

IFOAM (International Federation of Organic Agriculture Movements). (2005), *IFOAM Basic Standards*. Bonn: IFOAM.

Jia, L., et al. (2022), Keystone microbiome in the rhizosphere soil reveals the effect of long-term conservation tillage on crop growth in the Chinese Loess Plateau, *Plant Soil* 473: 457–472. https://doi.org/10.1007/s11104-022-05297-5

Jahangir, M. M. R., et al. (2021), Carbon and nitrogen accumulation in soils under conservation agriculture practices decreases with nitrogen application rates, *App. Soil Ecol.* 168: 104178. https://doi.org/10.1016/j.apsoil.2021.104178

Jat, H. S., et al. (2019), Climate smart agriculture practices improve soil organic carbon pools, biological properties and crop productivity in cereal-based systems of North-West India, *Catena*, 181: 104059. https://doi.org/10.1016/j.catena.2019.05.005

Joergensen, R. G. and Mueller, T. (1996), The fumigation-extraction method to estimate soil microbial biomass: calibration of the kEC value. *Soil Biol. Biochem.* 28: 33–37. https://doi.org/10.1016/0038-0717(95)00101-8

Kassam, A., et al. (2019). Global spread of conservation agriculture, *Int, J. Environ. Stud.* 76(1): 29–51. https://doi.org/10.1080/00207233.2018.1494927

Kaur, M. and Karnwal, A. (2023), Screening of endophytic Bacteria from stress-tolerating plants for abiotic stress tolerance and plant growth-promoting properties: Identification of potential strains for bioremediation and crop enhancement, *J. Agr. Food Res.* 14: 100723. https://doi.org/10.1016/j.jafr.2023.100723

Kelly, C., et al. (2020), The effects of dryland cropping system intensity on soil function and associated changes in macrofauna communities, *Soil Sci. Soc. Am. J.* 84(6): 1854–1870. https://doi.org/10.1002/saj2.20133

Kelly, C., et al. (2021), Winter cover crops and no-till promote soil macrofauna communities in irrigated, Mediterranean cropland in California, USA, *Appl. Soil Ecol.* 166: 104068. https://doi.org/10.1016/j.apsoil.2021.104068

Kiboi, M. N., et al. (2021), Soil nutrients and crop yield response to conservation-effective management practices in the sub-humid highlands agro-ecologies of Kenya, *Heliyon* 7: e07156. https://doi.org/10.1016/j.heliyon.2021.e07156

King, A. E. and Hofmockel, K. S. (2017), Diversified cropping systems support greater microbial cycling and retention of carbon and nitrogen, *Agr. Ecosyst. Environ.* 240: 66–76. https://doi.org/10.1016/j.agee.2017.01.040

Knapp, S. and van der Heijden, M. G. A. (2018), A global meta-analysis of yield stability in organic and conservation agriculture. *Nat. Commun.* 9:3632 https://doi.org/10.1038/s41467-018-05956-1

Kopittke, P. M., et al. (2024), Healthy soil for healthy humans and a healthy planet, *Crit. Rev. En. Sci. Tech.* 54(3): 210–221. https://doi.org/10.1080/10643389.2023.2228651

Krause, H. M., et al. (2023), Organic and conventional farming systems shape soil bacterial community composition in tropical arable farming, *Appl. Soil Ecol.* 191: 105054. https://doi.org/10.1016/j.apsoil.2023.105054

Lee, S. Y., et al. (2023), Inoculation effect of heavy metal tolerant and plant growth promoting rhizobacteria for rhizoremediation, *Int. J. Environ. Sci. Technol.* 21: 1419–1434. https://doi.org/10.1007/s13762-023-05078-2

Lehmann, J., et al. (2020), The concept and future prospects of soil health, *Nat. Rev. Earth Environ.* 1(10): 544–553. https://doi.org/10.1038/s43017-020-0080-8

Liu, S., et al. (2024a), Conservation tillage impacts on soil biodiversity: additional insights from the Collembola-associated bacteria, *Agr. Ecosyst. Environ.* 362: 108827. https://doi.org/10.1016/j.agee.2023.108827

Liu, B., et al. (2024b), Impact of diversified cropping systems and fertilization strategies on soil microbial abundance and functional potentials for nitrogen cycling, *Sci. Total Environ.* 932: 172954. https://doi.org/10.1016/j.scitotenv.2024.172954

Lopes, L. D. and Fernandes, M. F. (2020), Changes in microbial community structure and physiological profile in a kaolinitic tropical soil under different conservation agricultural practices, *Appl. Soil Ecol.* 152: 103545. https://doi.org/10.1016/j.apsoil.2020.103545

Mackay, J. E., et al. (2023), Tillage and pesticide seed treatments have distinct effects on soil microbial diversity and function, *Soil Biol. Biochem.* 176: 108860. https://doi.org/10.1016/j.soilbio.2022.108860

Mamabolo, E., et al. (2024), Conventional agricultural management negatively affects soil fauna abundance, soil physicochemical quality and multifunctionality, *Pedobiologia*, 104: 150961. https://doi.org/10.1016/j.pedobi.2024.150961

Mangalassery, S., et al. (2015), Impacts of zero tillage on soil enzyme activities, microbial characteristics and organic matter functional chemistry in temperate soils, *Eur. J. Soil Biol.* 68: 9–17. https://doi.org/10.1016/j.ejsobi.2015.03.001

Mathew, P. R. (2012), *Characterization of Soil Microbial Communities in Agricultural and Forest Ecosystems* (Doctoral dissertation).

McDaniel, M. D., et al. (2023), Diversifying and perennializing plants in agroecosystems alters retention of new C and N from crop residues, *Ecol. Appl.* 33(2): e2784. https://doi.org/10.1002/eap.2784

Mirzavand, J., et al. (2022), Biological indicators of soil quality under conventional, reduced, and no-tillage systems, *Arch. Agron. Soil Sci.* 68(3): 311–324. https://doi.org/10.1080/03650340.2020.1832656

Muoni, T., et al. (2019), Tillage and crop rotations enhance populations of earthworms, termites, dung beetles and centipedes: evidence from a long-term trial in Zambia, *J. Agr. Sci.* 157(6): 504–514. https://doi.org/10.1017/S002185961900073X

Mutema, M., et al. (2013), Effects of crop residues and reduced tillage on macrofauna abundance, *J. Organic Syst.* 8(1): 5–16.

Neogi, S., et al. (2014), Soil respiration, labile carbon pools, and enzyme activities as affected by tillage practices in a tropical rice–maize–cowpea cropping system, *Environ. Monit. Assess.* 186: 4223–4236. https://doi.org/10.1007/s10661-014-3693-x

Nunes, M. R., et al. (2020), Biological soil health indicators respond to tillage intensity: a US meta-analysis, *Geoderma*, 369: 114335 https://doi.org/10.1016/j.geoderma.2020.114335

Nyawade, S. O., et al. (2019), Short-term dynamics of soil organic matter fractions and microbial activity in smallholder potato-legume intercropping systems, *Appl. Soil Ecol.* 142: 123–135, https://doi.org/10.1016/j.apsoil.2019.04.015.

Oliveira, E. M., et al. (2024), Effects of conventional, organic and conservation agriculture on soil physical properties, root growth and microbial habitats in a long-term field experiment. *Geoderma* 447: 116927. https://doi.org/10.1016/j.geoderma.2024.116927

Ologunde, O. H., et al. (2024), Integrated agricultural system: a dynamic concept for improving soil quality, *J. Saudi Soc. Agr. Sci.* 23: 352–360. https://doi.org/10.1016/j.jssas.2024.03.002

Or, D., et al. (2021), Natural and managed soil structure: on the fragile scaffolding for soil functioning. *Soil Till. Res.* 208: 104912. https://doi.org/10.1016/j.still.2020.104912

Patra, S., et al. (2019), Soil hydraulic response to conservation agriculture under irrigated intensive cereal-based cropping systems in a semiarid climate, *Soil Till. Res.* 192: 151–163, https://doi.org/10.1016/j.still.2019.05.003

Petersen, H. (2002), Effects of non-inverting deep tillage vs. conventional ploughing on collembolan populations in an organic wheat field, *Eur. J. Soil Biol.* 38(2): 177–180. https://doi.org/10.1016/S1164-5563(02)01145-7

Philippot, L., et al. (2024), The interplay between microbial communities and soil properties, *Nat. Rev. Microbiol.* 22(4): 226–239. https://doi.org/10.1038/s41579-023-00980-5

Potapov, A. M., et al. (2022), Feeding habits and multifunctional classification of soil-associated consumers from protists to vertebrates, *Biol Rev*, 97: 1057–1117. https://doi.org/10.1111/brv.12832

Ramos, R. F., et al. (2023), Edaphic fauna in soil profile after three decades of different soil management and cover crops in a subtropical region, *An. Acad. Bras. Cienc.* 95: e20201667. https://doi.org/10.1590/0001-3765202320201667

Ren, H., et al. (2024), Green manure provides growth benefits for soil mesofauna by promoting soil fertility in agroecosystems, *Soil Till. Res.* 238: 106006. https://doi.org/10.1016/j.still.2024.106006

Rieff, G. G., et al. (2020), Impact of no-tillage versus conventional maize plantation on soil mesofauna with and without the use of a lambda-cyhalothrin based insecticide: a terrestrial model ecosystem experiment, *Appl. Soil Ecol.* 147: 103381. https://doi.org/10.1016/j.apsoil.2019.103381

Rocco, S., et al. (2024), Long-term soil quality and C stock effects of tillage and cover cropping in a conservation agriculture system. *Soil Till. Res.* 241: 106129. https://doi.org/10.1016/j.still.2024.106129

Rui, Y., et al. (2022), Persistent soil carbon enhanced in Mollisols by well-managed grasslands but not annual grain or dairy forage cropping systems, *P. Natl. Acad. Sci.* 119(7): e2118931119. https://doi.org/10.1073/pnas.2118931119

Sahu, P. K., et al. (2019), Connecting microbial capabilities with the soil and plant health: options for agricultural sustainability, *Ecol. Indic.* 105: 601–612. https://doi.org/10.1016/j.ecolind.2018.05.084

Samal, S. K., et al. (2017), Evaluation of long-term conservation agriculture and crop intensification in rice-wheat rotation of Indo-Gangetic Plains of South Asia: carbon dynamics and productivity, *Eur. J. Agron.* 90: 198–208. https://doi.org/10.1016/j.eja.2017.08.006

Sarker, R. R., et al. (2023), Tillage and Residue Management Impact on Microbial and Nematode Abundance Under Diverse Rice-Based Cropping Systems in Calcareous and Non-calcareous Floodplain Soils, *J. Soil Sci. Plant Nut.* 23(2): 2138–2151. https://doi.org/10.1007/s42729-023-01168-9

Sharma, S., et al. (2022), Total and labile pools of organic carbon in relation to soil biological properties under contrasting land-use systems in a dry mountainous region, *Carbon Manag.* 13(1): 352–371. https://doi.org/10.1080/17583004.2022.2089236

Shi, G., et al. (2021), Bacterial communities as indicators of soil health under a continuous cropping system, *Land Degerad. Dev.* 32(7): 2393–2408. https://doi.org/10.1002/ldr.3919

Simmons, B. L. and Coleman, D. C. (2008), Microbial community response to transition from conventional to conservation tillage in cotton fields, *Appl. soil Ecol.* 40(3): 518–528. https://doi.org/10.1016/j.apsoil.2008.08.003

Singh, J., et al. (2021), Comparison of soil carbon and nitrogen pools among poplar and eucalyptus based agroforestry systems in Punjab, India, *Carbon Manag.* 12(6): 693–708. https://doi.org/10.1080/17583004.2021.2011787

Sithole, N. J., et al. (2018), Long-term impact of no-till conservation agriculture on abundance and order diversity of soil macrofauna in continuous maize monocropping system, *Acta Agr. Scand. B–Soil Plant Sci.* 68(3): 220–229. https://doi.org/10.1080/09064710.2017.1381276

Steward, P. R., et al. (2018), The adaptive capacity of maize-based conservation agriculture systems to climate stress in tropical and subtropical environments: a meta-regression of yields, *Agri. Ecosyst. Environ.* 251: 194–202. https://doi.org/10.1016/j.agee.2017.09.019

Stefan, G., et al. (2014), Soil water availability strongly alters the community composition of soil protists, *Pedobiologia* 57: 205–213. https://doi.org/10.1016/j.pedobi.2014.10.001

Sultana, R., et al. (2024), Endophytic bacterium Sphingomonas panaciterrae NB5 influences soil properties and improves growth, nutrient contents, and yield of red amaranth (*Amaranthus tricolor* L.), *Curr. Plant Biol.*, 39: 100372. https://doi.org/10.1016/j.cpb.2024.100372

Takam Fongang, G. M., et al. (2024), What determines the adoption of conservation agriculture? Evidence from Quebec. *Environ. Manage.* 74: 775–789. https://doi.org/10.1007/s00267-024-02024-x

Tautges, N. E., et al. (2016), Soil microbial diversity and activity linked to crop yield and quality in a dryland organic wheat production system, *Appl. Soil Ecol.* 108: 258–268. https://doi.org/10.1016/j.apsoil.2016.09.003

Tisdall, J. M. and Oades, J. M. (1982), Organic matter and water stable aggregates in soils. *J. Soil Sci.* 33: 141–163. https://doi.org/10.1111/j.1365-2389.1982.tb01755.x

Tonon-Debiasi, B. C., et al. (2024), Microbial attributes as structural quality index for physical health of an Oxisol under compaction levels, *Soil Till. Res.* 235: 105872. https://doi.org/10.1016/j.still.2023.105872

Topalovic, O. and Geisen, S. (2023), Nematodes as suppressors and facilitators of plant performance, *New Phytol.* 238: 2305–2312. https://doi.org/10.1111/nph.18925

Torppa, K. A. and Taylor, A. R. (2022), Alternative combinations of tillage practices and crop rotations can foster earthworm density and bioturbation, *Appl. Soil Ecol.* 175: 104460. https://doi.org/10.1016/j.apsoil.2022.104460

Tschanz, P., et al. (2024), A review of soil tillage impacts on ground-nesting wild bees–mechanisms, implications, and future research perspectives. *Agr. Ecosyst. Environ.* 375: 109224. https://doi.org/10.1016/j.agee.2024.109224

Van Den Hoogen, J., et al. (2019), Soil nematode abundance and functional group composition at a global scale, *Nature* 572(7768): 194–198. https://doi.org/10.1038/s41586-019-1418-6

Vance, E. D., et al. (1987), An extraction method for measuring soil microbial biomass C. *Soil boil. Biochem.* 19(6): 703–707. https://doi.org/10.1016/0038-0717(87)90052-6

Velasquez, E. and Lavelle, P. (2019), Soil macrofauna as an indicator for evaluating soil based ecosystem services in agricultural landscapes, *Acta Oecologica* 100: 103446. https://doi.org/10.1016/j.actao.2019.103446

Vidaller, C. and Dutoit, T. (2022), Ecosystem services in conventional farming systems: a review, *Agron. Sustain. Dev.* 42: 22. https://doi.org/10.1007/s13593-021-00740-w

Wang, Y., et al.. (2011), Long-term impact of farming practices on soil organic carbon and nitrogen pools and microbial biomass and activity, *Soil Till. Res.* 117: 8–16. https://doi.org/10.1016/j.still.2011.08.002

Wang, J., et al. (2021), Intercropping perennial aquatic plants with rice improved paddy field soil microbial biomass, biomass carbon and biomass nitrogen to facilitate soil sustainability, *Soil Till. Res.* 208: 104908. https://doi.org/10.1016/j.still.2020.104908

Whalen, J. K. and Gul, S. (2023), Root Interactions with the Microbiome from the Rhizoplane to the Bulk Soil: an Overview. In M. J. Goss and M. Oliver (Eds), *Encyclopedia of Soils in the Environment*, 2nd ed. Academic Press, pp. 357–368. https://doi.org/10.1016/B978-0-12-822974-3.00243-3

Xu, L., et al. (2024), Combined metagenomics and metabolomic analysis of microbial community structure and metabolic function in continuous soybean cropping soils of Songnen Plain, China, *Chem. Biol. Tech. Agr.* 11(1): 46. https://doi.org/10.1186/s40538-024-00569-x

Yang, B., et al. (2021), Impact of land use type and organic farming on the abundance, diversity, community composition and functional properties of soil nematode communities in vegetable farming, *Agr. Ecosyst. Environ.* 318: 107488. https://doi.org/10.1016/j.agee.2021.107488

Zhang, H., et al. (2022), Subsoiling and conversion to conservation tillage enriched nitrogen cycling bacterial communities in sandy soils under long-term maize monoculture, *Soil Till. Res.* 215: 105197. https://doi.org/10.1016/j.still.2021.105197

Zhang, J., et al. (2024), Assessment of soil health and identification of key soil health indicators for five long-term crop rotations with varying fertility management, *Geoderma* 443: 116836. https://doi.org/10.1016/j.geoderma.2024.116836.

Zhang, S., et al. (2015), Conservation tillage positively influences the microflora and microfauna in the black soil of Northeast China, *Soil Till. Res.* 149: 46–52. https://doi.org/10.1016/j.still.2015.01.001

Zhang, X., et al. (2017), Ecosystem services of the soil food web after long-term application of agricultural management practices, *Soil Biol. Biochem.* 111: 36–43. https://doi.org/10.1016/j.soilbio.2017.03.017

Chapter 13

Using biological indicators to assess the health of forest soils

Weixin Zhang and Cancan Zhao, Henan University, China; Bin Wang, Xin Sun and Zhipeng Li, Chinese Academy of Sciences, China; Tao Liu, National Engineering Laboratory for Applied Technology of Forestry and Ecology in South China and Central South University of Forestry and Technology, China; Xiao Zhang, Northwest A&F University, China; Yuanhu Shao, Zhifeng Shen and Shenglei Fu, Henan University, China; and Yiqing Li, University of Hawaii at Hilo, USA

1 Introduction

One third of the Earth's land surface is covered by forests (Xing et al., 2023). The health of forest ecosystems is vital for sustaining essential ecological services such as biodiversity, clean water, soil conservation, timber production, and carbon sequestration (von Wilpert, 2022). Forest soils typically display a highly differentiated, hierarchical structure, characterized by high porosity, biological activity, and chemical reactivity. However, these soils are susceptible

http://dx.doi.org/10.19103/AS.2025.0159.16

to environmental changes and management practices (von Wilpert, 2022). Comprehensively and accurately assessing forest soil health has not only become a crucial foundation for the sustainable development of forests but also presents a significant challenge.

The concept of soil health originated from agricultural systems (Lehmann et al., 2020; Jian et al., 2020), and the U.S. Department of Agriculture Natural Resource Conservation Service (USDA-NRCS) has developed a set of criteria for evaluating soil health (Stott, 2019). Compared to agricultural fields, forests have their own unique attributes, so the definition of forest soil health also differs.

Several features of forest soils are particularly noteworthy:

- High spatial heterogeneity: Forests are highly heterogeneous ecosystems that comprise a multitude of reactive interfaces (Li et al., 2023a), which are more complex than those in grasslands and farmlands.
- Well-developed O horizon and relatively acidic conditions: Forest soils typically have a well-developed O horizon (organic layer) and throughout the profile tend to have pH values that are neutral to strongly acidic (Ranger, 2018; DeLuca et al., 2019).
- Relatively fungi-enriched microbial community structure: The microbial community structure of forest soils is relatively dominated by fungi (DeLuca et al., 2019; Marx, 2024). Note that, although forest soils are often characterized by a strong fungal food web and energy flow, and many studies emphasize the critical role of fungi in forest soil health, this does not mean that bacteria are unimportant. In fact, in most ecosystems, the absolute biomass of bacteria is higher than that of fungi (Wang et al., 2019), and bacteria do not always play a lesser role in the decomposition of surface litter compared to fungi (Cao et al., 2024).

Forests at different latitudes and elevations have their own unique biotic and abiotic characteristics in their soils. Theoretically, there is no basis to claim that the soils of certain forests are healthier than others. Only when forests are subjected to human or natural disturbances and the original ecological balances of the forest ecosystem are disrupted, do forest soils exhibit various sub-healthy or degraded states, accompanied by varying degrees of decline in the ecological service functions of forest soils. Therefore, we define forest soil health as follows:

> the continued capacity of soil to function as a vital living ecosystem that sustains native biodiversity, ecological balances, and ecosystem services in forests.

Previous studies have systematically reviewed the topic of 'Assessing soil biological health in forest soils,' by comparisons with grasslands and farmlands in order to highlight the distinctive architecture and functional roles of forest soil organisms (DeLuca et al., 2019).

2 Soil health and forest resilience

Soils are subject to numerous degradative processes (e.g. erosion, compaction, salinization, and acidification), which result in physical, chemical, and biological imbalances (Table 1); climatic fluctuations including frequent droughts, floods, and heat waves, intense rainstorms, and other extremes are the leading stressors of soil (Smith et al., 2022), while deforestation often exacerbates soil degradation (Veldkamp et al., 2020).

Furthermore, assessing forest health and resilience is an urgent task that requires multiple methods, global collaboration, data collection at multiple scales, and a willingness to be surprised (Marx, 2024). The ability of soil to continue providing all essential ecosystem services despite degradative processes is of paramount importance to ecosystem health; this inherent characteristic of soil is termed soil resilience (Blanco and Lal, 2023).

It is important to measure the capacity of forests to withstand and recover from natural and anthropogenic perturbations (i.e. their resilience) (Forzieri et al., 2022). Tropical, arid, and temperate forests are experiencing a significant decline in resilience, which is likely related to increased water limitations and climate variability; reductions in resilience are statistically linked to sudden increases in tree mortality and abrupt declines in forest primary productivity.

However, the role of soil health in this context is not yet clear, but soil can influence forest resilience, at least from the perspective of water supply. The resistance and resilience of forest soil structure, soil organic matter,

Table 1 The main health issues of forest soils

Health issue category	Health issue	References
Soil physical imbalance	Soil erosion Soil compaction Soil water imbalance Soil thermal imbalance	Veldkamp et al., 2020; Smith et al., 2022; Blanco and Lal, 2023
Soil chemical imbalance	Soil acidification Soil salinization Organic matter loss	
Soil biological imbalance	Biodiversity loss Imbalance of beneficial and harmful soil biota	Zhao et al., 2019; Li et al., 2023b
Reduced sustainability	Decline in resistance and resilience	Forzieri et al., 2022; Marx, 2024

and soil biota to both anthropogenic and natural disturbances, as well as their contributions to the overall resistance and resilience of the forest, are beginning to receive attention. For example, it has been recognized that the stability of soil organic carbon (SOC) depends not only on the physiochemical characteristics of SOC but also on the soil microenvironment, soil microbial community composition, and types of perturbations (Yang et al., 2020). Studies have also shown that the fungal-based food web, and the processes of C and N loss it governs, in undisturbed grassland soil was more resistant, although not resilient, and better able to adapt to drought than the bacterial-based food web of wheat soil (de Vries et al., 2012; Six, 2012). Considering that forest soils are often less disturbed by human activities and most plants have a symbiotic fungal network that assists them in absorbing nutrients and water (Pennisi and Cornwall, 2020), it is reasonable to assume that forests characterized by a relatively strong fungal-based food web would also be more resistant to drought.

Soil health, a cornerstone of forest ecosystem health, has seen its assessment methods evolve from traditional chemical and physical property measurements to include comprehensive assessments that incorporate biological indicators (Alkorta et al., 2003; Cardoso et al., 2013). The purpose of this chapter is to illustrate the challenges facing forest soil health, how to evaluate forest soil health based on biological indicators, and the issues that need to be considered, with the aim of providing valuable insights for ultimately constructing a simple and effective biological-based forest soil health indicator system.

3 How to characterize forest soil health

Similar to agricultural soil health, forest soil health also needs to consider at least three fundamental elements (Zhang et al., 2020a):

- soil organic matter;
- soil biota; and
- soil structure.

These elements regulate key ecological processes, such as the transformation and accumulation of organic matter, the maintenance of soil water and thermal balance, and the balance of biological interactions. Considering these three elements is not only due to their crucial roles in forest ecosystems but also because their recovery is very challenging once they are damaged or threatened by human or natural disturbances.

Among these, soil organic matter has long been a key indicator of agricultural soil health (Nunes et al., 2021; Martin et al., 2022; Sprunger and Martin, 2023). The accumulation of soil organic matter is also fundamentally important for forest soil health (Pouyat et al., 2020; Marx, 2024), but it often requires a long time (Devine et al., 2011; Chen et al., 2017). Soil biota are the most dynamic components of the soil. Traditional soil health assessment methods rely more on monitoring soil physical and chemical properties, which have significant limitations. Biological traits might be better predictors of soil health metrics than physical or chemical soil properties (Wilhelm et al., 2022). Of course, emphasizing the inclusion of soil biota in soil health assessment does not exclude the importance of non-biological factors. Soil structure not only directly reflects the apparent characteristics of the soil but also closely related to soil organic matter accumulation and soil biota activity, and together they influence soil chemical properties. Well-developed aggregate structures in forest soils enhance their ability to regulate water and heat, preventing severe drought or flooding, and help maintain soil temperature at a relatively stable level, thus providing a stable space and environment for soil biota and their regulated biochemical processes (Zhang et al., 2024a).

It is important to note that investigating forest soil health requires integrating vegetation, climate, and ecosystem functions within a unified One Health framework (Fig. 1a). Specifically, assessing soil health should not be limited to the soil itself. It is necessary to compare changes in key ecological processes, such as carbon and nutrient cycling and biological interactions, in soils of different health levels, and link these to important ecosystem services such as plant productivity, carbon sequestration (van Bruggen and Semenov, 2000; Cai and Wu, 2015), and the regulation of surface heat balance (Zhang et al., 2024a) (Fig. 1b).

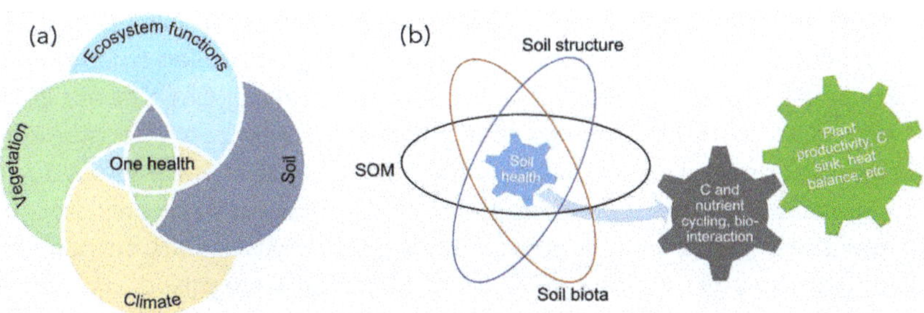

Figure 1 The conceptual model for understanding (a) one health approach and (b) the three major components of forest soil health and their influences on key ecological processes in forests. C: carbon; SOM: soil organic matter.

4 Assessing soil health through biological indicators: ecosystem functions and plant characteristics

Forest soil health can be shown in different dimensions by examining several aspects:

- ecosystem functions;
- physiological and ecological characteristics of plants (at community, population, and individual levels); and
- soil fauna (at community, population, and individual levels), soil microorganisms (at community, major groups, and individual levels), biochemical indicators, key ecological processes, and the resistance and resilience of these parameters (Table 2).

This section reviews the first two, whilst soil biota and other parameters are discussed in the following section.

4.1 Ecosystem functions and soil health

At the ecosystem level, the connection between vegetation productivity and soil can reflect the soil's ability to supply key resources such as nutrients and water. The coupling of the process by which plants form organic matter through photosynthesis and the process mediated by soil biota for the decomposition, transformation, and release of organic matter into nutrients, as well as the coupling of carbon, nutrients, and water, are key determinants of the resource use efficiency of the entire ecosystem.

However, the relationship between comprehensive indicators like vegetation productivity and soil health is highly uncertain. Our previous research found that during the early development stage of subtropical plantations, the increase in plantation productivity was not necessarily related to the initial soil organic matter stocks (Chen et al., 2017), and that the accumulation of soil organic matter was primarily controlled by understory shrubs and herbs rather than changes in tree biomass (Chen et al., 2019). Recently, a study in Europe also found that soil health (a composite index based on soil properties, biodiversity, and plant disease control) in woodlands was significantly higher than in croplands and grasslands, but only cropland and grassland productivity, not woodland productivity, could be positively linked to soil health (Romero et al., 2024). Similarly, when mixed forests were converted to pure stands for several years, plant-parasitic nematodes became more abundant in the pure stands, but the productivity of the pure stands did not decline (Zhao et al., 2019).

Table 2 Biological indicators that can characterize the health of forest soils

Indicator category	Indicators	Indicative significance	Uncertainties	References
Comprehensive ecosystem indicators	Vegetation productivity, vegetation carbon sink function	Soil's ability to supply key resources such as nutrients and water	Decoupling of plant carbon sink function and soil health	Cai and Wu, 2015; Liu et al., 2023a; Shang et al., 2023
Plant community indicators	Plant health status, plant diversity	Severity of soil-borne diseases; soil microbial support for coexistence of multiple plant species	Difficulty distinguishing the impact of other factors on what constitutes a healthy soil food web	Xi et al., 2021; Wang et al., 2021; Dai et al., 2021; Wang et al., 2024
Plant population indicators	Indicator plants, such as *Dicranopteris linearis*	Acidic soil conditions	Unable to capture slow soil acidification processes	Lin et al., 1957; Yang et al., 2021
	Indicator plants, such as ground cover plants	Soil organic matter and nutrient status	General applicability needs further validation	Ou et al., 2020
Plant physiological and ecological indicators	Plant stoichiometry	Soil's ability to supply nutrients	Influenced by plant intrinsic properties	Yan et al., 2017
	Root: shoot ratio, fine root diameter, and specific root length	Soil's ability to supply nutrients	Influenced by plant intrinsic properties	Zhang et al., 2018; Dai et al., 2021
	Leaf stomatal conductance and root system configuration changes	Soil's ability to supply water, i.e., degree of drought stress	The relationship is not strictly corresponding	Gupta et al., 2020
	Leaf yellowing and wilting, root rot	Excessive soil moisture, poor drainage	Difficulty distinguishing the impact of other factors	Nie et al., 2021
	Fine root distribution and biomass	Development of soil aggregate structure	Difficulty distinguishing the impact of other factors	Xiao et al., 2024
Soil biota community indicators	Complexity and stability of soil food web	Reflects the overall condition of soil health	Simple soil food webs can also be healthy	Wang et al., 2023

(*Continued*)

Table 2 (*Continued*)

Indicator category	Indicators	Indicative significance	Uncertainties	References
Soil fauna community indicators	Soil fauna community index	Reflects the overall condition of soil quality	Effect varies depending on the selected soil fauna groups	Yan et al., 2012
	Soil biological quality-earthworms (QBS-e)	Direct evaluation of soil quality in the field	Method developed for evaluating agricultural soil quality may need optimization for forests	Fusaro et al., 2018
	Soil biological quality-arthropod (QBS-ar)	Degree of soil environmental degradation	Primarily used for agricultural soils, its indication of forest soil health is pending	Menta et al., 2018
	Nematode-related indices	Soil fertility, degree of disturbance, soil food web energy characteristics, and organic carbon dynamics	Relationship with productivity is unclear	Bongers and Ferris, 1999; Ferris and Bongers, 2006; Ferris, 2010; Zhang et al., 2024b
Soil fauna population indicators	Earthworm population characteristics	Soil fertility, severity of soil-borne diseases	Effective primarily at local scales	Yasir et al., 2009; van Groenigen et al., 2014
	Population Ddensity of plant-parasitic nematodes	Severity of soil-borne diseases	Parasitic organisms are not always detrimental to ecosystem health	Zhao et al., 2019
	Soil fauna parasites	Infestation by parasitic pests	An important component of ecosystems, not always harmful	Poulin, 2014
Soil fauna physiological and ecological indicators	Soil fauna stoichiometric	Carbon and nutrient limitation in soil animals	Potential misjudgment, best combined with population change data	Zhang et al., 2023a
	Earthworm aestivation	Soil water and heat stress conditions	Difficulty in distinguishing between drought stress and cold stress	Edwards and Arancon, 2022
	Antioxidant enzyme activity in earthworms	Toxic environments in soil, such as aluminum toxicity due to acidification	Difficulty in distinguishing the impact of other stress factors	Velki and Hackenberger, 2013; Li et al., 2014; Hattab et al., 2015; Song et al., 2019

Soil microbial community indicators	Fungal vs. bacterial energy flow channels balance	Balance between organic matter decomposition and accumulation, degree of disturbance	Fungal: bacterial ratios can vary for many reasons	Wang et al., 2019; Fierer et al., 2021; Cao et al., 2024
	Core soil microbial populations	Ability of plants and soil to resist pathogenic organisms	Difficulty in determining core populations, requires further research	Marx, 2024
Soil microbial population indicators	Types and hyphal density of plant-symbiotic fungi in soil	Plant's ability to utilize soil nutrients, an important foundation for vegetation recovery	Mycorrhizal fungi do not always play a role	Marx, 2024; Wang et al., 2024
	Population density of soil-borne diseases	Disease pressure faced by plants	Diseases are part of the ecosystem and are not always detrimental	Kasanen et al., 2022
Physiological and ecological indicators of soil microorganisms	Microbe stoichiometric	Indicate the carbon and nutrient balance status of soil biota	Thresholds may be ecosystem specific	Cui et al., 2024a
	Phospholipid fatty acid (PLFA) – bacterial stress indicator	Activity of soil microorganisms, severity of adverse conditions such as drought or resource limitations	Difficulty in distinguishing the effects of different stress factors	Shen et al., 2024
Soil biochemical indicators	Proportion of readily available sugars in soil	Energy supply status of the soil, activity level of the soil system	Appropriate sugar content in different ecosystems needs to be determined	Gunina and Kuzyakov, 2015
	Enzyme activity	Enzyme stoichiometric indicating the balance of carbon and nutrients in the soil	Inconsistencies in inferring nutrient limitations	Mori et al., 2023

(Continued)

Table 2 (*Continued*)

Indicator category	Indicators	Indicative significance	Uncertainties	References
Key process indicators	Resource utilization efficiency indices	Efficiency of carbon, nutrient, and water utilization	The relationship between resource utilization efficiency and productivity in soil systems may decouple	Zhou and Ding, 2007; Feng et al., 2021
	Soil organic matter accumulation	Carbon sequestration coefficient (SQ), potential for soil carbon storage	Vegetation carbon sink function and soil carbon storage may not be synchronized	Zhang et al., 2013
Resistance and resilience index	Resistance and resilience of various biological indicators	Testing the sustainability of soil health	Difficulty in comprehensively evaluating the resistance and resilience of different indicators	Blanco and Lal, 2023

4.2 Plant communities and soil health

At the plant dimension, soil health can be reflected through various aspects, including plant community health indicators, changes in indicator plant populations, and physiological and ecological indicators of individual plants.

How do plant communities reflect soil health? Many useful insights can be gained from studies on plant-soil feedbacks (PSF). The weaker the negative PSF, where soil microbes and abiotic properties reduce plant performance, the higher the theoretical soil health level of the ecosystem (Xi et al., 2021; Wang et al., 2021). For example, if the same plant species grows faster in a given ecosystem, with a higher proportion of older and taller individuals, it may indicate that the plant is receiving more adequate nutrients from the soil and does not need to allocate additional resources to resist soil-borne diseases. Additionally, the coexistence patterns within and between plant species often reflect the characteristics of the soil biotic community, as maintaining plant diversity, especially rare species, often requires the support of soil microorganisms (Dai et al., 2021; Wang et al., 2024).

The distribution of plant populations can also reflect the physicochemical environment of the soil. Some plants in forests are particularly sensitive to changes in soil pH and nutrient status, making their distribution useful for indicating soil conditions. The role of understory shrubs and herbs is especially noteworthy. For example, the fern *Dicranopteris linearis*, widely distributed in red soil areas of subtropical and tropical regions, can serve as an indicator of acidic soils (Lin et al., 1957; Yang et al., 2021). *Alpinia chinensis* can indicate phosphorus limitation in forest soils (available P $\leq$ 2 mg/kg) (Ou et al., 2020).

Physiological and ecological indicators of individual plants can also reflect the physicochemical environment of the soil. For example, plant element stoichiometry (Yan et al., 2017), root: shoot ratio, fine root diameter, and specific root length can reflect the soil's ability to supply nutrients (Zhang et al., 2017; Dai et al., 2021). Stomatal conductance, root architecture, and visible symptoms on leaves and roots can reflect the degree of soil drought or waterlogging (Gupta et al., 2020; Nie et al., 2021). Root biomass and distribution can reflect the development of soil aggregate structure (Xiao et al., 2024).

5 Assessing soil health through biological indicators: soil fauna

Soil fauna refers to eukaryotic and heterotrophic organisms, with the exception of fungi, that live temporarily or permanently in the soil. These organisms contribute millions of species and account for up to 60% of all terrestrial animal biomass (Bar-On et al., 2018). They play crucial roles in processes such as litter decomposition, aggregate formation, energy fluxes, and structural maintenance, making them essential indicators of soil health in the face of environmental changes (Sun et al., 2021; Potapov et al., 2024).

Healthy forest soil typically serves as a biodiversity reservoir for soil fauna, which can be broadly categorized into three groups: microfauna, mesofauna, and macrofauna, based on their body size (Fenton, 1947; Wang et al., 2023):

- Microfauna, such as protozoa and nematodes, regulate microbial communities and influence soil microstructure, nutrient mineralization, and carbon sequestration (Schmidt et al., 2000; Liu et al., 2023b).
- Mesofauna, including springtails (collembolans) and mites, contribute to the breakdown of organic matter and act as intermediaries between soil micro-resources and larger consumers (George et al., 2017).
- Macrofauna, such as earthworms and termites, are recognized as soil engineers.

They not only facilitate litter decomposition and aggregate formation but also modify soil structure and nutrient dynamics through their gut passage and burrowing activities. Predators, like spiders and ground beetles, shape soil ecosystems by controlling smaller organisms through soil food webs (Gongalsky, 2021). Note that soil-dwelling mammals, although their activities may alter the air and water flow through the soil, are not typically included in the term 'soil fauna' (Yarwood et al., 2020; Coleman et al., 2024).

Forests are composed of diverse woody plants with a complex vertical structure, providing tons of living and dead materials rich in recalcitrant organic components such as lignin, tannins, and phenols (Bonanomi et al., 2021). To decompose these resources and maintain energy fluxes in food webs, forests require a significant number of specialist primary consumers, such as detritivores and phytophages, which are rarely found in grassland and farmland ecosystems. As a result, healthy forests exhibit a pyramidal structure and a highly complex composition of soil food webs (Wang et al., 2023), sustaining long food chains, high energy fluxes, and nutrient cycling processes (Fig. 2).

However, the composition of each trophic group changes in unhealthy forest soils. For example, one common forest management practice, litter removal from the forest floor, reduces the feeding and sheltering resources for litter-dwelling soil fauna, which primarily consist of primary consumers such as Diplopoda and Isopoda (Inkotte et al., 2023). Due to the decline in primary consumers, the food web structure becomes less complex, and less energy is transferred to animals in higher trophic levels (Fig. 2). Therefore, the structure of the soil food web should be considered an important indicator of soil health.

Soil fauna communities are highly sensitive to environmental disturbances, such as logging, fires, and land-use changes, and are increasingly recognized as key indicators of soil health in forest ecosystems (Marshall, 2000; Frouz et al.,

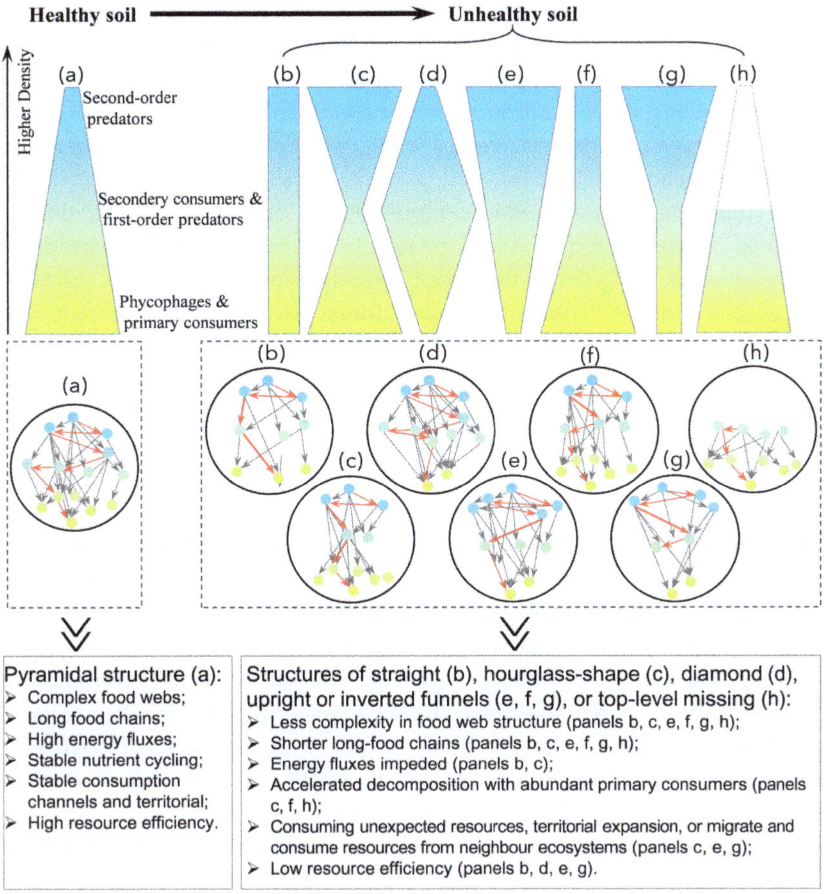

Figure 2 A conceptual scheme illustrating the contrasting characteristics of soil food webs in healthy and unhealthy soils. The changes in trophic groups and overall shapes are based on studies such as O'Gorman et al. (2019), Nagelkerken et al. (2020), and Wang et al. (2024). The green-blue gradient represents the trophic level of soil fauna. Arrows denote the predator–prey relationships, and red arrows highlight the longest food chain in the food web.

2006). These disturbances affect soil characteristics and biodiversity, often leading to structural changes in the composition of soil fauna communities, which in turn impact ecosystem functions. Soil arthropods, such as collembolans and mites, which play essential roles in both above- and belowground processes, respond quickly to changes in soil health. The QBS-ar (Qualità Biologica del Suolo – arthropod) index, which focuses on arthropod morphological traits, has been proposed to assess soil health by measuring the adaptation of soil fauna to varying land-use and environmental changes (Menta et al., 2018). Additionally, an abundance-based Fauna Index (FAI) has been designed to incorporate the diversity of the soil faunal community, its functional traits, and the abundance of its members. Given that the abundance and diversity of soil fauna may integrate physical, chemical, and microbiological properties of soil, FAI values show a strong correlation with soil quality (Yan et al., 2012), providing a more comprehensive approach to monitoring soil biodiversity and ecosystem functions.

In general, the soil fauna community can be classified into different functional groups, including decomposers, herbivores, scavengers, predators, and parasites. Among these, the relationship between herbivorous soil fauna, such as plant-parasitic nematodes, and forest health is relatively apparent; however, parasitic organisms within soil fauna, like parasitic protists, also play a crucial role in shaping ecological patterns by influencing host behavior, performance, and even speciation processes (Poulin, 2014; Li et al., 2023a), yet they are often overlooked. Contrary to the common perception that parasites negatively impact animal health and economic interests, they are integral components of natural communities (Kuris et al., 2008; Wood and Johnson, 2015). Research on aboveground and aquatic systems has shown that healthy or minimally disturbed ecosystems typically harbor highly diverse parasite communities, while human-induced changes often lead to significant declines in local parasite species diversity (Li et al., 2023b; Wood et al., 2018).

Following this logic, parasitic organisms hosted by soil fauna are likely sensitive to human activities and hold significant potential as ecological indicators of ecosystem health and human impacts. Recent advances in high-throughput technologies have greatly expanded our understanding of parasites in soil animals. Mounting evidence suggests that soils are inhabited by high biomass and diversity of protistan parasites, particularly Apicomplexa (Geisen et al., 2015; Mahé et al., 2017; Oliverio et al., 2020). Additionally, there is strong evidence of a significant decline in local Apicomplexa biodiversity in soil systems associated with land-use changes (Li et al., 2023b; Schulz et al., 2019). This decline is attributed to both a reduction in host species diversity and the hotter, drier conditions prevalent in human-modified systems (Li et al., 2023b).

While indices beyond species richness are still lacking, the use of animal parasites in soil ecosystems as ecological indicators of forest system health holds great promise, especially with ongoing advancements in molecular techniques and database development (Bass and del Campo, 2020; Del Campo et al., 2019).

A healthy soil generally provides multiple ecosystem functions, and the functional traits of soil fauna offer a direct representation of their response and adaptation to macro- and microenvironmental conditions (Qiao et al., 2022). These traits have significant potential for characterizing communities, quantifying food web structures, and understanding their contributions to ecosystem functions. For instance, characteristics such as body size, feeding habits, and dispersal capabilities in organisms like collembolans and mites reflect soil fertility and microclimate stress (Pan et al., 2022). Functional trait-based indices are often more accurate than biodiversity-based indices in reflecting soil health, although their application is still limited by our understanding of species traits.

Soil fauna of varying sizes perform different ecological functions in the soil, yet they are intricately linked through complex interactions with each other, microorganisms, and plants. These interactions form the soil food web, which is critical for sustaining ecosystem structure and function. The connectivity, complexity, and stability of the soil food web serve as key indicators of soil quality and ecosystem services. Monitoring shifts in functional groups, biomass, and energy flow within the two main food webs – the detrital and rhizosphere webs – helps assess soil health (Robinson et al., 2024).

In the detrital food web, energy flows from plant residues to a series of decomposers: micro-decomposers (including bacteria, fungi, herbivorous, and omnivorous nematodes), meso-decomposers (such as springtails and oribatid mites), and macro-decomposers (like earthworms, woodlice, and millipedes). From there, energy is transferred to organisms occupying higher trophic levels. The detrital food web is highly depended on the dominant decomposers, which not only assimilate nutrients but also fragment organic matter into smaller particles and produce fecal pellets that serve as substrates for other soil microbial decomposers. While, the rhizosphere food web is centered on plant roots and the herbivorous organisms, including herbivorous nematodes and hemiedaphic and euedaphic springtails, channeling energy through pathways associated with root systems. Due to the physical structure of soil, the diversity of consumers feeding on roots is more limited compared to those consuming plant residues on the surface. The stability of these energy transfers across trophic levels is essential for maintaining the health and resilience of the soil ecosystem.

6 Using representative taxa: earthworm communities and soil health

Earthworms can indicate soil quality in several ways (Fründ et al., 2011):

- the abundance and species composition of the earthworm community;
- the behavior of individual earthworms in contact with soil substrate, such as preference or avoidance activities;
- the accumulation of chemicals from the soil into their bodies; and
- the biochemical or cytological stress biomarkers in the earthworm tissues.

At the local scale, in forest ecosystems at the same successional stage, habitats with higher soil organic matter content often have greater earthworm biomass. Earthworm activity is commonly found to enhance soil fertility and promote plant growth (van Groenigen et al., 2014; He et al., 2018). Earthworms have also been shown to suppress fungal diseases by increasing the diversity of rhizosphere bacteria and related chitinase genes (Yasir et al., 2009). Additionally, earthworm activity can significantly inhibit plant-parasitic nematodes (Shao et al., 2017).

Furthermore, in an old beech forest, middens of the earthworm *Lumbricus terrestris* were found to contribute to the patchiness and heterogeneity of the forest floor and serve as important microhabitats for some soil animals, especially bacterial-feeding and omnivorous nematodes (Maraun et al., 1999). Clearly, earthworms have a strong positive association with soil fertility, soil-borne diseases, and soil biodiversity, all of which are key elements of soil health. Tools like the QBS-e (Qualità Biologica del Suolo – earthworm) index, although initially proposed for assessing farmland soil quality (Fusaro et al., 2018), can be used to evaluate forest soil health based on earthworm populations without requiring extensive taxonomic expertise. Note that the relationship between earthworm population size and forest soil health may change with forest types, e.g. the acid conifer soils will usually be low in earthworm populations even when the soils are healthy.

Moreover, earthworms are sensitive to environmental stress and are generally considered a biological indicator of soil abiotic conditions. When earthworms are exposed to stress, they often respond by altering their biomarkers. Measuring changes in malondialdehyde (MDA), glutathione peroxidase (GSH-Px), superoxide dismutase (SOD), and catalase (CAT) in earthworm tissues can reflect the health status of the soil environment (Velki and Hackenberger, 2013; Li et al., 2014; Hattab et al., 2015; Song et al., 2019).

However, on a global scale, the relationship between earthworm abundance, biomass, and diversity and soil parameters is weak (Phillips et al., 2019). This suggests that the relationship between earthworms and soil health is more complex than previously thought.

7 Using representative taxa: nematode communities and soil health

Soil nematodes are ubiquitous and well-adapted to a wide range of environmental conditions. There are an estimated $4.4 \pm 0.64 \times 10^{20}$ nematodes (with a total biomass of approximately 0.3 gigatonnes) inhabiting surface soils worldwide, making them the most abundant animals on Earth and filling all trophic levels in the soil food web (van den Hoogen et al., 2019). Among biological indicators, soil nematodes stand out due to their pivotal role in ecosystem functions and their rapid response to disturbance, making them an important tool for assessing forest soil health. Their transparency allows for the observation of diagnostic internal features without dissection, and their short life cycles enhance their potential as bioindicators for environmental health assessments (Bongers and Ferris, 1999; Ferris and Bongers, 2006).

Ecological indices using nematode community characteristics to describe soil health can be categorized into four types: species diversity, trophic group diversity, life history diversity, and functional diversity indices (Li et al., 2007). Not only does the species diversity of nematode communities indicate soil health, but specific groups can also serve as indicator organisms. For example, omnivorous-predatory nematodes are generally considered indicators of structurally stable and well-developed soil food webs, while the family Rhabditidae, as r-strategists, are typically dominant in disturbed habitats.

Nematode-based indices have become increasingly important for gaining insights into soil ecosystem health and functioning. The Maturity Index (MI), introduced by Bongers (1990), calculates the weighted mean frequency of nematodes across five colonizer-persister (c-p) classes. These classes reflect the functional responses of non-parasitic soil nematodes to changes in resources and the environment. The Plant-Parasitic Index (PPI) was developed to account for plant-parasitic nematodes; a higher PPI value signifies a greater abundance of K-strategy plant-parasitic nematodes, which is generally more detrimental to plant growth. Ferris et al. (2001) introduced a comprehensive system that integrates c-p scaling with feeding groups to assess soil food webs, categorizing them as basal, enriched, structured, or a combination thereof. This system led to the development of four diagnostic indices: the Basal Index (BI), Enrichment Index (EI), Structure Index (SI), and Channel Index (CI), providing a nuanced understanding of soil food web dynamics.

In addition to feeding habits and c-p values, nematode functional traits include body length, body width, stylet length, and oesophagus length, among others (Hou et al., 2023; Xue et al., 2024). Studies that combine multiple functional traits of nematodes can provide more comprehensive insights into nematode ecological strategies and their implications for soil health. Recently, trade-offs between nematode traits have been proposed to explain their ecological strategies, ranging from fast-growing, high-metabolism nematodes to slow-growing but larger-bodied nematodes, described as the nematode economics spectrum (Zhang et al., 2024b). The nematode economics spectrum, which influences SOC through its regulation of microbial traits such as microbial carbon use efficiency and microbial biomass, offers a new perspective for enhancing the predictive power of SOC dynamics (Zhang et al., 2024b).

Furthermore, the nematode metabolic footprint was developed using carbon as a metric for assessing the response of nematodes to resource distribution (Ferris, 2010). This study marked an important step for soil ecologists, transitioning from using nematode community structure to indicate soil (structural) health to using nematode biomass and energy flow to reflect soil (functional) health. The nematode metabolic footprint includes two key processes: carbon stored for growth and carbon used for respiration. Thus, it can serve as a valuable tool for linking soil nematode communities with carbon cycling. However, the nematode metabolic footprint does not account for the predation relationships of predatory and omnivorous nematodes at the highest trophic level on nematodes at lower trophic levels. Therefore, the dynamics of energy flux across multiple trophic levels within the soil food web cannot be fully understood (Barnes et al., 2018).

The soil food web energy model was refined by introducing energy conversion rates between different trophic levels (Barnes et al., 2014). Barnes et al. (2018) further enhanced this model by incorporating individual metabolic requirements, body size, phylogenetic differences, and the effects of temperature on metabolic rates into the assessment of energy flow through network nodes. Energy flow in the soil food web considers multiple functional traits, focusing not only on the soil organisms themselves but also on linking biodiversity to ecosystem functionality. The energy flow within the nematode food web makes nematodes an excellent bioindicator for assessing soil health (Zhang et al., 2024b).

It is important to note that the aforementioned calculation methods still fail to account for the turnover rates of soil animal populations (Wang et al., 2019). For example, if in habitat A, nematodes complete one life cycle in a week, and in habitat B, nematodes complete ten life cycles in a week, then the actual accumulated nematode biomass in habitat B over 1 week (relative to standing stock) would be ten times that of habitat A. This scenario cannot be accurately captured by considering only nematode respiration and predation.

Finally, scientists attempt to understand the complexity and stability of food webs by analyzing the topology of co-occurrence networks, thereby indicating SOC sequestration and soil multifunctionality (Wang et al., 2024; Cao et al., 2024). Previous studies have found that the stability of soil food webs is well-approximated by the cascade model. The results strongly support the general hypothesis that the stability of rich and complex soil food webs is primarily driven by correlations in interaction strength and the organization of the soil food web into functional groups (de Castro et al., 2021). Therefore, the complexity and stability of food webs are better predictors of soil functions. However, whether more complex soil food webs necessarily indicate healthier soils still requires further data to substantiate this notion.

8 Soil microbial communities and soil health

Microorganisms play fundamental ecological roles in forests. Forests are highly heterogeneous ecosystems that encompass a multitude of microbial-mediated reactive interfaces, such as the root–soil interface, litter–soil interface, root–root interface, and plant–atmosphere interface (Li et al., 2023a). Therefore, soil microbial diversity is a critical driving force of the soil processes essential for maintaining soil health (Wall et al., 2015). However, there is a lack of mechanistic understanding of how soil microbes relate to soil health, with much of the focus being on agricultural ecosystems. Research on the effects of microbial communities on forest soil health lags behind that of agricultural microbiomes, as the development of a forest ecosystem is a complex systematic process influenced not only by the soil but also by roots and the rhizosphere, litter, and the phyllosphere. We propose the following possible pathways by which soil microbes may influence soil health in forest ecosystems.

Firstly, sustaining soil functions over the long term is crucial for forest soil health. Soil microbes play a fundamental role in maintaining multiple forest soil functions, such as nutrient cycling, primary productivity, decomposition of plant inputs, and climate regulation (Banerjee and van der Heijden, 2023), which may be closely related to forest soil health (Fig. 3a). Numerous studies

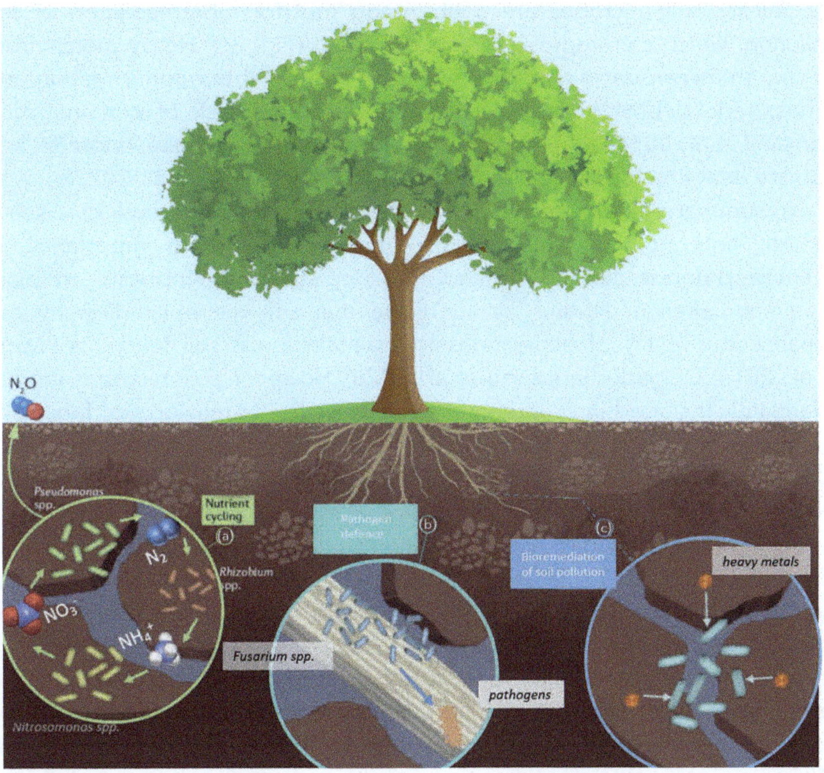

Figure 3 The three pathways by which soil microbes may influence soil health in forest ecosystems, i.e. (a) nutrient cycling; (b) pathogen defence; and (c) bioremediation of soil pollution.

have demonstrated that soil microbes decompose organic material to release tree-available inorganic nutrients, alter nutrient availability through oxidation, reduction, solubilization, and chelation processes, and store and release nutrients in necromass (Marschner and Rengel, 2007). These processes drive forest soil nutrient cycling and regulate the energy flux in the soil.

Secondly, soil microbes serve as the first line of defense against soil-borne pathogens, which can improve forest soil health (Fig. 3b). For example, kauri (*Agathis australis*) trees, one of New Zealand's most ancient native species, are currently threatened by dieback disease caused by the invasive soil-borne pathogen *Phytophthora agathidicida* (Black et al., 2018). Soil microbial communities, such as *Penicillium*, *Trichoderma*, *Enterobacteriacae*, *Actinetobacter,* and *Pseudomonas*, may be biocontrol agents against kauri dieback that can support kauri health in the face of pathogen attack (Byers et al., 2020). Soil microbes can compete with pathogens for physical space and resources, induce systemic resistance and priming for enhanced defense, and engage in direct defense responses such as antibiosis (production of antibiotics, lytic enzymes, and volatiles) and hyperparasitism (Du Jardin, 2015).

Thirdly, soil microbes are a promising tool for bioremediation of soil pollution, which can improve forest soil health (Fig. 3c). Heavy metals from mining areas can enter and accumulate in forest soil through irrigation and atmospheric deposition, significantly increasing the risk of contamination (Esmaeili et al., 2014). Soil microbes can assimilate, accumulate, degrade, and stabilize heavy metals, further reducing soil pollution (Priya et al., 2022).

In summary, soil microbes play a major role in nutrient cycling, disease control, and pollutant remediation, underscoring their importance in quantifying forest soil health and the long-standing emphasis on plant production. Recent studies also indicate that soil microbial diversity can enhance the ability of ecosystems to maintain forest soil functions (Wagg et al., 2014; Delgado-Baquerizo et al., 2020). However, soil microbes are not adequately included in or integrated into analytical frameworks of forest soil health due to the lack of mechanistic understanding of how soil microbes relate to soil functions (Bünemann et al., 2018). Advancing our understanding of the causal relationships between soil microbial indicators and soil functions is therefore critically needed and represents the next frontier in forest soil health research. Additionally, the relationships between soil health and microbial survival and functioning are poorly understood, particularly because most microbial studies fail to investigate soil health at the genetic level, which is necessary to understand the interactions between soil microbes and forest soil health. Multi-omics methods, such as metagenomics and transcriptomics, have been successfully used to study microbial communities at the genetic level (Wubs et al., 2016). Combining these multi-omics methods with soil health assessments will enable the analysis and tracking of the expression of microbial functional genes, thereby providing deep insights into the effects of the overall microbiome function on forest soil health.

On the other hand, investigating the composition and activity of microorganisms related to soil health processes can, in turn, indicate the status of soil health. For instance, measuring soil heterotrophic respiration (primarily driven by soil microbes) can provide insight into overall soil vitality. The qCO_2, calculated from microbial biomass and respiration data, can reflect microbial carbon use efficiency. Measuring the abundance and expression of biological nitrogen fixation genes can indicate the intensity of biological nitrogen fixation. Assessing the infection rate and biomass of arbuscular mycorrhizal fungi (AMF) can reflect the ability of the plant–AMF symbiosis to absorb nutrients and resist adverse conditions. Determining the ratio of beneficial rhizosphere microorganisms to pathogenic organisms can reflect plant disease resistance (Anthony et al., 2022; Kasanen et al., 2022).

In fact, numerous microbial-based indicators of soil health, such as the fungal: bacterial ratio, microbial enzyme activities, mycorrhizal abundance, and microbial pathogens, have been proposed, and some of these tests are commercially available and used routinely in farmlands (Fierer et al., 2021). However, these microbial metrics for soil health assessment should be interpreted with caution, as interpretations will be highly context dependent (Fierer et al., 2021). For instance, fungal:bacterial ratios can vary for many reasons such as forest type (broadleaf vs. coniferous), hydrothermal conditions, natural disturbances, higher enzyme activities do not necessarily indicate greater nutrient availability, and increased root colonization by AMF may not always correlate with enhanced plant growth (Fierer et al., 2021). Similarly, machine learning approaches using characteristics of microbial communities (such as 16S rRNA gene-based surveys of the soil microbiome) can effectively predict soil health metrics in agricultural fields (Wilhelm et al., 2022).

In agricultural systems, there has been considerable research on the resistance and resilience of soil microbial communities (Zhang et al., 2023b). Studies have found that soil microbial communities are generally not resistant to perturbations, and a full recovery when the perturbations are removed is rare. However, the input of organic materials into soils usually increases resilience. At the individual level, dormancy and stress tolerance are key factors that influence microbial resistance and resilience. At the population level, life-history strategies (such as copiotrophs/oligotrophs and r/K strategists) also play a significant role. At the community level, factors such as microbial diversity, community composition (the relative abundance of fungi vs. bacteria, and rare vs. dominant species), and the interactions (networks) among microbes are suggested to be crucial for influencing microbial resistance and resilience. These intrinsic factors, while derived from studies in farmlands, can provide valuable insights into the resistance and resilience of forest soil microbial communities to disturbances.

Similar to the recently reported core microbiota in the human gut that affects human health (Wu et al., 2024), healthy forests may also possess their own core microbiota. In healthy forest soils, besides the fungi that significantly influence the forest (Marx, 2024), what other components make up the core microbiota of forest soils? What are their characteristics? We postulate that forest soil health may be associated with core microbiota across multiple dimensions, such as the rhizosphere soil core microbiota, non-rhizosphere soil core microbiota, and the gut core microbiota of soil fauna (e.g. earthworms).

9 Biochemical indicators of forest soil health

Changes in the chemical stoichiometry of plants, soil animals, and microorganisms, as well as enzymes, can indicate the balance of carbon and nutrients in ecosystems (Yan et al., 2017; Zhang et al., 2023a; Mori et al., 2023; Cui et al., 2024a). These indicators can help determine whether the primary biological groups in the ecosystem are carbon-limited, nitrogen-limited, phosphorus-limited, or limited by multiple elements simultaneously.

Additionally, the PLFA-based bacterial stress index can reflect the stress conditions of bacteria due to resource limitations or water and heat stress (Shen et al., 2024).

Furthermore, from the perspective of energy supply, one can understand the 'intensity' of material cycling and energy transfer in ecosystems. Healthy soil should have sufficient vitality but not be overly active. This soil vitality can be characterized by the content of readily available sugars and their proportion of total organic carbon (Gunina and Kuzyakov, 2015). The appropriate scenario is that organic matter decomposition and transformation should neither be too fast nor too slow, and respiration should neither be too strong nor too weak. Rapid decomposition can support high productivity in the short term but consumes resources quickly and often with low efficiency. Slow decomposition leads to more resource accumulation but results in low productivity. This also involves the balance between the bacterial and fungal energy flow channels. A healthy forest should possess a self-regulating mechanism to maintain a balanced state between bacterial and fungal energy flow channels. However, the optimal equilibrium may vary and requires further explored.

10 Ecological processes, forest soil health, and resilience

Healthy soil should be able to provide relatively adequate nutrients and water, coupled with the plant photosynthetic carbon-fixing process, while maintaining relatively high resource use efficiency to support relatively high and stable productivity.

Soil organic matter accumulation is the ecological process most closely related to soil health. Gradual accumulation rather than depletion of soil organic matter should be an important indicator of healthy forest soils. Since changes in soil organic matter typically occur slowly, it can be challenging to determine whether forest soils are accumulating organic matter. For example, in subtropical Eucalyptus plantations, soil organic matter only begins to accumulate significantly after about 10 years following afforestation (Chen et al., 2017). However, since newly input organic matter typically exhibits lower natural abundances of ^{13}C and ^{15}N, combining the analysis of soil organic matter content with changes in the natural abundances of ^{13}C and ^{15}N can be useful for assessing the potential and capacity of soil organic matter accumulation (Chen et al., 2019).

The resource use efficiency of plants and soil biota is a touchstone for forest soil health. Resource use efficiency is a comprehensive reflection of the stoichiometric ratios in forest ecosystems, the balance between fungal and bacterial energy flow channels, and the equilibrium between organic matter decomposition and accumulation. In an ideal forest ecosystem, light, carbon, nutrients, and water are all adequately supplied and well-coupled, leading to high productivity with minimal resource waste and supporting long food chains. In reality, forest ecosystems often face limitations of one or more resources, leading to decoupling of resources and subsequent resource wastage.

For example, severe nutrient limitation in soils can significantly reduce the carbon use efficiency of soil microorganisms, as indicated by a marked increase in the metabolic quotient (microbial respiration per unit biomass, qCO_2) (Feng et al., 2021). In the early stages of subtropical plantation development, maintaining productivity often involves higher soil microbial qCO_2, meaning that a greater proportion of energy is invested to obtain nutrients (Ding et al., 1992; Zhou and Ding, 2007). It has also been found that the carbon use efficiency of forest soil microorganisms increases with higher plant diversity, thereby establishing a direct link between plant diversity and soil organic matter accumulation (Duan et al., 2023).

Notably, the earthworm-mediated carbon trap (Zhang et al., 2013) can lead to the formation of "sponge-like resource islands," providing a large amount of available yet relatively stable and less leachable resources for plants and microorganisms (He et al., 2018; Zhang et al., 2020b). This can alleviate competition for nutrients between plants and bacteria (Lv et al., 2016; He et al., 2020). The role of earthworms in the sustainable development of forest ecosystems warrants further investigation.

Additionally, while theoretically healthy soil can support higher vegetation productivity and aboveground carbon sequestration, these two aspects may also become decoupled. For example, as forests age, the carbon sequestration capacity of the vegetation itself (excluding soil carbon sequestration processes) may decline (Shang et al., 2023). At this point, stating that older forests have less healthy soil compared to younger forests may not be accurate; moreover, the soil carbon sequestration capacity of mature forests may significantly increase (Zhou et al., 2006). Finally, soil drought has become a key factor causing tree mortality and reduced carbon sequestration in forests (Liu et al., 2023b). As the impacts of global change has become increasingly severe, maintaining the coupling of carbon, nutrients, and water has become more urgent.

Nevertheless, the rate of soil organic matter accumulation and resource use efficiency are not necessarily indicative of healthier forest soil when they are higher. Determining the specific levels at which soil can be considered 'healthy' requires further research.

10.1 The resistance and resilience of soil health

How the biological, biochemical, and process indicators of soil health change under anthropogenic and natural disturbances is a critical measure of forest soil health. Two decades ago, a study proposed indices for quantifying the resistance and resilience of soil biota to exogenous disturbances, such as soil drought (Orwin and Wardle, 2004). Theoretically, any soil health indicator can be quantified using these indices to measure its resistance and resilience.

Currently, soil microorganisms are commonly used biological indicators in the calculation of resistance and resilience. A global-scale study demonstrated strong connections between microbial community composition and the resistance of multiple ecosystem functions linked to C, N, and P cycling (multifunctionality resistance) in natural dryland ecosystems (Delgado-Baquerizo et al., 2017). Soil microbial species richness, as well as the microbial biomasses of C, N, and P, has been widely used as a state variable to study soil resilience (Yi and Jackson, 2021). Additionally, changes in key ecological processes that are simple to measure, such as soil respiration, are frequently used to characterize the resistance of soils to external disturbances (Todman et al., 2016). Clearly, combining biological, biochemical, and process indicators can provide a more comprehensive understanding of the resistance and resilience of soil health.

11 Challenges and future trends

11.1 Establishing reference systems for healthy soils across different spatiotemporal scales to assess forest soil health

Evaluating forest soil health using a single standard is unreasonable. Soil health should be aligned with the forest and soil development stage or the soil development characteristics of the climatic zone in which it is located. From a temporal perspective, forest soils are continuously developing and changing. The level of soil health must match the developmental stage of the vegetation and soil. Each developmental stage has its optimal forest soil condition. Allowing forests at different ages or developmental stages to reach their natural states ensures that the soil is in the healthy condition appropriate for that stage.

For example, the soil biota in healthy soils are also continually evolving. Abnormal changes in the biotic community that deviate from natural patterns may signal issues with soil health. To determine whether the evolution of a forest soil is abnormal, a reference system for comparison is essential. However, a significant proportion of previous studies neither considered reference ecosystems (41%) nor provided basic information about soil types (<21%), highlighting the urgent need to standardize assessment methods to ensure data comparability (Gatica-Saavedra et al., 2023).

As forests succeed, does the soil become increasingly healthy? If forests continue to develop, will they eventually decline? Does the forest ecosystem have some negative feedback mechanisms to curb processes or factors that are detrimental to the overall health of the ecosystem? These questions are worth deep consideration. Ultimately, forests or forest soils themselves are not inherently healthy or unhealthy; the distinction arises only when human expectations are applied.

From a spatial perspective, the standards for soil health in forest ecosystems should differ across different climatic zones. For example, in southern regions with favorable water and heat conditions, organic matter decomposes rapidly, resulting in relatively low soil organic matter content. However, due to the fast turnover, the flux of organic matter over a given period is not low. Therefore, comparing the absolute content of existing soil organic matter between southern forests and northern cold forests, where organic matter decomposes more slowly, is not only poorly comparable but can also be misleading.

Another example is nematodes, particularly those at higher trophic levels, which are highly sensitive to changes in water and heat conditions (Nielsen et al., 2014; Li et al., 2023c). Drought can cause nematodes to become smaller (Lu et al., 2023; Cui et al., 2024b), and nematode diversity and density tend to increase in wet, cold ecosystems (Procter, 1990). Therefore, soil health indicators based on nematodes should ideally be evaluated separately for different climatic zones.

11.2 How to construct a simple and effective comprehensive index for forest soil health?

Theoretically, to comprehensively reflect the health status of forest soils, it is necessary to effectively integrate biological indicators with physical and chemical indicators, thereby developing a more comprehensive soil health assessment system. However, more complex soil health indicators are not always better. A focus on the synergistic changes in soil structure, soil organic matter, and soil biota, as well as their regulatory effects on key ecological processes such as soil organic matter accumulation, can provide a robust framework for assessing soil health. The specific construction of a soil health index should be tailored to the 'pain points' of the forest ecosystem being assessed.

For example, in tropical plantations during the early stages of development, the contradiction between the high nutrient demand of vegetation and the limited nutrient supply from the soil is a major bottleneck limiting plantation development. The more this contradiction is alleviated, the healthier the plantation becomes. Therefore, when assessing such plantations, the index can be designed to focus on quantifying the balance between nutrient supply and demand. For relatively impoverished tropical plantations, the nutrients released from the decomposition of organic matter are a crucial source of plant nutrition. Thus, the point at which soil organic matter transitions from depletion to significant accumulation marks the beginning of improved soil health in these plantations.

However, accurately quantifying the accumulation rate of soil organic matter in forest soils is quite challenging. As vegetation and soil develop, the soil volume often expands, leading to poor comparability of organic matter stocks calculated from soil samples of apparently the same volume at different times and locations. We found that the fundamental reason for the calculation bias in organic matter stocks due to changes in soil volume lies in the non-conservative nature of the soil reference frame. To address this, we proposed the Equivalent Mineral-Matter Volume (EMMV) approach, which can more accurately calculate and integrate soil organic matter data across different spatiotemporal scales (Zhang et al., 2019).

Furthermore, the size of soil biotic communities directly affects their ability to influence organic matter transformation processes. However, quantifying the accumulation of soil biotic community biomass over a given time period also poses significant challenges. When comparing a key process (such as organic matter transformation or changes in soil biotic populations), it is not sufficient to look only at current stocks or instantaneous fluxes; instead, cumulative amounts over a unit time should be considered. Using the same approach, the calculation of energy flow through different trophic levels in the soil food web can be improved. This involves not only calculating the biomass stocks of organisms at each trophic level but also considering their respiration and population turnover (Wang et al., 2019).

11.3 Strengthen research on mechanisms for maintaining forest soil health

Further investigation of diverse negative feedback mechanisms related to soil health is needed to explore how ecosystems maintain soil health through self-regulation. Given that the weakest area of research is the connection between soil biodiversity, soil food web characteristics, and forest ecosystem services, efforts can be focused on the following three aspects.

First, strengthen research on the spatial and temporal distribution patterns of soil biodiversity and its functions across multiple scales. Forests are the terrestrial ecosystems with the highest biodiversity, and soil is a critical component of forest biodiversity. The maintenance and stability of forest ecosystem functions largely depend on the composition of soil biotic functional groups. However, studies have shown that globally, only 0.3% of sampling sites simultaneously focus on soil biodiversity and ecosystem functions (Guerra et al., 2020). Moreover, the focus on soil biotic groups is primarily on microbial communities, such as bacteria and fungi, with a lack of research on soil fauna. Both the temporal scale (e.g. few sampling occasions) and the horizontal spatial scale (uneven distribution of sampling points) limit our understanding of forest soil biodiversity (Shao et al., 2023). Notably, most studies on the vertical distribution of forest biodiversity focus on changes along elevation gradients and mainly concern larger animals (such as birds) or insects, with fewer studies on soil invertebrates (Sun et al., 2020). Reports on the vertical distribution of soil biota within an ecosystem are scarce (soil biota do not always remain in the soil). If the research scope is expanded to climate zones or even a global scale, these 'hidden' biodiversity hotspots (blind hot spots) may play a significant role in forest soil health.

Second, shift the focus and content of soil food web research. This involves two main aspects:

- **From single trophic levels to food web cascading effects.** Extend the focus from the response of a single group to disturbances to the feedbacks of multi-trophic level interactions, and elucidate the mechanisms of changes in food web complexity and cascading effects among organisms.
- **From structure to function.** Past research on food web structure has mostly concentrated on 'abundance structure,' such as the individual density of soil biota at the species and trophic group levels. Currently, the 'energy structure,' which characterizes soil biota biomass based on functional traits like body length and width, should receive more attention. Compared to abundance structure, energy structure provides a more accurate reflection of ecosystem functions. Therefore, research on the functional traits of key

groups is urgently needed, which is also a crucial foundation for using soil biota to indicate forest soil health.

Third, enhance research on the coupling processes of 'biota-water-heat' in forest soils. The direct role of forests in regulating surface heat balance and thus influencing global climate change is increasingly recognized (Tang et al., 2024). However, most existing studies focus on how aboveground vegetation regulates surface temperature through changes in photosynthesis, radiation reflection, and evapotranspiration, while neglecting the impact of biotic activity in the soil on the water–heat coupling processes that regulate surface heat balance. In fact, soil biotic activity can alter soil structure and the soil CO_2 reservoir, thereby regulating soil heat balance, which is crucial for maintaining soil biotic activity (Zhang et al., 2024a).

In summary, the assessment of forest soil health constitutes a complex, integrative endeavor. A comprehensive evaluation necessitates the consideration of aboveground vegetation and soil biota, alongside their interrelated ecological processes. This holistic approach not only serves as a pivotal entry point for soil health assessment but also represents a significant challenge and bottleneck in the field. Forest soil health is inherently dynamic, evolving through different stages of forest development and varying with geographical context; thus, assessments devoid of these specific considerations are inadequate. Furthermore, the definition of forest soil health is influenced by varying human expectations regarding the ecosystem services provided by forests. Future research must focus on identifying clear, relevant indicators of forest soil health that closely align with human well-being, thereby providing a robust foundation for sustainable forest management.

12 References

Alkorta, I., et al. (2003), Soil enzyme activities as biological indicators of soil health, *Reviews on Environmental Health* 18(1), 65–73.

Anthony, M.A., et al. (2022), Forest tree growth is linked to mycorrhizal fungal composition and function across Europe, *The ISME Journal* 16(5), 1327–1336.

Banerjee, S. and van der Heijden, M.G.A. (2023), Soil microbiomes and one health, *Nature Reviews Microbiology* 21(1), 6–20.

Barnes, A.D., et al. (2014), Consequences of tropical land use for multitrophic biodiversity and ecosystem functioning, *Nature Communications* 5(1), 5351.

Barnes, A.D., et al. (2018), Energy flux: the link between multitrophic biodiversity and ecosystem functioning, *Trends in Ecology & Evolution* 33(3), 186–197.

Bar-On, Y.M., et al. (2018), The biomass distribution on Earth, *Proceedings of the National Academy of Sciences of the United States of America* 115(25), 6506–6511.

Bass, D. and del Campo, J. (2020), Microeukaryotes in animal and plant microbiomes: Ecologies of disease? *European Journal of Protistology* 76, 125719.

Black, A., et al. (2018), Calling time on New Zealand's oldest tree species, *Nature* 561(7722), 177–178.

Blanco, H. and Lal, R. (2023), Soil Resilience. In H. Blanco and R. Lal (Eds), *Soil Conservation and Management*, Springer, Cham, Switzerland, pp. 525–544.

Bonanomi, G., et al. (2021), Decomposition of woody debris in Mediterranean ecosystems: the role of wood chemical and anatomical traits, *Plant and Soil* 460(1–2), 263–280.

Bongers, T. (1990), The maturity index: an ecological measure of environmental disturbance based on nematode species composition, *Oecologia* 83(1), 14–19.

Bongers, T. and Ferris, H. (1999), Nematode community structure as a bioindicator in environmental monitoring, *Trends in Ecology and Evolution* 14(6), 224–228.

Bünemann, E.K., et al. (2018), Soil quality: a critical review, *Soil Biology and Biochemistry* 120, 105–125.

Byers, A.K., et al. (2020), Soil microbial community restructuring and functional changes in ancient kauri (*Agathis australis*) forests impacted by the invasive pathogen *Phytophthora agathidicida*, *Soil Biology and Biochemistry* 150, 108016.

Cai, X.X. and Wu, J.Z. (2015), Research progress on the assessment of forest soil health, *Forest Engineering* 31, 37–41.

Cao, T.T., et al. (2024), Enlarging interface reverses the dominance of fungi over bacteria in litter decomposition, *Soil Biology and Biochemistry* 198, 109543.

Cardoso, E., et al. (2013), Soil health: looking for suitable indicators. What should be considered to assess the effects of use and management on soil health? *Scientia Agricola* 70(4), 274–289.

Chen, Y.Q., et al. (2017), Reforestation makes a minor contribution to soil carbon accumulation in the short term: evidence from four subtropical plantations, *Forest Ecology and Management* 384, 400–405.

Chen, Y.Q., et al. (2019), Labile C dynamics reflect soil organic carbon sequestration capacity: understory plants drive topsoil c process in subtropical forests, *Ecosphere* 10(6), e02784.AAA

Coleman, D.C., et al. (2024), Soil Fauna: Occurrence, Biodiversity, and Roles in Ecosystem .Function In E.A. Paul and S.D. Frey (Eds.), *Soil Microbiology, Ecology and Biochemistry* (5th ed.), Elsevier, Amsterdam, Netherlands, pp. 131–159.

Cui, Y., et al. (2024a). Predicting microbial nutrient limitations from a stoichiometry-based threshold framework. *The Innovation Geoscience* 2(1), 100048.

Cui, S., et al. (2024b), Body size induced changes in metabolic carbon of soil nematodes under N deposition and precipitation regime change in a temperate grassland, *Ecological Processes* 13, 9.

Dai, D., et al. (2021), Advances in mechanisms of rare species maintenance and plant-soil feedback in plant communities, *Biodiversity Science* 29(12), 1687–1699.

de Castro, F., et al. (2021), Local stability properties of complex, species-rich soil food webs with functional block structure, *Ecology and Evolution* 11(22), 16070–16081.

de Vries, F., et al. (2012), Land use alters the resistance and resilience of soil food webs to drought, *Nature Climate Change* 2, 276–280.

del Campo, J., et al. (2019), Assessing the diversity and distribution of apicomplexans in host and free-living environments using high-throughput amplicon data and a phylogenetically informed reference framework, *Frontiers in Microbiology* 10, 1–15.

Delgado-Baquerizo, M., et al. (2017), Soil microbial communities drive the resistance of ecosystem multifunctionality to global change in drylands across the globe, *Ecology Letters* 20(10), 1295–1305.

Delgado-Baquerizo, M., et al. (2020), Multiple elements of soil biodiversity drive ecosystem functions across biomes. *Nature Ecology & Evolution* 4(2), 210–220.

DeLuca, T.H., et al. (2019), Assessing Soil Biological Health in Forest Soils. In M. Busse, et al. (Eds), *Developments in Soil Science, Global Change and Forest Soils*, Elsevier, Amsterdam, Netherlands, pp. 397–426.

Devine, S., et al. (2011), Soil carbon change through two meters during forest succession alongside a 30-yr agroecosystem experiment, *Forest Science* 57, 36–50.

Ding, M.M., et al. (1992), Effect of afforestation on microbial biomass and activity in soils of tropical China, *Soil Biology and Biochemistry* 24(9), 865–872.

du Jardin, P. (2015), Plant biostimulants: definition, concept, main categories and regulation, *Scientia Horticulturae* 196, 3–14.

Duan, P., et al. (2023), Tree species diversity increases soil microbial carbon use efficiency in a subtropical forest, *Global Change Biology* 29, 7131–7144.

Edwards, C.A. and Arancon, N.Q. (2022), *Biology and Ecology of Earthworms* (4th ed.). Springer, New York.

Esmaeili, A., et al. (2014), A geochemical survey of heavy metals in agricultural and background soils of the Isfahan industrial zone, Iran, *Catena* 121, 88–98.

Feng, J., et al. (2021), Soil microbial trait-based strategies drive metabolic efficiency along an altitude gradient, *ISME Communications* 1, 71.

Fenton, G.R. (1947), The soil fauna: with special reference to the ecosystem of forest soil, *Journal of Animal Ecology* 16(1), 76.

Ferris, H. (2010), Form and function: metabolic footprints of nematodes in the soil food web, *European Journal of Soil Biology* 46(2), 97–104.

Ferris, H. and Bongers, T. (2006), Nematode indicators of organic enrichment, *Journal of Nematology* 38(1), 3–12.

Ferris, H., et al. (2001), A framework for soil food web diagnostics: extension of the nematode faunal analysis concept, *Applied Soil Ecology* 18(1), 13–29.

Fierer, N., et al. (2021), How microbes can, and cannot, be used to assess soil health, *Soil Biology and Biochemistry* 153, 108111.

Forzieri, G., et al. (2022), Emerging signals of declining forest resilience under climate change, *Nature* 608, 534–539.

Frouz, J., et al. (2006), Effects of soil macrofauna on other soil biota and soil formation in reclaimed and unreclaimed post mining sites: results of a field microcosm experiment, *Applied Soil Ecology* 33, 308–320.

Fründ, H.C., et al. (2011). Earthworms as Bioindicators of Soil Quality. In A. Karaca (Eds), *Biology of Earthworms. Soil Biology*, Springer, Berlin, Heidelberg, Vol. 24. https://doi.org/10.1007/978-3-642-14636-7_16

Fusaro, S., et al. (2018), Soil Biological Quality Index based on earthworms (QBS-e): a new way to use earthworms as bioindicators in agroecosystems, *Ecological Indicators* 93, 1276–1292.

Gatica-Saavedra, P., et al. (2023), Soil health indicators for monitoring forest ecological restoration: a critical review, *Restoration Ecology* 31, e13836.

Geisen, S., et al. (2015), Not all are free-living: high-throughput DNA metabarcoding reveals a diverse community of protists parasitizing soil metazoa, *Molecular Ecology* 24, 4556–4569.

George, P.B.L., et al. (2017), Evaluation of mesofauna communities as soil quality indicators in a national-level monitoring programme, *Soil Biology and Biochemistry* 115, 537–546.

Gongalsky, K.B. (2021), Soil macrofauna: Study problems and perspectives, *Soil Biology and Biochemistry* 159, 108281.

Guerra, C.A., et al. (2020), Blind spots in global soil biodiversity and ecosystem function research, *Nature Communications* 11(1), 3870.

Gunina, A. and Kuzyakov, Y. (2015), Sugars in soil and sweets for microorganisms: Review of origin, content, composition and fate, *Soil Biology & Biochemistry* 90, 87–100.

Gupta A., et al. (2020), The physiology of plant responses to drought, *Science* 368, 266–269.

Hattab, S., et al. (2015), Transcriptional expression levels and biochemical markers of oxidative stress in the earthworm *Eisenia Andrei* after exposure to 2,4-dichlorophenoxyacetic acid (2,4-D), *Ecotoxicology and Environmental Safety* 122, 76–82.

He, X.X., et al. (2018), Cooperation of earthworm and arbuscular mycorrhizae enhanced plant N uptake by balancing absorption and supply of ammonia, *Soil Biology & Biochemistry* 116, 351–359.

He, X.X., et al. (2020), Earthworms negate the adverse effect of arbuscular mycorrhizae on living bacterial biomass and bacterial necromass accumulation in a subtropical soil, *Soil Biology and Biochemistry* 151, 108052.

Hou, W., et al. (2023), Functional traits of soil nematodes define their response to nitrogen fertilization, *Functional Ecology* 37, 1197–1210.

Inkotte, J., et al. (2023), Litter removal impacts on soil biodiversity and eucalypt plantation development in the seasonal tropics, *Journal of Forestry Research (Harbin)* 34, 735–748.

Jian, J., et al. (2020), A database for global soil health assessment, *Scientific Data* 7, 16.

Kasanen, R., et al. (2022), Forest Tree Disease Control and Management. In F. O. Asiegbu and A. Kovalchuk (Eds), *Forest Microbiology*, Academic Press, Oxford, UK, pp. 425–462.

Kuris, A.M., et al. (2008), Ecosystem energetic implications of parasite and free-living biomass in three estuaries, *Nature* 454, 515–518.

Lehmann, J., et al. (2020), The concept and future prospects of soil health, *Nature Reviews Earth & Environment* 1, 544–553.

Li, F.Q., et al. (2023a), Microbiome sustains forest ecosystem functions across hierarchical scales, *Eco-Environment & Health* 2, 24–31.

Li, G.X., et al. (2023c), Nematode biomass changes along an elevational gradient are trophic group dependent but independent of body size, *Global Change Biology* 29(17), 4898–4909.

Li, Z., et al. (2023b), Land use decouples parasite–metazoan host biodiversity associations in soils across subtropical and temperate zones in China, *Global Ecology and Biogeography* 32(12), 2164–2176.

Li, Q., et al. (2007), Present situation and prospect of soil nematode diversity in farmland ecosystems, *Biodiversity Science* 15, 134–141.

Li, Y., et al. (2014), Responses of the earthworm *Eisenia Andrei* exposed to sublethal aluminium levels in an artificial soil substrate, *Chemistry and Ecology* 30, 611–621.

Lin, H.H., et al. (1957), The chemical compostion of certain oxylophytes, calciphytes, and halophytes, *Acta Pedologica Sinica* 5(3), 247–270.

Liu, X., et al. (2023a), Microfauna community assembly and cascading relationship with microflora in cropland ecosystems along a latitudinal gradient, *Agriculture, Ecosystems & Environment* 357, 108678.

Liu, H., et al. (2023b), From tree to forest: multiple carbon sink constraints, *The Innovation* 4(4), 100463.
Lu, L.L., et al. (2023), Drought shifts soil nematodes to smaller size across biological scales, *Soil Biology and Biochemistry* 184, 109099.
Lv, M., et al. (2016), Plants modify the effects of earthworms on the soil microbial community and its activity in a subtropical ecosystem, *Soil Biology and Biochemistry* 103, 446–451.
Mahé, F., et al. (2017), Parasites dominate hyperdiverse soil protist communities in Neotropical rainforests, *Nature Ecology and Evolution* 1, 1–8.
Maraun, M., et al. (1999), Middens of the earthworm *Lumbricus terrestris* (Lumbricidae): microhabitats for micro- and mesofauna in forest soil, *Pedobiologia* 43, 276–287.
Marschner, P. and Rengel, Z.(2007), *Nutrient Cycling in Terrestrial Ecosystems*, Springer Science & Business Media, Leipzig, Germany.
Marshall, V.G. (2000), Impacts of forest harvesting on biological processes in northern forest soils, *Forest Ecology and Management* 133, 43–60.
Martin, T., et al. (2022), The integration of nematode communities into the soil biological health framework by factor analysis, *Ecological Indicators* 136, 108676.
Marx, V. (2024), How's your health, forests? *Nature Methods* 21, 1145–1148.
Menta, C., et al. (2018), Microarthropods biodiversity in natural, seminatural and cultivated soils-QBS-ar approach, *Applied Soil Ecology* 123, 740–743.
Mori, T., et al. (2023), Enzymatic C:N:P stoichiometry: questionable assumptions and inconsistencies to infer soil microbial nutrient limitation, *Geoderma* 429, 116242. https://doi.org/10.1016/j.geoderma.2022.116242.
Nagelkerken I., et al. (2020), Trophic pyramids reorganize when food web architecture fails to adjust to ocean change, *Science* 369, 829–832.
Nie, G.P., et al. (2021), Plant response to waterlogging stress: research progress, *Chinese Agricultural Science Bulletin* 37(18), 57–64.
Nielsen, U.N., et al. (2014), Global-scale patterns of assemblage structure of soil nematodes in relation to climate and ecosystem properties, *Global Ecology & Biogeography* 23(9), 968–978.
Nunes, M.R., et al. (2021), The soil health assessment protocol and evaluation applied to soil organic C, *Soil Science Society of America Journal* 85, 1196–1213.
O'Gorman, E.J., et al. (2019), A simple model predicts how warming simplifies wild food webs, *Nature Climate Change* 9, 611–616.
Oliverio, A.M., et al. (2020), The global-scale distributions of soil protists and their contributions to belowground systems, *Science Advances* 6, 1–10.
Orwin, K.H. and Wardle, D.A. (2004), New indices for quantifying the resistance and resilience of soil biota to exogenous disturbances, *Soil Biology and Biochemistry* 36, 1907–1912.
Ou, Y.D., et al. (2020), Indicative effect of ground vegetation on soil nutrient status in evergreen broad-leaved forest of Guangdong. *Journal of Tropical and Subtropical Botany* 28(2), 115–123.
Pan, S., et al. (2022), Effect of urban greening and afforestation on soil microarthropod communities in coastal reclaimed land: insights from functional traits, *Applied Soil Ecology* 173, 104391.
Pennisi, E. and Cornwall, W. (2020), Hidden web of fungi could shape the future of forests, *Science* 369, 1042–1043.

Phillips, H.R.P., et al. (2019), Global distribution of earthworm diversity, *Science* 366, 480–485.
Potapov, A.M., et al. (2024), Rainforest transformation reallocates energy from green to brown food webs, *Nature* 627, 116–122.
Poulin, R. (2014), Parasite biodiversity revisited: frontiers and constraints, *International Journal for Parasitology* 44, 581–589.
Pouyat, R., et al. (2020), *Forest and Rangeland Soils of the United States Under Changing Conditions*, Springer Nature, Cham, Switzerland.
Priya, A.K., et al. (2022), Biosorption of heavy metals by microorganisms: evaluation of different underlying mechanisms, *Chemosphere* 307, 135957.
Procter, D.L.C. (1990), Global overview of the functional roles of soil-living nematodes in terrestrial communities and ecosystems, *Journal of Nematology* 22(1), 1–7.
Qiao, Z., et al. (2022), Urbanization and greenspace type as determinants of species and functional composition of collembolan communities, *Geoderma* 428, 116175.
Ranger, J. (2018), Forest Soils: Characteristics and Sustainability. In J. Berthelin, et al. (Eds), *Soils as a Key Component of the Critical Zone 1*, John Wiley & Sons, London, UK.
Robinson, J.M., et al. (2024), Food webs in food webs: the micro–macro interplay of multilayered networks, *Trends in Ecology & Evolution* 39(10), 913–922.
Romero, F., et al. (2024), Soil health is associated with higher primary productivity across Europe, *Nature Ecology and Evolution* 8, 1847–1855.
Schmidt, I.K., et al. (2000), Long-term manipulation of the microbes and microfauna of two subarctic heaths by addition of fungicide, bactericide, carbon and fertilizer, *Soil Biology and Biochemistry* 32, 707–720.
Schulz, G., et al. (2019), Changes in trophic groups of protists with conversion of rainforest into rubber and oil palm plantations, *Frontiers in Microbiology* 10, 1–14.
Shang R., et al. (2023), China's current forest age structure will lead to weakened carbon sinks in the near future, *The Innovation* 4(5), 100515.
Shao, Y., et al. (2017), Nitrogen deposition cancels out exotic earthworm effects on plant-feeding nematode communities, *Journal of Animal Ecology* 6, 708–717.
Shao, Y., et al. (2023), Drivers of nematode diversity in forest soils across climatic zones, *Proceedings of the Royal Society B* 290, 20230107.
Shen, Z.F., et al. (2024), Fourteen-year field experiment reveals neutral effects of N and P deposition on abundance and stoichiometric traits of the earthworm *Pontoscolex corethrurus* in tropical plantations, *Soil Biology & Biochemistry* 197, 109540.
Six, J. (2012), Fungal friends against drought, *Nature Climate Change* 2, 234–235.
Smith, C., et al. (2022), Using soil sustainability and resilience concepts to support future land management practice: a case study of mt Grand Station, Hāwea, New Zealand, *Sustainability* 14(3), 1808.
Song, P., et al. (2019), Phthalate induced oxidative stress and DNA damage in earthworms (*Eisenia Fetida*), *Environment International* 129, 10–17.
Sprunger, C.D. and Martin, T.K. –(2023), An integrated approach to assessing soil biological health. *Advances in Agronomy* 182, 131–168.
Stott, D.E. (2019), *Recommended Soil Health Indicators and Associated Laboratory Procedures*, Soil Health Technical Note No. 450-03. U.S. Department of Agriculture, Natural Resources Conservation Service, Washington, DC.

Sun, X., et al. (2020), Changes in diversity and body size of Onychiurinae (Collembola: Onychiuridae) along an altitudinal gradient in Changbai Mountain, China, *Soil Ecology Letters* 2(3), 230–239.

Sun, X., et al. (2021), Soil Fauna and Soil Health, *Acta Pedologica Sinica* 58(5), 1073–1083.

Tang, T., et al. (2024), Historical land use and land cover change reduced the temperature of cold extremes more than that of hot extremes, *The Innovation Geoscience* 2(3), 100079.

Todman, L., et al. (2016), Defining and quantifying the resilience of responses to disturbance: a conceptual and modelling approach from soil science, *Scientific Reports* 6, 28426.

van Bruggen, A.H.C. and Semenov, A.M. (2000), In search of biological indicators for soil health and disease suppression, *Applied Soil Ecology* 15, 13–24.

van den Hoogen, J., et al. (2019), Soil nematode abundance and functional group composition at a global scale, *Nature* 572, 194–198.

van Groenigen, J., et al. (2014), Earthworms increase plant production: a meta-analysis, *Scientific Reports* 4, 6365.

Veldkamp, E., et al. (2020), Deforestation and reforestation impacts on soils in the tropics, *Nature Reviews Earth & Environment* 1, 590–605.

Velki, M. and Hackenberger, B. K. (2013), Inhibition and recovery of molecular biomarkers of earthworm *Eisenia Andrei* after exposure to organophosphate dimethoate, *Soil Biology and Biochemistry* 57, 100–108.

Wagg, C., et al. (2014), Soil biodiversity and soil community composition determine ecosystem multifunctionality, *Proceedings of the National Academy of Sciences* 111(14), 5266–5270.

Wall, D.H., et al. (2015), Soil biodiversity and human health, *Nature* 528(7580), 69–76.

Wang, G.Z., et al. (2021), Plant soil feedback theory and its applications and prospects in natural and agricultural ecosystems, *Acta Ecologica Sinica* 41(23), 9130–9143.

Wang, B., et al. (2023), Small urban fragments maintain complex food webs of litter-dwelling arthropods in a subtropical city in China, *Soil Biology and Biochemistry* 185, 109150.

Wang, X.L., et al. (2019), Fungi to bacteria ratio: Historical misinterpretations and potential implications, *Acta Oecologica* 95, 1–11.

Wang, W.T., et al. (2024), Soil microbial influences over coexistence potential in multispecies plant communities in a subtropical forest, *Ecology* 105, e4415.

Wilhelm, R.C., et al. (2022), Predicting measures of soil health using the microbiome and supervised machine learning, *Soil Biology and Biochemistry* 164, 108472.

von Wilpert, K. (2022), Forest soils–what's their peculiarity? *Soils* 6(1), 1–25.

Wood, C.L. and Johnson, P.T.J. (2015), A world without parasites: exploring the hidden ecology of infection, *Frontiers in Ecology and the Environment* 13, 425–434.

Wood, C.L., et al. (2018), Human impacts decouple a fundamental ecological relationship–The positive association between host diversity and parasite diversity, *Global Change Biology* 24, 3666–3679.

Wu, G.J., et al. (2024), A core microbiome signature as an indicator of health, *Cell* 187(23): 6550–6565.

Wubs, E.R., et al. (2016), Soil inoculation steers restoration of terrestrial ecosystems, *Nature Plants* 2(8), 1–5.

Xi, N., et al. (2021), Relationships between plant–soil feedbacks and functional traits, *Journal of Ecology* 109, 3411–3423.

Xiao, T., et al. (2024), Effects of vegetation roots on the structure and hydraulic properties of soils: a perspective review, *Science of The Total Environment* 906, 167524.

Xing, C., et al. (2023), A review of forest height inversion by PolInSAR: theory, advances, and perspectives, *Remote Sens* (Basel) 15, 1–29.

Xue, Q., et al. (2024), Soil nematode community profiling using reference-free mito-metagenomics, *Soil Biology and Biochemistry* 199, 109613.

Yan, S.K., et al. (2012), A soil fauna index for assessing soil quality, *Soil Biology and Biochemistry* 47, 158–165.

Yan, Z.B., et al. (2017), An assessment on the uncertainty of the nitrogen to phosphorus ratio as a threshold for nutrient limitation in plants, *Annals of Botany* 120, 937–942.

Yang, J.J., et al. (2020), Soil organic carbon stability under natural and anthropogenic-induced perturbations, *Earth-Science Reviews* 205, 103199.

Yang, L., et al. (2021), Rethinking the ecosystem functions of *Dicranopteris*, a widespread genus of ferns. *Frontiers in Plant Science* 11, 581513.

Yarwood, S.A., et al. (2020), Forest and Rangeland Soil Biodiversity. In R. Pouyat et al. (Eds), *Forest and Rangeland Soils of the United States Under Changing Conditions*, Springer, Cham, Switzerland.

Yasir, M., et al. (2009), Bacterial community composition and chitinase gene diversity of vermicompost with antifungal activity, *Bioresource Technology* 100(19), 4396–4403.

Yi CX and Jackson N. (2021), A review of measuring ecosystem resilience to disturbance. *Environ. Res. Lett.*, 16, 053008. DOI 10.1088/1748-9326/abdf09

Zhang, W.X., et al. (2013), Earthworms facilitate carbon sequestration through unequal amplification of carbon stabilization compared with mineralization, *Nature Communications* 4, 2576.

Zhang, K., et al. (2018), Global-scale patterns of nutrient density and partitioning in forests in relation to climate, *Global Change Biology* 24(1), 536–551.

Zhang, B., et al. (2023a), Multidimensional stoichiometric mismatch explains differences in detritivore biomass across three forest types, *Journal of Animal Ecology* 92(2), 454–465.

Zhang, B., et al. (2023b), Resistance and resilience of soil microbial communities: progress and perspective, *Acta Ecologica Sinica* 43(14), 5674–5685.

Zhang, W.X., et al. (2024a), Fluctuating "soil CO_2-lake" is key for understanding global climate change, *The Innovation* 5(4), 100642.

Zhang, C., et al. (2024b), An integrated fast–slow plant and nematode economics spectrum predicts soil organic carbon dynamics during natural restoration, *New Phytologist* 245, 13. https://doi.org/10.1111/nph.20166.

Zhang, W.X., et al. (2019), An alternative approach to reduce algorithm-derived biases in monitoring soil organic carbon changes, *Ecology and Evolution* 9, 7586–7596.

Zhang, W.X., et al. (2020a), Soil biota and sustainable agriculture: a review, *Acta Ecologica Sinica* 40, 3183–3206.

Zhang, W.X., et al. (2020b), An ignored key link in greenhouse effect: soil and soil CO_2 slow heat loss, *Soil Ecology Letters* 2(4), 308–316.

Zhao, C.C., et al. (2019), Bamboo forest management leads to a shift in the soil energy channel, *Geoderma* 353, 201–203.

Zhou, G.Y., et al. (2006), Old-growth forests can accumulate carbon in soils, *Science* 314, 1417–1417.

Zhou, L.X. and Ding, M.M. (2007), Soil microbial characteristics as bioindicators of soil health, *Biodiversity Science* 15, 162–171.

Index

www.ingramcontent.com/pod-product-compliance
Ingram Content Group UK Ltd.
Pitfield, Milton Keynes, MK11 3LW, UK
UKHW021824100926
8556IPUK00003B/151

9 781835 450758